D0585501

Basic VLSI Design

Third Edition

To Ella, Douglas, Christopher, and Nicholas
and
Michelle, Kylie, Natasha, and Kamran Jr

Basic VLSI Design

Third Edition

Douglas A. Pucknell
Kamran Eshraghian
Department of Electrical
and Electronic Engineering
University of Adelaide

Prentice Hall

Sydney New York London Toronto Tokyo Singapore Mexico City Amsterdam

Acquisitions Editor: Andrew Binnie.
Production Editor: Katie Millar.
Cover design: The Modern Art Production Group, Prahran, Victoria.
Typeset by Keyboard Wizards, Allambie Heights, NSW.

Printed in Australia by McPherson's Printing Group.

1 2 3 4 5 98 97 96 95 94

ISBN 0 13 079153 9

National Library of Australia
Cataloguing-in-Publication Data

Pucknell, Douglas A. (Douglas Albert)
Basic VLSI design.

3rd ed.
Includes bibliographies and index.
ISBN 0 13 079153 9.

1. Integrated circuits – Very large scale integration – Design and construction. 2. Bipolar integrated circuits – Design and construction. 3. Metal oxide semiconductors. 4. Gallium arsenide semiconductors. I. Eshraghian, Kamran. II. Title. (Series: Silicon systems engineering series).

621.395

Library of Congress
Cataloging-in-Publication Data

Pucknell, Douglas A. (Douglas Albert)
Basic VLSI design / by Doug Pucknell and Kamran Eshraghian — 3rd ed.

p. cm.
Includes bibliographical references and index.
ISBN 0-13-079153-9:

1. Integrated circuits – Very large scale integration – Design and construction. 2. Metal oxide semiconductors – Design and construction. I. Eshraghian, Kamran. II. Title.

TK7874.P82 1994 93–50694
621.39'5--dc20

Prentice Hall, Inc., Englewood Cliffs, New Jersey
Prentice Hall Canada, Inc., Toronto
Prentice Hall Hispanoamericana, SA, Mexico
Prentice Hall of India Private Ltd, New Delhi
Prentice Hall International, Inc., London
Prentice Hall of Japan, Inc., Tokyo
Prentice Hall of Southeast Asia Pty Ltd, Singapore
Editora Prentice Hall do Brasil Ltda, Rio de Janeiro

PRENTICE HALL

A division of Simon & Schuster

Contents

List of color plates xiii
Preface xv
Acknowledgments xix
About the Authors xxi

1 A review of microelectronics and an introduction to MOS technology 1
1.1 Introduction to integrated circuit technology 2
1.2 The integrated circuit (IC) era 4
1.3 Metal-oxide-semiconductor (MOS) and related VLSI technology 6
1.4 Basic MOS transistors 6
1.5 Enhancement mode transistor action 9
1.6 Depletion mode transistor action 10
1.7 nMOS fabrication 10
1.7.1 Summary of an nMOS process 14
1.8 CMOS fabrication 15
1.8.1 The p-well process 15
1.8.2 The n-well process 17
1.8.3 The twin-tub process 19
1.9 Thermal aspects of processing 19
1.10 BiCMOS technology 21
1.10.1 BiCMOS fabrication in an n-well process 24
1.10.2 Some aspects of bipolar and CMOS devices 24
1.11 Production of E-beam masks 25
1.12 Observations 26

2 Basic electrical properties of MOS and BiCMOS circuits 28

2.1 Drain-to-source current I_{ds} versus voltage V_{ds} relationships 29
2.1.1 The non-saturated region 30
2.1.2 The saturated region 32
2.2 Aspects of MOS transistor threshold voltage V_t 34
2.3 MOS transistor transconductance g_m and output conductance g_{ds} 35
2.4 MOS transistor figure of merit ω_0 37
2.5 The pass transistor 38
2.6 The nMOS inverter 38
2.7 Determination of pull-up to pull-down ratio ($Z_{p.u.}/Z_{p.d.}$) for an nMOS inverter driven by another nMOS inverter 40
2.8 Pull-up to pull-down ratio for an nMOS inverter driven through one or more pass transistors 42
2.9 Alternative forms of pull-up 45
2.10 The CMOS inverter 47
2.11 MOS transistor circuit model 50
2.12 Some characteristics of npn bipolar transistors 52
2.12.1 Transconductance g_m — bipolar 52
2.12.2 Comparative aspects of key parameters of CMOS and bipolar transistors 53
2.12.3 BiCMOS inverters 54
2.13 Latch-up in CMOS circuits 57
2.14 BiCMOS latch-up susceptibility 59
2.15 Observations 59
2.16 Tutorial exercises 60

3 MOS and BiCMOS circuit design processes 61

3.1 MOS layers 62
3.2 Stick diagrams 62
3.2.1 nMOS design style 67
3.2.2 CMOS design style 68
3.3 Design rules and layout 72
3.3.1 Lambda-based design rules 73
3.3.2 Contact cuts 75
3.3.3 Double metal MOS process rules 77
3.3.4 CMOS lambda-based design rules 78
3.4 General observations on the design rules 79
3.5 2 μm double metal, double poly. CMOS/BiCMOS rules 83
3.6 1.2 μm double metal, single poly. CMOS rules 84
3.7 Layout diagrams — a brief introduction 84
3.8 Symbolic diagrams — translation to mask form 90
3.9 Observations 90
3.10 Tutorial exercises 92

4 Basic circuit concepts 94

4.1 Sheet resistance R_s 95
4.2 Sheet resistance concept applied to MOS transistors and inverters 96
4.2.1 Silicides 98
4.3 Area capacitances of layers 99
4.4 Standard unit of capacitance $\square C_g$ 100
4.5 Some area capacitance calculations 100
4.6 The delay unit τ 102
4.7 Inverter delays 104
4.7.1 A more formal estimation of CMOS inverter delay 105
4.8 Driving large capacitive loads 107
4.8.1 Cascaded inverters as drivers 108
4.8.2 Super buffers 110
4.8.3 BiCMOS drivers 111
4.9 Propagation delays 114
4.9.1 Cascaded pass transistors 114
4.9.2 Design of long polysilicon wires 115
4.10 Wiring capacitances 116
4.10.1 Fringing fields 116
4.10.2 Interlayer capacitances 117
4.10.3 Peripheral capacitance 117
4.11 Choice of layers 118
4.12 Observations 119
4.13 Tutorial exercises 120

5 Scaling of MOS circuits 123

5.1 Scaling models and scaling factors 124
5.2 Scaling factors for device parameters 125
5.2.1 Gate area A_g 125
5.2.2 Gate capacitance per unit area C_0 or C_{ox} 126
5.2.3 Gate capacitance C_g 126
5.2.4 Parasitic capacitance C_x 126
5.2.5 Carrier density in channel Q_{on} 126
5.2.6 Channel resistance R_{on} 126
5.2.7 Gate delay T_d 127
5.2.8 Maximum operating frequency f_0 127
5.2.9 Saturation current I_{dss} 127
5.2.10 Current density J 127
5.2.11 Switching energy per gate E_g 128
5.2.12 Power dissipation per gate P_g 128
5.2.13 Power dissipation per unit area P_a 128
5.2.14 Power-speed product P_T 128
5.2.15 Summary of scaling effects 128
5.3 Some discussion on and limitations of scaling 129
5.3.1 Substrate doping 129

5.3.2 Limits of miniaturization 132
5.3.3 Limits of interconnect and contact resistance 134
5.4 Limits due to subthreshold currents 139
5.5 Limits on logic levels and supply voltage due to noise 139
5.7 Limits due to current density 142
5.8 Observations 144
5.9 References 145

6 Subsystem design and layout 146
6.1 Some architectural issues 147
6.2 Switch logic 148
6.2.1 Pass transistors and transmission gates 148
6.3 Gate (restoring) logic 149
6.3.1 The inverter 150
6.3.2 Two-input nMOS, CMOS and BiCMOS *Nand* gates 150
6.3.3 Two-input nMOS, CMOS and BiCMOS *Nor* gates 156
6.3.4 Other forms of CMOS logic 159
6.4 Examples of structured design (combinational logic) 165
6.4.1 A parity generator 165
6.4.2 Bus arbitration logic for n-line bus 167
6.4.3 Multiplexers (data selectors) 171
6.4.4 A general logic function block 174
6.4.5 A four-line Gray code to binary code converter 175
6.4.6 The programmable logic array (PLA) 176
6.5 Some clocked sequential circuits 176
6.5.1 Two-phase clocking 176
6.5.2 Charge storage 181
6.5.3 Dynamic register element 182
6.5.4 A dynamic shift register 183
6.6 Other system considerations 184
6.6.1 Bipolar drivers for bus lines 184
6.6.2 Basic arrangements for bus lines 186
6.6.3 The precharged bus concept 186
6.6.4 Power dissipation for CMOS and BiCMOS circuits 188
6.6.5 Current limitations for V_{DD} and GND (V_{SS}) rails 189
6.6.6 Further aspects of V_{DD} and V_{SS} rail distribution 190
6.7 Observations 192
6.8 Tutorial exercises 192

7 Subsystem design processes 196
7.1 Some general considerations 197
7.1.1 Some problems 198
7.2 An illustration of design processes 198

7.2.1 The general arrangement of a 4-bit arithmetic processor 199
7.2.2 The design of a 4-bit shifter 203
7.3 Observations 207
7.4 Tutorial exercises 209

8 Illustration of the design process — computational elements 210

8.1 Some observations on the design process 211
8.2 Regularity 211
8.3 Design of an ALU subsystem 212
8.3.1 Design of a 4-bit adder 213
8.3.2 Implementing ALU functions with an adder 224
8.4 A further consideration of adders 226
8.4.1 The Manchester carry-chain 226
8.4.2 Adder enhancement techniques 228
8.4.3 A comparison of adder enhancement techniques 237
8.5 Multipliers 240
8.5.1 The serial-parallel multiplier 240
8.5.2 The Braun array 242
8.5.3 Twos complement multiplication using the Baugh-Wooley method 242
8.5.4 A pipelined multiplier array 244
8.5.5 The modified Booth's algorithm 248
8.5.6 Wallace tree multipliers 251
8.5.7 Recursive decomposition of the multiplication 251
8.5.8 Dadda's method 253
8.6 Observations 253
8.7 Tutorial exercises 254
8.8 References 255

9 Memory, registers, and aspects of system timing 256

9.1 System timing considerations 257
9.2 Some commonly used storage/memory elements 257
9.2.1 The dynamic shift register stage 257
9.2.2 A three-transistor dynamic RAM cell 259
9.2.3 A one-transistor dynamic memory cell 261
9.2.4 A pseudo-static RAM/register cell 263
9.2.5 Four-transistor dynamic and six-transistor static CMOS memory cells 266
9.2.6 JK flip-flop circuit 269
9.2.7 D flip-flop circuit 273
9.3 Forming arrays of memory cells 273
9.3.1 Building up the floor plan for a 4 × 4-bit register array 274
9.3.2 Selection and control of the 4 × 4-bit register array 276

9.3.3 Random access memory (RAM) arrays 278
9.4 Observations 283
9.5 Tutorial exercises 283

10 Practical aspects and testability 285

10.1 Some thoughts on performance 286
10.1.1 Optimization of nMOS and CMOS inverters 287
10.1.2 Noise margins 292
10.2 Further thoughts on floor plans/layout 294
10.3 Floor plan layout of the 4-bit processor 298
10.4 Input/output (I/O) pads 298
10.5 'Real estate' 301
10.6 Further thoughts on system delays 303
10.6.1 Buses 303
10.6.2 Control paths, selectors, and decoders 303
10.6.3 Use of an asymmetric two-phase clock 305
10.6.4 More nasty realities 306
10.7 Ground rules for successful design 307
10.8 The real world of VLSI design 316
10.9 Design styles and philosophy 316
10.10 The interface with the fabrication house 318
10.10.1 CIF (Caltech. Intermediate Form) code 319
10.11 CAD tools for design and simulation 324
10.12 Aspects of design tools 324
10.12.1 Graphical entry layout 324
10.12.2 Design verification prior to fabrication 327
10.12.3 Design rule checkers (DRC) 328
10.12.4 Circuit extractors 330
10.12.5 Simulators 330
10.13 Test and testability 332
10.13.1 System partitioning 333
10.13.2 Layout and testability 334
10.13.3 Reset/initialization 334
10.13.4 Design for testability 334
10.13.5 Testing combinational logic 336
10.13.6 Testing sequential logic 339
10.13.7 Practical design for test (DFT) guidelines 341
10.13.8 Scan design techniques 349
10.13.9 Built-in-self-test (BIST) 353
10.13.10 Future trends 358
10.14 References 359

11 Some CMOS design projects 362

11.1 Introduction to project work 363
11.2 CMOS project 1 — an incrementer/decrementer 363

11.2.1 Behavioral description 363
11.2.2 Structural description 364
11.2.3 Physical description 365
11.2.4 Design verification 367
11.3 CMOS project 2 — left/right shift serial/parallel register 367
11.3.1 Behavioral description 367
11.3.2 Structural description 367
11.3.3 Physical description 371
11.3.4 Design verification 372
11.4 CMOS project 3 — a comparator for two *n*-bit numbers 372
11.4.1 Behavioral description 373
11.4.2 Structural description 376
11.4.3 Physical description 377
11.4.4 Symbolic or stick representation to mask transformation 378
11.4.5 Design verification 379
11.5 CMOS/BiCMOS project 4 — a two-phase non-overlapping clock generator with buffered output on both phases 381
11.5.1 Behavioral description 384
11.5.2 Structural description 384
11.5.3 Design process 385
11.5.4 Final test (simulation) results 391
11.5.5 Further thoughts 391
11.6 CMOS project 5 — design of a ∂*latch* — an event-driven latch element for EDL systems 396
11.6.1 A brief overview of event-driven logic (EDL) concepts (Pucknell 1993) 396
11.6.2 Behavioral description of a ∂*latch* 399
11.6.3 Structural description 399
11.6.4 Circuit action 401
11.6.5 Mask layout and performance simulation 401
11.7 Observations 405
11.8 References 405

12 Ultra-fast VLSI circuits and systems — introduction to GaAs technology 406

12.1 Ultra-fast systems 406
12.1.1 Submicron CMOS technology 406
12.1.2 Gallium arsenide VLSI technology 407
12.2 Gallium arsenide crystal structure 408
12.2.1 A compound semiconductor 410
12.2.2 Doping process 411
12.2.3 Channeling effect 412
12.2.4 Energy band structure 412
12.2.5 Electron velocity-field behavior 414

12.3 Technology development 414
12.3.1 Gallium arsenide devices 418
12.3.2 Metal semiconductor FET (MESFET) 418
12.3.3 GaAs fabrication 420
12.4 Device modeling and performance estimation 435
12.4.1 Device characterization 435
12.4.2 Drain to source current derivation 435
12.4.3 Transconductance and output conductance 442
12.4.4 Logic voltage swing 445
12.4.5 Direct-coupled FET logic (DCFL) inverter 446
12.5 MESFET-based design 451
12.5.1 MESFET design methodology 451
12.5.2 Gallium arsenide layer representations 451
12.5.3 Design methodology and layout style 453
12.5.4 Layout design rules 458
12.5.5 Symbolic approach to layout for GaAs MESFETs 463
12.6 GaAs MESFET classes of logic 465
12.6.1 Normally-on logic gates 465
12.6.2 Normally-off logic gates 465
12.7 VLSI design — the final ingredients 468
12.8 Tutorial exercises 468

Appendix A 470
Appendix B 474
Appendix C 483
Further reading 489
Index 491

List of color plates

Color plates

1 (a) Encodings for a simple single metal nMOS process
(b) Color encodings for a double metal CMOS p-well process
(c) Additional encodings for a double metal double poly. BiCMOS n-well process
(d) Color stick diagram examples

2 Example layout encodings

3 ORBIT™ 2μm design rules (a)(b)

4 ORBIT™ 2μm design rules (c)

5 ORBIT™ 2μm design rules (d)(e)

6 ORBIT™ 2μm design rules (f)

7 1-bit CMOS shift register cell

8 (a) A BiCMOS 2 input nand gate
(b) A BiCMOS 2 input nor gate

9 (a) 3I/P nMOS nor gate
(b) 2I/P CMOS (p-well) nor gate

10 n-type pass transistor based 4-way MUX

11 CMOS transmission gate based 4-way MUX

12 Mask layout for two-phase (and complements) clock generator

Preface

The microscopic dimensions of current silicon-integrated circuitry make possible the design of digital circuits which may be very complex and yet extremely economical in space, power requirements and cost, and potentially very fast. The space, power and cost aspects have made silicon the dominant fabrication technology for electronics in very wide ranging areas of application. The combination of complexity and speed is finding ready applications for VLSI systems in digital processing, and particularly in those application areas requiring sophisticated high speed digital processing. Although silicon MOS-based circuitry will meet most requirements in such systems and the technology is still being enhanced by ongoing improvements in fabrication, there are ultimate limitations associated with the velocity of electrons (and holes) in silicon which will make MOS circuitry unsuitable for some ultra fast systems that are now being contemplated. Thus, other techniques are being actively investigated to complement silicon technology, including the use of materials other than silicon for the production of integrated circuits. One promising technology is the production of very fast circuits in gallium arsenide.

The overwhelming majority of VLSI systems in silicon utilize nMOS, CMOS or BiCMOS technology and, although nMOS designs are now mostly outmoded, it is advantageous to understand the processes and to be able to design or analyze nMOS circuits as the need arises. This added learning load is no real burden since the three technologies are closely interrelated and design is based on common concepts. It is further possible to relate some silicon design methodology and nMOS circuit concepts to the design of gallium arsenide circuits.

A significant feature of this edition is the expansion of CMOS circuitry to include bipolar transistors — the BiCMOS process. This is of particular interest, for example, where larger capacitive loads must be driven. The nature and characteristics of the relevant bipolar devices are dealt with and design rules for an n-well BiCMOS process appear in the text and are used in design examples.

Essential matters for nMOS, CMOS and BiCMOS digital circuit design are covered in Chapters 1 to 9, including numerous illustrative design exercises over Chapters 6 to 9. Learning to design circuits in silicon is essentially a 'hands-on' process and further exercises are set as tutorial work in Chapters 2, 3, 4, 6, 7, 8 and 9, and more demanding work is set out in the six CMOS design projects that comprise Chapter 11. Lambda-based design rules (Mead and Conway style*) are used in most design exercises since they are easily understood, easily remembered and applied, and can be used for fabrication. The rule set given covers both nMOS and p-well-based CMOS. However, more effective designs may be based on 'real-world' micron-based rule sets and two such rule sets are included in this text. Both are from Orbit Semiconductor Inc. of California, USA: one rule set covers a 2 micron double metal, double polysilicon n-well BiCMOS process and the second set is for a 1.2 micron n-well double metal, single polysilicon CMOS process.

An extended coverage of digital arithmetic circuitry is now included in Chapter 8 and a large part of Chapter 10 is devoted to an expanded treatment of testability considerations.

Chapter 11 is devoted entirely to project work designed to illustrate typical approaches to the design of a variety of system requirements.

Chapter 12 builds on the earlier chapters to introduce gallium arsenide (GaAs) technology and establishes suitable encoding, notation, design rules and basic design methodology. Some GaAs logic circuit arrangements are examined.

We have been particularly careful to ensure that this third edition does not omit essential material from the second edition and maintains the format and approach with which users of the earlier editions have become familiar.

Although much new material has been included, all the core material from the second edition appears in roughly the same order as before, with the exception of the chapter dealing with PLAs and finite state machines. This subject, rightly or wrongly, we now feel belongs more appropriately in texts covering digital logic in general and combinational and sequential logic in particular. One such text now forms part of this Silicon Engineering Series of texts**. However, in order to maintain a coverage of the PLA, we have included the essential material as Appendix C.

Thus, those who have based coursework on the first or second editions of this text should not be seriously affected by any omissions and should benefit substantially from the additional material presented in this new edition.

---

* Mead, C. A. & Conway, L. A. *Introduction to VLSI systems*, Addison-Wesley, USA, 1980.

** Pucknell, D. A. *Fundamentals of Digital Logic Design*, Prentice Hall, Australia, 1990.

In conclusion, the authors have set out to present a balanced and structured course covering the 'technologies of the nineties'. The book covers the design of circuits in silicon and introduces the newer GaAs technology in a way that is compatible with design in silicon. We have set out to present this text in a form that is readily used and easily assimilated. Most of the material presented is based on coursework taught over a number of years and is therefore 'tried and trusted'.

Douglas Pucknell and Kamran Eshraghian
Adelaide, February 1994

Acknowledgments

First, we must acknowledge valuable and valued contributions from Dr B. Hochet of the Swiss Federal Institute of Technology, Lausanne, on adder enhancement techniques (Chapter 8), Professor K. Trivedi of Duke University on inverter optimization (Chapter 10), and Dr A. Osseiran, also of the Swiss Federal Institute of Technology, on design for testability (Chapter 10).

Any teaching program must inevitably benefit greatly from interaction between the instructors and the students. This has certainly been the case for the coursework on which this text is based and we owe much to our undergraduate students, our postgraduate research colleagues, and fellow members of the academic staff. It is therefore virtually impossible to name all those who have made useful contributions.

However, we can at least pay tribute to those who have directly contributed material that appears in this text, including an extract from a final year student project report by Brenton Cooper and design work contributed by postgraduate researcher Shannon Morton. Both contributions appear in Chapter 11.

Our VLSI work has also benefited both directly and indirectly from our interaction with Orbit Semiconductor Inc. of Sunnyvale, California, with Orbit's Gary Kennedy through Laird Varzaly and Alf Grasso of Integrated Silicon Design Pty Ltd (ISD) of Adelaide, who represent Orbit in Australia. We have also made reference in the text to the use of ISD software in teaching and in design work generally, and we are most grateful for the way in which ISD has always supported our work.

In a global sense, like that of most others in the VLSI world, our approach has been significantly influenced by the pioneering work of Carver Mead and Lynn Conway, who were largely responsible for bringing VLSI technology within the scope of the ordinary electronics engineer and engineering student.

Finally, we acknowledge with gratitude the unflagging support and patience of our families and our colleagues, and the editorial and production team of our publishers, Prentice Hall of Sydney, who have endured yet again the emergence of this latest edition of *Basic VLSI Design*.

About the authors

Douglas Pucknell — PhD, BSc, BE, CEng, MIEE.
Douglas Pucknell gained his initial experience in engineering while serving in the Electrical Branch of the Royal Navy. On leaving the RN in 1954, he studied for a BSc degree in Edinburgh, Scotland, where he subsequently joined the staff of Ferranti Ltd as a design engineer. After eleven years with Ferranti, and having become Chief Engineer (Logic Circuits), he took up an appointment at the University of Adelaide, South Australia, and obtained his PhD degree in 1973. He is now an Associate Professor in the department of Electrical and Electronic Engineering.

His teaching and research interests lie in the fields of Digital Systems, Microcomputer based Engineering and in VLSI design and he is a co-founder of the high-tech company Integrated Silicon Design Pty Ltd (ISD). He has a long-term interest in technology transfer to industry and has been an organiser and presenter of many post-experience courses in Australia and overseas.

Kamran Eshraghian — PhD, MEng, BTech, FIE (Aust), SMIREE, CPeng.
Kamran Eshraghian obtained his PhD, MEngSc, and BTech degrees from the University of Adelaide, South Australia. In 1977 he joined the Department of Electrical and Electronic Engineering at the University of Adelaide after spending some 10 years with Philips Research. In 1987, he became the Director of the Centre for Gallium Arsenide VLSI Technology, University of Adelaide, the very first centre established in Australia to pursue research in the general area of ultra high speed VLSI systems. He spent a period of time at the Department of Computer Science, Duke University, North Carolina and at the Department of Electronic Engineering, EPFL, Lausanne, Switzerland. He co-founded two successful Australian high technology companies, providing intimate links between university research and industry. At present he is the Foundation Professor of Electronic, Computer and Communications Engineering at Edith Cowan University, Perth, Western Australia.

1 A review of microelectronics and an introduction to MOS technology

If you would have the kindness to begin at the beginning, I should be vastly obliged; all these stories that begin in the middle simply fog my wit.

Count Anthony Hamilton

Objectives

This chapter 'sets the scene' by reviewing the evolution of integrated circuits (ICs) and comparing the general characteristics of currently available technologies, including BiCMOS and GaAs as well as nMOS and CMOS.

Basic MOS transistor action is briefly reviewed and an overview of fabrication processes is given to help appreciate the nature of the technologies.

1.1 Introduction to integrated circuit technology

There is no doubt that our daily lives are significantly affected by electronic engineering technology. This is true on the domestic scene, in our professional disciplines, in the workplace, and in leisure activities. Indeed, even at school, tomorrow's adults are exposed to and are coming to terms with quite sophisticated electronic devices and systems. There is no doubt that revolutionary changes have taken place in a relatively short time and it is also certain that even more dramatic advances will be made in the next decade.

Electronics as we know it today is characterized by reliability, low power dissipation, extremely low weight and volume, and low cost, coupled with an ability to cope easily with a high degree of sophistication and complexity. Electronics, and in particular the integrated circuit, has made possible the design of powerful and flexible processors which provide highly intelligent and adaptable devices for the user. Integrated circuit memories have provided the essential elements to complement these processors and, together with a wide range of logic and analog integrated circuitry, they have provided the system designer with components of considerable capability and extensive application. Furthermore, the revolutionary advances in technology have not yet by any means run their full course and the potential for future developments is exciting to say the least.

Up until the 1950s, electronic active device technology was dominated by the vacuum tube and, although a measure of miniaturization and circuit integration did take place, the technology did not lend itself to miniaturization as we have come to accept it today. Thus the vast majority of present-day electronics is the result of the invention of the transistor in 1947.

The invention of the transistor by William B. Shockley, Walter H. Brattain and John Bardeen of Bell Telephone Laboratories was followed by the development of the Integrated Circuit (IC). The very first IC emerged at the beginning of 1960 and since that time there have already been four generations of ICs: SSI (small scale integration), MSI (medium scale integration), LSI (large scale integration), and VLSI (very large scale integration). Now we are beginning to see the emergence of the fifth generation, ULSI (ultra large scale integration), which is characterized by complexities in excess of 3 million devices on a single IC chip. Further miniaturization is still to come and more revolutionary advances in the application of this technology must inevitably occur.

Over the past several years, Silicon CMOS technology has become the dominant fabrication process for relatively high performance and cost effective VLSI circuits. The revolutionary nature of this development is indicated by the way in which the number of transistors integrated in circuits on a single chip has grown, as indicated in Figure 1–1. Such progress is highlighted by recent products such as RISC chips in which it is possible to process some 35 million instructions per second. In order to improve on this throughput rate it will be necessary to improve

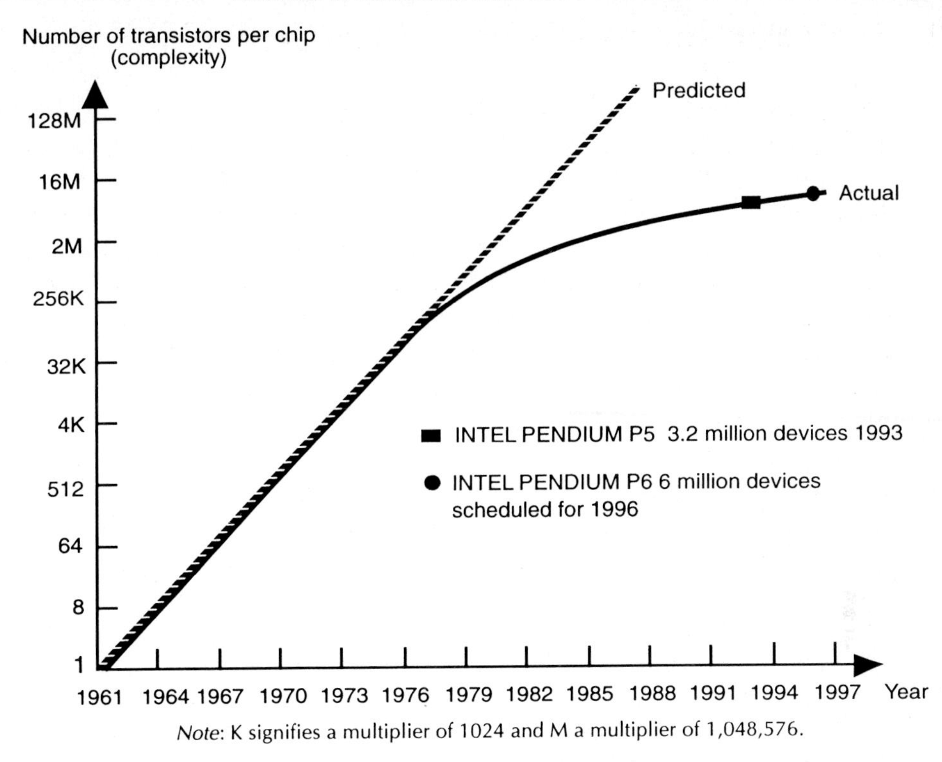

Figure 1–1 Moore's first law: transistors integrated on a single chip (commercial products)

the technology, both in terms of scaling and processing, and through the incorporation of other enhancements such as BiCMOS. The implication of this approach is that existing silicon technology could effectively facilitate the tripling of rate. Beyond this, that is, above 100 million instructions per second, one must look to other technologies. In particular, the emerging gallium arsenide (GaAs) based technology will be most significant in this area of ultra high speed logic/fast digital processors. GaAs also has further potential as a result of its photo-electronic properties, both as a receiver and as a transmitter of light. GaAs in combination with silicon will provide the designer with some very exciting possibilities.

It is most informative in assessing the role of the currently available technologies to review their speed and power performance domains. This has been set out as Figure 1–2 and the potential presented by each may be readily assessed.

This text deals mostly with silicon-based VLSI, including BiCMOS, but also introduces GaAs-based technology. ECL-based technology is not covered here, but much of the material given is relevant to the general area of the design of digital integrated circuits.

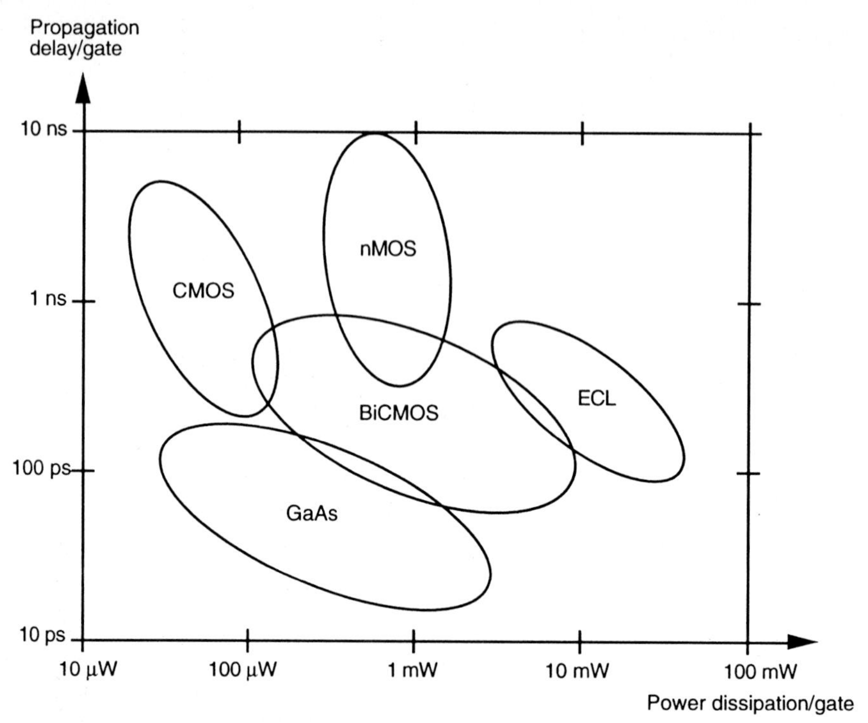

Figure 1–2 Speed/power performance of available technologies

1.2 The integrated circuit (IC) era

Such has been the potential of the silicon integrated circuit that there has been an extremely rapid growth in the number of transistors (as a measure of complexity) being integrated into circuits on a single silicon chip. In less than three decades, this number has risen from tens to millions as can be seen in Figure 1–1. The figure sets out what has become known as 'Moore's first law' after predictions made by Gordon Moore (of Intel) in the 1960s. It may be seen that his predictions have largely come true except for an increasing divergence between 'predicted' and 'actual' over the last few years due to problems associated with the complexities involved in designing and testing such very large circuits.

Such has been the impact of this revolutionary growth that IC technology now affects almost every aspect of our lives. More is still to come since we have not yet reached the limits of miniaturization and there is no doubt that tens of millions of transistors will be readily integrated onto a single chip in the future. This evolutionary process is reflected in Table 1–1.

Truly the 1970s, the 1980s and now the 1990s may well be described as the integrated circuit era.

Table 1–1 Microelectronics evolution

Year	1947	1950	1961	1966	1971	1980	1990	2000
Technology	*Invention of the transistor*	*Discrete components*	*SSI*	*MSI*	*LSI*	*VLSI*	*ULSI**	*GSI†*
Approximate numbers of transistors per chip in commercial products	1	1	10	100–1000	1000–20,000	20,000–1,000,000	1,000,000–10,000,000	>10,000,000
Typical products	—	Junction Transistor and diode	Planar devices Logic gates Flip-flops	Counters Multiplexers Adders	8 bit micro-processors ROM RAM	16 and 32 bit micro-processors Sophisticated peripherals GHM Dram	Special processors, Virtual reality machines, smart sensors	

* Ultra large-scale integration
† Giant-scale integration

Note: The boundary lines between technologies in the table are *not* artificially created. Crossing each boundary requires new design methodology, simulation approaches, and new methods for determining and routing communications and for handling complexity.

1.3 Metal-oxide-semiconductor (MOS) and related VLSI technology

Within the bounds of MOS technology, the possible circuit realizations may be based on pMOS, nMOS, CMOS and now BiCMOS devices.

However, this text will deal with nMOS, then with CMOS (which includes nMOS and pMOS transistors) and BiCMOS, and finally with GaAs technology, all of which may be classed as leading integrated circuit technologies.

Although CMOS is the dominant technology, some of the examples used to illustrate the design processes will be presented in nMOS form. The reasons for this are as follows:

- For nMOS technology, the design methodology and the design rules are easily learned, thus providing a simple but excellent introduction to structured design for VLSI.
- nMOS technology and design processes provide an excellent background for other technologies. In particular, some familiarity with nMOS allows a relatively easy transition to CMOS technology and design.
- For GaAs technology some arrangements in relation to logic design are similar to those employed in nMOS technology. Therefore, understanding the basics of nMOS design will assist in the layout of GaAs circuits.

Not only is VLSI technology providing the user with a new and more complex range of 'off the shelf' circuits, but VLSI design processes are such that system designers can readily design their own special circuits of considerable complexity. This provides a new degree of freedom for designers and it is probable that some very significant advances will result. Couple this with the fact that integration density is increasing rapidly, as advances in technology shrink the feature size for circuits integrated in silicon. Typical manufacturers' commercial IC products have shown this trend quite clearly as indicated in Figure 1–3 and, simultaneously, the effectiveness of the circuits produced has increased with scaling down. A common measure of effectiveness is the speed power product of the basic logic gate circuit of the technology (for nMOS, the *Nor* gate; with *Nand* and *Nor* gates for CMOS). Speed power product is measured in picojoules (pJ) and is the product of the gate switching delay in nanoseconds and the gate power dissipation in milliwatts. Typical figures are implied in Figure 1–2.

1.4 Basic MOS transistors

Having now established some background, let us turn our attention to basic MOS processes and devices. In particular, let us examine the basic nMOS enhancement and depletion mode transistors as shown in Figures 1–4 (a) and (b).

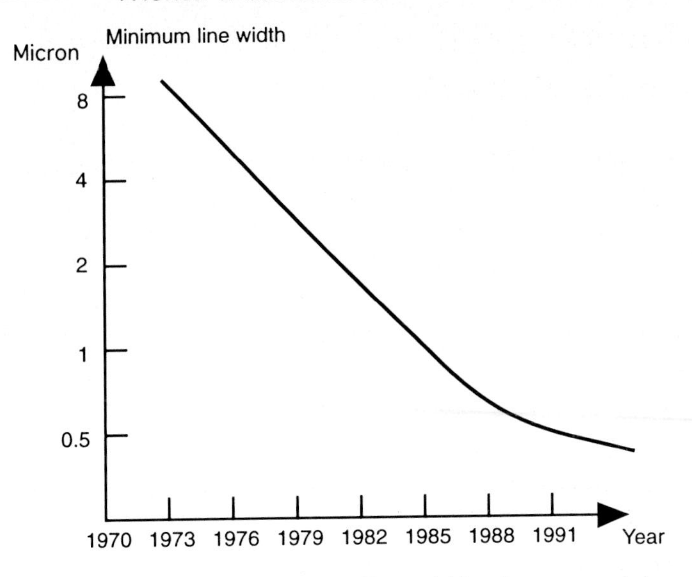

Figure 1–3 Approximate minimum line width of commercial products versus year

nMOS devices are formed in a p-type substrate of moderate doping level. The source and drain regions are formed by diffusing n-type impurities through suitable masks into these areas to give the desired n-impurity concentration and give rise to depletion regions which extend mainly in the more lightly doped p-region as shown. Thus, source and drain are isolated from one another by two diodes. Connections to the source and drain are made by a deposited metal layer. In order to make a useful device, there must be the capability for establishing and controlling a current between source and drain, and this is commonly achieved in one of two ways, giving rise to the enhancement mode and depletion mode transistors.

Consider the enhancement mode device first, shown in Figure 1–4(a). A polysilicon gate is deposited on a layer of insulation over the region between source and drain. Figure 1–4(a) shows a basic enhancement mode device in which the channel is not established and the device is in a non-conducting condition, $V_D = V_S = V_{gs} = 0$. If this gate is connected to a suitable positive voltage with respect to the source, then the electric field established between the gate and the substrate gives rise to a charge inversion region in the substrate under the gate insulation and a conducting path or channel is formed between source and drain.

The channel may also be established so that it is present under the condition $V_{gs} = 0$ by implanting suitable impurities in the region between source and drain during manufacture and prior to depositing the insulation and the gate. This arrangement is shown in Figure 1–4(b). Under these circumstances, source and drain are connected by a conducting channel, but the channel may now be closed by applying a suitable negative voltage to the gate.

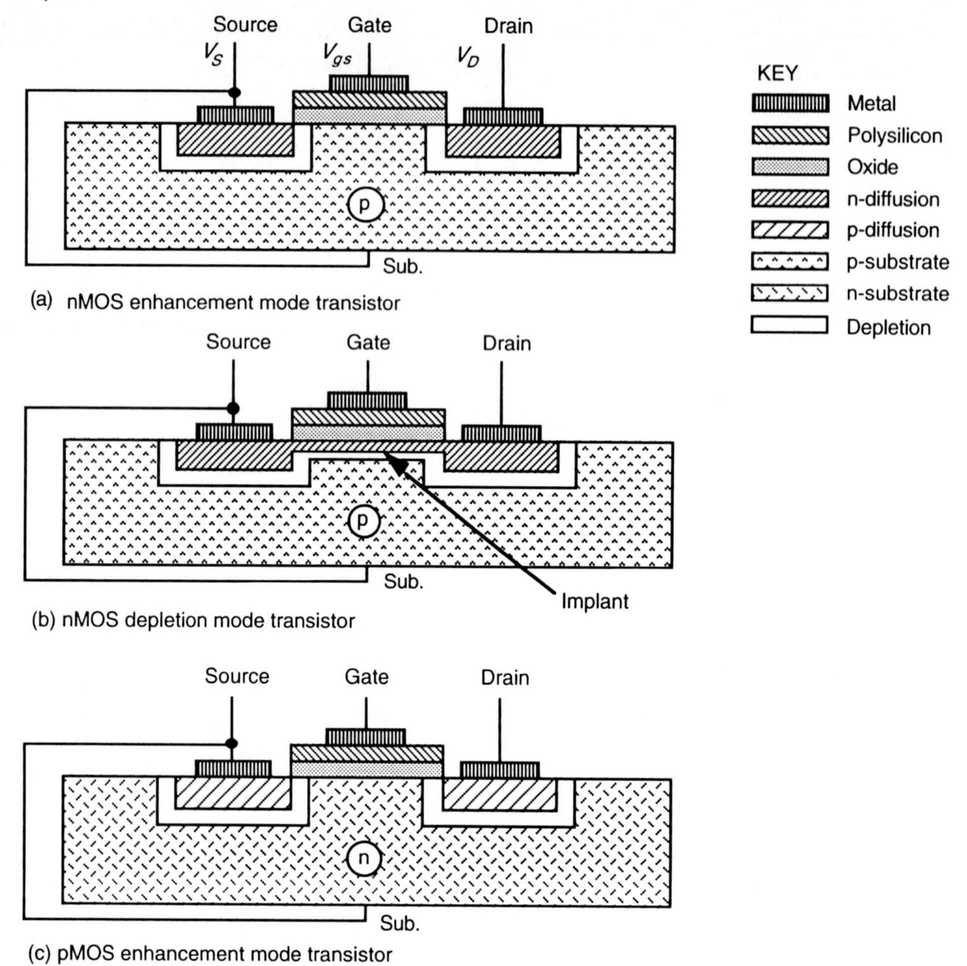

Figure 1–4 MOS transistors (V_D = 0V. Source gate and substrate to 0V)

In both cases, variations of the gate voltage allow control of any current flow between source and drain.

Figure 1–4(c) shows the basic pMOS transistor structure for an enhancement mode device. In this case the substrate is of n-type material and the source and drain diffusions are consequently p-type. In the figure, the conditions shown are those for an unbiased device; however, the application of a *negative* voltage of suitable magnitude ($> |V_t|$) between gate and source will give rise to the formation of a channel (p-type) between the source and drain and current may then flow if the drain is made negative with respect to the source. In this case the current is carried by holes as opposed to electrons (as is the case for nMOS devices). In consequence, pMOS transistors are inherently slower than nMOS, since hole mobility μ_p is less, by a factor of approximately 2.5, than electron mobility μ_n.

However, bearing these differences in mind, the discussions of nMOS transistors which follow relate equally well to pMOS transistors.

1.5 Enhancement mode transistor action

To gain some understanding of this mechanism, let us further consider the enhancement mode device, as in Figure 1–5, under three sets of conditions. It must first be recognized that in order to establish the channel in the first place, a minimum voltage level of *threshold voltage* V_t must be established between gate

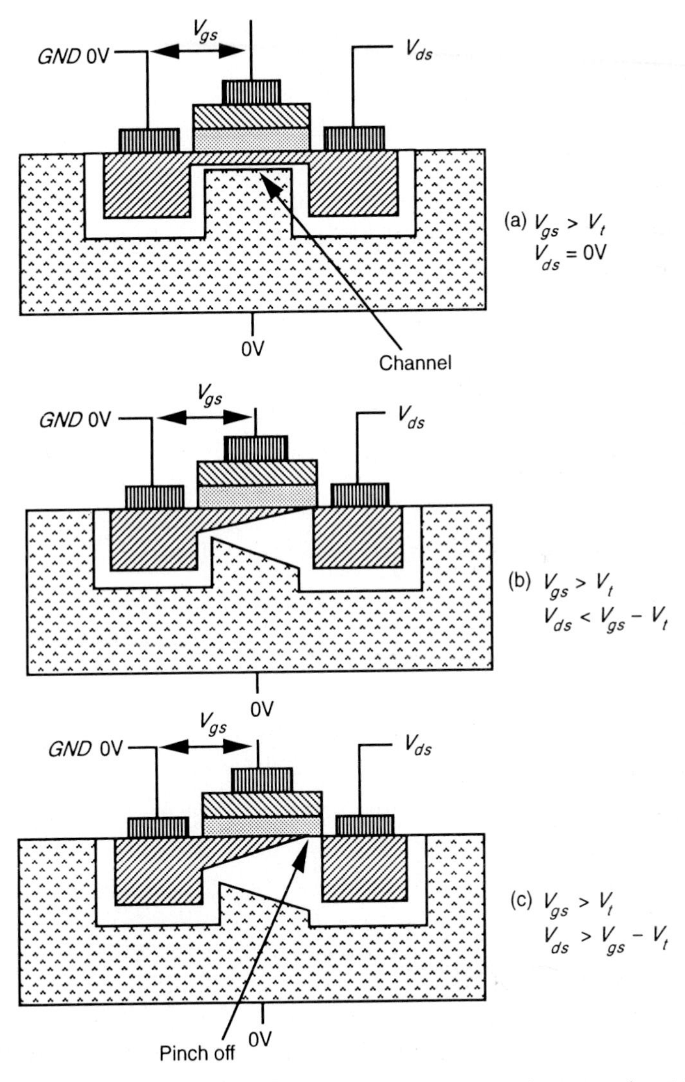

Note: V_{ds} is the drain-to-source voltage. Substrate assumed connected to 0V.

Figure 1–5 Enhancement mode transistor for particular values of V_{ds} with ($V_{gs} > V_t$)

and source (and of course between gate and substrate as a result). Figure 1–5(a) then indicates the conditions prevailing with the channel established but no current flowing between source and drain ($V_{ds} = 0$). Now consider the conditions prevailing when current flows in the channel by applying a voltage V_{ds} between drain and source. There must, of course, be a corresponding IR drop = V_{ds} along the channel. This results in the voltage between gate and channel varying with distance along the channel with the voltage being a maximum of V_{gs} at the source end. Since the effective gate voltage is $V_g = V_{gs} - V_t$ (no current flows when $V_{gs} < V_t$), there will be voltage available to invert the channel at the drain end so long as $V_{gs} - V_t \geqslant V_{ds}$. The limiting condition comes when $V_{ds} = V_{gs} - V_t$. For all voltages $V_{ds} < V_{gs} - V_t$, the device is in the non-saturated region of operation which is the condition shown in Figure 1–5(b).

Consider now what happens when V_{ds} is increased to a level greater than $V_{gs} - V_t$. In this case, an IR drop = $V_{gs} - V_t$ takes place over less than the whole length of the channel so that over part of the channel, near the drain, there is insufficient electric field available to give rise to an inversion layer to create the channel. The channel is therefore 'pinched off', as indicated in Figure 1–5(c). Diffusion current completes the path from source to drain in this case, causing the channel to exhibit a high resistance and behave as a constant current source. This region, known as *saturation,* is characterized by almost constant current for increase of V_{ds} above $V_{ds} = V_{gs} - V_t$. In all cases, the channel will cease to exist and no current will flow when $V_{gs} < V_t$. Typically, for enhancement mode devices, $V_t = 1$ volt for $V_{DD} = 5$ volt or, in general terms, $V_t = 0.2\ V_{DD}$.

1.6 Depletion mode transistor action

For depletion mode devices the channel is established, because of the implant, even when $V_{gs} = 0$, and to cause the channel to cease to exist a negative voltage V_{td} must be applied between gate and source.

V_{td} is typically $< -0.8\ V_{DD}$, depending on the implant and substrate bias, but, threshold voltage differences aside, the action is similar to that of the enhancement mode transistor.

Commonly used symbols for nMOS and pMOS transistors are set out in Figure 1–6.

1.7 nMOS fabrication

A brief introduction to the general aspects of the polysilicon gate self-aligning nMOS fabrication process will now be given. As well as being relevant in their own right, the fabrication processes used for nMOS are relevant to CMOS and

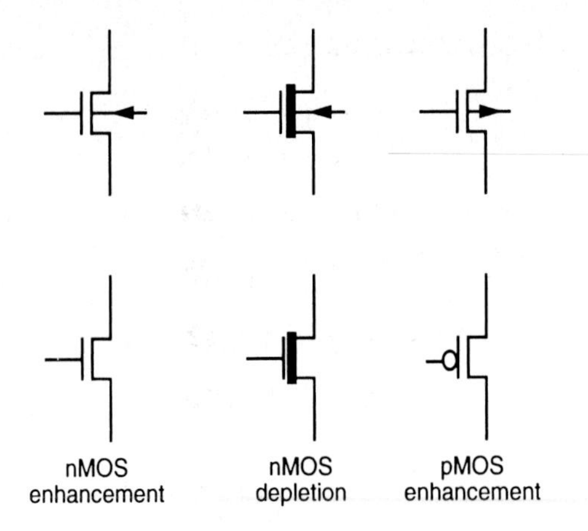

Figure 1–6 Transistor circuit symbols

BiCMOS which may be viewed as involving additional fabrication steps. Also, it is clear that an appreciation of the fabrication processes will give an insight into the way in which design information must be presented and into the reasons for certain performance characteristics and limitations. An nMOS process is illustrated in Figure 1–7 and may be outlined as follows:

1. Processing is carried out on a thin wafer cut from a single crystal of silicon of high purity into which the required p-impurities are introduced as the crystal is grown. Such wafers are typically 75 to 150 mm in diameter and 0.4 mm thick and are doped with, say, boron to impurity concentrations of $10^{15}/cm^3$ to $10^{16}/cm^3$, giving resistivity in the approximate range 25 ohm cm to 2 ohm cm.
2. A layer of silicon dioxide (SiO_2), typically 1 μm thick, is grown all over the surface of the wafer to protect the surface, act as a barrier to dopants during processing, and provide a generally insulating substrate onto which other layers may be deposited and patterned.
3. The surface is now covered with a photoresist which is deposited onto the wafer and spun to achieve an even distribution of the required thickness.
4. The photoresist layer is then exposed to ultraviolet light through a mask which defines those regions into which diffusion is to take place together with transistor channels. Assume, for example, that those areas exposed to ultraviolet radiation are polymerized (hardened), but that the areas required for diffusion are shielded by the mask and remain unaffected.
5. These areas are subsequently readily etched away together with the underlying silicon dioxide so that the wafer surface is exposed in the window defined by the mask.

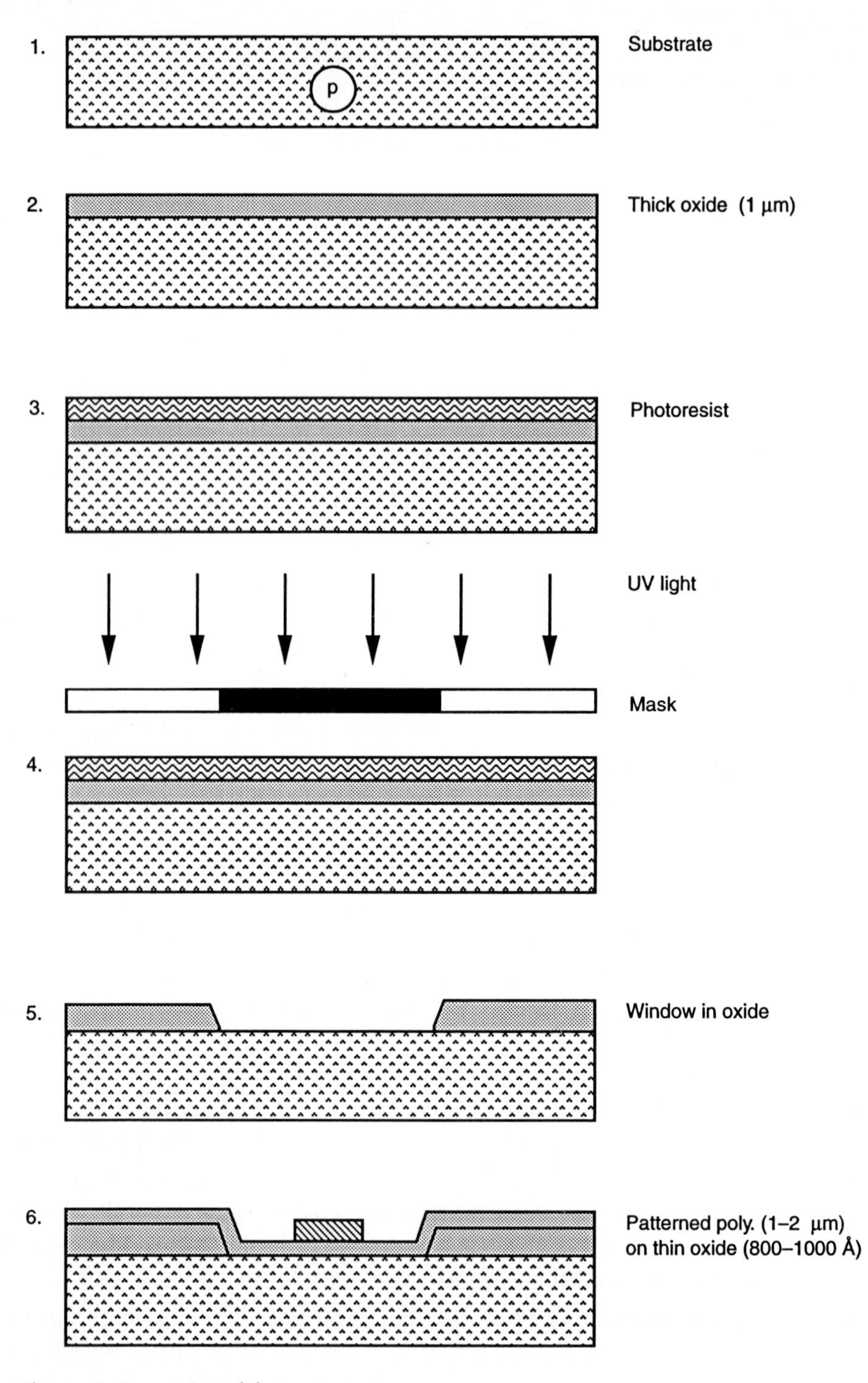

Figure 1–7 nMOS fabrication process

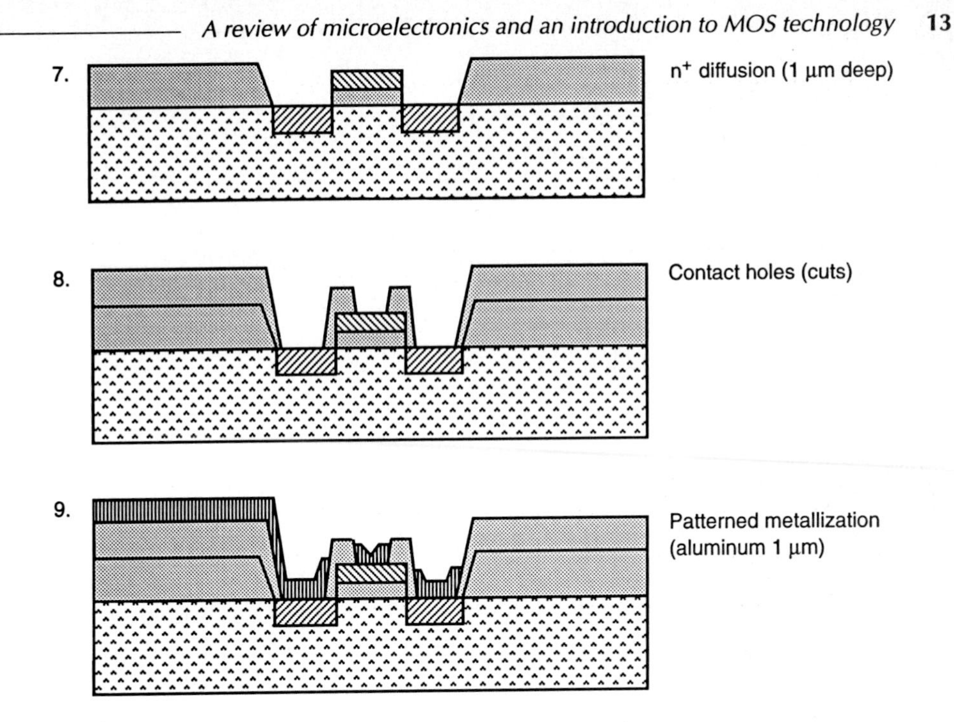

Figure 1–7 continued

6. The remaining photoresist is removed and a thin layer of SiO_2 (0.1 μm typical) is grown over the entire chip surface and then polysilicon is deposited on top of this to form the gate structure. The polysilicon layer consists of heavily doped polysilicon deposited by chemical vapor deposition (CVD). In the fabrication of fine pattern devices, precise control of thickness, impurity concentration, and resistivity is necessary.
7. Further photoresist coating and masking allows the polysilicon to be patterned (as shown in Step 6), and then the thin oxide is removed to expose areas into which n-type impurities are to be diffused to form the source and drain as shown. Diffusion is achieved by heating the wafer to a high temperature and passing a gas containing the desired n-type impurity (for example, phosphorus) over the surface as indicated in Figure 1–8. Note that the polysilicon with underlying thin oxide and the thick oxide act as masks during diffusion — the process is self-aligning.
8. Thick oxide (SiO_2) is grown over all again and is then masked with photoresist and etched to expose selected areas of the polysilicon gate and the drain and source areas where connections (i.e. contact cuts) are to be made.
9. The whole chip then has metal (aluminum) deposited over its surface to a thickness typically of 1 μm. This metal layer is then masked and etched to form the required interconnection pattern.

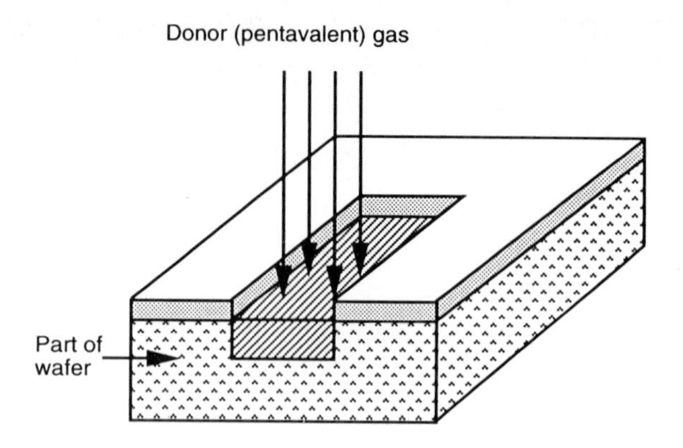

Figure 1–8 Diffusion process

It will be seen that the process revolves around the formation or deposition and patterning of three layers, separated by silicon dioxide insulation. The layers are diffusion within the substrate, polysilicon on oxide on the substrate, and metal insulated again by oxide.

To form depletion mode devices it is only necessary to introduce a masked ion implantation step between Steps 5 and 6 in Figure 1–7. Again, the thick oxide acts as a mask and this process stage is also self-aligning.

Consideration of the processing steps will reveal that relatively few masks are needed and the self-aligning aspects of the masking processes greatly ease the problems of mask registration. In practice, some extra process steps are necessary, including the overglassing of the whole wafer, except where contacts to the outside world are required. However, the process is basically straightforward to envisage and circuit design eventually comes down to the business of delineating the masks for each stage of the process. The essence of the process may be reiterated as follows.

1.7.1 Summary of an nMOS process

- Processing takes place on a p-doped silicon crystal wafer on which is grown a 'thick' layer of SiO_2.
- *Mask* 1 — Pattern SiO_2 to expose the silicon surface in areas where paths in the diffusion layer or source, drain or gate areas of transistors are required. Deposit thin oxide over all. For this reason, this mask is often known as the *'thinox' mask* but some texts refer to it as the *diffusion mask.*
- *Mask* 2 — Pattern the ion implantation within the thinox region where depletion mode devices are to be produced — *self-aligning.*

- *Mask* 3 — Deposit polysilicon over all (1.5 μm thick typically), then pattern using Mask 3. Using the same mask, remove thin oxide layer where it is not covered by polysilicon.
- Diffuse n^+ regions into areas where thin oxide has been removed. Transistor drains and sources are thus self-aligning with respect to the gate structures.
- *Mask* 4 — Grow thick oxide over all and then etch for contact cuts.
- *Mask* 5 — Deposit metal and pattern with Mask 5.
- *Mask* 6 — Would be required for the overglassing process step.

1.8 CMOS fabrication

There are a number of approaches to CMOS fabrication, including the p-well, the n-well, the twin-tub, and the silicon-on-insulator processes. In order to introduce the reader to CMOS design we will be concerned mainly with well-based circuits. The p-well process is widely used in practice and the n-well process is also popular, particularly as it was an easy retrofit to existing nMOS lines.

For the lambda-based rules set out later, we will assume a p-well process.

1.8.1 The p-well process

A brief overview of the fabrication steps may be obtained with reference to Figure 1–9, noting that the basic processing steps are of the same nature as those used for nMOS.

In primitive terms, the structure consists of an n-type substrate in which p-devices may be formed by suitable masking and diffusion and, in order to accommodate n-type devices, a deep p-well is diffused into the n-type substrate as shown.

This diffusion must be carried out with special care since the p-well doping concentration and depth will affect the threshold voltages as well as the breakdown voltages of the n-transistors. To achieve low threshold voltages (0.6 to 1.0 V), we need either deep well diffusion or high well resistivity. However, deep wells require larger spacing between the n- and p-type transistors and wires because of lateral diffusion and therefore a larger chip area.

The p-wells act as substrates for the n-devices within the parent n-substrate, and, provided that voltage polarity restrictions are observed, the two areas are electrically isolated. However, since there are now in effect two substrates, two substrate connections (V_{DD} and V_{SS}) are required, as shown in Figure 1–10.

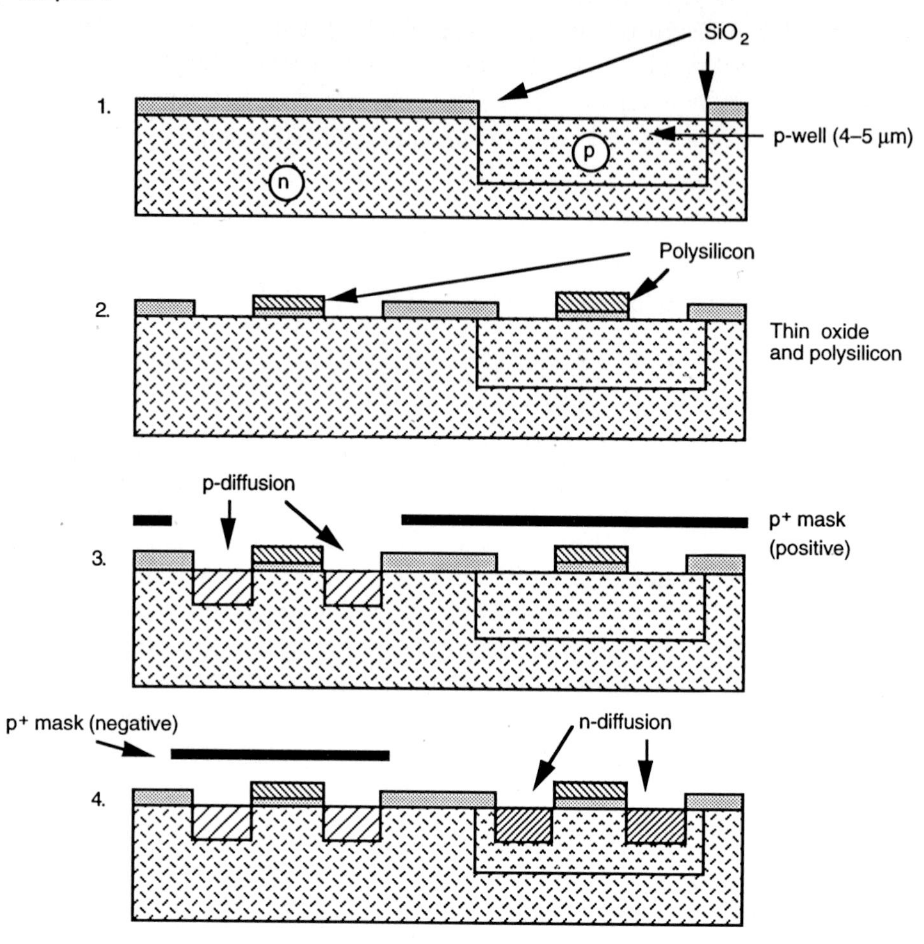

Figure 1–9 CMOS p-well process steps

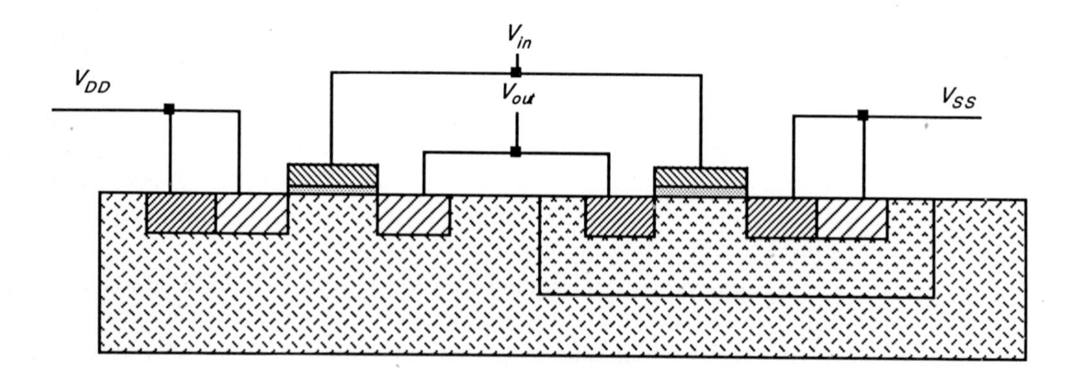

Figure 1–10 CMOS p-well inverter showing V_{DD} and V_{SS} substrate connections

In all other respects — masking, patterning, and diffusion — the process is similar to nMOS fabrication. In summary, typical processing steps are:

- *Mask* 1 — defines the areas in which the deep p-well diffusions are to take place.
- *Mask* 2 — defines the thinox regions, namely those areas where the thick oxide is to be stripped and thin oxide grown to accommodate p- and n-transistors and diffusion wires.
- *Mask* 3 — used to pattern the polysilicon layer which is deposited after the thin oxide.
- *Mask* 4 — A p-plus mask is now used (to be in effect 'Anded' with Mask 2) to define all areas where p-diffusion is to take place.
- *Mask* 5 — This is usually performed using the negative form of the p-plus mask and, with Mask 2, defines those areas where n-type diffusion is to take place.
- *Mask* 6 — Contact cuts are now defined.
- *Mask* 7 — The metal layer pattern is defined by this mask.
- *Mask* 8 — An overall passivation (overglass) layer is now applied and Mask 8 is needed to define the openings for access to bonding pads.

1.8.2 The n-well process

As indicated earlier, although the p-well process is widely used, n-well fabrication has also gained wide acceptance, initially as a retrofit to nMOS lines.

N-well CMOS circuits are also superior to p-well because of the lower substrate bias effects on transistor threshold voltage and inherently lower parasitic capacitances associated with source and drain regions.

Typical n-well fabrication steps are illustrated in Figure 1–11. The first mask defines the n-well regions. This is followed by a low dose phosphorus implant driven in by a high temperature diffusion step to form the n-wells. The well depth is optimized to ensure against p-substrate to p^+ diffusion breakdown without compromising the n-well to n^+ mask separation. The next steps are to define the devices and diffusion paths, grow field oxide, deposit and pattern the polysilicon, carry out the diffusions, make contact cuts, and finally metallize as before.

It will be seen that an n^+ mask and its complement may be used to define the n- and p-diffusion regions respectively. These same masks also include the V_{DD} and V_{SS} contacts (respectively). It should be noted that, alternatively, we could have used a p^+ mask and its complement, since the n^+ and p^+ masks are generally complementary.

By way of illustration, Figure 1–12 shows an inverter circuit fabricated by the n-well process, and this may be directly compared with Figure 1–10.

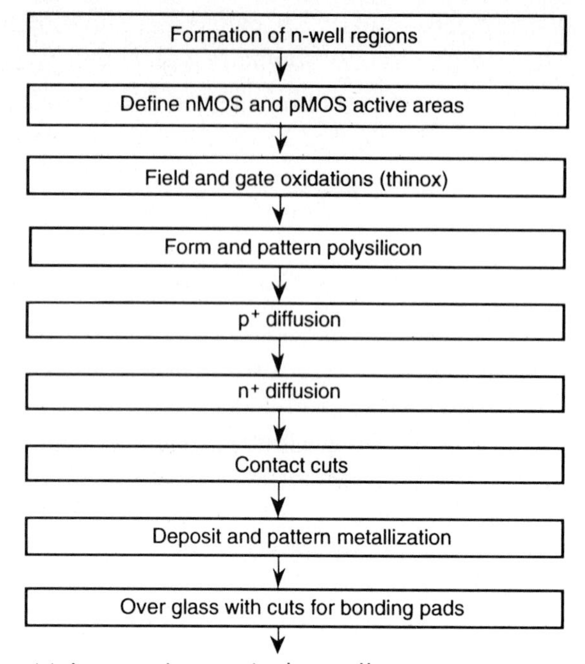

Figure 1–11 Main steps in a typical n-well process

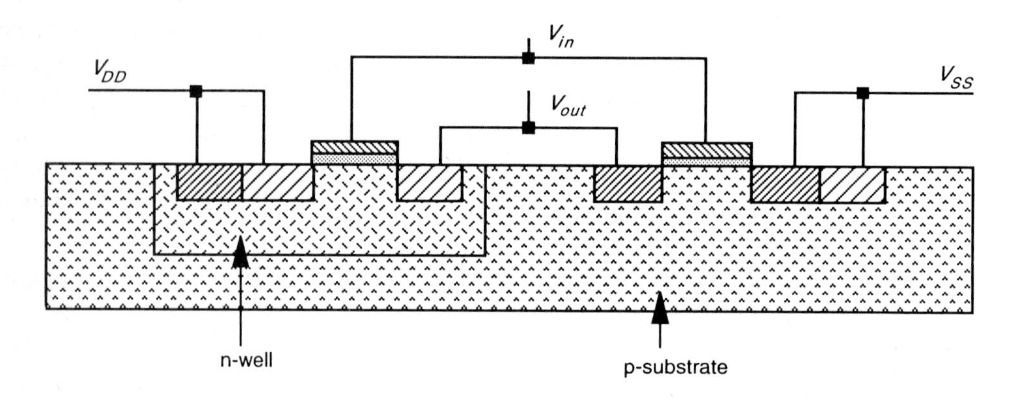

Figure 1–12 Cross-sectional view of n-well CMOS inverter

Owing to differences in charge carrier mobilities, the n-well process creates non-optimum p-channel characteristics. However, in many CMOS designs (such as domino-logic and dynamic-logic structures), this is relatively unimportant since they contain a preponderance of n-channel devices. Thus the n-channel transistors are mainly those used to form logic elements, providing speed and high density of elements.

Latch-up problems can be considerably reduced by using a low-resistivity epitaxial p-type substrate as the starting material, which can subsequently act as a very low resistance ground-plane to collect substrate currents.

However, a factor of the n-well process is that the performance of the already poorly performing p-transistor is even further degraded. Modern process lines have come to grips with these problems, and good device performance may be achieved for both p-well and n-well fabrication.

The design rules which are presented for 1.2 μm and 2 μm technologies in this text are for Orbit™ n-well processes.

1.8.2.1 The Berkeley n-well process

There are a number of p-well and n-well fabrication processes, and, in order to look more closely at typical fabrication steps, we will use the Berkeley n-well process as an example. This process is illustrated in Figure 1–13.

1.8.3 The twin-tub process

A logical extension of the p-well and n-well approaches is the twin-tub fabrication process.

Here we start with a substrate of high resistivity n-type material and then create both n-well and p-well regions. Through this process it is possible to preserve the performance of n-transistors without compromising the p-transistors. Doping control is more readily achieved and some relaxation in manufacturing tolerances results. This is particularly important as far as latch-up is concerned.

In general, the twin-tub process allows separate optimization of the n- and p-transistors. The arrangement of an inverter is illustrated in Figure 1–14, which may in turn be compared with Figures 1–10 and 1–12.

1.9 Thermal aspects of processing

The processes involved in making nMOS and CMOS devices have differing high temperature sequences as indicated in Figure 1–15.

The CMOS p-well process, for example, has a high temperature p-well diffusion process (1100 to 1250°C), the nMOS process having no such requirement. Because of the simplicity, ease of fabrication, and high density per unit area of nMOS circuits, many of the earlier IC designs, still in current use, have been fabricated using nMOS technology, and it is likely that nMOS and CMOS system designs will continue to coexist for some time to come.

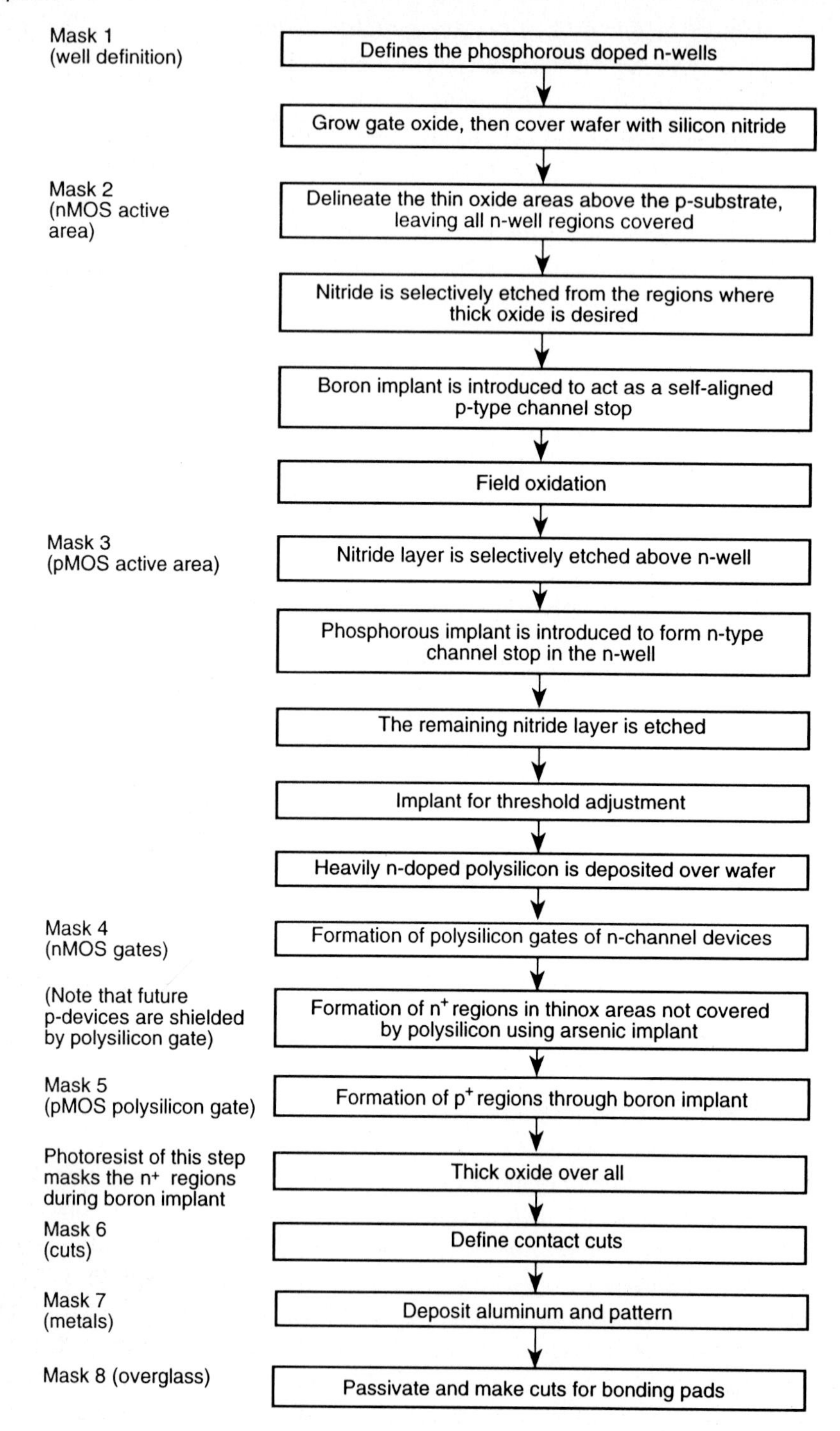

Figure 1–13 Flow diagram of Berkeley n-well fabrication

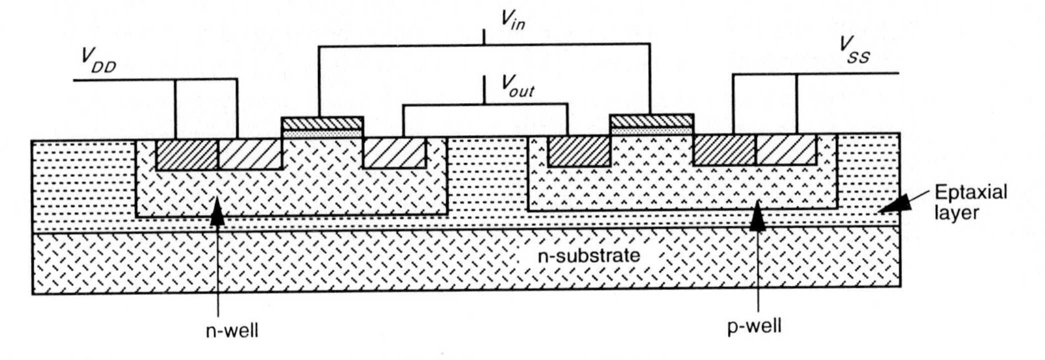

Figure 1–14 Twin-tub structure

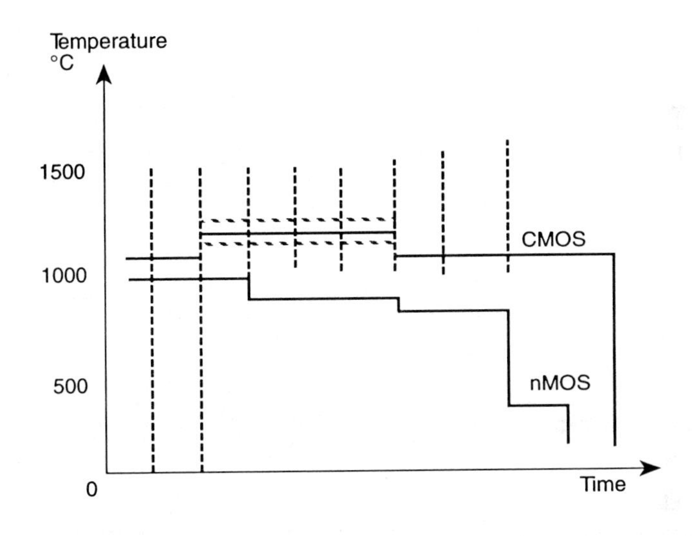

Figure 1–15 Thermal sequence difference between nMOS and CMOS processes

1.10 BiCMOS technology

A known deficiency of MOS technology lies in the limited load driving capabilities of MOS transistors. This is due to the limited current sourcing and current sinking abilities associated with both p- and n-transistors and although it is possible, for example, to design so-called super buffers using MOS transistors alone, such arrangements do not always compare well with the capabilities of bipolar transistors. Bipolar transistors also provide higher gain and have generally better noise and high frequency characteristics than MOS transistors and it may be seen (Figure 1–2) that using BiCMOS gates may be an effective way of speeding up VLSI circuits. However, the application of BiCMOS in subsystems such as ALU, ROM, a register-file, or, for that matter, a barrel shifter is not always an effective way of improving speed. This is because most gates in such structures do not have to drive large capacitive loads so that the BiCMOS arrangements give no speed advantage. To take advantage of BiCMOS, the whole functional entity, not just the logic gates, must be considered. A comparison between the characteristics of CMOS and bipolar circuits is set out in Table 1–2 and the differences are self-evident. BiCMOS technology goes some way toward combining the virtues of both technologies.

When considering CMOS technology, it becomes apparent that theoretically there should be little difficulty in extending the fabrication processes to include bipolar as well as MOS transistors. Indeed, a problem of p-well and n-well CMOS processing is that parasitic bipolar transistors are inadvertently formed as part of the outcome of fabrication. The production of npn bipolar transistors with good performance characteristics can be achieved, for example, by extending the standard n-well CMOS processing to include further masks to add two additional layers — the n^+ subcollector and p^+ base layers. The npn transistor is formed in an n-well and the additional p^+ base region is located in the well to form the p-base region of the transistor. The second additional layer, the buried n^+ subcollector (BCCD), is added to reduce the n-well (collector) resistance and thus improve the quality of the bipolar transistor. The simplified general arrangement of such a bipolar npn transistor may be appreciated with regard to Figure 1–16. Bipolar transistor characteristics will follow in Chapter 2 and the relevant design rules

Table 1–2 Comparisons between CMOS and bipolar technologies

CMOS technology	*Bipolar technology*
• Low static power dissipation	• High power dissipation
• High input impedance (low drive current)	• Low input impedance (high drive current)
• Scalable threshold voltage	
• High noise margin	• Low voltage swing logic
• High packing density	• Low packing density
• High delay sensitivity to load (fan-out limitations)	• Low delay sensitivity to load
• Low output drive current	• High output drive current
• Low g_m ($g_m \alpha V_{in}$)	• High g_m ($g_m \alpha e^{V_{in}}$)
	• High f_t at low currents
• Bidirectional capability (drain and source are interchangeable)	• Essentially unidirectional
• A near ideal switching device	

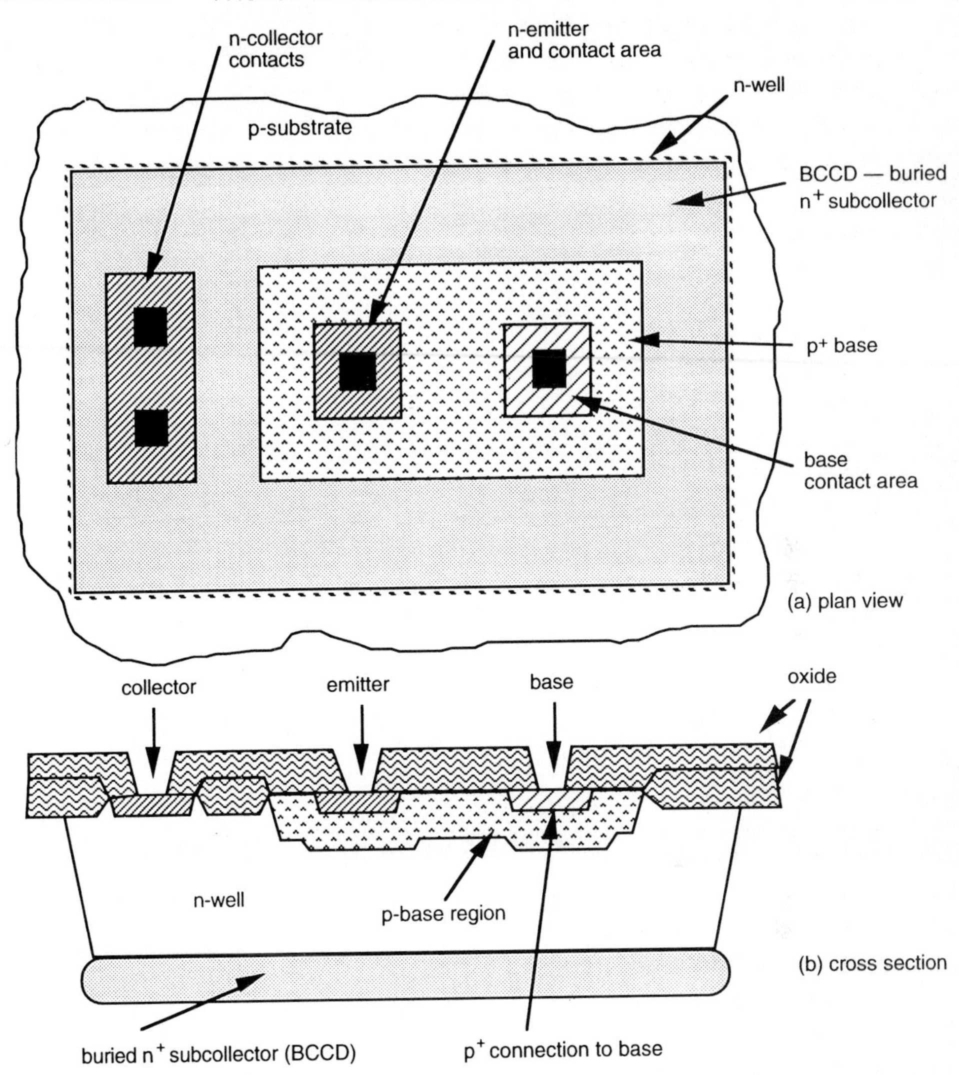

Note: For clarity, the layers have not been drawn transparent but BCCD underlies the entire area and the p^+ base underlies all within its boundary.

Figure 1–16 Arrangement of BiCMOS npn transistor (Orbit 2 μm CMOS)

are dealt with in Chapter 3. A quick appraisal of Figure 3–13(f) will serve to further illustrate the actual geometry of a BiCMOS bipolar transistor in n-well technology. Since extra design and processing steps are involved, there is an inevitable increase in cost and this is reflected in Figure 1–17, which also includes ECL and GaAs gates for cost comparison.

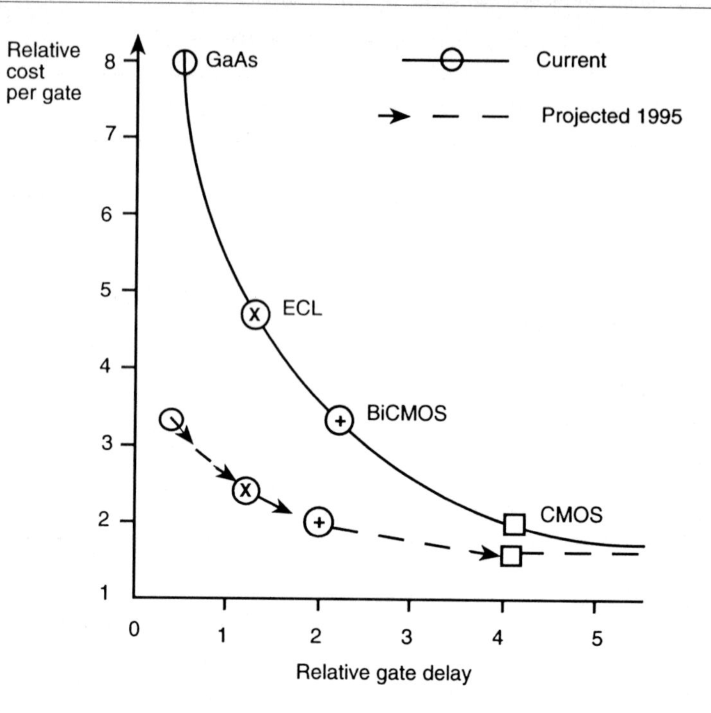

Figure 1–17 Cost versus delay for logic gate

1.10.1 BiCMOS fabrication in an n-well process

The basic process steps used are those already outlined for CMOS but with additional process steps and additional masks defining (i) the p^+ base region; (ii) n^+ collector area; and (iii) the buried subcollector (BCCD).

Table 1–3 sets out the process steps for a single poly. single metal CMOS n-well process, showing the additional process steps for the bipolar devices.

1.10.2 Some aspects of bipolar and CMOS devices

Clearly there are relative advantages and disadvantages when comparing bipolar technology with CMOS technology. A readily assimilated comparison of some key features was set out in Table 1–2.

It will be seen that there are several advantages if the properties of CMOS and bipolar technologies could be combined. This is achieved to a significant extent in the BiCMOS technology. As in all things, there is a penalty which, in this case, arises from the additional process steps, some loss of packing density and thus higher cost.

Table 1–3 n-well BiCMOS fabrication process steps

Single poly. single metal CMOS	*Additional steps for bipolar devices*
• Form n-well	
	• Form buried n^+ layer (BCCD)
• Delineate active areas	
• Channel stop	
	• Form deep n^+ collector
• Threshold V_t adjustment	
• Delineate poly./gate areas	
• Form n^+ active areas	
• Form p^+ active areas	
	• Form p^+ base for bipolars
• Define contacts	
• Delineate the metal areas	

A cost comparison of all current high speed technologies may be assessed from Figure 1–17.

A further advantage which arises from BiCMOS technology is that analog amplifier design is facilitated and improved. High impedance CMOS transistors may be used for the input circuitry while the remaining stages and output drivers are realized using bipolar transistors.

To take maximum advantage of available silicon technologies one might envisage the following mix of technologies in a silicon system:

CMOS	for logic
BiCMOS	for I/O and driver circuits
ECL	for critical high speed parts of the system

However, in this text we will not be dealing with the ECL technology.

1.11 Production of E-beam masks

All the processes discussed have made use of masks at various stages of fabrication. In many processes, the masks are produced by standard optical techniques and much has been written on the photolithographic processes involved. However, as geometric dimensions shrink, and also to allow for the processing of a number of different chip designs on a single wafer, other techniques are evolving. One popular process employed for this purpose uses an E-beam machine. A rough outline of this type of mask making follows:

1. The starting material consists of chrome-plated glass plates which are coated with an E-beam sensitive resist.
2. The E-beam machine is loaded with the mask description data (MEBES).
3. Plates are loaded into the E-beam machine, where they are exposed with the patterns specified by the customer's mask data.
4. After exposure to the E-beam, the plates are introduced into a developer to bring out the patterns left by the E-beam in the resist coating.
5. The cycle is followed by a bake cycle and a plasma de-summing, which removes the resist residue.
6. The chrome is then etched and the plate is stripped of the remaining E-beam resist.

The advantages of E-beam masks are:

- tighter layer to layer registration;
- smaller feature sizes.

There are two approaches to the design of E-beam machines:

- raster scanning;
- vector scanning.

In the first case, the electron beam scans all possible locations (in a similar fashion to a television display), and a bit map is used to turn the E-beam on and off, depending on whether the particular location being scanned is to be exposed or not.

For vector scanning, the beam is directed only to those locations which are to be exposed. Although this is inherently faster, the data handling involved is more complex.

1.12 Observations

This chapter has set the scene by introducing the basically simple MOS transistor structures and the relatively straightforward fabrication processes used in the manufacture of nMOS, CMOS and BiCMOS circuits. We have also attempted to emphasize the revolutionary spread of integrated circuit technology which has, in the short space of 30 years, advanced to a point where we now see highly complex systems completely integrated onto a single chip.

Although this text concentrates on digital circuits and systems, similar techniques can be applied to the design and fabrication of analog devices. Indeed, the trends are toward systems of VLSI (and beyond) complexity which will in future include, on single chips, significant analog interfaces and other appropriate circuitry. This higher level of integration will lead to fewer packages and interconnections and to more complex systems than today. There will be a marked beneficial effect on cost and reliability of the systems that will be available to all professions and disciplines and in most aspects of everyday life.

Our discussions of fabrication have in some instances simplified the processes used in order to reveal or emphasize the essential features. Indeed, the fabrication of similar devices by different fabricators may vary considerably in detail. This is also the case with the design rules (see Chapter 3) which are specified by the fabricator. Design rules will be introduced via the concept of 'lambda-based' rules, which are a result of the work of Mead and Conway, and although not producing the tightest layouts, these rules are acceptable to many fabricators. A study of lambda-based rules also provides a good way of absorbing the essential concepts underlying any set of design rules. However, the text also gives an up-to-date set of real world 'micron-based' rules for 2 μm and for 1.2 μm n-well CMOS technologies which may be used when the designer reaches an acceptable level of competence. The 2 μm rule set is for a BiCMOS process and thus also provides for bipolar npn transistors. It must be noted here that '2 μm technology', for example, means that the minimum line width (and, consequently, the typical feature size of the geometry) of the chip layout will be 2 μm.

In order to understand the basic features MOS and BiCMOS IC technologies, we must now look into the basic electrical properties.

2 Basic electrical properties of MOS and BiCMOS circuits

There is no virtue in not knowing what can be known.

Aldous Huxley

Objectives

If design is to be effectively carried out, or indeed if the performance of circuits realized in MOS technology is to be properly understood, then the practitioner must have a sound understanding of the MOS active devices.

This chapter establishes the basic characteristics of MOS transistor and examines various possibilities for configuring inverter circuits. In the case of nMOS circuits the need for and values of the ratio rules are established.

Discussion then extends to the characteristics of BiCMOS transistors and the ensuing inverter circuitry.

Finally, aspects of latch-up are considered for CMOS and BiCMOS devices. Having introduced the MOS transistor and the processes used to produce it, we are now in a position to gain some understanding of the electrical characteristics of the basic MOS circuits — enhancement and depletion mode transistors and inverters. Our considerations will be based on reasonable approximations so that the essential features can be evaluated and illustrated in a concise and easily absorbed manner. VLSI designers should have a good knowledge of the behavior of the circuits they are designing or designing with. Even if large systems are being designed, using computer-aided design processes, it is essential that the designs be based on a sound foundation of understanding if those systems are to meet performance specifications.

The following expressions and discussion relate directly to nMOS transistors, but pMOS expressions are also given where appropriate and, generally, a reversal of voltage and current polarities of nMOS expressions and the exchange of μ_n for μ_p and electrons for holes will yield pMOS from nMOS expressions.

We will then briefly discuss some bipolar transistor characteristics which are relevant to an understanding of BiCMOS circuits. Bipolar transistor parameters are also compared with comparable parameters for CMOS transistors.

2.1 Drain-to-source current I_{ds} versus voltage V_{ds} relationships

The whole concept of the MOS transistor evolves from the use of a voltage on the gate to induce a charge in the channel between source and drain, which may then be caused to move from source to drain under the influence of an electric field created by voltage V_{ds} applied between drain and source. Since the charge induced is dependent on the gate to source voltage V_{gs}, then I_{ds} is dependent on both V_{gs} and V_{ds}. Consider a structure, as in Figure 2–1, in which electrons will flow source to drain:

$$I_{ds} = -I_{sd} = \frac{\text{Charge induced in channel } (Q_C)}{\text{Electron transit time } (\tau)} \tag{2.1}$$

First, transit time:

$$\tau_{sd} = \frac{\text{Length of channel } (L)}{\text{Velocity } (v)}$$

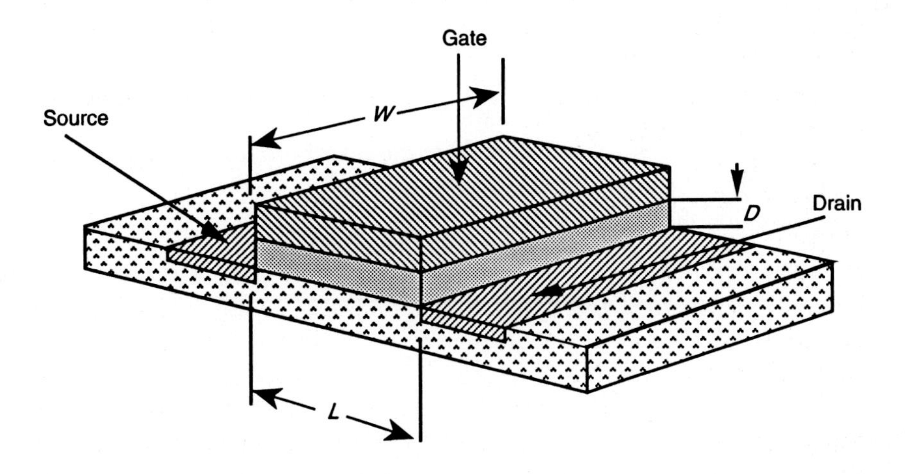

Figure 2–1 nMOS transistor structure

but velocity

$$v = \mu E_{ds}$$

where

$$\mu = \text{electron or hole mobility (surface)}$$
$$E_{ds} = \text{electric field (drain to source)}$$

Now

$$E_{ds} = \frac{V_{ds}}{L}$$

so that

$$v = \frac{\mu V_{ds}}{L}$$

Thus

$$\tau_{sd} = \frac{L^2}{\mu V_{ds}} \tag{2.2}$$

Typical values of μ at room temperature are:

$$\mu_n \div 650 \text{ cm}^2/\text{V sec (surface)}$$
$$\mu_p \div 240 \text{ cm}^2/\text{V sec (surface)}$$

2.1.1 The non-saturated region

Charge induced in channel due to gate voltage is due to the voltage difference between the gate and the channel, V_{gs} (assuming substrate connected to source). Now note that the voltage along the channel varies linearly with distance X from the source due to the IR drop in the channel (see Figure 1–5) and assuming that the device is not saturated then the average value is $V_{ds}/2$. Furthermore, the effective gate voltage $V_g = V_{gs} - V_t$ where V_t is the threshold voltage needed to invert the charge under the gate and establish the channel.

Note that the charge/unit area = $E_g \varepsilon_{ins} \varepsilon_0$. Thus induced charge

$$Q_c = E_g \varepsilon_{ins} \varepsilon_0 WL$$

where

$$E_g = \text{average electric field gate to channel}$$
$$\varepsilon_{ins} = \text{relative permittivity of insulation between gate and channel}$$
$$\varepsilon_0 = \text{permittivity of free space}$$

(*Note*: $\varepsilon_0 = 8.85 \times 10^{-14}$F cm^{-1}; $\varepsilon_{ins} \div 4.0$ for silicon dioxide)

Now

$$E_g = \frac{\left((V_{gs} - V_t) - \frac{V_{ds}}{2}\right)}{D}$$

where D = oxide thickness.

Thus

$$Q_c = \frac{WL\varepsilon_{ins}\varepsilon_0}{D}\left((V_{gs} - V_t) - \frac{V_{ds}}{2}\right) \tag{2.3}$$

Now, combining equations (2.2) and (2.3) in equation (2.1), we have

$$I_{ds} = \frac{\varepsilon_{ins}\varepsilon_0\mu}{D}\frac{W}{L}\left((V_{gs} - V_t) - \frac{V_{ds}}{2}\right)V_{ds}$$

or

$$I_{ds} = K\frac{W}{L}\left((V_{gs} - V_t)V_{ds} - \frac{V_{ds}^2}{2}\right) \tag{2.4}$$

in the non-saturated or resistive region where $V_{ds} < V_{gs} - V_t$ and

$$K = \frac{\varepsilon_{ins}\varepsilon_0\mu}{D}$$

The factor W/L is, of course, contributed by the geometry and it is common practice to write

$$\beta = K\frac{W}{L}$$

so that

$$I_{ds} = \beta\left((V_{gs} - V_t)V_{ds} - \frac{V_{ds}^2}{2}\right) \tag{2.4a}$$

which is an alternative form of equation 2.4.

Noting that gate/channel capacitance

$$C_g = \frac{\varepsilon_{ins}\varepsilon_0 WL}{D} \text{ (parallel plate)}$$

we also have

$$K = \frac{C_g\mu}{WL}$$

so that

$$I_{ds} = \frac{C_g \mu}{L^2}\left((V_{gs} - V_t)V_{ds} - \frac{V_{ds}^{\ 2}}{2}\right) \tag{2.4b}$$

which is a further alternative form of equation 2.4.

Sometimes it is convenient to use *gate capacitance per unit area* C_0 (which is often denoted C_{ox}) rather than C_g in this and other expressions. Noting that

$$C_g = C_0\, WL$$

we may also write

$$I_{ds} = C_0 \mu \frac{W}{L}\left((V_{gs} - V_t)V_{ds} - \frac{V_{ds}^{\ 2}}{2}\right) \tag{2.4c}$$

2.1.2 The saturated region

Saturation begins when $V_{ds} = V_{gs} - V_t$ since at this point the IR drop in the channel equals the effective gate to channel voltage at the drain and we may assume that the current remains fairly constant as V_{ds} increases further. Thus

$$I_{ds} = K\frac{W}{L}\frac{(V_{gs} - V_t)^2}{2} \tag{2.5}$$

or, we may write

$$I_{ds} = \frac{\beta}{2}(V_{gs} - V_t)^2 \tag{2.5a}$$

or

$$I_{ds} = \frac{C_g \mu}{2L^2}(V_{gs} - V_t)^2 \tag{2.5b}$$

We may also write

$$I_{ds} = C_0 \mu \frac{W}{2L}\left(V_{gs} - V_t\right)^2 \tag{2.5c}$$

The expressions derived for I_{ds} hold for both enhancement and depletion mode devices, but it should be noted that the threshold voltage for the nMOS depletion mode device (denoted as V_{td}) is *negative.*

Typical characteristics for nMOS transistors are given in Figure 2–2. pMOS transistor characteristics are similar, with suitable reversal of polarity.

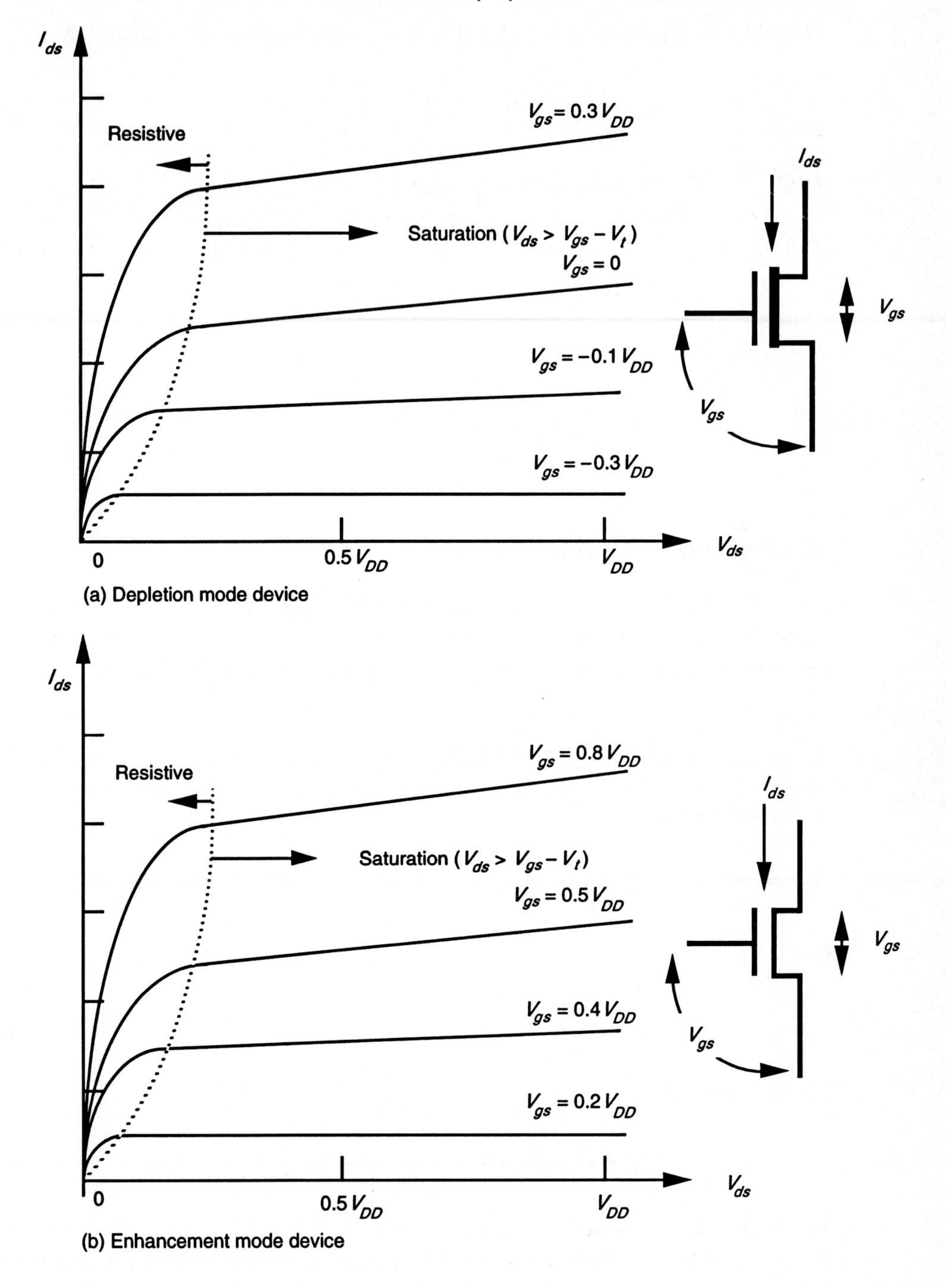

Figure 2–2 MOS transistor characteristics

2.2 Aspects of MOS transistor threshold voltage V_t

The gate structure of a MOS transistor consists, electrically, of charges stored in the dielectric layers and in the surface to surface interfaces as well as in the substrate itself.

Switching an enhancement mode MOS transistor from the off to the on state consists in applying sufficient gate voltage to neutralize these charges and enable the underlying silicon to undergo an inversion due to the electric field from the gate.

Switching a depletion mode nMOS transistor from the on to the off state consists in applying enough voltage to the gate to add to the stored charge and invert the 'n' implant region to 'p'.

The threshold voltage V_t may be expressed as:

$$V_t = \phi_{ms} \frac{Q_B - Q_{SS}}{C_0} + 2\phi_{fN} \tag{2.6}$$

where

Q_B = the charge per unit area in the depletion layer beneath the oxide
Q_{SS} = charge density at Si:SiO_2 interface
C_0 = capacitance per unit gate area
ϕ_{ms} = work function difference between gate and Si
ϕ_{fN} = Fermi level potential between inverted surface and bulk Si.

Now, for polysilicon gate and silicon substrate, the value of ϕ_{ms} is negative but negligible, and the magnitude and sign of V_t are thus determined by the balance between the remaining negative term $\frac{-Q_{SS}}{C_0}$ and the other two terms, both of which are positive. To evaluate V_t, each term is determined as follows:

$$Q_B = \sqrt{2\varepsilon_0 \varepsilon_{Si} q N (2\phi_{fN} + V_{SB})} \text{ coulomb/m}^2$$

$$\phi_{fN} = \frac{kT}{q} \ln \frac{N}{n_i} \text{ volts}$$

$$Q_{SS} = (1.5 \text{ to } 8) \times 10^{-8} \text{ coulomb/m}^2$$

depending on crystal orientation, and where

V_{SB} = substrate bias voltage (negative w.r.t. source for nMOS, positive for pMOS)
$q = 1.6 \times 10^{-19}$ coulomb
N = impurity concentration in the substrate (N_A or N_D as appropriate)
ε_{Si} = relative permittivity of silicon $\doteqdot$ 11.7)
n_i = intrinsic electron concentration ($1.6 \times 10^{10}/cm^3$ at 300°K)
k = Boltzmann's constant = 1.4×10^{-23} joule/°K

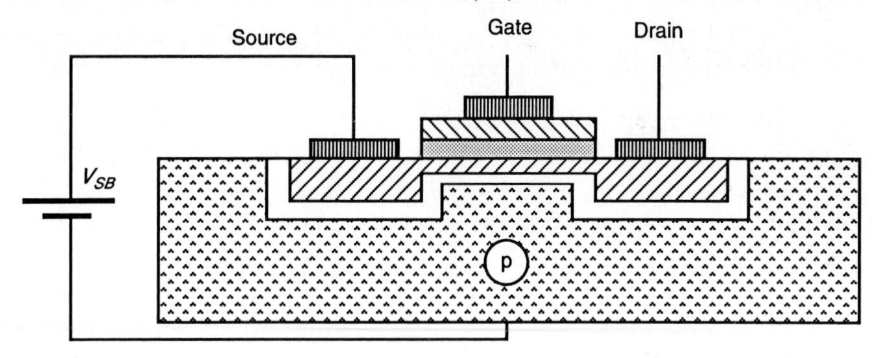

Figure 2–3 Body effect (nMOS device shown)

The *body effects* may also be taken into account since the substrate may be biased with respect to the source, as shown in Figure 2–3.

Increasing V_{SB} causes the channel to be depleted of charge carriers and thus the threshold voltage is raised.

Change in V_t is given by $\Delta V_t \doteqdot \gamma (V_{SB})^{1/2}$ where γ is a constant which depends on substrate doping so that the more lightly doped the substrate, the smaller will be the body effect.

Alternatively, we may write

$$V_t = V_t(0) + \left(\frac{D}{\varepsilon_{ins}\varepsilon_0} \right) \sqrt{2\varepsilon_0 \varepsilon_{Si} QN}.(V_{SB})^{1/2}$$

where $V_t(0)$ is the threshold voltage for $V_{SB} = 0$.

To establish the magnitude of such effects, typical figures for V_t are as follows:
For nMOS enhancement mode transistors:

$V_{SB} = 0$ V; $V_t = 0.2V_{DD}$ (= + 1 V for V_{DD} = + 5 V)
$V_{SB} = 5$ V; $V_t = 0.3V_{DD}$ (= + 1.5 V for V_{DD} = + 5 V)
} Similar but negative values for pMOS

For nMOS depletion mode transistors:

$V_{SB} = 0$ V; $V_{td} = -0.7V_{DD}$ (= − 3.5 V for V_{DD} = + 5 V)
$V_{SB} = 5$ V; $V_{td} = -0.6V_{DD}$ (= − 3.0 V for V_{DD} = + 5 V)

2.3 MOS transistor transconductance g_m and output conductance g_{ds}

Transconductance expresses the relationship between output current I_{ds} and the input voltage V_{gs} and is defined as

$$g_m = \frac{\delta I_{ds}}{\delta V_{gs}} \Big| V_{ds} = \text{constant}$$

To find an expression for g_m in terms of circuit and transistor parameters, consider that the charge in channel Q_c is such that

$$\frac{Q_c}{I_{ds}} = \tau$$

where τ is transit time. Thus change in current

$$\delta I_{ds} = \frac{\delta Q_c}{\tau_{ds}}$$

Now

$$\tau_{ds} = \frac{L^2}{\mu V_{ds}} \qquad \text{(from 2.2)}$$

Thus

$$\delta I_{ds} = \frac{\delta Q_c V_{ds} \mu}{L^2}$$

but change in charge

$$\delta Q_c = C_g \delta V_{gs}$$

so that

$$\delta I_{ds} = \frac{C_g \delta V_{gs} \mu V_{ds}}{L^2}$$

Now

$$g_m = \frac{\delta I_{ds}}{\delta V_{gs}} = \frac{C_g \mu V_{ds}}{L^2}$$

In saturation

$$V_{ds} = V_{gs} - V_t$$

$$g_m = \frac{C_g \mu}{L^2}(V_{gs} - V_t) \qquad (2.7)$$

and substituting for $C_g = \frac{\varepsilon_{ins} \varepsilon_0 WL}{D}$

$$g_m = \frac{\mu \varepsilon_{ins} \varepsilon_0}{D} \frac{W}{L}(V_{gs} - V_t) \qquad (2.7a)$$

Alternatively,

$$g_m = \beta(V_{gs} - V_t)$$

It is possible to increase the g_m of a MOS device by increasing its width. However, this will also increase the input capacitance and area occupied.

A reduction in the channel length results in an increase in ω_0 owing to the higher g_m. However, the gain of the MOS device decreases owing to the strong degradation of the output resistance = $1/g_{ds}$.

The output conductance g_{ds} can be expressed by

$$g_{ds} = \frac{\delta I_{ds}}{\delta V_{gs}} = \lambda . I_{ds} \, \alpha \left(\frac{1}{L} \right)^2$$

Here the strong dependence on the channel length is demonstrated as

$$\lambda \alpha \left(\frac{1}{L} \right) \text{ and } I_{ds} \, \alpha \left(\frac{1}{L} \right)$$

for the MOS device.

2.4 MOS transistor figure of merit ω_0

An indication of frequency response may be obtained from the parameter ω_0 where

$$\omega_0 = \frac{g_m}{C_g} = \frac{\mu}{L^2}(V_{gs} - V_t)\left(= \frac{1}{\tau_{sd}} \right) \tag{2.8}$$

This shows that switching speed depends on gate voltage above threshold and on carrier mobility and inversely as the square of channel length. A fast circuit requires that g_m be as high as possible.

Electron mobility on a (100) oriented n-type inversion layer surface (μ_n) is larger than that on a (111) oriented surface, and is in fact about three times as large as hole mobility on a (111) oriented p-type inversion layer. Surface mobility is also dependent on the effective gate voltage ($V_{gs} - V_t$).

For faster nMOS circuits, then, one would choose a (100) oriented p-type substrate in which the inversion layer will have a surface carrier mobility $\mu_n \doteqdot 650$ cm²/V sec at room temperature.

Compare this with the typical bulk mobilities

$$\mu_n = 1250 \text{ cm}^2/\text{V sec}$$
$$\mu_p = 480 \text{ cm}^2 \text{ V sec}$$

from which it will be seen that $\frac{\mu_s}{\mu} \doteqdot 0.5$ (where μ_s = surface mobility and μ = bulk mobility).

2.5 The pass transistor

Unlike bipolar transistors, the isolated nature of the gate allows MOS transistors to be used as switches in series with lines carrying logic levels in a way that is similar to the use of relay contacts. This application of the MOS device is called the *pass transistor* and switching logic arrays can be formed — for example, an *And* array as in Figure 2–4.

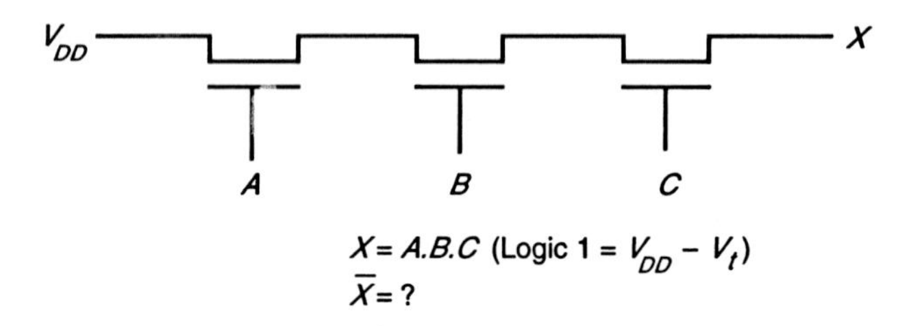

Note: Means must exist so that X assumes ground potential when $A + B + C = 0$.

Figure 2–4 Pass transistor *And* gate

2.6 The nMOS inverter

A basic requirement for producing a complete range of logic circuits is the inverter. This is needed for restoring logic levels, for *Nand* and *Nor* gates, and for sequential and memory circuits of various forms. In the treatment of the inverter used in this section, the authors wish to acknowledge the influence of material previously published by Mead and Conway.

The basic inverter circuit requires a transistor with source connected to ground and a load resistor of some sort connected from the drain to the positive supply rail V_{DD}. The output is taken from the drain and the input applied between gate and ground.

Resistors are not conveniently produced on the silicon substrate; even modest values occupy excessively large areas so that some other form of load resistance is required. A convenient way to solve this problem is to use a depletion mode transistor as the load, as shown in Figure 2–5.

Now:

- With no current drawn from the output, the currents I_{ds} for both transistors must be equal.
- For the depletion mode transistor, the gate is connected to the source so it is always on and only the characteristic curve $V_{gs} = 0$ is relevant.
- In this configuration the depletion mode device is called the pull-up (p.u.) and the enhancement mode device the pull-down (p.d.) transistor.

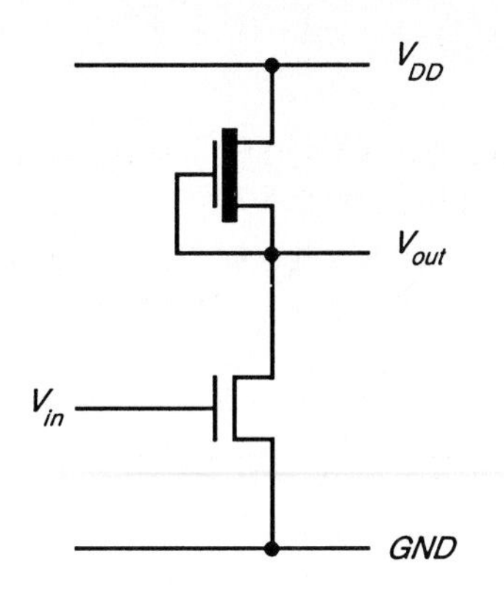

Figure 2–5 nMOS inverter

- To obtain the inverter transfer characteristic we superimpose the $V_{gs} = 0$ depletion mode characteristic curve on the family of curves for the enhancement mode device, noting that maximum voltage across the enhancement mode device corresponds to minimum voltage across the depletion mode transistor.
- The points of intersection of the curves as in Figure 2–6 give points on the transfer characteristic, which is of the form shown in Figure 2–7.

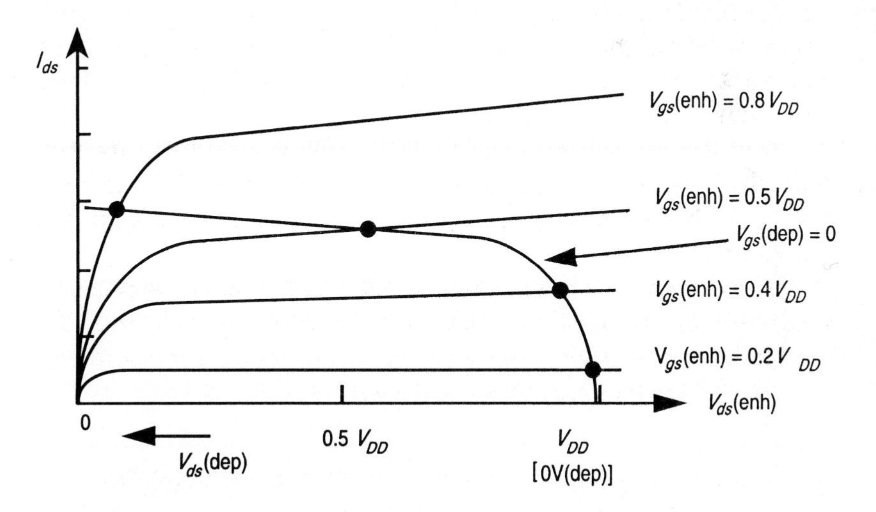

$V_{ds}(\text{enh}) = V_{DD} - V_{ds}(\text{dep}) = V_{out}$

$V_{gs}(\text{enh}) = V_{in}$ . . . intersection points give transfer characteristic

Figure 2–6 Derivation of nMOS inverter transfer characteristic

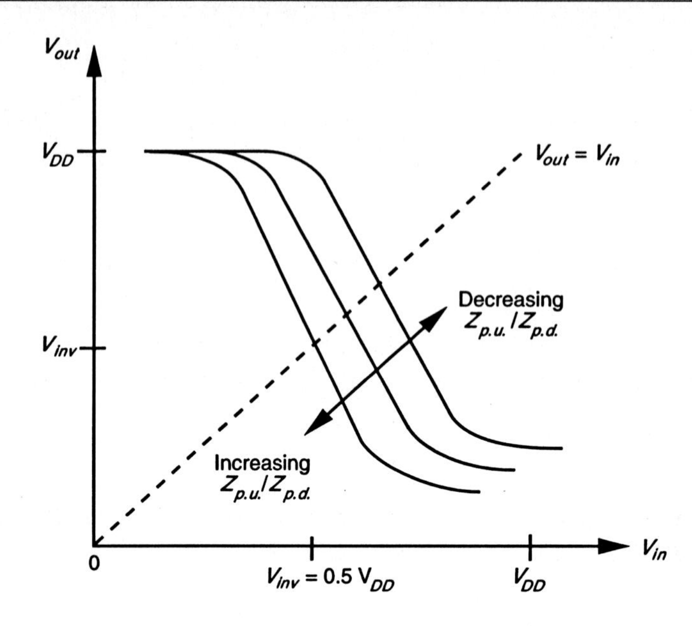

Figure 2–7 nMOS inverter transfer characteristic

- Note that as V_{in} (= V_{gs} p.d. transistor) exceeds the p.d. threshold voltage current begins to flow. The output voltage V_{out} thus decreases and the subsequent increases in V_{in} will cause the p.d. transistor to come out of saturation and become resistive. Note that the p.u. transistor is initially resistive as the p.d. turns on.
- During transition, the slope of the transfer characteristic determines the gain:

$$\text{Gain} = \frac{\delta V_{out}}{\delta V_{in}}$$

- The point at which $V_{out} = V_{in}$ is denoted as V_{inv} and it will be noted that the transfer characteristics and V_{inv} can be shifted by variation of the ratio of pull-up to pull-down resistances (denoted $Z_{p.u.}/Z_{p.d.}$ where Z is determined by the length to width ratio of the transistor in question).

2.7 Determination of pull-up to pull-down ratio ($Z_{p.u.}/Z_{p.d.}$) for an nMOS inverter driven by another nMOS inverter

Consider the arrangement in Figure 2–8 in which an inverter is driven from the output of another similar inverter. Consider the depletion mode transistor for

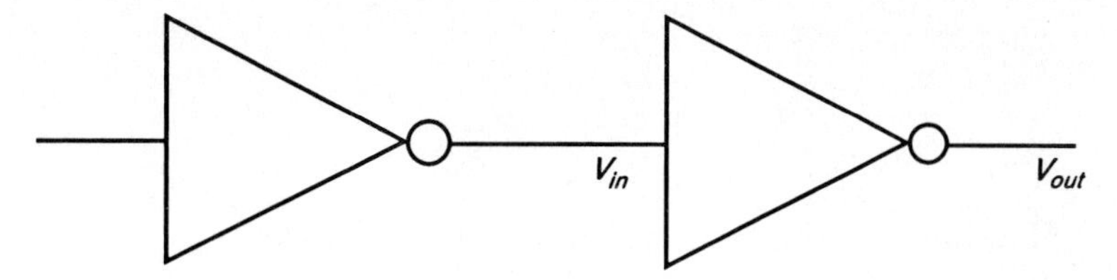

Figure 2–8 nMOS inverter driven directly by another inverter

which $V_{gs} = 0$ under all conditions, and further assume that in order to cascade inverters without degradation of levels we are aiming to meet the requirement

$$V_{in} = V_{out} = V_{inv}$$

For equal margins around the inverter threshold, we set $V_{inv} = 0.5V_{DD}$. At this point both transistors are in saturation and

$$I_{ds} = K\frac{W}{L}\frac{(V_{gs} - V_t)^2}{2}$$

In the depletion mode

$$I_{ds} = K\frac{W_{p.u.}}{L_{p.u.}}\frac{(-V_{td})^2}{2} \quad \text{since } V_{gs} = 0$$

and in the enhancement mode

$$I_{ds} = K\frac{W_{p.d.}}{L_{p.d.}}\frac{(V_{inv} - V_t)^2}{2} \quad \text{since } V_{gs} = V_{inv}$$

Equating (since currents are the same) we have

$$\frac{W_{p.d.}}{L_{p.d.}}(V_{inv} - V_t)^2 = \frac{W_{p.u.}}{L_{p.u.}}(-V_{td})^2$$

where $W_{p.d.}$, $L_{p.d.}$, $W_{p.u.}$, and $L_{p.u.}$ are the widths and lengths of the pull-down and pull-up transistors respectively.

Now write

$$Z_{p.d.} = \frac{L_{p.d.}}{W_{p.d.}}; Z_{p.u.} = \frac{L_{p.u.}}{W_{p.u.}}$$

we have

$$\frac{1}{Z_{p.d.}}(V_{inv} - V_t)^2 = \frac{1}{Z_{p.u.}}(-V_{td})^2$$

whence

$$V_{inv} = V_t - \frac{V_{td}}{\sqrt{Z_{p.u.}/Z_{p.d.}}} \qquad (2.9)$$

Now we can substitute typical values as follows

$$V_t = 0.2V_{DD};\ V_{td} = -0.6V_{DD}$$

$$V_{inv} = 0.5V_{DD} \text{ (for equal margins)}$$

thus, from equation (2.9)

$$0.5 = 0.2 + \frac{0.6}{\sqrt{Z_{p.u.}/Z_{p.d.}}}$$

whence

$$\sqrt{Z_{p.u.}/Z_{p.d.}} = 2$$

and thus

$$Z_{p.u.}/Z_{p.d.} = 4/1$$

for an inverter directly driven by an inverter.

2.8 Pull-up to pull-down ratio for an nMOS inverter driven through one or more pass transistors

Now consider the arrangement of Figure 2–9 in which the input to inverter 2 comes from the output of inverter 1 but passes through one or more nMOS transistors used as switches in series (called *pass transistors*).

We are concerned that connection of pass transistors in series will degrade the logic 1 level into inverter 2 so that the output will not be a proper logic 0 level. The critical condition is when point A is at 0 volts and *B* is thus at V_{DD}, but the voltage into inverter 2 at point *C* is now reduced from V_{DD} by the threshold

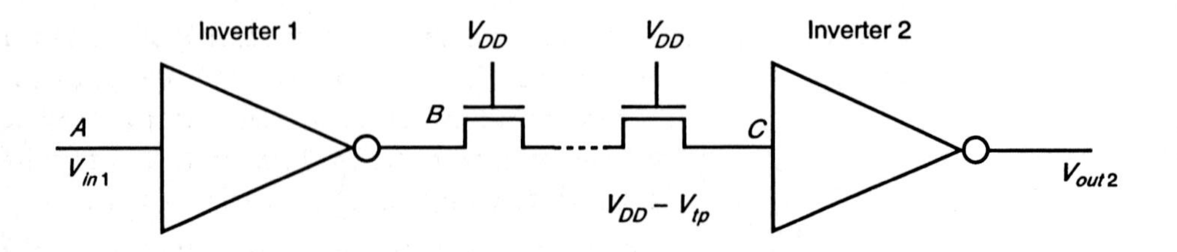

Figure 2–9 Pull-up to pull-down ratios for inverting logic coupled by pass transistors

voltage of the series pass transistor. With all pass transistor gates connected to V_{DD} (as shown in Figure 2–8), there is a loss of V_{tp}, however many are connected in series, since no static current flows through them and there can be no voltage drop in the channels. Therefore, the input voltage to inverter 2 is

$$V_{in2} = V_{DD} - V_{tp}$$

where

$$V_{tp} = \text{threshold voltage for a pass transistor}$$

We must now ensure that for this input voltage we get out the same voltage as would be the case for inverter 1 driven with input = V_{DD}.

Consider inverter 1 (Figure 2–10(a)) with input = V_{DD}. If the input is at V_{DD}, then the p.d. transistor T_2 is conducting but with a low voltage across it; therefore, it is in its resistive region represented by R_1 in Figure 2–10. Meanwhile, the p.u. transistor T_1 is in saturation and is represented as a current source.

For the p.d. transistor

$$I_{ds} = K\frac{W_{p.d.1}}{L_{p.d.1}}\left((V_{DD} - V_t)V_{ds1} - \frac{V_{ds1}^2}{2}\right) \qquad \text{(from 2.4)}$$

Therefore

$$R_1 = \frac{V_{ds1}}{I_{ds}} = \frac{1}{K}\frac{L_{p.d.1}}{W_{p.d.1}}\left(\frac{1}{V_{DD} - V_t - \frac{V_{ds1}}{2}}\right)$$

Note that V_{ds1} is small and $\frac{V_{ds1}}{2}$ may be ignored.

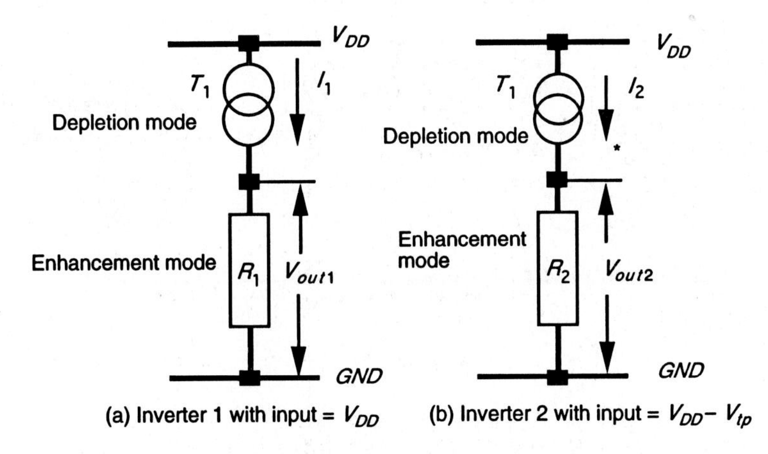

Figure 2–10 Equivalent circuits of inverters 1 and 2

Thus

$$R_1 \doteqdot \frac{1}{K} Z_{p.d.1}\left(\frac{1}{V_{DD} - V_t}\right)$$

Now, for depletion mode p.u. in saturation with $V_{gs} = 0$

$$I_1 = I_{ds} = K \frac{W_{p.u.1}}{L_{p.u.1}} \frac{(-V_{td})^2}{2} \qquad \text{(from 2.5)}$$

The product

$$I_1 R_1 = V_{out\,1}$$

Thus

$$V_{out1} = I_1 R_1 = \frac{Z_{p.d.1}}{Z_{p.u.1}}\left(\frac{1}{V_{DD} - V_t}\right)\frac{(V_{td})^2}{2}$$

Consider inverter 2 (Figure 2–10(b)) when input = $V_{DD} - V_{tp}$. As for inverter 1

$$R_2 \doteqdot \frac{1}{K} Z_{p.d.2} \frac{1}{((V_{DD} - V_{tp}) - V_t)}$$

$$I_2 = K \frac{1}{Z_{p.u.2}} \frac{(-V_{td})^2}{2}$$

whence

$$V_{out2} = I_2 R_2 = \frac{Z_{p.d.2}}{Z_{p.u.2}}\left(\frac{1}{V_{DD} - V_{tp} - V_t}\right)\frac{(-V_{td})^2}{2}$$

If inverter 2 is to have the same output voltage under these conditions then $V_{out\,1} = V_{out\,2}$. That is

$$I_1 R_1 = I_2 R_2$$

Therefore

$$\frac{Z_{p.u.2}}{Z_{p.d.2}} = \frac{Z_{p.u.1}}{Z_{p.d.1}} \frac{(V_{DD} - V_t)}{(V_{DD} - V_{tp} - V_t)}$$

Taking typical values

$$V_t = 0.2\ V_{DD}$$

$$V_{tp} = 0.3\ V_{DD}{}^*$$

$$\frac{Z_{p.u.2}}{Z_{p.d.2}} = \frac{Z_{p.u.1}}{Z_{p.d.1}} \frac{0.8}{0.5}$$

Therefore

$$\frac{Z_{p.u.2}}{Z_{p.d.2}} \div 2\frac{Z_{p.u.1}}{Z_{p.d.1}} = \frac{8}{1}$$

Summarizing for an nMOS inverter:

- An inverter driven directly from the output of another should have a $Z_{p.u.}/Z_{p.d.}$ ratio of $\geqslant 4/1$.
- An inverter driven through one or more pass transistors should have a $Z_{p.u.}/Z_{p.d.}$ ratio of $\geqslant 8/1$.

Note: It is the driven, *not* the driver, whose ratio is affected.

2.9 Alternative forms of pull-up

Up to now we have assumed that the inverter circuit has a depletion mode pull-up transistor as its load. There are, however, at least four possible arrangements:

1. *Load resistance R_L* (Figure 2–11). This arrangement is not often used because of the large space requirements of resistors produced in a silicon substrate.
2. *nMOS depletion mode transistor pull-up* (Figure 2–12).
   (a) Dissipation is high since rail to rail current flows when V_{in} = logical 1.
   (b) Switching of output from 1 to 0 begins when V_{in} exceeds V_t of p.d. device.
   (c) When switching the output from 1 to 0, the p.u. device is non-saturated initially and this presents lower resistance through which to charge capacitive loads.

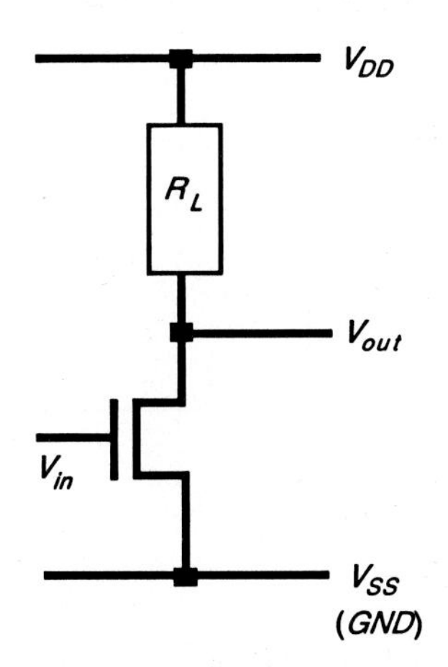

Figure 2–11 Resistor pull-up

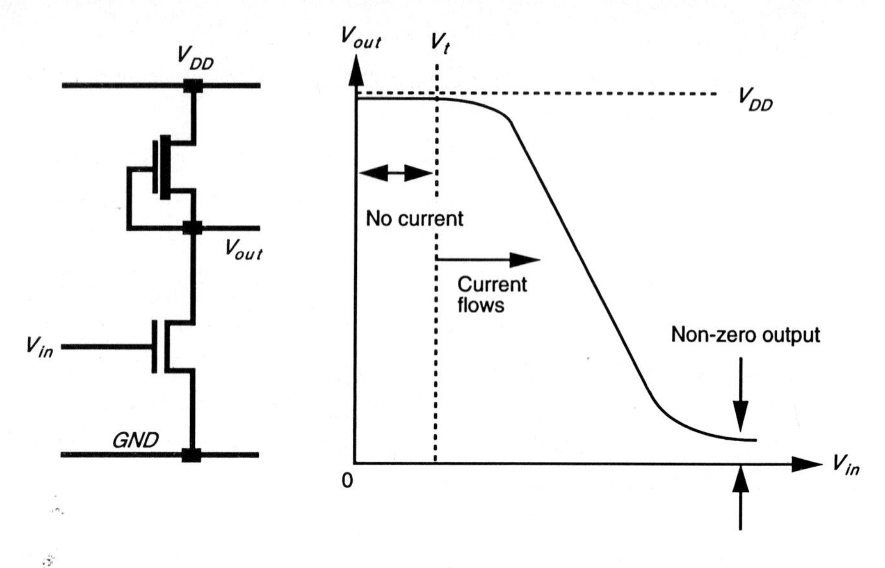

Figure 2–12 nMOS depletion mode transistor pull-up and transfer characteristic

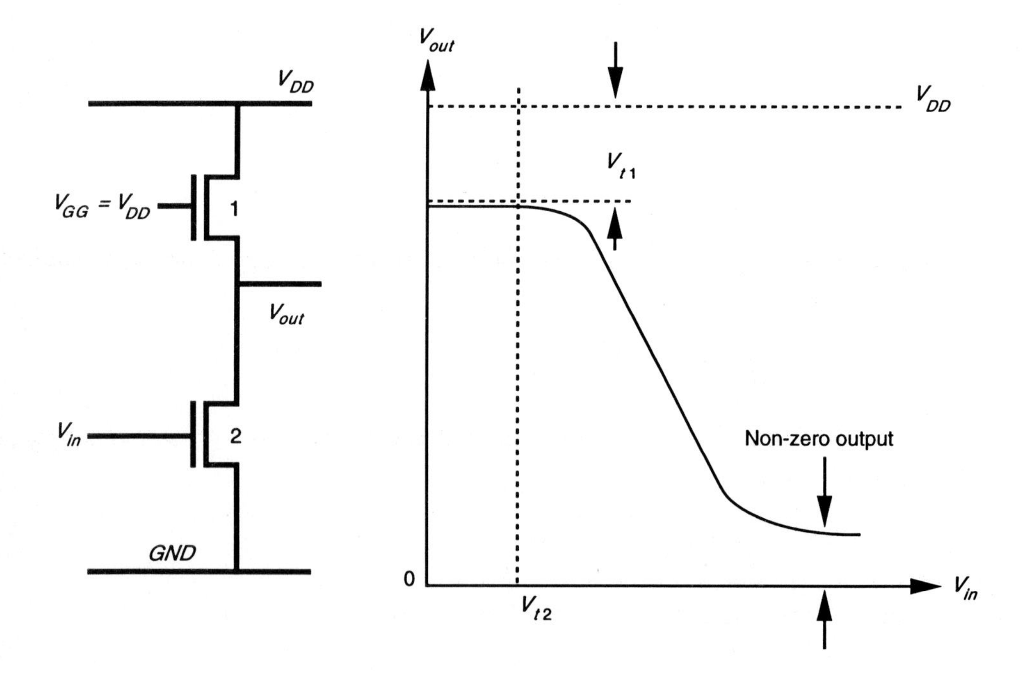

Figure 2–13 nMOS enhancement mode pull-up and transfer characteristic

3. *nMOS enhancement mode pull-up* (Figure 2–13).
   (a) Dissipation is high since current flows when V_{in} = logical 1 (V_{GG} is returned to V_{DD}).
   (b) V_{out} can never reach V_{DD} (logical 1) if $V_{GG} = V_{DD}$ as is normally the case.
   (c) V_{GG} may be derived from a switching source, for example, one phase of a clock, so that dissipation can be greatly reduced.
   (d) If V_{GG} is higher than V_{DD} then an extra supply rail is required.
4. *Complementary transistor pull-up* (CMOS) (Figure 2–14).
   (a) No current flow either for logical 0 or for logical 1 inputs.
   (b) Full logical 1 and 0 levels are presented at the output.
   (c) For devices of similar dimensions the p-channel is slower than the n-channel device.

2.10 The CMOS inverter

The general arrangement and characteristics are illustrated in Figure 2–14. We have seen (equations 2.4 and 2.5) that the current/voltage relationships for the MOS transistor may be written

$$I_{ds} = K\frac{W}{L}\left((V_{gs} - V_t\right)V_{ds} - \frac{V_{ds}^{\ 2}}{2}$$

in the resistive region, or

$$I_{ds} = K\frac{W}{L}\frac{(V_{gs} - V_t)^2}{2}$$

in saturation. In both cases the factor K is a technology-dependent parameter such that

$$K = \frac{\varepsilon_{ins}\varepsilon_0\mu}{D}$$

The factor W/L is, of course, contributed by the geometry and it is common practice to write

$$\beta = K\frac{W}{L}$$

so that, for example

$$I_{ds} = \frac{\beta}{2}(V_{gs} - V_t)^2$$

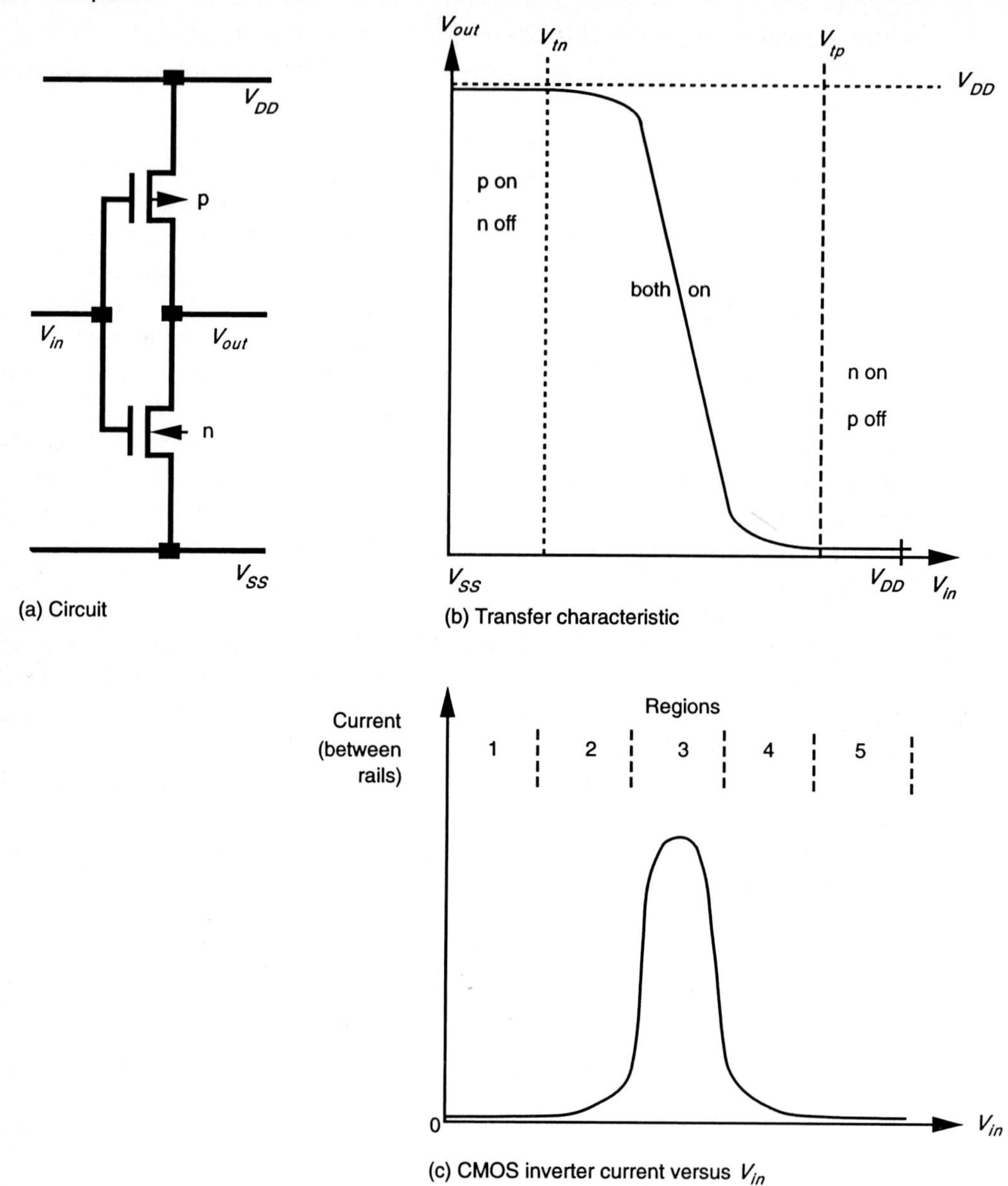

Figure 2–14 Complementary transistor pull-up (CMOS)

in saturation, and where β may be applied to both nMOS and pMOS transistors as follows

$$\beta_n = \frac{\varepsilon_{ins}\varepsilon_0\mu_n}{D}\frac{W_n}{L_n}$$

$$\beta_p = \frac{\varepsilon_{ins}\varepsilon_0\mu_p}{D}\frac{W_p}{L_p}$$

where W_n and L_n, W_p and L_p are the n- and p-transistor dimensions respectively. With regard to Figures 2–14(b) and 2–14(c), it may be seen that the CMOS inverter has five distinct regions of operation.

Considering the static conditions first, it may be seen that in *region 1* for which V_{in} = logic 0, we have the p-transistor fully turned on while the n-transistor is fully turned off. Thus no current flows through the inverter and the output is directly connected to V_{DD} through the p-transistor. A good logic 1 output voltage is thus present at the output.

In *region 5* V_{in} = logic 1, the n-transistor is fully on while the p-transistor is fully off. Again, no current flows and a good logic 0 appears at the output.

In *region 2* the input voltage has increased to a level which just exceeds the threshold voltage of the n-transistor. The n-transistor conducts and has a large voltage between source and drain; so it is in saturation. The p-transistor is also conducting but with only a small voltage across it, it operates in the unsaturated resistive region. A small current now flows through the inverter from V_{DD} to V_{SS}. If we wish to analyze the behavior in this region, we equate the p-device resistive region current with the n-device saturation current and thus obtain the voltage and current relationships.

Region 4 is similar to region 2 but with the roles of the p- and n-transistors reversed. However, the current magnitudes in regions 2 and 4 are small and most of the energy consumed in switching from one state to the other is due to the larger current which flows in region 3.

Region 3 is the region in which the inverter exhibits gain and in which both transistors are in saturation.

The currents (with regard to Figure 2–14(c) in each device must be the same since the transistors are in series, so we may write

$$I_{dsp} = -I_{dsn}$$

where

$$I_{dsp} = \frac{\beta_p}{2}(V_{in} - V_{DD} - V_{tp})^2$$

and

$$I_{dsn} = \frac{\beta_n}{2}(V_{in} - V_{tn})^2$$

from whence we can express V_{in} in terms of the β ratio and the other circuit voltages and currents

$$V_{in} = \frac{V_{DD} + V_{tp} + V_{tn}(\beta_n / \beta_p)^{1/2}}{1 + (\beta_n / \beta_p)^{1/2}} \tag{2.10}$$

Since both transistors are in saturation, they act as current sources so that the equivalent circuit in this region is two current sources in series between V_{DD} and

V_{SS} with the output voltage coming from their common point. The region is inherently unstable in consequence and the changeover from one logic level to the other is rapid.

If $\beta_n = \beta_p$ and if $V_{tn} = -V_{tp}$, then from equation 2.10

$$V_{in} = 0.5\ V_{DD}$$

This implies that the changeover between logic levels is symmetrically disposed about the point at which

$$V_{in} = V_{out} = 0.5\ V_{DD}$$

since only at this point will the two β factors be equal. But for $\beta_n = \beta_p$ the device geometries must be such that

$$\mu_p W_p / L_p = \mu_n W_n / L_n$$

Now the mobilities are inherently unequal and thus it is necessary for the width to length ratio of the p-device to be two to three times that of the n-device, namely

$$W_p / L_p \doteqdot 2.5\ W_n / L_n$$

However, it must be recognized that mobility μ is affected by the transverse electric field in the channel and is thus dependent on V_{gs} (and thus on V_{in} in this case). It has been shown empirically that the actual mobility is

$$\mu = \mu_z (1 - \phi(V_{gs} - V_t))^{-1}$$

ϕ is a constant approximately equal to 0.05, V_t includes any body effect, and μ_z is the mobility with zero transverse field. Thus a β ratio of 1 will only hold good around the point of symmetry when $V_{out} = V_{in} = 0.5\ V_{DD}$.

The β ratio is often unimportant in many configurations and in most cases minimum size transistor geometries are used for both n- and p-devices. Figure 2–15 indicates the trends in the transfer characteristic as the ratio is varied. The changes indicated in the figure would be for quite large variations in β ratio (e.g. up to 10:1) and the ratio is thus not too critical in this respect.

2.11 MOS transistor circuit model

The MOS transistor can be modeled with varying degrees of complexity. However, a consideration of the actual physical construction of the device (as in Figure 2–16) leads to some understanding of the various components of the model.

Notes: C_{GC} = gate to channel capacitance

C_{GS} = gate to source capacitance } Small for self-aligning nMOS process

C_{GD} = gate to drain capacitance }

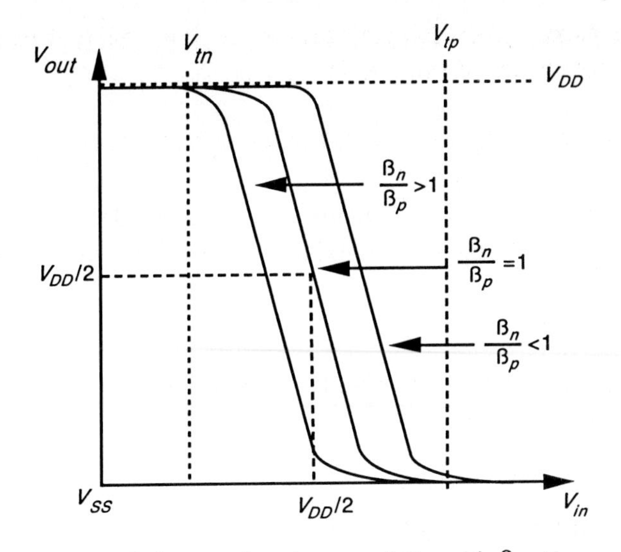

Figure 2–15 Trends in transfer characteristic with β ratio

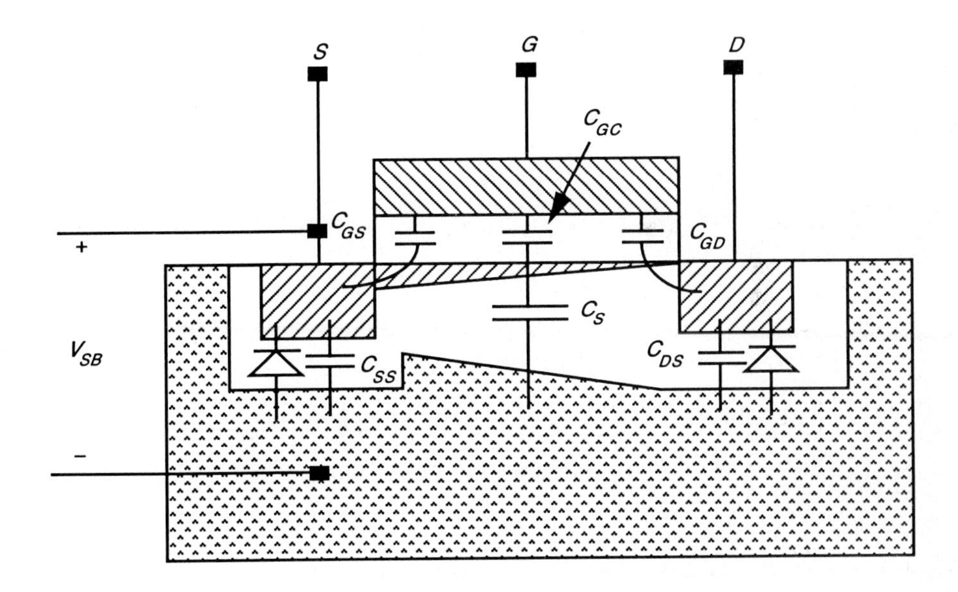

Figure 2–16 nMOS transistor model

Remaining capacitances are associated with the depletion layer and are voltage dependent. Note that C_{SS} indicates source-to-substrate, C_{DS} drain-to-substrate, and C_S channel-to-substrate capacitances.

2.12 Some characteristics of npn bipolar transistors

The key properties of MOS transistors and MOS inverters having been covered, it is now desirable to extend our thoughts into some properties of bipolar transistors and into BiCMOS inverters.

In dealing with bipolar transistor characteristics, it will be assumed that the reader is familiar with the basic operation and the fundamental aspects of bipolar transistors.

2.12.1 Transconductance g_m — bipolar

The transconductance of a bipolar transistor is commonly presented as

$$g_m = I_c / \frac{kT}{q}$$

where

I_c = collector current
q = electron charge
k = Boltzmann's constant
T = temperature °K

The expression can be rewritten in the form

$$g_m \, \alpha \, A_E \, e^{V_{be}(q/kT)}$$

where

V_{be} is the base to emitter voltage

and

A_E is the emitter area.

Note that the following factors may be deduced

- $g_m \, \alpha \, e^{V_{be}}$, that is, exponentially dependent on input voltage V_{be}
- $g_m \, \alpha \, I_c$
- g_m is independent of process
- g_m is a weak function of transistor size.

Remembering that, for MOS transistors

$$g_m = \frac{\mu \varepsilon_{ins} \varepsilon_0}{D} \frac{W}{L} (V_{gs} - V_t)$$

where

D = oxide thickness (often denoted t_{ox})

Comparisons can be made between MOS and bipolar transistor g_m as follows:

1. For $I_c = I_{ds}$ the difference between the thermal voltage (kT/q) and the effective gate voltage ($V_{gs} - V_t$) introduces a large difference in transconductance.
2. If inputs are controlled by equal amounts of charge

   that is

   $$C_g \text{ (MOS)} = C_{base} \text{ (bipolar)}$$

   then

   $$g_m \text{ (bipolar)} >> g_m \text{ (MOS)}$$

   noting that

   $$C_{base} = \tau_F I_c (q/kT)$$
   $$C_g = C_0 A$$

where C_0 (*often denoted as* C_{ox}) is the gate to channel capacitance per unit area and $A = W.L.$ τ_F is the forward transit time.

2.12.2 Comparative aspects of key parameters of CMOS and bipolar transistors

In order to put matters in perspective, a comparison of key parameters follows in Table 2–1.

Table 2–1 A comparison of some parameters

CMOS	*Bipolar*
1. $I_{ds} = \frac{(\mu C_0)}{2} \frac{W}{L} (V_{gs} - V_t)^2$ $= \frac{\beta}{2} (V_{gs} - V_t)^2$ [In saturation]	$I_c = I_s \exp(qV_{be}/kT)$
2. $g_m = (2\beta)^{1/2} (I_{ds})^{1/2}$ [expressions given can be put in this form]	$g_m = I_c / \frac{kT}{q}$
3. $I_{ds}/A = (\mu C_0/2L^2)(V_{gs} - V_t)^2$	$I_c/A = 1/(R_B \mu \tau_B)$

where I_{ds}/A and I_c/A are current/area and R_B is base resistance and τ_B is the base transit time (usually in the order of 10–30ps).

Evaluating, we may see that *I/A* for bipolar is five times better than that for CMOS. A discussion of the current drive aspects of BiCMOS circuits will be found in Chapter 4 (section 4.8.3).

2.12.3 BiCMOS inverters

As in nMOS and CMOS logic circuitry, the basic logic element is the inverter circuit.

When designing with BiCMOS in mind, the logical approach is to use MOS switches to perform the logic function and bipolar transistors to drive the output loads. The simplest logic function is that of inversion, and a simple BiCMOS inverter circuit is readily set out as shown in Figure 2–17.

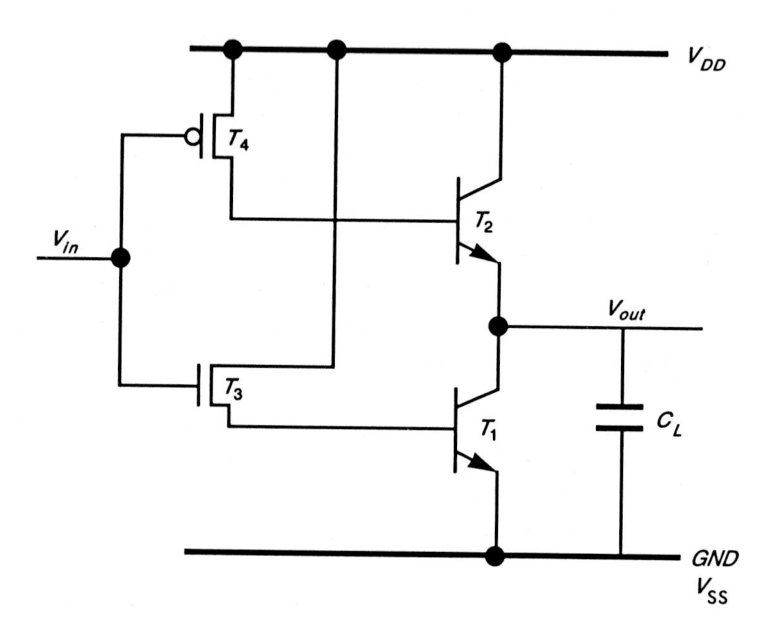

Figure 2–17 A simple BiCMOS Inverter

It consists of two bipolar transistors T_1 and T_2 with one nMOS transistor T_3 and one pMOS transistor T_4, both being enhancement mode devices. The action of the circuit is straightforward and may be described as follows:

- With $V_{in} = 0$ volts(*GND*) T_3 is off so that T_1 will be non-conducting. But T_4 is on and supplies current to the base of T_2 which will conduct and act as a current source to charge the load C_L toward +5 volts(V_{DD}). The output of the inverter will rise to +5 volts less the base to emitter voltage V_{BE} of T_2.
- With $V_{in} = +5$ volts(V_{DD}) T_4 is off so that T_2 will be non-conducting. But T_3 will now be on and will supply current to the base of T_1 which will conduct and act as a current sink to the load C_L discharging it toward 0 volts(*GND*). The output of the inverter will fall to 0 volts plus the saturation voltage V_{CEsat} from the collector to the emitter of T_1.
- T_1 and T_2 will present low impedances when turned on into saturation and the load C_L will be charged or discharged rapidly.

- The output logic levels will be good and will be close to the rail voltages since V_{CEsat} is quite small and V_{BE} is approximately +0.7 volts.
- The inverter has a high input impedance.
- The inverter has a low output impedance.
- The inverter has a high current drive capability but occupies a relatively small area.
- The inverter has high noise margins.

However, owing to the presence of a DC path from V_{DD} to GND through T_3 and T_1, this is not a good arrangement to implement since there will be a significant static current flow whenever V_{in} = logic 1. There is also a problem in that there is no discharge path for current from the base of either bipolar transistor when it is being turned off. This will slow down the action of this circuit.

An improved version of this circuit is given in Figure 2–18, in which the DC path through T_3 and T_1 is eliminated, but the output voltage swing is now reduced, since the output cannot fall below the base to emitter voltage V_{BE} of T_1.

An improved inverter arrangement, using resistors, is shown in Figure 2–19. In this circuit resistors provide the improved swing of output voltage when each bipolar transistor is off, and also provide discharge paths for base current during turn-off.

The provision of on chip resistors of suitable value is not always convenient and may be space-consuming, so that other arrangements — such as in Figure 2–20 — are used. In this circuit, the transistors T_5 and T_6 are arranged to turn on when T_2 and T_1 respectively are being turned off.

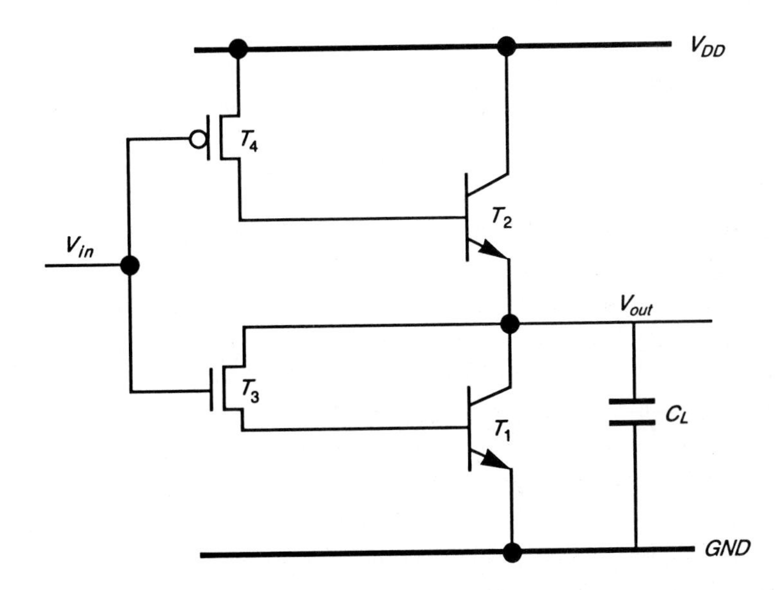

Figure 2–18 An alternative BiCMOS inverter with no static current flow

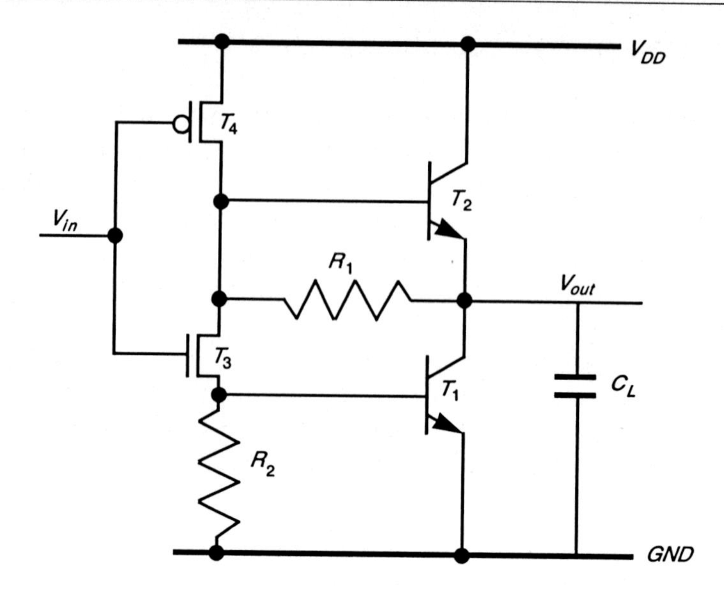

Figure 2–19 An improved BiCMOS inverter with better output logic levels

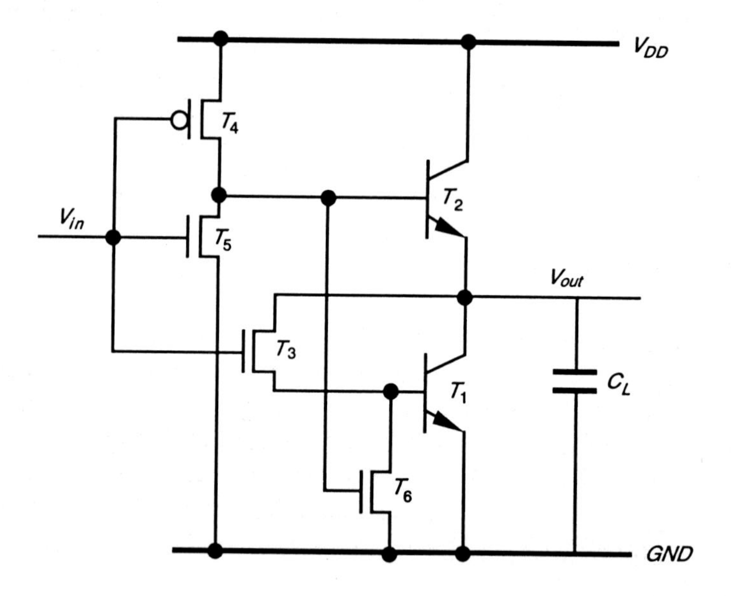

Figure 2–20 An improved BiCMOS inverter using MOS transistors for base current discharge

In general, BiCMOS inverters offer many advantages where high load current sinking and sourcing is required. The arrangements lead on to the BiCMOS gate circuits which will be dealt with in Chapter 5.

2.13 Latch-up in CMOS circuits

A problem which is inherent in the p-well and n-well processes is due to the relatively large number of junctions which are formed in these structures and, as mentioned earlier, the consequent presence of parasitic transistors and diodes. Latch-up is a condition in which the parasitic components give rise to the establishment of low-resistance conducting paths between V_{DD} and V_{SS} with disastrous results. Careful control during fabrication is necessary to avoid this problem.

Latch-up may be induced by glitches on the supply rails or by incident radiation. The mechanism involved may be understood by referring to Figure 2–21, which shows the key parasitic components associated with a p-well structure in which an inverter circuit (for example) has been formed.

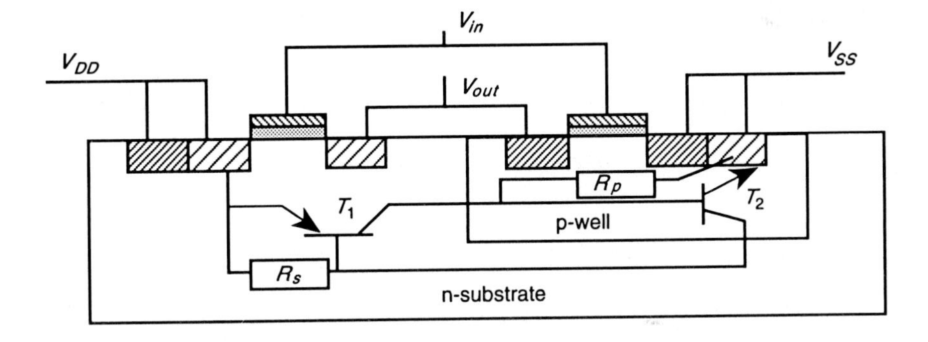

Figure 2–21 Latch-up effect in p-well structure

There are, in effect, two transistors and two resistances (associated with the p-well and with regions of the substrate) which form a path between V_{DD} and V_{SS}. If sufficient substrate current flows to generate enough voltage across R_s to turn on transistor T_1, this will then draw current through R_p and, if the voltage developed is sufficient, T_2 will also turn on, establishing a self-sustaining low-resistance path between the supply rails. If the current gains of the two transistors are such that $\beta_1 \times \beta_2 > 1$, latch-up may occur. Equivalent circuits are given in Figure 2–22.

With no injected current, the parasitic transistors will exhibit high resistance, but sufficient substrate current flow will cause switching to the low-resistance state as already explained. The switching characteristic of the arrangement is outlined in Figure 2–23.

Once latched-up, this condition will be maintained until the latch-up current drops below I_l. It is thus essential for a CMOS process to ensure that V_l and I_l are not readily achieved in any normal mode of operation.

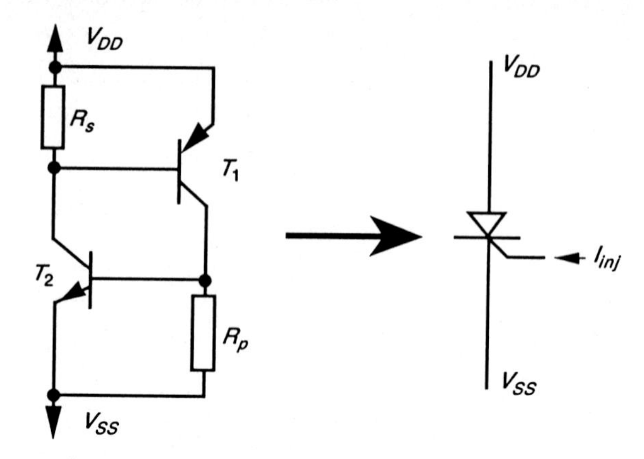

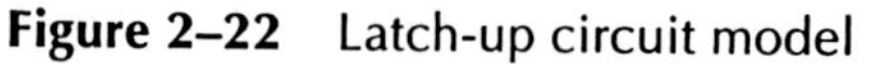

Figure 2–22 Latch-up circuit model

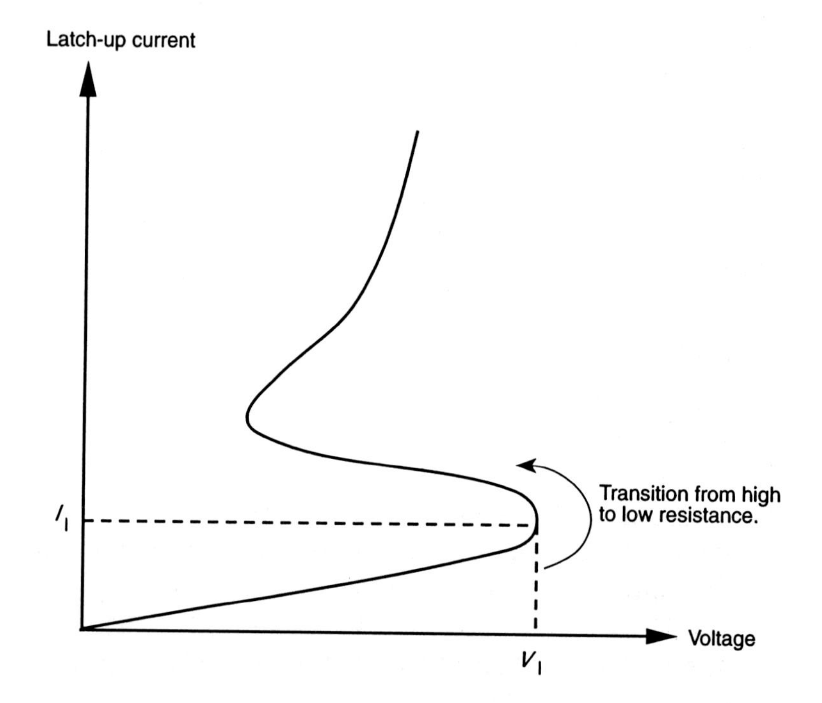

Figure 2–23 Latch-up current versus voltage

Remedies for the latch-up problem include:

1. an increase in substrate doping levels with a consequent drop in the value of R_s ;
2. reducing R_p by control of fabrication parameters and by ensuring a low contact resistance to V_{SS} ;
3. other more elaborate measures such as the introduction of guard rings.

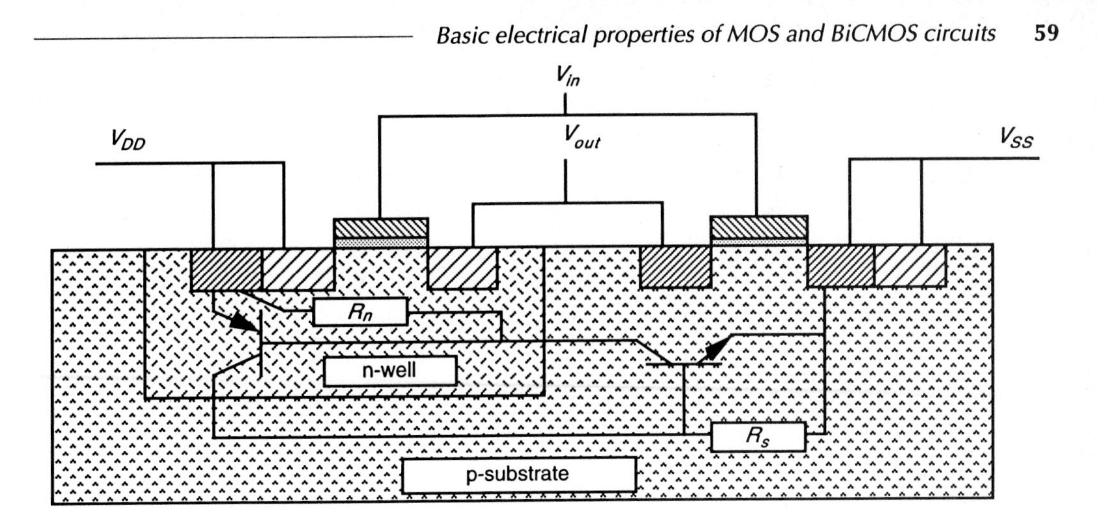

Figure 2–24 Latch-up circuit for n-well process

For completeness, the latch-up configuration for an n-well structure is given in Figure 2–24.

2.14 BiCMOS Latch-up susceptibility

One benefit of the BiCMOS process is that it produces circuits which are less likely to suffer from latch-up problems. This is due to several factors:

- A reduction of substrate resistance R_s.
- A reduction of n-well resistance $R_{w.}$
- A reduction of R_s and R_w means that a larger lateral current is necessary to invite latch-up and a higher value of holding current is also required.
- The parasitic (vertical) pnp transistor which is part of the n-well latch-up circuit has its beta reduced owing to the presence of the buried n+ layer. This has the effect of reducing carrier lifetime in the n-base region and this contributes the reduction in beta.

2.15 Observations

This chapter has established the underlying properties of MOS active devices and simple circuits configured when using them. The reason for such encumbrances as ratio rules has been explained and it is now appropriate to discuss the means by which circuits can be interconnected in silicon.

2.16 Tutorial exercises

1. Compare the relative merits of three different forms of pull-up for an inverter circuit. What is the best choice for realization in (a) nMOS technology? (b) CMOS technology?
2. In the inverter circuit, what is meant by $Z_{p.u.}$ and $Z_{p.d.}$? Derive the required ratio between $Z_{p.u.}$ and $Z_{p.d.}$ if an nMOS inverter is to be driven from another nMOS inverter.
3. For a CMOS inverter, calculate the shift in the transfer characteristic (Figure 2–15) when the β_n/β_p ratio is varied from 1/1 to 10/1.

3 MOS and BiCMOS circuit design processes

The artist must understand that he does not (only) create — he materializes.
Horia Bernea

Objectives

The purpose of this chapter is to provide an insight into the methods and means for materializing circuit designs in silicon.

Design processes are aided by simple concepts such as stick and symbolic diagrams but the key element is a set of design rules. Design rules are the communication link between the designer specifying requirements and the fabricator who materializes them. Design rules are used to produce workable mask layouts from which the various layers in silicon will be formed or patterned.

The first set of design rules introduced here are 'lambda-based'. These rules are straightforward and relatively simple to apply. However, they are 'real' and chips can be fabricated from mask layouts using the lambda-based rule set.

Tighter and faster designs will be realized if a fabricator's line is used to its full advantage and such rule sets are generally particular not only to the fabricator but also to a specific technology.

Two such design rule sets, from Orbit*, are also introduced in this chapter.

*Orbit Semiconductor Inc., California.

3.1 MOS layers

MOS design is aimed at turning a specification into masks for processing silicon to meet the specification. We have seen that MOS circuits are formed on four basic layers — *n-diffusion*, *p-diffusion*, *polysilicon*, and *metal*, which are isolated from one another by thick or thin (thinox) silicon dioxide insulating layers. The thin oxide (thinox) mask region includes n-diffusion, p-diffusion, and transistor channels. Polysilicon and thinox regions interact so that a transistor is formed where they cross one another. In some processes, there may be a second metal layer and also, in some processes, a second polysilicon layer. Layers may be deliberately joined together where contacts are formed. We have also seen that the basic MOS transistor properties can be modified by the use of an implant within the thinox region and this is used in nMOS circuits to produce depletion mode transistors.

We have also seen that bipolar transistors can be included in this design process by the addition of extra layers to a CMOS process. This is referred to as BiCMOS technology, and in this text it is dealt with in an n-well CMOS environment.

We must find a way of capturing the topology and layer information of the actual circuit in silicon so that we can set out simple diagrams which convey both *layer* information and *topology.*

3.2 Stick diagrams

Stick diagrams may be used to convey layer information through the use of a color code — for example, in the case of nMOS design, green for n-diffusion, red for polysilicon, blue for metal, yellow for implant, and black for contact areas. In this text the color coding has been complemented by monochrome encoding of the lines so that black and white copies of stick diagrams do not lose the layer information. The encodings chosen are shown and illustrated in color as Color plates 1(a)–(d) and in monochrome form as Figures 3–1(a)–(d). When you are drawing your own stick diagrams you should use single lines in the appropriate colors, as in Color plate 1(d) noting that yellow lines are outlined in green for clarity only.

Note that mask layout information, which is also color coded, may also be hatched for monochrome encoding, also shown in Figures 3–1(a)–(c). Monochrome encoding schemes are widely illustrated throughout the text, and it will be noted that diagrams and mask layouts in this form are readily reproduced by copying machines.

The color and monochrome encoding scheme used has been evolved to cover nMOS, CMOS, and BiCMOS processes and to be compatible with the design processes of gallium arsenide. The color encoding is compatible with color terminals, printers, and plotters having quite simple color palettes. Using color workstations,

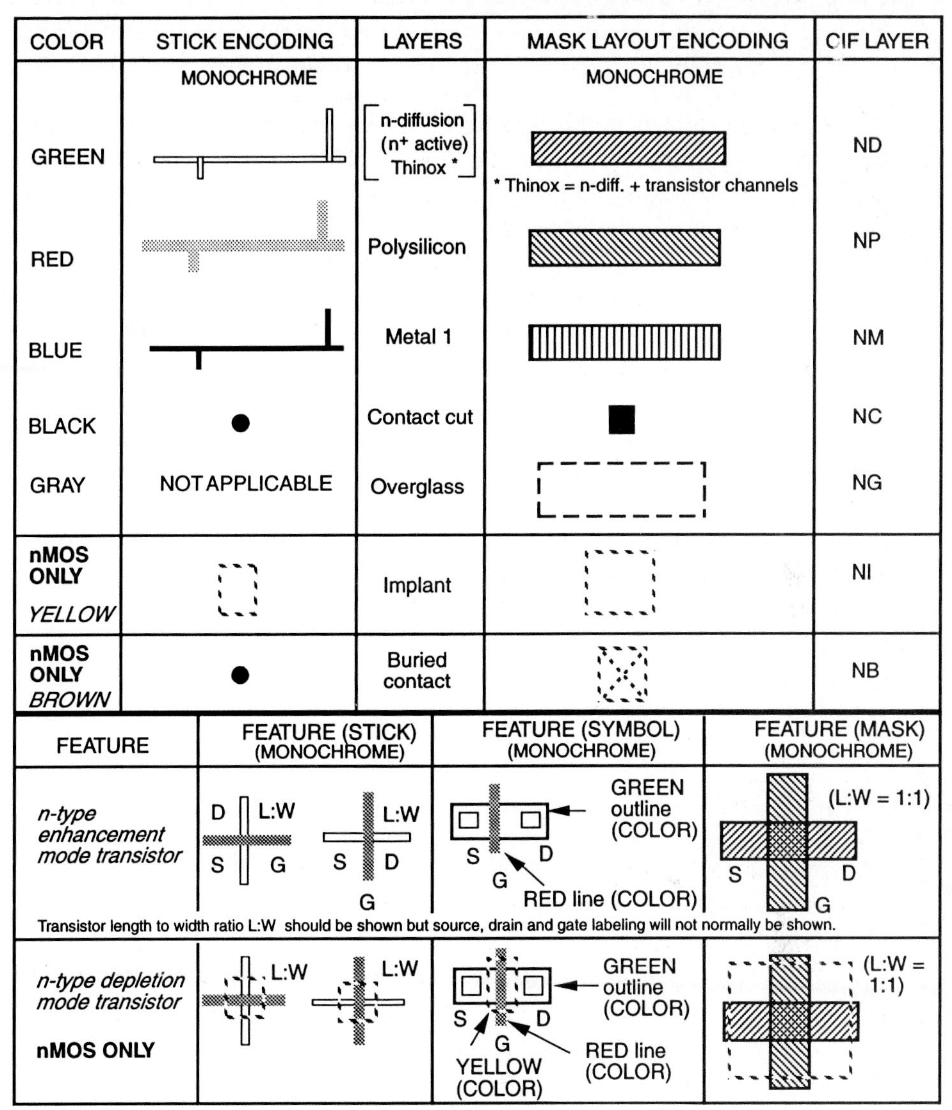

Figure 3–1(a) Encodings for a simple single metal nMOS process (see Color plate 1(a) for nMOS color encoding details)

the mask areas are usually color filled while pen plotters produce color outlines only. In this text, most color diagrams incorporate color outlines and color hatching (hatching as for the monochrome encoding) so that the detail of underlying areas may be easily discerned where layers intersect or are superimposed. This form of color representation is acceptable for those with color vision difficulties and may also be copied by a monochrome copier without losing the encoding. The various representations are indicated in Color plate 2.

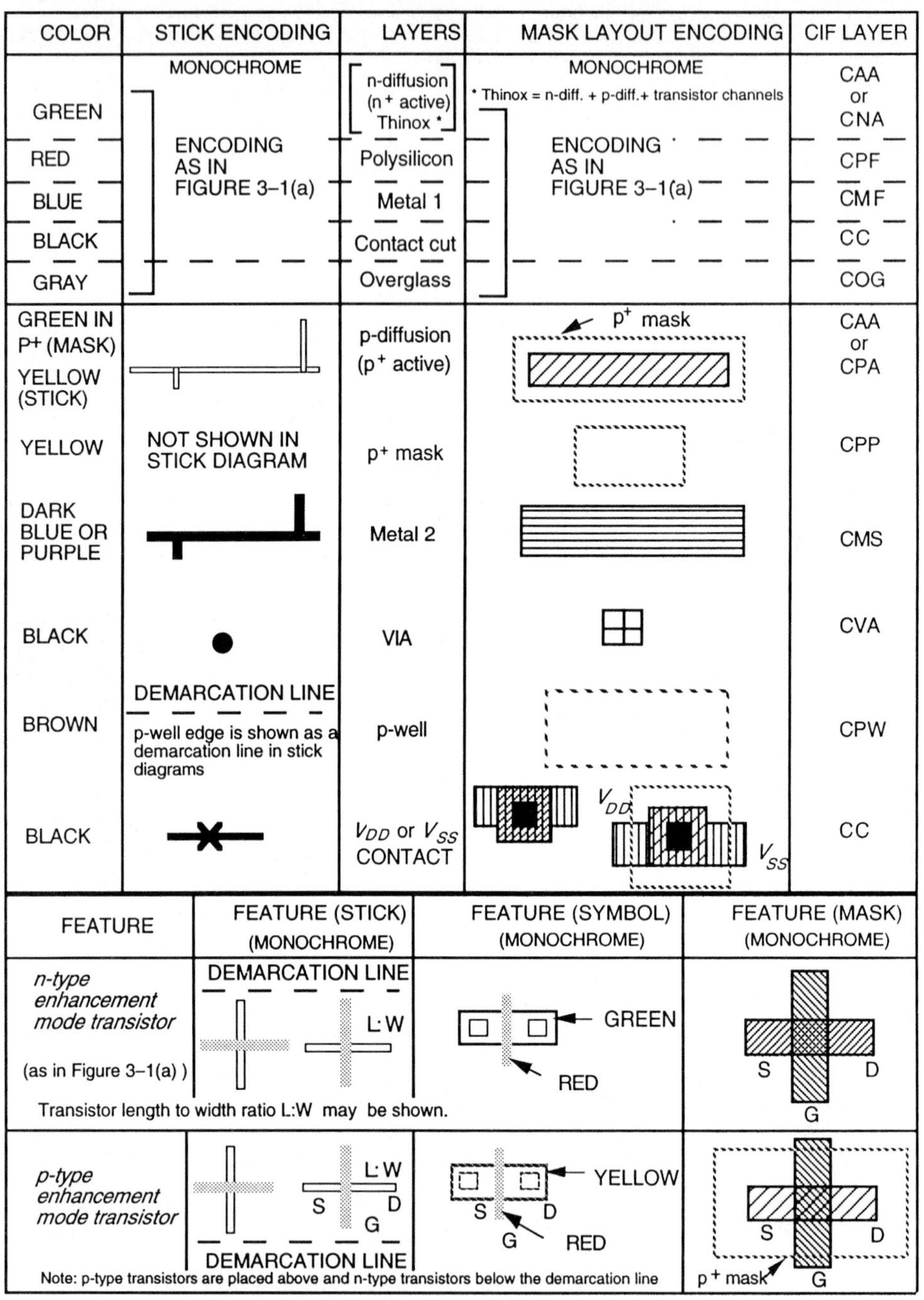

COLOR	STICK ENCODING	LAYERS	MASK LAYOUT ENCODING	CIF LAYER
GREEN	MONOCHROME ENCODING AS IN FIGURE 3–1(a)	n-diffusion (n$^+$ active) Thinox *	MONOCHROME * Thinox = n-diff. + p-diff.+ transistor channels ENCODING AS IN FIGURE 3–1(a)	CAA or CNA
RED		Polysilicon		CPF
BLUE		Metal 1		CMF
BLACK		Contact cut		CC
GRAY		Overglass		COG
GREEN IN P+ (MASK) YELLOW (STICK)		p-diffusion (p$^+$ active)	p$^+$ mask	CAA or CPA
YELLOW	NOT SHOWN IN STICK DIAGRAM	p$^+$ mask		CPP
DARK BLUE OR PURPLE		Metal 2		CMS
BLACK		VIA		CVA
BROWN	DEMARCATION LINE p-well edge is shown as a demarcation line in stick diagrams	p-well		CPW
BLACK		V_{DD} or V_{SS} CONTACT	V_{DD} V_{SS}	CC

FEATURE	FEATURE (STICK) (MONOCHROME)	FEATURE (SYMBOL) (MONOCHROME)	FEATURE (MASK) (MONOCHROME)
n-type enhancement mode transistor (as in Figure 3–1(a))	DEMARCATION LINE L:W	GREEN RED	S D G
Transistor length to width ratio L:W may be shown.			
p-type enhancement mode transistor	L:W S D G DEMARCATION LINE	YELLOW S D G RED	S D G p$^+$ mask
Note: p-type transistors are placed above and n-type transistors below the demarcation line			

The same well encoding and demarcation line are used for an n-well process.
For p-well process, the n features are in the well. For an n-well process, the p features are in the well.

Figure 3–1(b) Encodings for a double metal CMOS p-well process (see Color plate 1(b) for CMOS color encoding details)

COLOR	STICK ENCODING	LAYERS	MASK LAYOUT ENCODING	CIF LAYER
	MONOCHROME		MONOCHROME	
ORANGE		Polysilicon 2		CPS
SEE COLOR PLATE 1(c)		Bipolar npn transistor	see Figure 3–13(f)	Not applicable
PINK	Not separately encoded	p-base of bipolar npn transistor	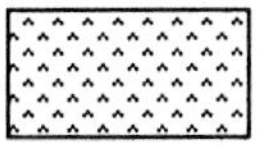	CBA
PALE GREEN	Not separately encoded	Buried collector of bipolar npn transistor	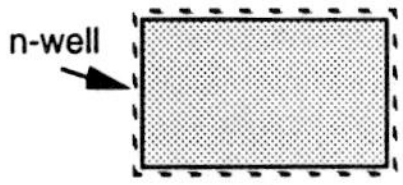 n-well	CCA

FEATURE	FEATURE (STICK) (MONOCHROME)	FEATURE (SYMBOL) (MONOCHROME)	FEATURE (MASK) (MONOCHROME)
n-type enhancement poly. 2 transistor	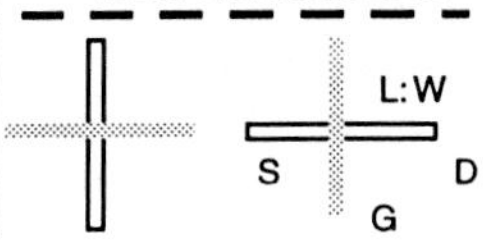 DEMARCATION LINE; L:W; S; D; G	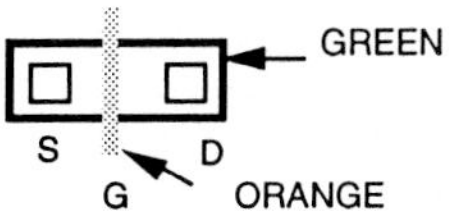 GREEN; S; D; G; ORANGE	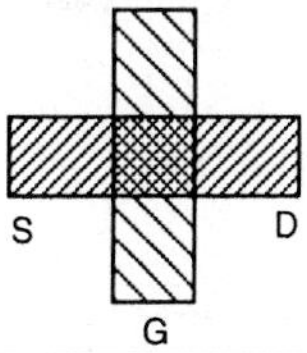 S; D; G
Transistor length to width ratio L:W may be shown.			
p-type enhancement poly. 2 transistor	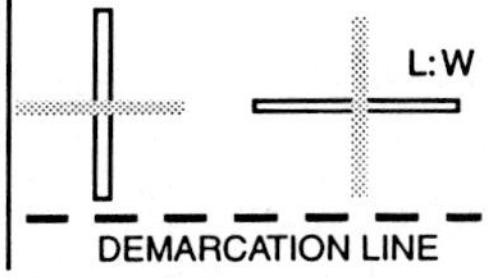 L:W; DEMARCATION LINE	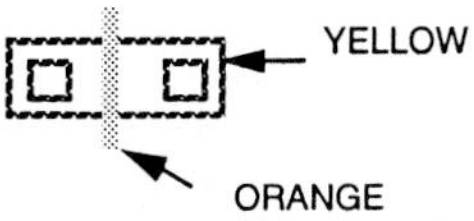 YELLOW; ORANGE	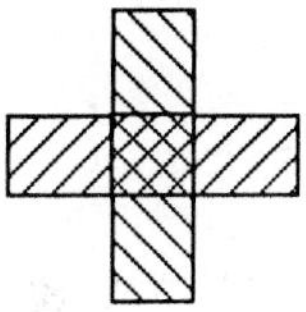
Note: p-type transistors are placed above and n-type transistors below the demarcation line.			
npn bipolar transistor			See Figure 3–13(f) and Color plate 6

The same well encoding and demarcation line as in Figure 3–1(b) are used for an n-well process. For a p-well process, the n features are in the well. For an n-well process, the p features are in the well.

Figure 3–1(c) Additional encodings for a double metal double poly. BiCMOS n-well process (see Color plates 1(c) and 6 for additional CMOS and BiCMOS color encoding details)

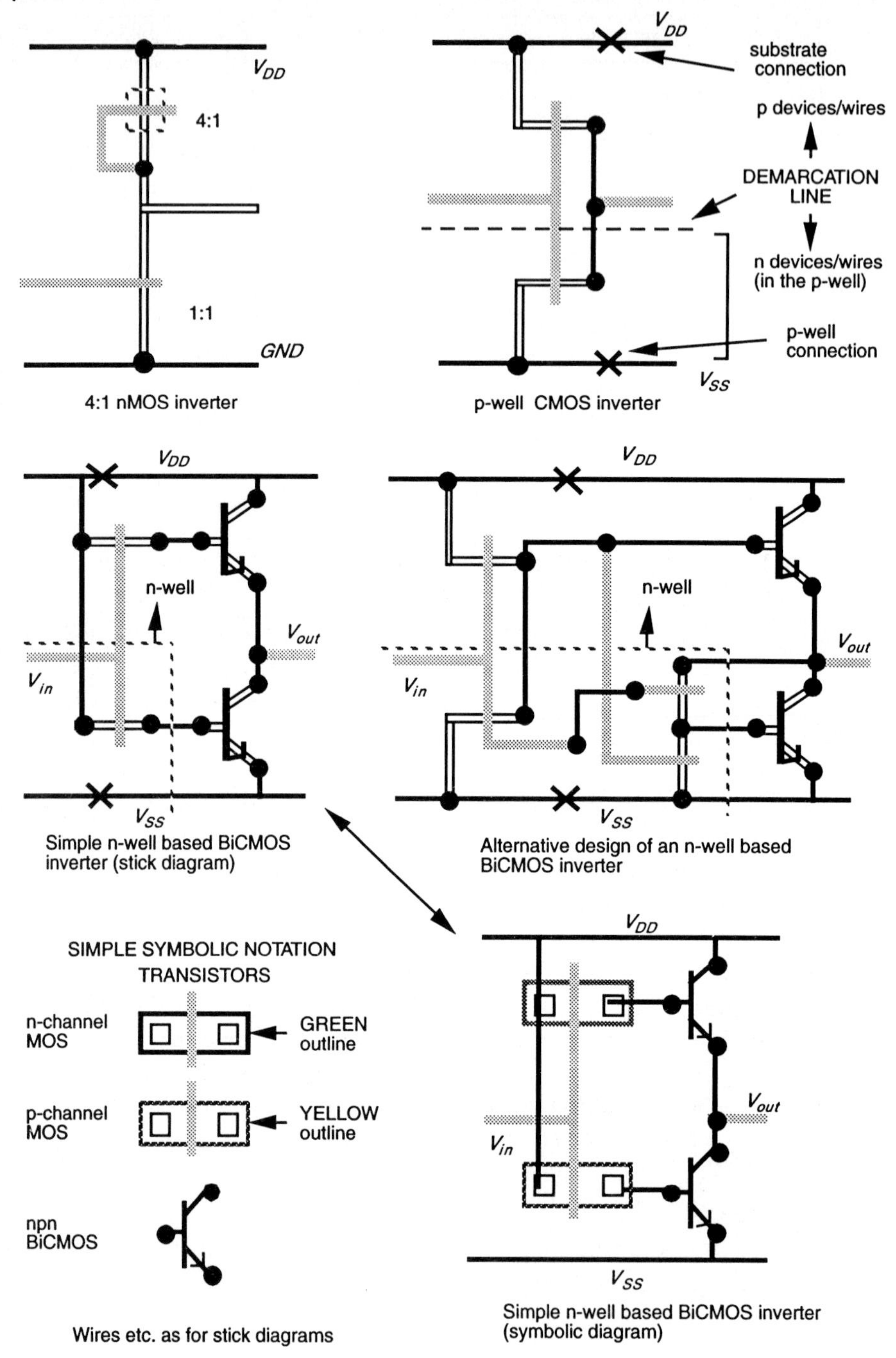

Monochrome stick diagram examples

Figure 3–1(d) Stick diagrams and simple symbolic encoding (see also Color plate 1(d))

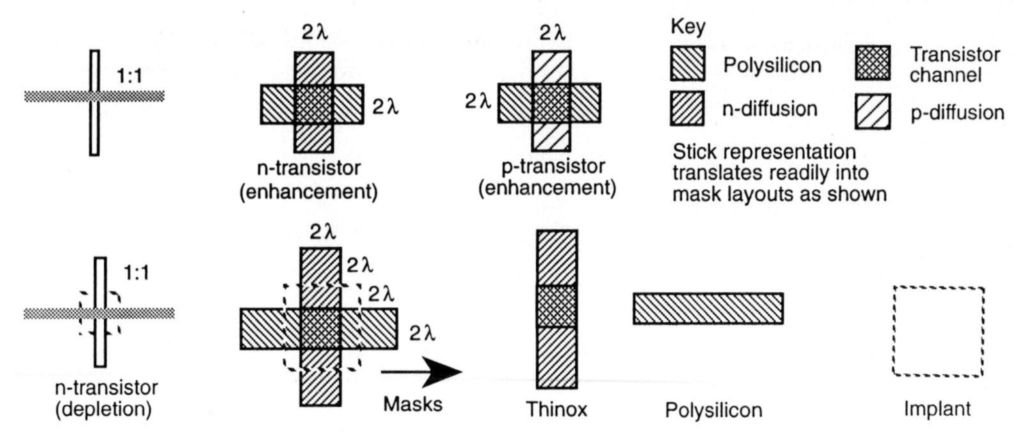

Figure 3–2 Stick diagrams and corresponding mask layout examples

In order to facilitate the learning and use of the encoding schemes, the simple set required for a single metal nMOS design is set out first as Figure 3–1(a) and Color plate 1(a); for a double metal CMOS p-well process the required encodings are extended by those given as Figure 3–1(b) and Color plate 1(b). Figure 3–1(c) and Color plate 1(c) further extend the representations to cover a second polysilicon layer and BiCMOS technology.

In this chapter we will see how basic circuits are represented in stick diagram and in symbolic form. We will be using stick representation quite widely throughout the text. The layout of stick diagrams faithfully reflects the topology of the actual layout in silicon. To illustrate stick diagrams, inverter circuits are presented in Figure 3–1(d) and in Color plate 1(d) — in nMOS, in p-well CMOS, and in n-well BiCMOS technology. A symbolic form of diagram is often most convenient and such diagrams are based on the simple symbol set included in Figures 3–1(a)–(c) and Color plates 1(a)–(c). The simplicity of symbolic form is illustrated in Figure 3–1(d), in Color plate 1(d), and in Color plate 7.

Having conveyed layer information and topology by using stick or symbolic diagrams, these diagrams are relatively easily turned into mask layouts as, for example, the transistor stick diagrams of Figure 3–2 stressing the ready translation into mask layout form.

In order that the mask layouts produced during design will be compatible with the fabrication processes, a set of design rules are set out for layouts so that, if obeyed, the rules will produce layouts which will work in practice.

3.2.1 nMOS design style

In order to start with a relatively simple process, we will consider single metal, single polysilicon nMOS technology (see Figure 3–1(a) and Color plate 1(a)).

A rational approach to stick diagram layout is readily adopted for such nMOS circuits and the approach recommended here is both easy to use and to turn into a mask layout. The layout of nMOS involves:

- n-diffusion [n-diff.] and other thinoxide regions [thinox] (green);
- polysilicon 1 [poly.] — since there is only one polysilicon layer here (red);
- metal 1 [metal] — since we use only one metal layer here (blue);
- implant (yellow);
- contacts (black or brown [buried]).

A transistor is formed wherever poly. crosses n-diff. (red over green) and all diffusion wires (interconnections) are n-type (green).

When starting a layout, the first step normally taken is to draw the metal (blue) V_{DD} and GND rails in parallel allowing enough space between them for the other circuit elements which will be required. Next, thinox (green) paths may be drawn between the rails for inverters and inverter-based logic as shown in Figure 3–3(a), not forgetting to make contacts as appropriate. Inverters and inverter-based logic comprise a pull-up structure, usually a depletion mode transistor, connected from the output point to V_{DD} and a pull-down structure of enhancement mode transistors suitably interconnected between the output point and GND. This step in the process is illustrated in Figure 3–3(b), remembering that poly. (red) crosses thinox (green) wherever transistors are required. Do not forget the implants (yellow) for depletion mode transistors and do not forget to write in the length to width (L:W) ratio for each transistor. Ratios are important, particularly in nMOS and nMOS-like circuits.

Signal paths may also be switched by pass transistors, and long signal paths may often require metal buses (blue). Allowing for the fact that the stick diagram may well represent only a small section of circuit which will be replicated many times, a convenient strategy is to run power rails and bus(es) in parallel in metal (blue) and then propagate control signals at right angles on poly. as shown. At this stage of design, 'leaf-cell' boundaries are conveniently shown on the stick diagram and these are placed so that replicated cells may be directly interconnected by direct abutment on a side-by-side and/or top-to-bottom basis.The aspects just discussed are illustrated in Figure 3–3(c).

From the very beginning a design style should encourage the concepts of 'regularity' (through the use of replication) and generality so that design effort can be minimized and the interconnection of leaf-cells, subsystems and systems is facilitated.

3.2.2 CMOS design style

The stick and layout representations for CMOS used in this text are a logical extension of the nMOS approach and style already outlined. They are based on the widely accepted work of Mead and Conway.

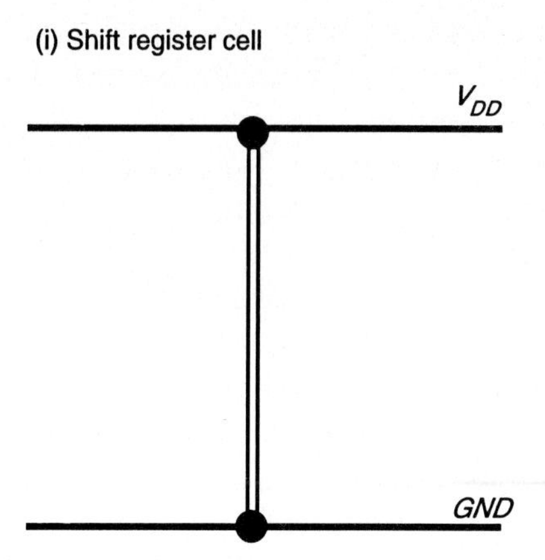

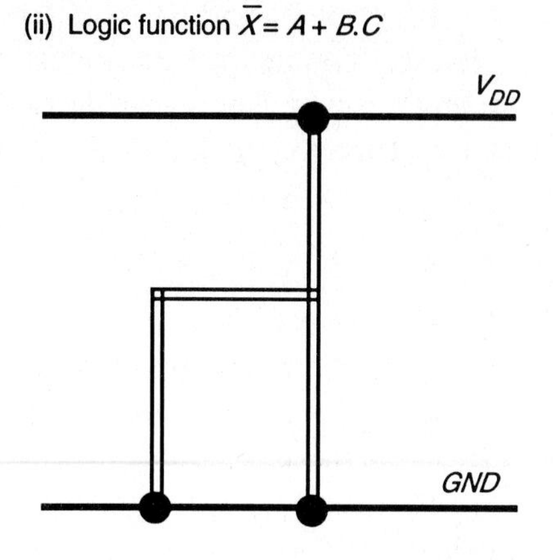

(a) Rails and thinox paths

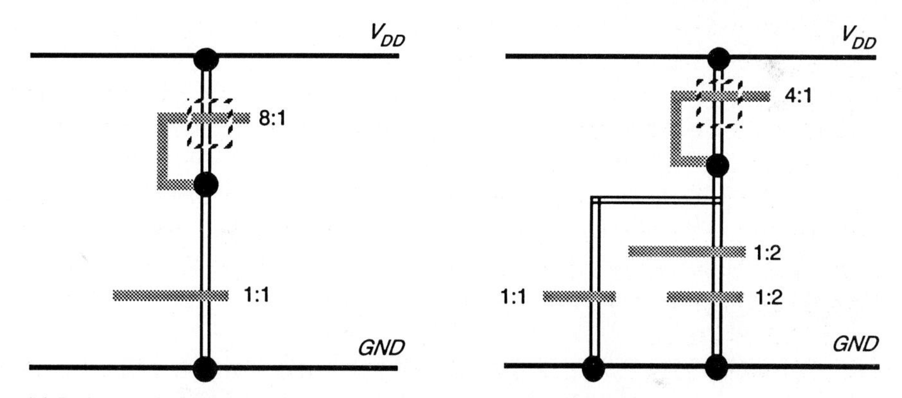

(b) Pull-up and pull-down structures (polysilicon), implants, and ratios

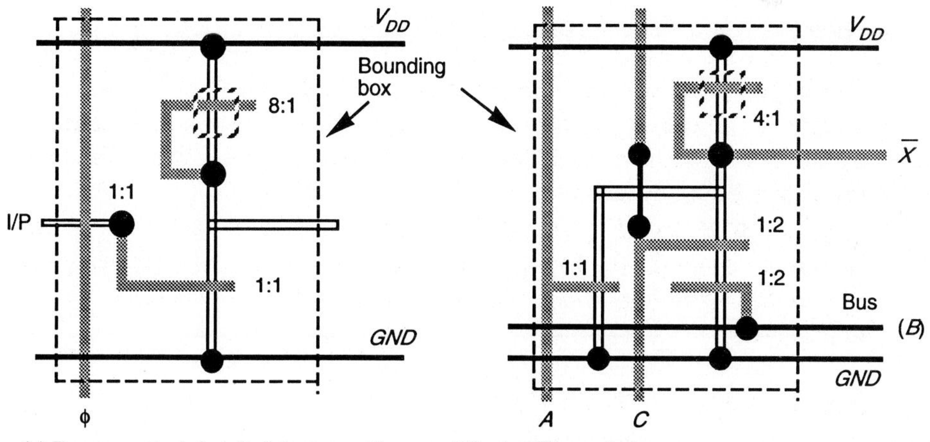

(c) Buses, control signals, interconnections, and 'leaf-cell' boundaries

Figure 3–3 Examples of nMOS stick layout design style

All features and layers defined in Figure 3–1, with the exception of implant (yellow) and the buried contact (brown), are used in CMOS design. Yellow in CMOS design is now used to identify p-transistors and wires, as depletion mode devices are not utilized. As a result, no confusion results from the allocation of the same color to two different features. The two types of transistor used, 'n' and 'p', are separated in the stick layout by the demarcation line (representing the p-well boundary) above which all p-type devices are placed (transistors and wires (yellow)). The n-devices (green) are consequently placed below the demarcation line and are thus located in the p-well. These factors are emphasized by Figure 3–4.

Diffusion paths must not cross the demarcation line and n-diffusion and p-diffusion wires must not join. The 'n' and 'p' features are normally joined by metal where a connection is needed. Apart from the demarcation line, there is no indication of the actual p-well topology at this (stick diagram) level of abstraction; neither does the p+ mask appear. Their geometry will appear when the stick diagram is translated to a mask layout. However, we must not forget to place crosses on V_{DD} and V_{SS} rails to represent the substrate and p-well connection respectively. The design style is illustrated simply by taking as an example the design of a single bit of a shift register. The design begins with the drawing of the V_{DD} and V_{SS} rails in parallel and in metal and the creation of an (imaginary) demarcation line in between, as in Figure 3–5(a). The n-transistors are then placed below this line and thus close to $V_{SS,}$ while p-transistors are placed above the line and below V_{DD}. In both cases, the transistors are conveniently placed with their diffusion paths parallel to the rails (horizontal in the diagram) as shown in Figure 3–5(b). A similar approach can be taken with transistors in symbolic form.

A sound approach is to now interconnect the n- with the p-transistors as required, using metal and connect to the rails as shown in Figure 3–5(c). It must be remembered that only metal and polysilicon can cross the demarcation line but with that restriction, wires can run in diffusion also. Finally, the remaining interconnections are made as appropriate and the control signals and data inputs are added. These steps are illustrated in Figure 3–5(d).

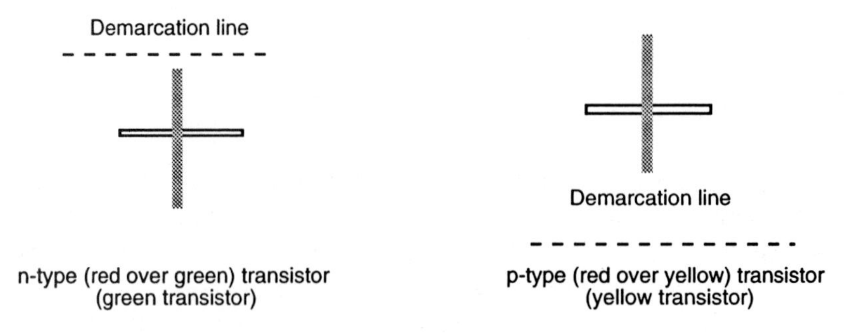

Figure 3–4 n-type and p-type transistors in CMOS design

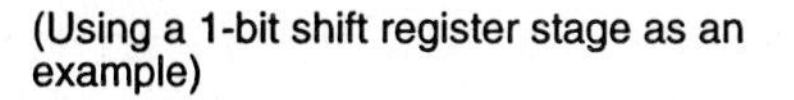

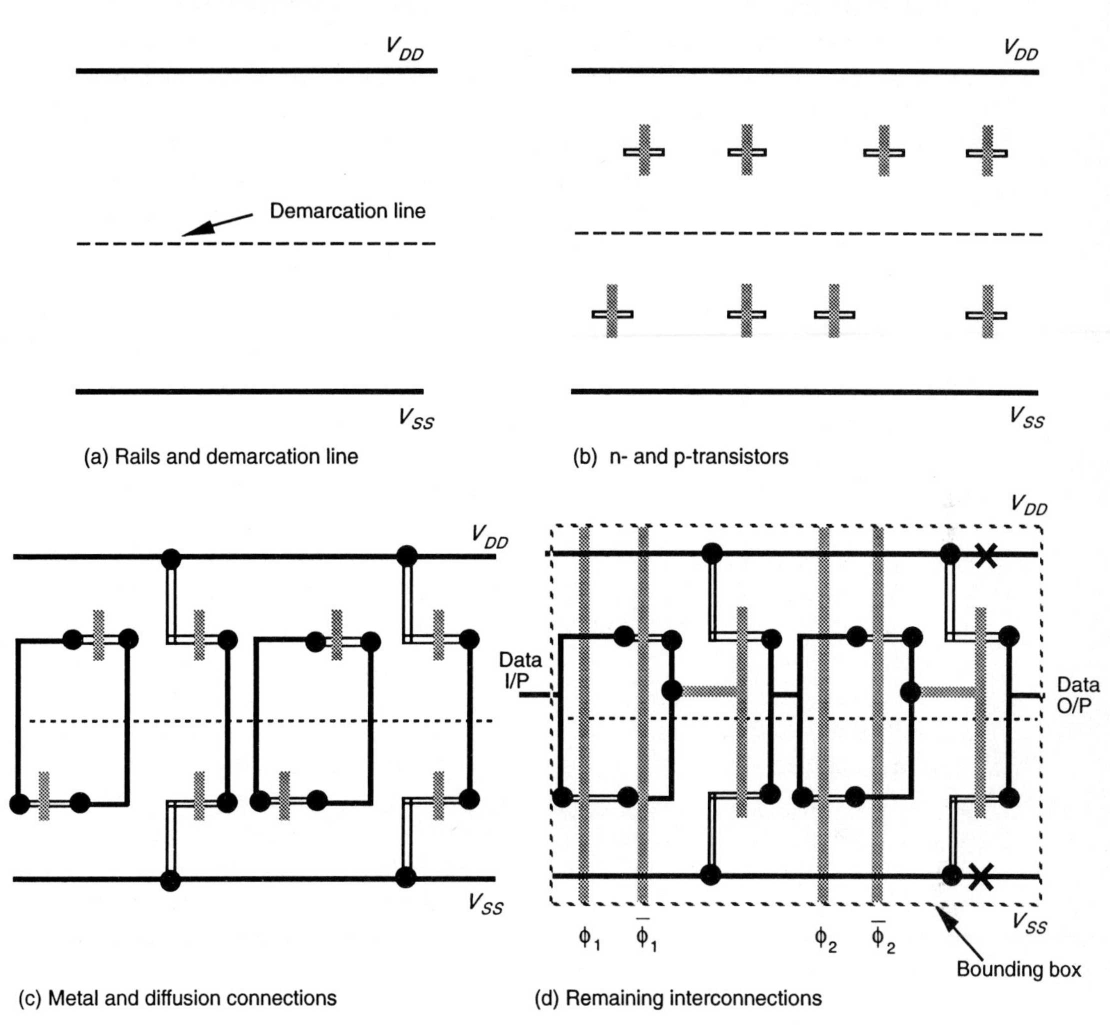

(a) Rails and demarcation line

(b) n- and p-transistors

(c) Metal and diffusion connections

(d) Remaining interconnections

Note: The contact crosses in (d) should represent one V_{DD} contact for every four p-transistors and one V_{SS} contact for every four n-transistors.

Figure 3–5 Example of CMOS stick layout design style

Although the circuit layout is now complete, we must not forget to represent the V_{ss} and V_{DD} contact crosses — one on the V_{DD} line for every four p-transistors and one on the V_{SS} line for every four n-transistors. The bounding box for the entire leaf-cell may also be shown if appropriate.

This design style is straightforward in application but later on we may recognize that sometimes transistors can be merged to advantage. We will also see how stick diagrams are turned into mask layouts, noting for CMOS layouts that the thinox mask includes all green features (n-devices) and all yellow features (p-devices) in the stick diagram.

An even simpler representation, which nevertheless carries much of the information present in a stick diagram, is to draw a symbolic diagram as in Figure 3–5(e). This diagram represents the same circuit as Figure 3–5(d) and the similarities are quite apparent. This form of diagram facilitates transistor merging, as shown, and is also readily translated to mask layouts.

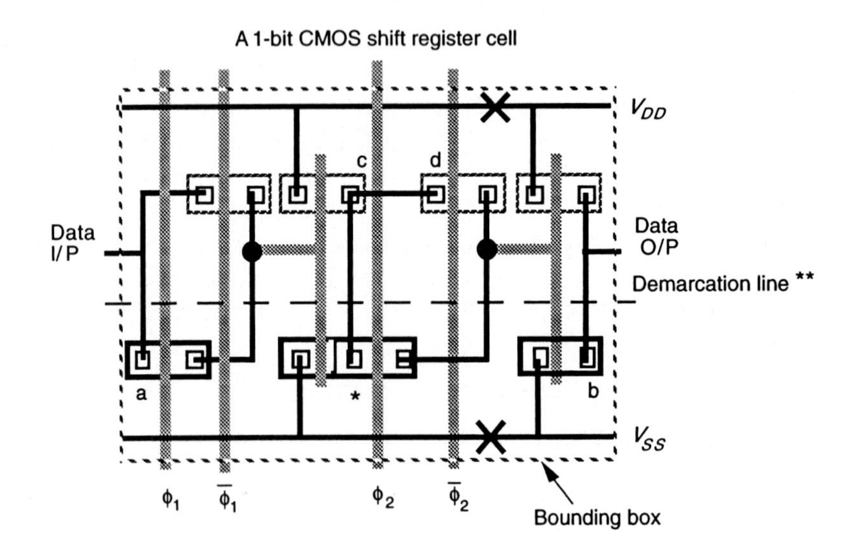

* Note that two transistors (n-type) are merged as shown. When abutting cells, transistors a and b could also be merged. It is also possible to merge p-type transistors c and d etc.

** Demarcation line may be shown but is not essential since transistor symbols are already encoded.

Figure 3–5(e) Symbolic form of diagram (CMOS shift register)

3.3 Design rules and layout

The object of a set of design rules is to allow a ready translation of circuit design concepts, usually in stick diagram or symbolic form, into actual geometry in silicon. The design rules are the effective interface between the circuit/system designer and the fabrication engineer. Clearly, both sides of the interface have a vested interest in making their own particular tasks as easy as possible and design rules usually attempt to provide a workable and reliable compromise that is friendly to both sides.

Circuit designers in general want tighter, smaller layouts for improved performance and decreased silicon area. On the other hand, the process engineer wants design rules that result in a *controllable and reproducible* process. Generally we find that there has to be a compromise for a competitive circuit to be produced at a reasonable cost.

One of the important factors associated with design rules is the achievable definition of the process line. Definition is determined by process line equipment and process design. For example, it is found that if a 10:1 wafer stepper is used instead of a 1:1 projection mask aligner, the level-to-level registration will be closer. Design rules can be affected by the maturity of the process line. For example, if the process is mature, then one can be assured of the process line capability, allowing tighter designs with fewer constraints on the designer.

The simple 'lambda (λ)-based' design rules set out first in this text are based on the invaluable work of Mead and Conway and have been widely used, particularly in the educational context and in the design of multiproject chips. The design rules are based on a single parameter λ which leads to a simple set of rules for the designer, and wide acceptance of the rules by a large cross-section of the fabrication houses and silicon brokers, and allows for scaling of the designs to a limited extent. This latter feature may help to give designs a longer lifetime. The simplicity of lambda-based rules also provides a simple introduction to design rules and to mask layout design in general and helps to set the scene for the 'micron-based' rule sets which follow.

3.3.1 Lambda-based design rules

In general, design rules and layout methodology based on the concept of λ provide a process and feature size-independent way of setting out mask dimensions to scale.

All paths in all layers will be dimensioned in λ units and subsequently λ can be allocated an appropriate value compatible with the feature size of the fabrication process. This concept means that the actual mask layout design takes little account of the value subsequently allocated to the feature size, but the design rules are such that, if correctly obeyed, the mask layouts will produce working circuits for a range of values allocated to λ. For example, λ can be allocated a value of 1.0 μm so that minimum feature size on chip will be 2 μm (2λ). Design rules, also due to Mead and Conway, specify line widths, separations, and extensions in terms of λ, and are readily committed to memory. Design rules can be conveniently set out in diagrammatic form as in Figure 3–6 for the widths and separation of conducting paths, and in Figure 3–7 for extensions and separations associated with transistor layouts.

The design rules associated with contacts between layers are set out in Figures 3–8 and 3–9 and it will be noted that connection can be made between two or, in the case of nMOS designs, three layers.

In all cases, the use of the design rules will be illustrated in layouts resulting from exercises worked through in the text.

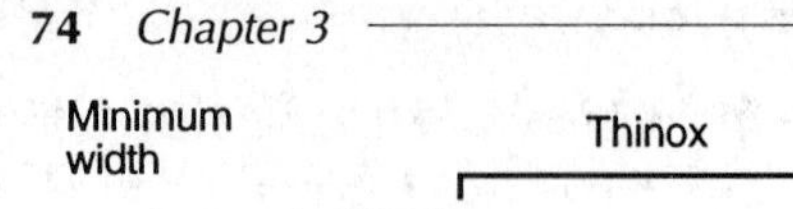

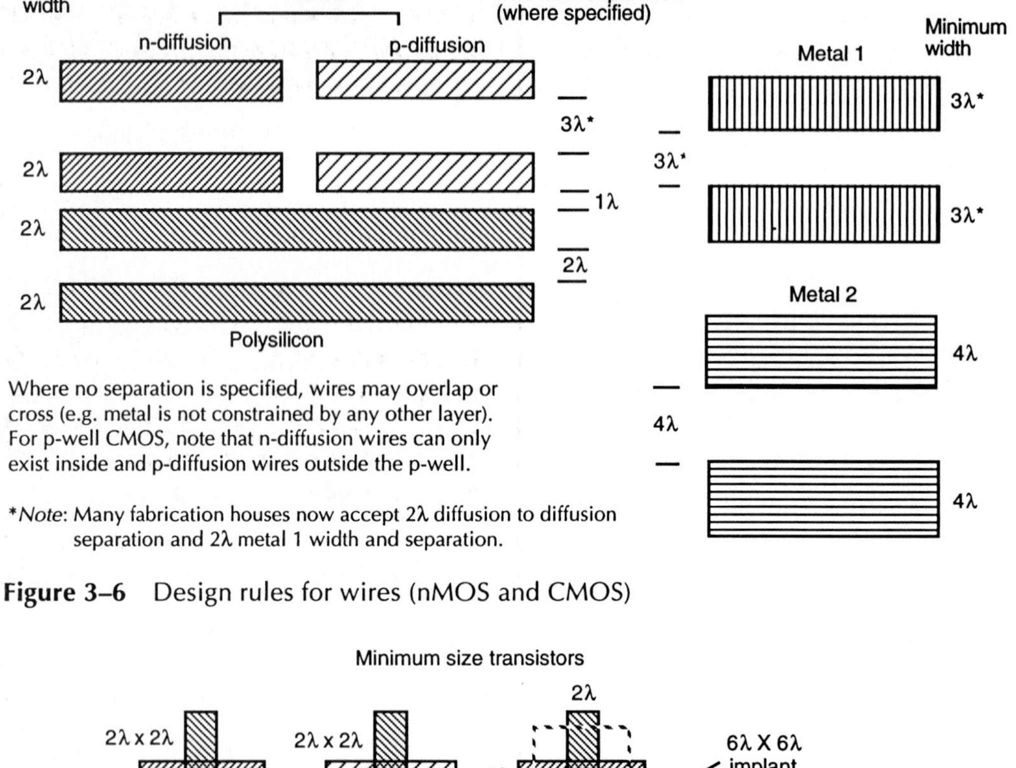

Where no separation is specified, wires may overlap or cross (e.g. metal is not constrained by any other layer). For p-well CMOS, note that n-diffusion wires can only exist inside and p-diffusion wires outside the p-well.

**Note*: Many fabrication houses now accept 2λ diffusion to diffusion separation and 2λ metal 1 width and separation.

Figure 3–6 Design rules for wires (nMOS and CMOS)

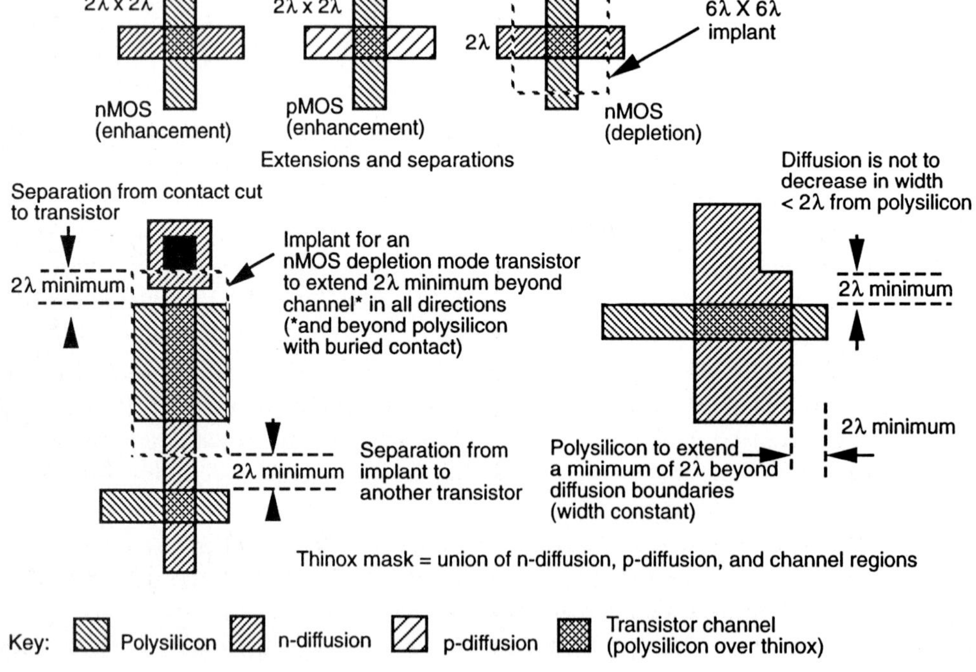

Figure 3–7 Transistor design rules (nMOS, pMOS, and CMOS)

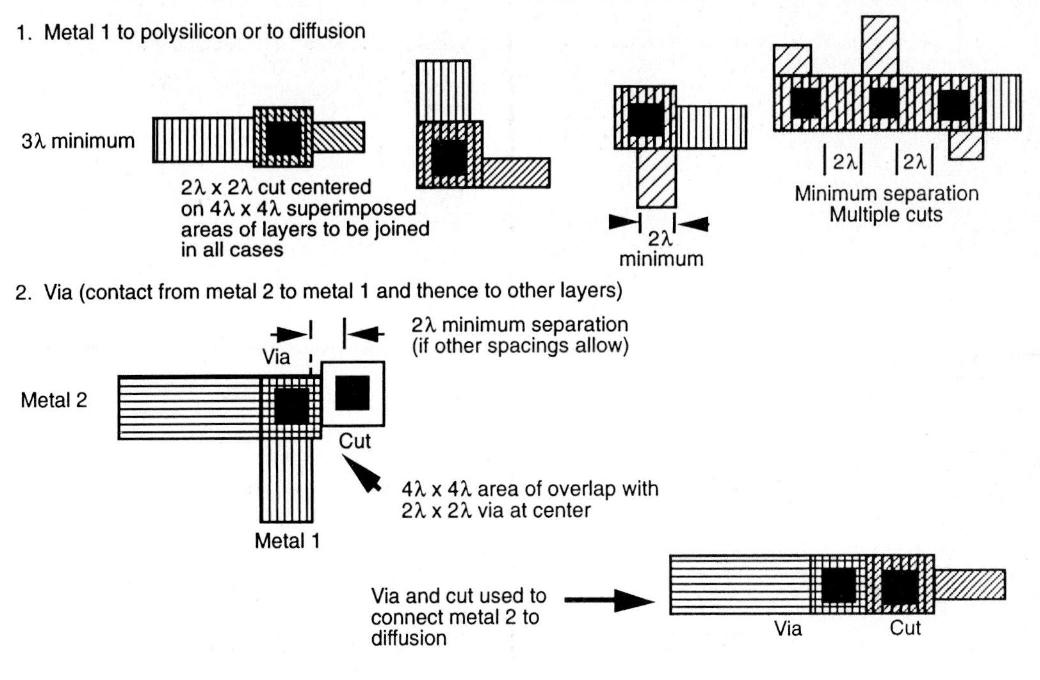

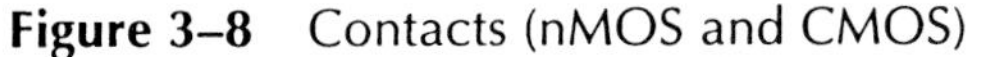
Figure 3–8 Contacts (nMOS and CMOS)

3.3.2 Contact cuts

When making contacts between polysilicon and diffusion in nMOS circuits it should be recognized that there are three possible approaches — poly. to metal then metal to diff., or a *buried contact* poly. to diff., or a *butting contact* (poly. to diff. using metal). Of the latter two, the buried contact is the most widely used, giving economy in space and a reliable contact. Butting contacts were widely used at one time but have been mostly superseded by buried contacts and have been included here and in the figures for the sake of completeness. In CMOS designs, poly. to diff. contacts are almost always made via metal.

When making connections between metal and either of the other two layers (as in Figure 3–8), the process is quite simple. The $2\lambda \times 2\lambda$ contact cut indicates an area in which the oxide is to be removed down to the underlying polysilicon or diffusion surface. When deposition of the metal layer takes place the metal is deposited through the contact cut areas onto the underlying area so that contact is made between the layers.

When connecting diffusion to polysilicon using the butting contact approach (see Figure 3–9), the process is rather more complex. In effect, a $2\lambda \times 2\lambda$ contact cut is made down to each of the layers to be joined. The layers are butted together in such a way that these two contact cuts become contiguous. Since the polysilicon and diffusion outlines overlap and thin oxide under polysilicon acts as a mask in

Buried contact: Basically, layers are joined over a 2λ x 2λ area with the buried contact cut extending by 1λ in all directions around the contact area except that the contact cut extension is increaed to 2λ in diffusion paths leaving the contact area. This is to avoid forming unwanted transistors (see following examples).

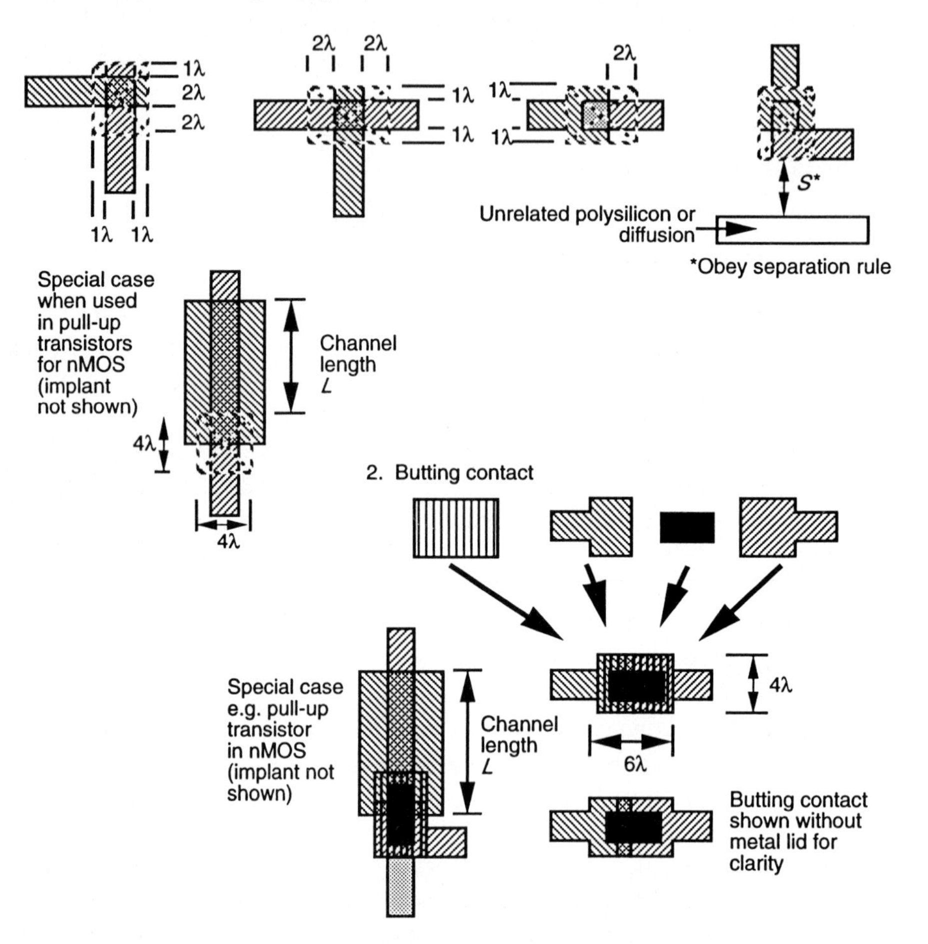

Figure 3–9 Contacts polysilicon to diffusion (nMOS only in the main text)

the diffusion process, the polysilicon and diffusion layers are also butted together. The contact between the two butting layers is then made by a metal overlay as shown in the figure. It is hoped that the cross-sectional view of the butting contact in Figure 3–10(b) helps to make the nature of the contact apparent.

The buried contact approach shown in Figures 3–9 and 3–10 is simpler, the contact cut (broken line) in this case indicating where the thin oxide is to be removed to reveal the surface of the silicon wafer before polysilicon is deposited. Thus, the polysilicon is deposited directly on the underlying crystalline wafer. When diffusion takes place, impurities will diffuse into the polysilicon as well as into the diffusion region within the contact area. Thus a satisfactory connection between polysilicon and diffusion is ensured. Buried contacts can be smaller in area than their butting contact counterparts and, since they use no metal layer, they are subject to fewer design rule restrictions in a layout.

The design rules in this case ensure that a reasonable contact area is achieved and that there will be no transistor formed unintentionally in series with the contact. The rules are such that they also avoid the formation of unwanted diffusion to polysilicon contacts and protect the gate oxide of any transistors in the vicinity of the buried contact cut area.

3.3.3 Double metal MOS process rules

A powerful extension to the process so far described is provided by a second metal layer. This gives a much greater degree of freedom, for example, in distributing global V_{DD} and V_{SS} (*GND*) rails in a system. Other processes also allow a second polysilicon layer and one such process will be introduced later.

From the overall chip interconnection aspect, the second metal layer in particular is important and, although the use of such a layer is readily envisaged, its disposition relative to (and details of) its connection to other layers using metal 1 to metal 2 contacts, called *vias*, can be readily established with reference to Figures 3–8 and 3–10(c).

Usually, second level metal layers are coarser than the first (conventional) layer and the isolation layer between the layers may also be of relatively greater thickness. To distinguish contacts between first and second metal layers, they are known as *vias* rather than contact cuts. The second metal layer representation is color coded dark blue (or purple). For the sake of completeness, the process steps for a two-metal layer process are briefly outlined as follows.

The oxide below the first metal layer is deposited by atmospheric chemical vapor deposition (CVD) and the oxide layer between the metal layers is applied in a similar manner. Depending on the process, removal of selected areas of the oxide is accomplished by plasma etching, which is designed to have a high level of vertical ion bombardment to allow for high and uniform etch rates.

Similarly, the bulk of the process steps for a double polysilicon layer process are similar in nature to those already described, except that a second thin oxide layer is grown after depositing and patterning the first polysilicon layer (Poly. 1) to isolate it from the now to be deposited second poly. layer (Poly. 2). The presence of a second poly. layer gives greater flexibility in interconnections and also allows Poly. 2 transistors to be formed by intersecting Poly. 2 and diffusion.

To revert to the double metal process it is convenient at this point to consider the layout strategy commonly used with this process. The approach taken may be summarized as follows:

1. Use the second level metal for the global distribution of power buses, that is, V_{DD} and *GND* (V_{SS}), and for clock lines.

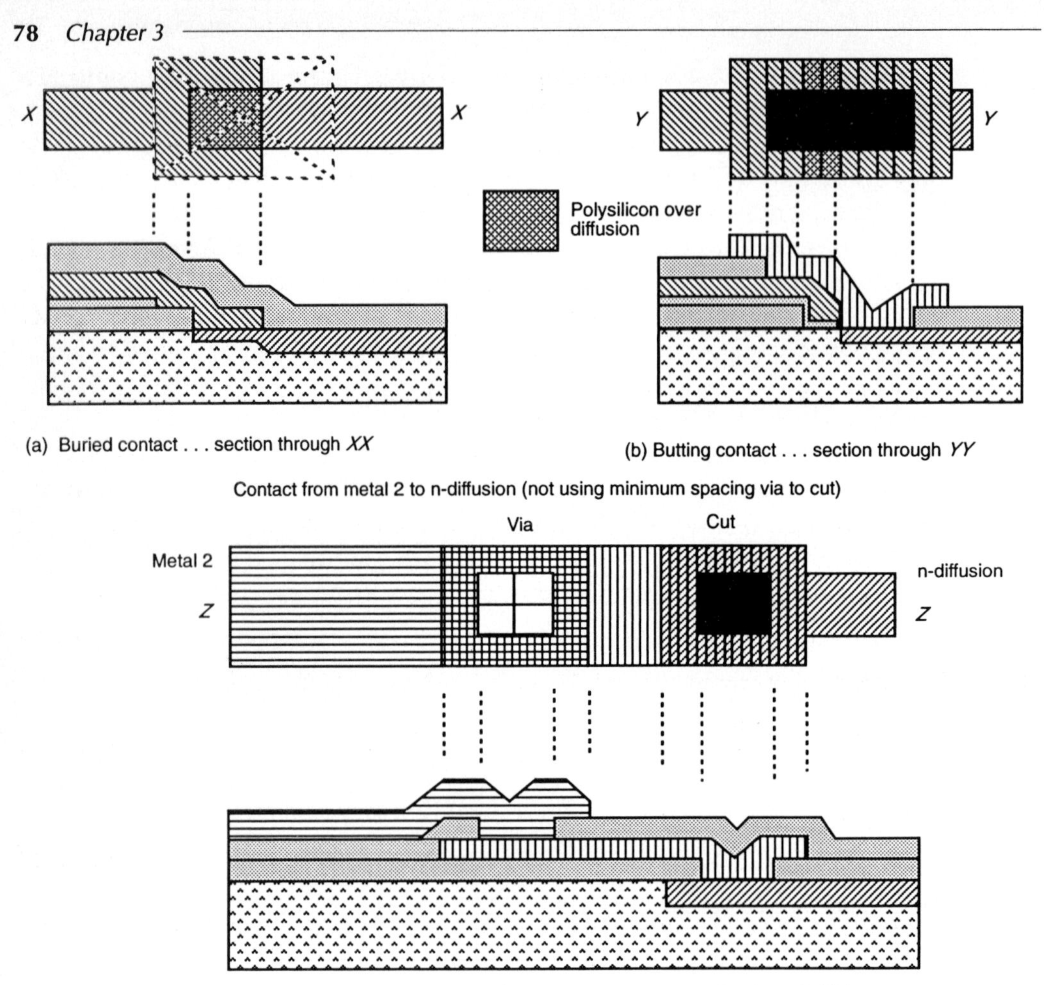

Figure 3–10 Cross-sections through some contact structures

2. Use the first level metal for local distribution of power and for signal lines.
3. Lay out the two metal layers so that the conductors are mutually orthogonal wherever possible.

3.3.4 CMOS lambda-based design rules

The CMOS fabrication process is much more complex than nMOS fabrication, which, in turn, has been simplified for ready presentation in this text. The new reader may well think that the design rules discussed here are complex enough, but in fact they constitute an abstract of the actual processing steps which are used to produce the chip. In a CMOS process, for example, the actual set of industrial design rules may well comprise more than 100 separate rules, the documentation for which spans many pages of text and/or many diagrams. Two such rule sets, micron-based, will be given in this text.

However, extending the Mead and Conway concepts, which we have already set out for nMOS designs, and noting the exclusion of butting and buried contacts, it is possible to add rules peculiar to CMOS (Figure 3–11) to those already set out in Figures 3–6 to 3–10. The additional rules are concerned with those features unique to p-well CMOS, such as the p-well and p+ mask and the special 'substrate' contacts. We have already provided for the p-transistors and p-wires in Figures 3–6 to 3–10. The rules given are also readily translated to an n-well process.

Although the CMOS rules in total may seem difficult to comprehend for the new designer, once use has been made of the simpler nMOS rules the transition to CMOS is not hard to achieve. The real key to success in VLSI design is to put it into practice, and this text attempts to encourage the reader to do just that.

3.4 General observations on the design rules

Owing to the microscopic nature of dimensions and features of silicon circuits, a major problem is presented by possible deviation in line widths and in interlayer registration.

If the line widths are too small, it is possible for lines thus defined to be discontinuous in places.

If separate paths in a layer are placed too close together, it is possible that they will merge in places or interfere with each other.

For the lambda-based rules discussed initially, the design rules are formulated in terms of a length unit λ which is related to the resolution of the process. λ may be viewed as a bound on the width deviation of a feature from its ideal 'as drawn' size and also as a bound on the maximum misalignment of any one mask. In the worst case, these effects may combine to cause the relative position of feature edges on different mask levels to deviate by as much as 2λ in their interrelationship. Inevitably, a consequence of using the lambda-based concept is that every dimension must be rounded up to whole λ values and this leads to layouts which do not fully exploit the capabilities of the process.

Similar concepts underlie the establishment of 'micron-based' rule sets, but actual dimensions are given so that full advantage can be taken of the fabrication line capabilities and tighter layouts result.

Layout rules, therefore, provide strict guidelines for preparing the geometric layouts which will be used to configure the actual masks used during fabrication and can be regarded as the main communication link between circuit/systems designers and the process engineers engaged in manufacture.

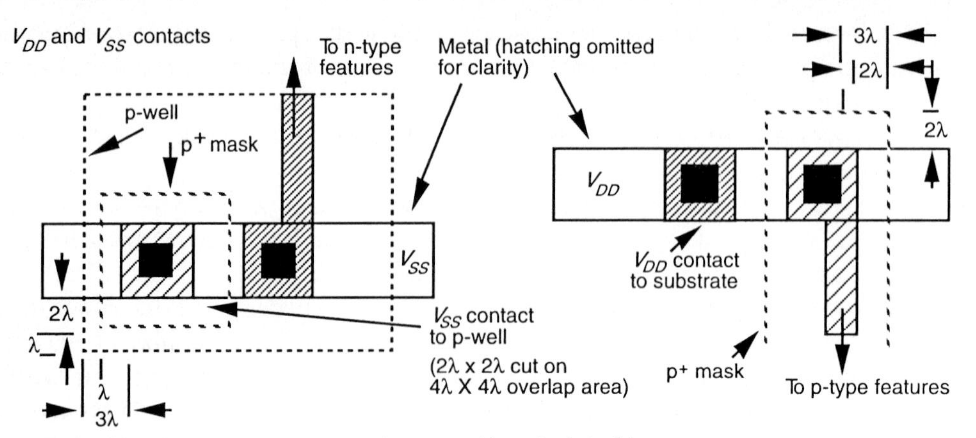

Each of the above arrangements can be merged into single 'split' contacts.

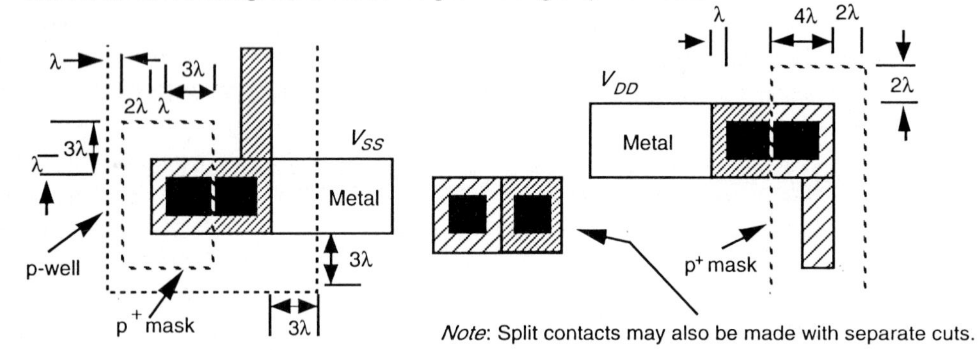

p-well and p^+ mask rules

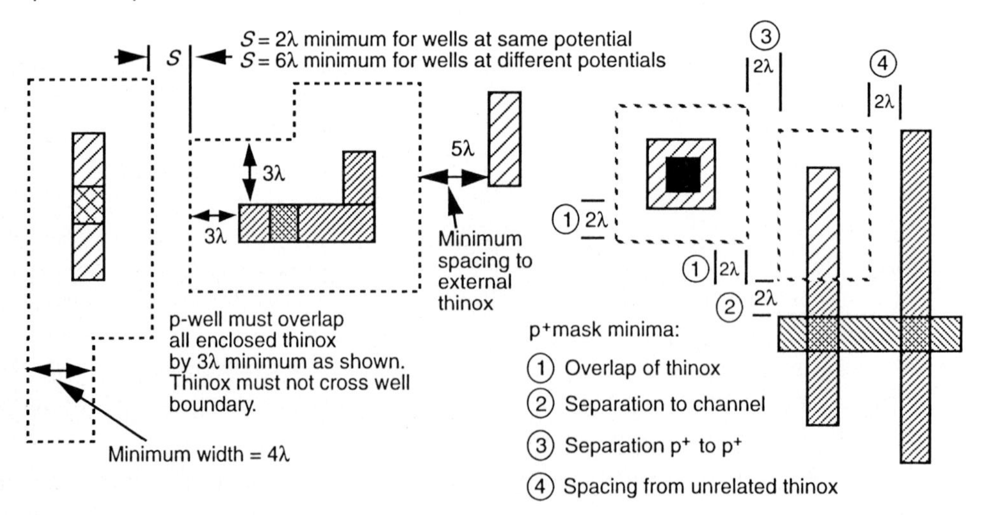

Figure 3–11 Particular rules for p-well CMOS process

The goal of any set of design rules should be to optimize yield while keeping the geometry as small as possible without compromising the reliability of the finished circuit.

On the questions of yield and reliability, even the conservative nature of the lambda-based rules can stand reevaluation when these two factors are of paramount importance. In particular, the rules associated with contacts can be improved upon in the light of experience. Figure 3–12 sets out aspects that may be observed for high yield and in high reliability situations.

In our proposed scheme of events in creating stick layouts for CMOS, we have assumed that poly. and metal can both freely cross well boundaries and this is indeed the case, but we should be careful to try to exclude poly. from areas which lie within p+ mask areas where possible. The reason for this is that the resistance of the poly. layer is reduced in current processes by n-type doping. Clearly the p+ doping which takes place inside the p+ mask will also dope the poly. which is already in place when the p+ doping step takes place. This results in an increase in the n-doping poly. resistance which may be significant in certain parts of a system.

The 3λ metal width rule is a conservative one but is implemented to allow for the fact that the metal layer is deposited after the others and on top of them and several layers of silicon dioxide, so that the surface on which it sits is quite 'mountainous'. The metal layer is also light-reflective and these factors combine to result in poor edge definition. In double metal the second layer of metal has an even more uneven terrain on which to be deposited and patterned. Hence metal 2 is often wider than metal 1.

Metal to metal separation is also large and is brought about mainly by difficulties in defining metal edges accurately during masking operations on the highly reflective metal.

All diffusion processes are such that lateral diffusion occurs as well as impurity penetration from the surface. Hence the separation rules for diffusion allow for this and relatively large separations are specified. This is particularly the case for the p-well diffusions which are deep diffusions and thus have considerable lateral spread.

Transitions from thin gate oxide to thick field oxide in the oxidation process also use up space and this is another reason why the lambda-based rules require a minimum separation between thinox regions of 3λ. In effect, this implies that the minimum feature size for thick oxide is 3λ.

The simplicity of the lambda-based rules makes this approach to design an appropriate one for the novice chip designer and also, perhaps, for those applications in which we are not trying to achieve the absolute minimum area and the absolute maximum performance. Because lambda-based rules try 'to be all things to all people', they do suffer from least common denominator effects and from the upward rounding of all process line dimension parameters into integer values of lambda.

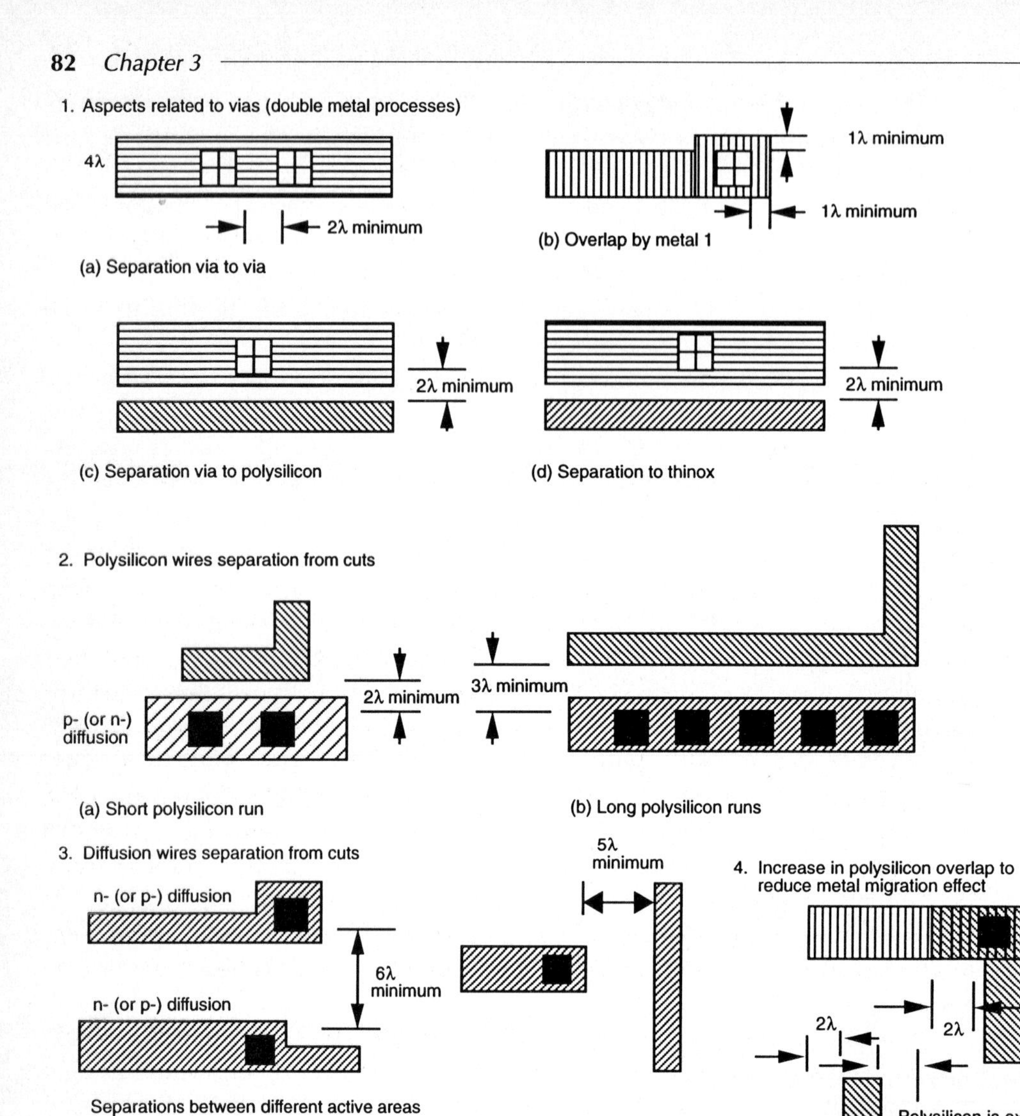

Figure 3–12 Further aspects of λ-based design rules for contacts, including some factors contributing to higher yield/reliability

The performance of any fabrication line in this respect clearly comes down to a matter of tolerances and definitions in terms of microns (or some other suitable unit of length). Thus, expanded sets of rules often referred to as micron-based rules are available to the more experienced designer to allow for the use of the full capability of any process. Also, many processes offer additional layers, which again adds to the possibilities presented to the designer.

In order to properly represent these important aspects, the next section introduces Orbit Semiconductor's 2 μm feature size double metal, double poly., n-well CMOS rules which also offer a BiCMOS capability.

3.5 2 μm double metal, double poly. CMOS/BiCMOS rules*

In order to accommodate the additional features present in this technology, it is necessary to extend the range of color and monochrome encodings previously used for double metal p-well CMOS. The encoding used is compatible with that already described, but as far as color assignments are concerned the following extension/additions are made: n-well — brown (same as p-well); Poly. 1 — red; Poly. 2 — orange; nDiff. (n-active) — green; pDiff. (p-active) — yellow (a green outline to the yellow may be used to show pDiff. clearly in color stick diagrams). Hatching, which is compatible with monochrome encoding, may also be added to color mask encoding, to distinguish underlying layers and to allow for ready copying of color diagrams on monochrome copying machines.

For BiCMOS the following are added: buried n^+ subcollector — pale green; p-base — pink. These extra features are set out in Figure 3–1(c) and in Color plate 1(c).

The use of color encoding is illustrated in the colorplates section of this book. The monochrome encoded rule set for the Orbit™ 2μm double metal double poly. BiCMOS process is given in Figures 3–13(a)–(f). The rule set is also presented in color as Color plates 3 to 6. Note the relative complexity of these rule sets. It must be further noted that an appropriate set of electrical parameters must accompany each set of design rules and the parameters for the Orbit™ 2 μm process are included in Appendix A.

* The rules and other details have been supplied by Orbit Semiconductors Inc. of Sunnyvale, California, through Integrated Silicon Design Pty Ltd of Adelaide, Australia. Their joint cooperation is gratefully acknowledged.

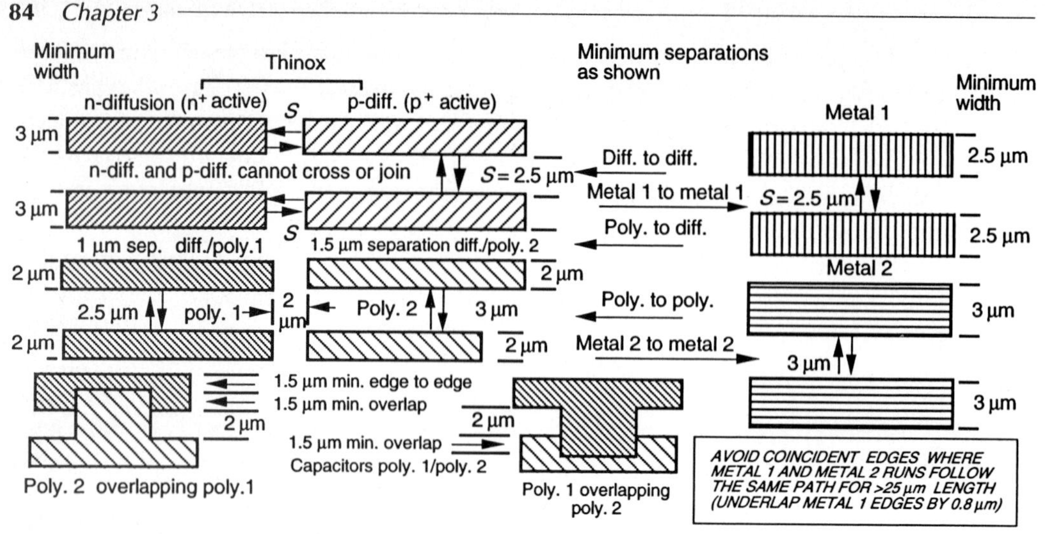

Note: *Where* no separation is specified, wires may overlap or cross (e.g. metal may cross any layer). For p-well CMOS, n-diff. wires can only exist inside and p-diff. wires outside the p-well. For n-well CMOS, p-diff. wires can only exist inside and n-diff. wires outside the n-well.

Figure 3–13(a) Design rules for wires (interconnects) (Orbit 2 µm CMOS)

3.6 1.2 µm double metal, single poly. CMOS rules*

As fabrication technology improves, so the feature size reduces and a separate set of micron-based design rules must accompany each new feature size. In order to open up the possibilities presented by this text, we have included the Orbit™ 1.2 µm rules in Appendix B together with the relevant electrical parameters.

3.7 Layout diagrams — a brief introduction

Mask layout diagrams may be hand-drawn on, say, 5 mm squared paper. In the case of lambda-based rules, the side of each square is taken to represent λ and, for micron-based rules, it will be taken to represent the least common factor associated with the rules (for example, 0.25 µm per side for the 2 µm process and

* The rules and other related details have been supplied by Orbit Semiconductors Inc. of Sunnyvale, California, through Integrated Silicon Design Pty Ltd of Adelaide, Australia. Their joint cooperation is gratefully acknowledged.

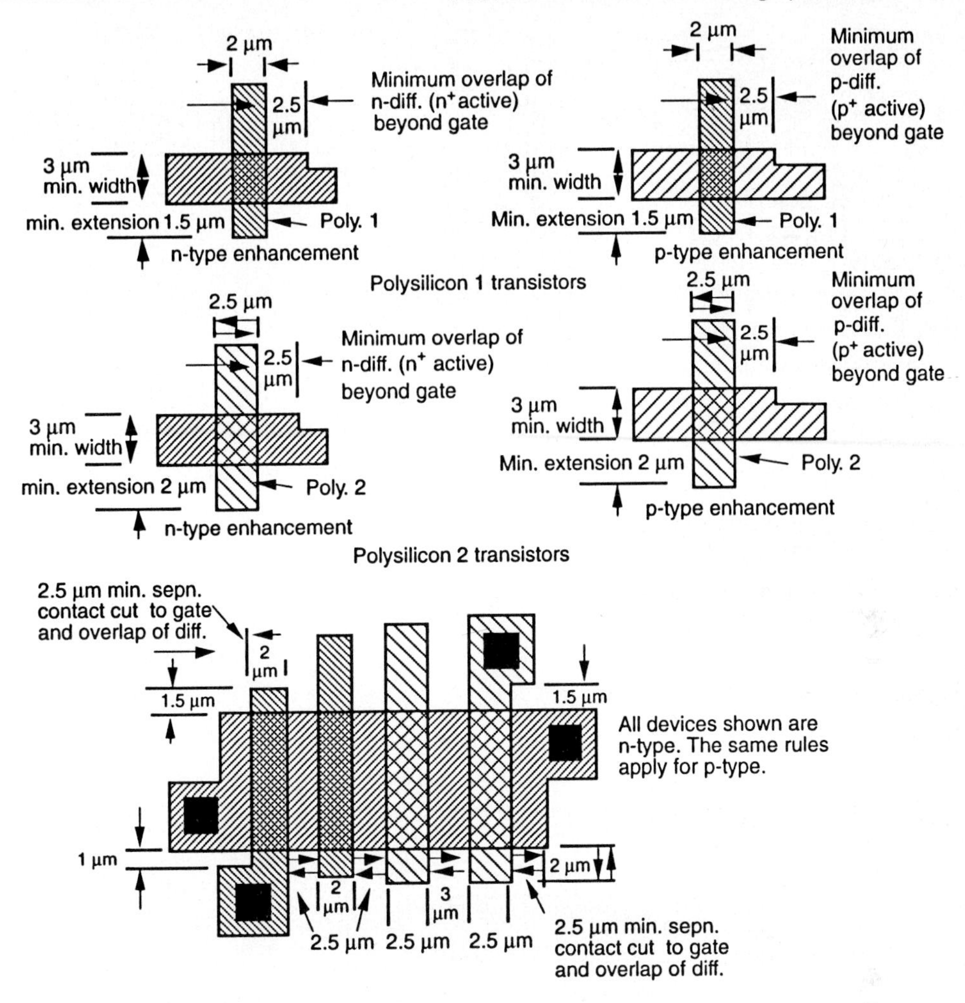

Figure 3–13(b) Transistor related design rules (Orbit 2 μm CMOS) minimum sizes and overlaps

0.2 μm per side for the 1.2 μm Orbit™ process layout). Most CAD VLSI tools also offer convenient facilities for mask level design.

The introductory layout diagrams which follow in Figures 3–14 to 3–17 inclusive have been included to illustrate the use of the lambda-based rule set and many more examples will appear later in the text.

The use of butting contacts has not been illustrated here as the reader is to be discouraged from using a facility which is not widely available now, but example layouts appear elsewhere for the sake of continuity with earlier designs and previous editions of this book.

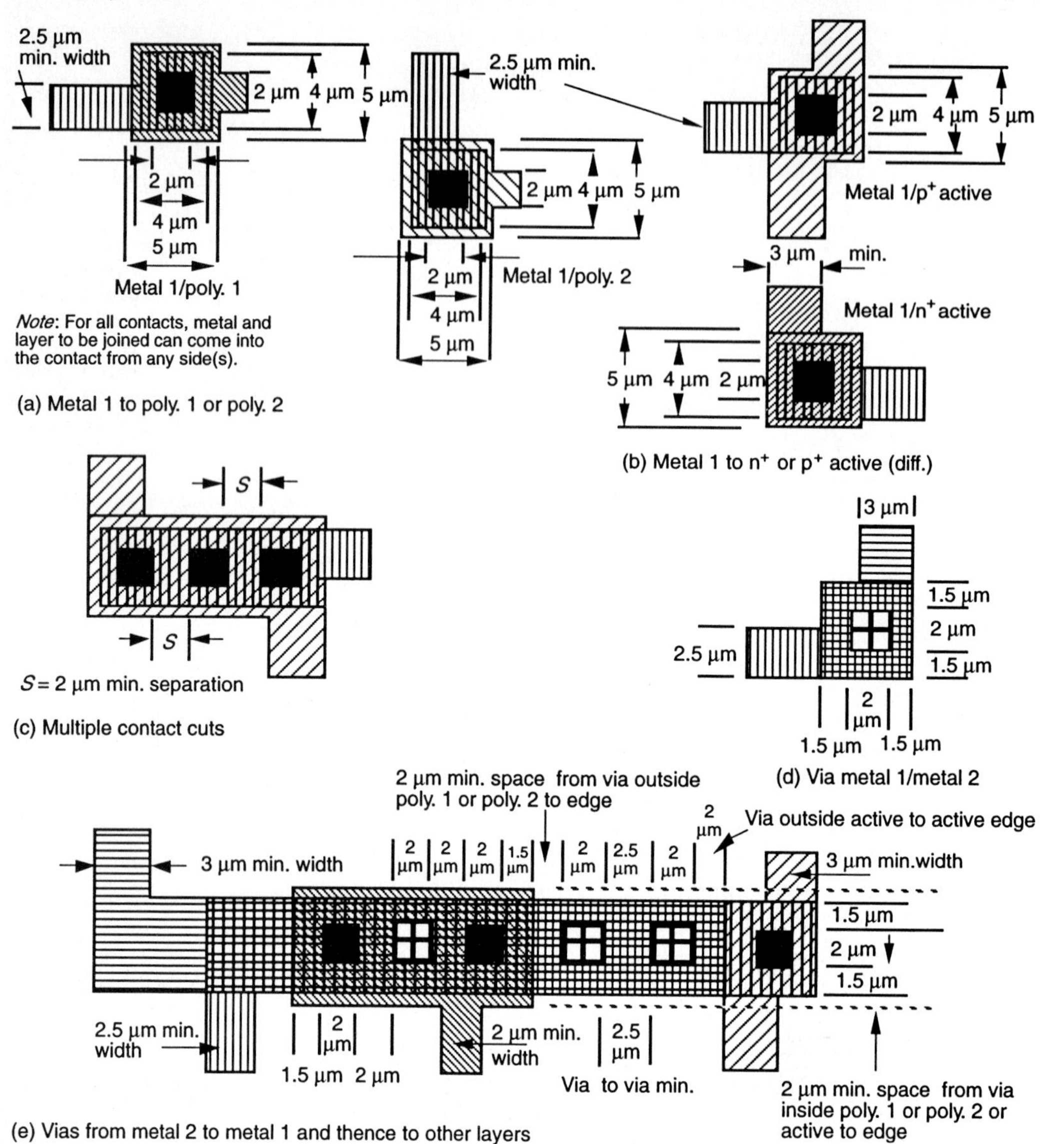

Figure 3–13(c) Rules for contacts and vias (Orbit 2 μm CMOS)

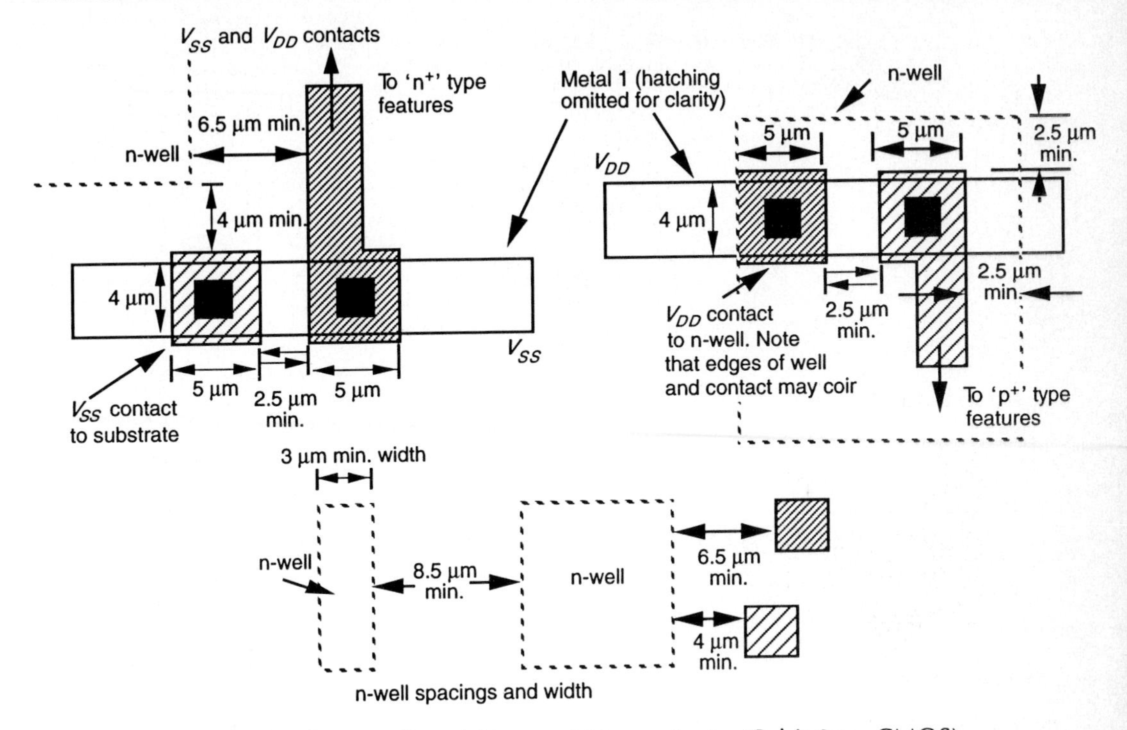

Figure 3–13(d) Rules for n-well and V_{DD} and V_{SS} contacts (Orbit 2μm CMOS)

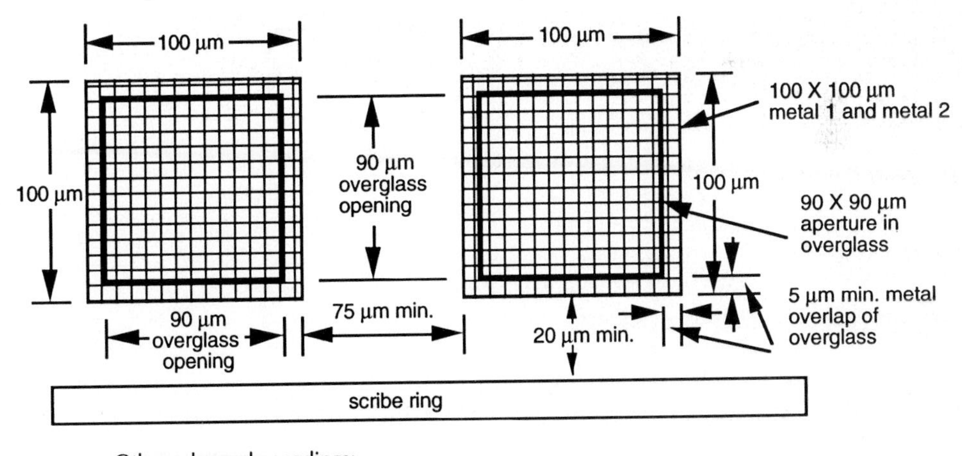

Other rules and encodings:
Via overlap of pad 2 μm.
Pad to active separation 20 μm min.
Color encoding for overglass mask . . . Gray

Figure 3–13(e) Rules for pad and overglass geometry (Orbit 2μm CMOS)

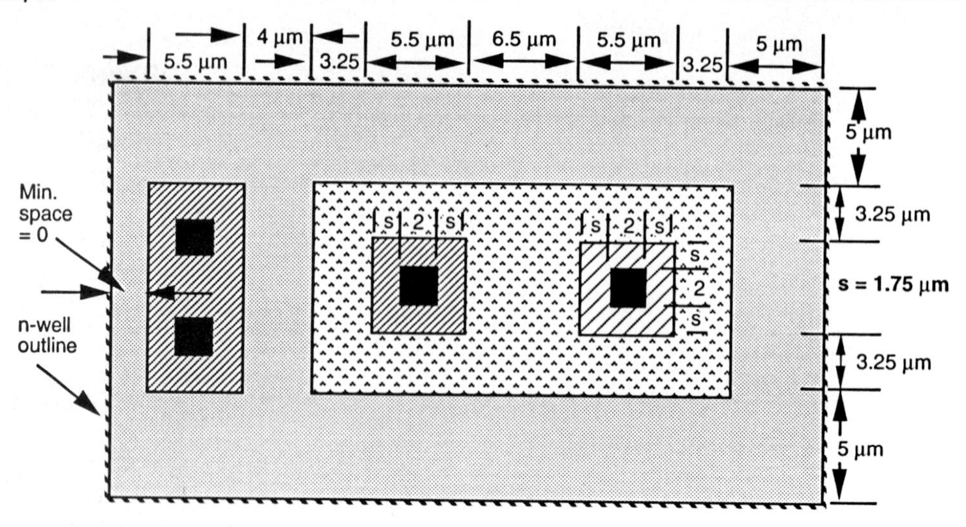

Note: For clarity, the layers have not been drawn transparent but BCCD underlies the entire area and the p-base underlies all within its boundary.

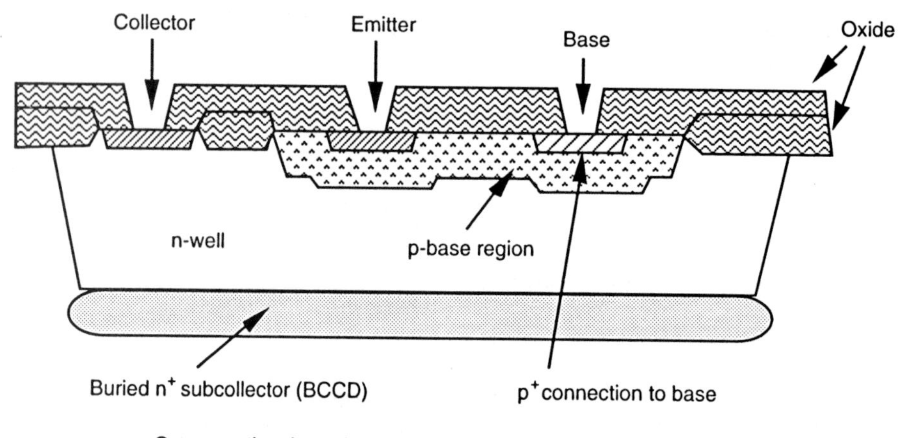

Cross-section through npn transistor (Orbit 2 μm BiCMOS)

Figure 3–13(f) Special rules for BiCMOS transistors (Orbit 2μm CMOS)

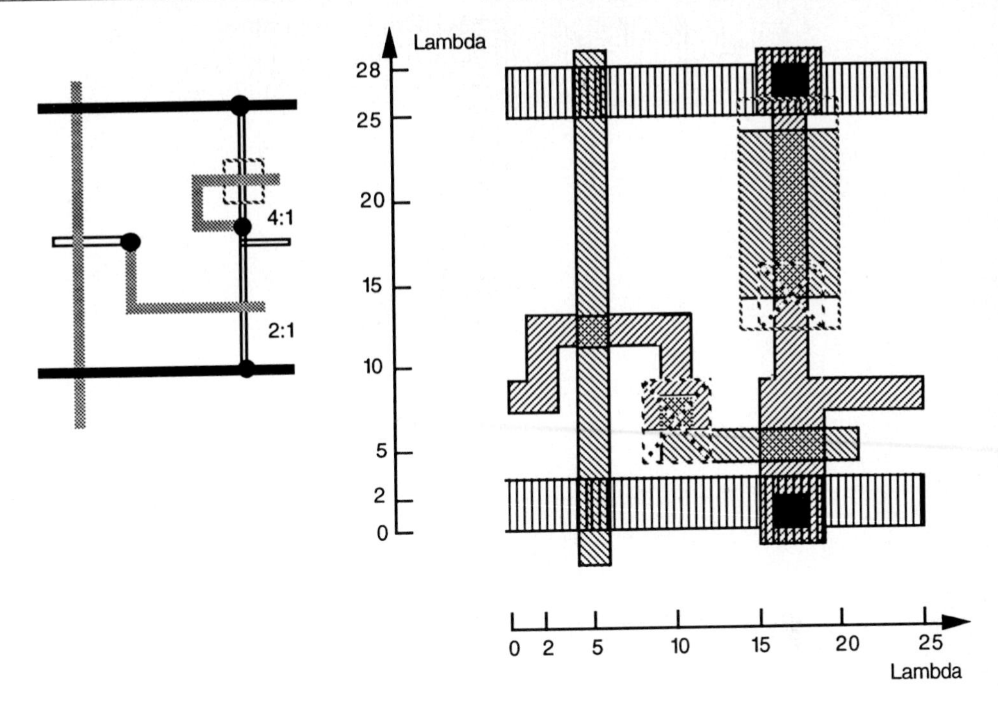

Figure 3–14 Stick diagram and layout for nMOS shift register cell

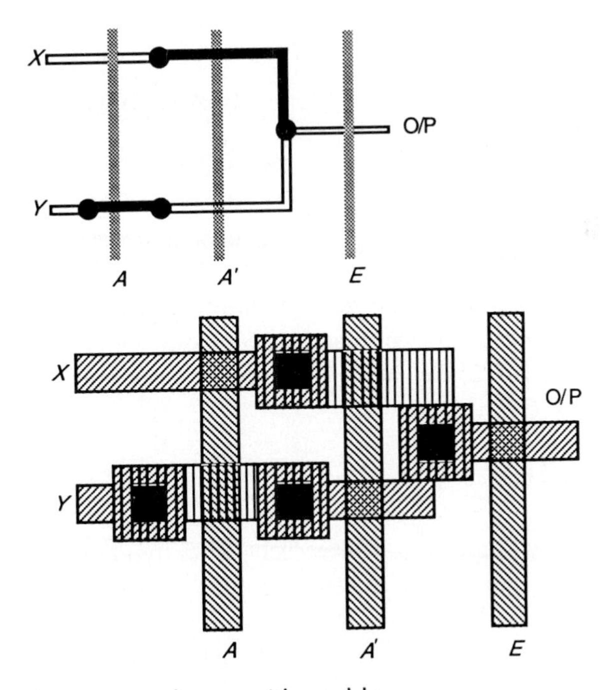

Figure 3–15 Two-way selector with enable

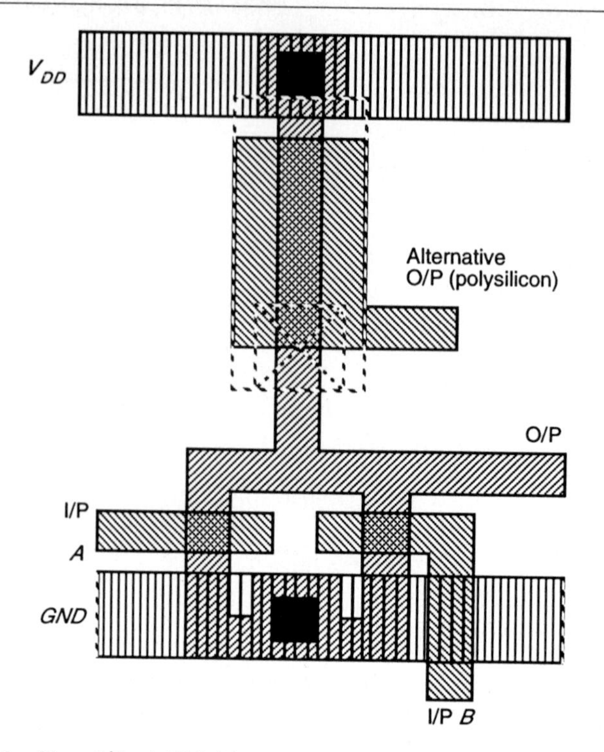

Figure 3–16 Two I/P nMOS *Nor* gate

3.8 Symbolic diagrams — translation to mask form

The symbolic form of diagram is also readily translated to mask layout form. Take, for example, the symbolic form of a 1-bit CMOS shift register cell given earlier in Figure 3–5(e). The symbolic form is reproduced in Figure 3–17(a) and the resultant mask layout is presented as Figure 3–17(b). This is also presented in color as Color plate 7. The translation process should be self-evident from the figures. Further examples of mask layout from symbolic diagram form follow in Chapter 5 (for BiCMOS gates) and in other parts of the text.

3.9 Observations

This chapter has introduced three sets of design rules with which nMOS and CMOS designs may be fabricated. Designs incorporating BiCMOS technology are covered by the 'Orbit' 2 μm double metal, double poly. n-well process rules. We are now in a position to use the design rules and, for simplicity, most design examples will use the lambda-based rules. As the budding designer becomes

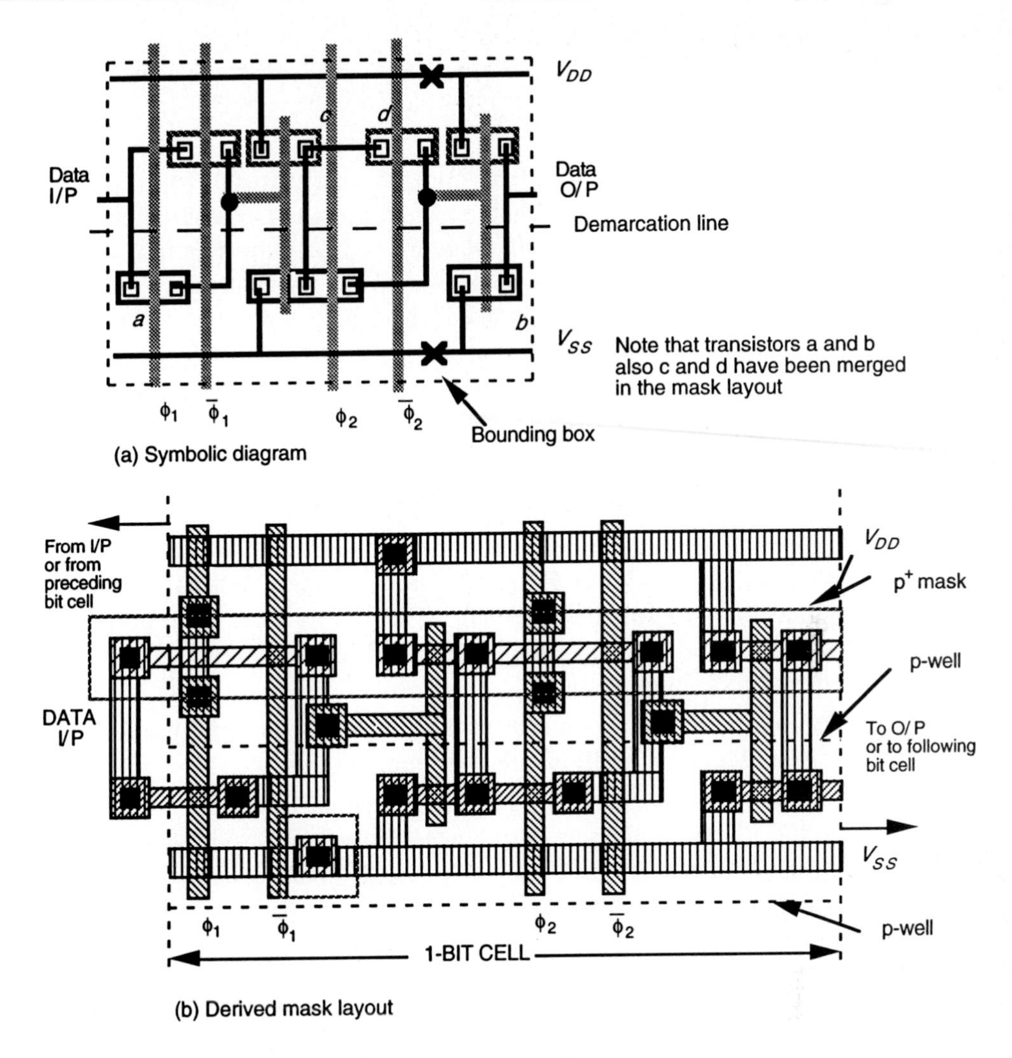

Figure 3–17 A 1-bit CMOS shift register cell

more proficient, designs are readily completed using one or other of the 'Orbit' rule sets.

Before we begin any design work, however, a further chapter is necessary to establish, explain and evaluate other key circuit parameters.

3.10 Tutorial exercises

Note: Use colors to represent layers.

1. First draw the *circuit diagrams* for each of Figures 3–14 to 3–16 and then, after closing this book, draw a stick diagram and a mask layout diagram for each. These efforts may then be compared with those in the book, although note that lack of conformity in detail may not mean that a layout, for example, is incorrect. Check your layouts against the design rules given in the text.
2. Draw the stick diagram and a mask layout for an 8:1 nMOS inverter circuit. Both the input and output points should be on the polysilicon layer.
3. With regard to Figure 3–15, what will be the state of the output (O/P) when control line E is at 0 volts? Could any simple modification to this circuit improve its operation? If so, set out a modified stick diagram and corresponding mask layout.
4. With regard to Figure 3–14, determine suitable left-hand and right-hand boundary lines for this leaf-cell, so that a series of such leaf-cells can be butted directly together side by side without violating any design rules, yet occupying minimum area.
5. Can you reduce the area occupied by the leaf-cell of Figure 3–14? Draw an alternative layout to illustrate your contention.
6. Figure 3–18 presents a simple CMOS layout. Study the layout, and from it produce a circuit diagram. Explain the nature and purpose of the circuit. Using this layout, explain how you could construct a four-way multiplexer (selector) circuit.

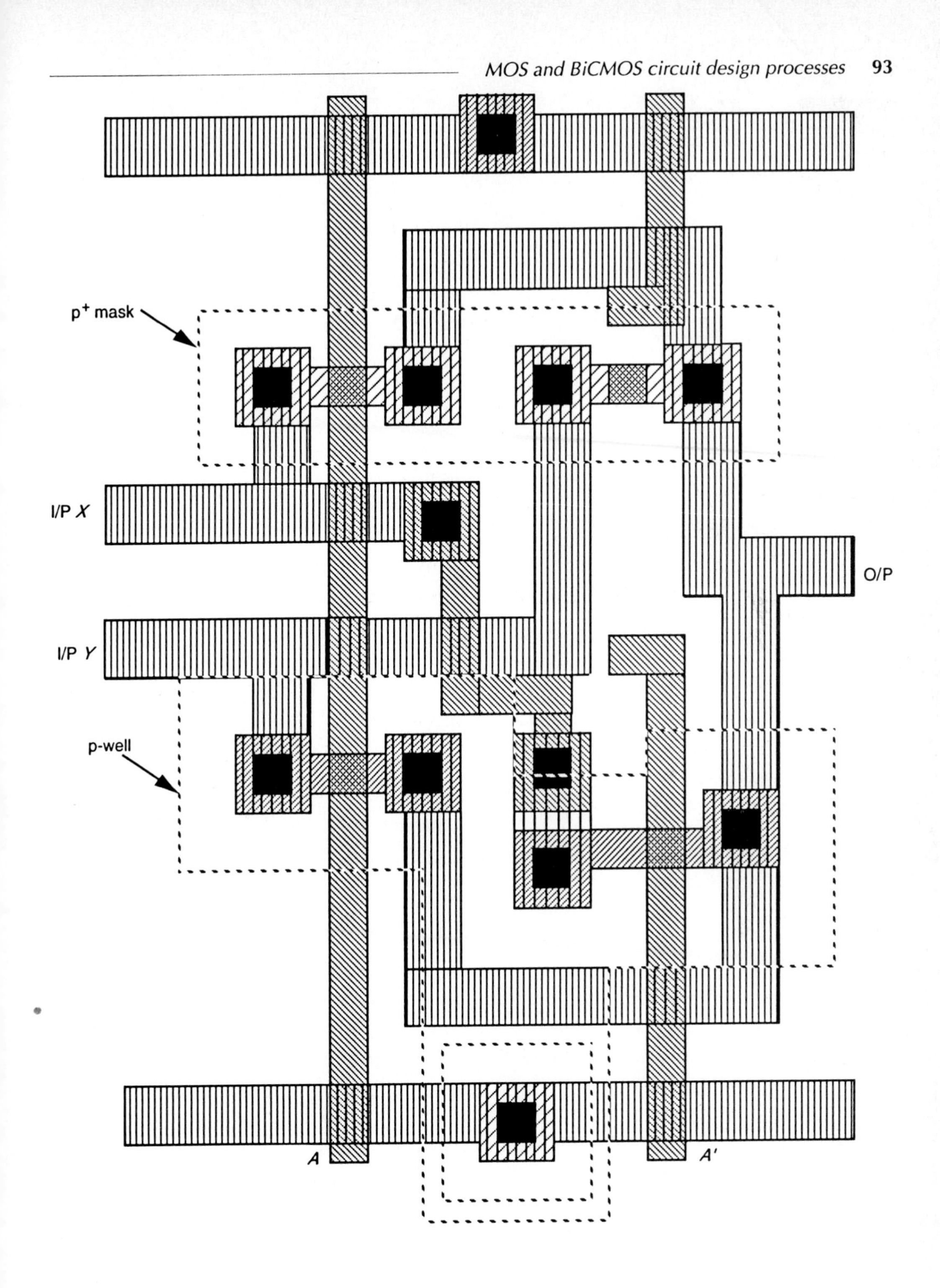

Figure 3–18 CMOS layout example

4 Basic circuit concepts

Education is a progressive discovery of our own ignorance.

Will Durant

Objectives

The active devices of MOS technology having been dealt with in some measure, it is now appropriate to consider their interconnection as circuits. The 'wiring-up' of circuits takes place through the various conductive layers which are produced by the MOS processing and it is therefore necessary to be aware of the resistive and capacitive characteristics of each layer.

Concepts such as sheet resistance R_s and a standard unit of capacitance $\square Cg$ help greatly in evaluating the effects of wiring and input and output capacitances. Further, the delays associated with wiring, with inverters and with other circuitry may be conveniently evaluated in terms of a delay unit τ.

Parameter values for the layers in 5 μm, 2 μm and 1.2 μm technologies are given in this chapter so that actual designs may be evaluated. Means of dealing with larger capacitive loads are also discussed, as are the factors affecting the choice of layer for various interconnection purposes.

So far we have established equations (Chapter 2) which characterize the behavior of MOS transistors, aspects of their use in both nMOS and CMOS circuits, and the pull-up to pull-down ratios which must be observed when nMOS inverters and pass transistors are interconnected. However, as yet we have not considered the actual resistance and capacitance values associated with transistors, nor have we considered circuit wiring and parasitics. In order to simplify the treatment of such components, there are basic circuit concepts which will now be introduced, and for particular MOS processes we can set out approximate circuit parameters which greatly ease the design process in allowing straightforward calculations. In order to take advantage of BiCMOS circuitry we must also examine some basic properties of bipolar transistors.

4.1 Sheet resistance R_s

Consider a uniform slab of conducting material of resistivity ρ, of width W, thickness t, and length between faces L. The arrangement is shown in Figure 4–1.

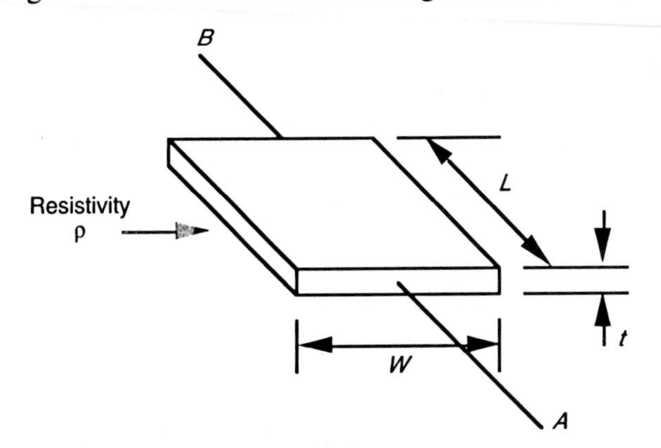

Figure 4–1 Sheet resistance model

With reference to Figure 4–1, consider the resistance R_{AB} between two opposite faces.

$$R_{AB} = \frac{\rho L}{A} \text{ ohm}$$

where

$$A = \text{cross-section area}$$

Thus

$$R_{AB} = \frac{\rho L}{tW} \text{ ohm}$$

Now, consider the case in which $L = W$, that is, a square of resistive material, then

$$R_{AB} = \frac{\rho}{t} = R_s$$

where

$$R_s = \text{ohm per square or sheet resistance}$$

Thus

$$R_s = \frac{\rho}{t} \text{ ohm per square}$$

Note that R_s is completely independent of the area of the square; for example, a 1 μm per side square slab of material has exactly the same resistance as a 1 cm per side square slab of the same material if the thickness is the same.

Thus the actual values associated with the layers in a MOS circuit depend on the thickness of the layer and the resistivity of the material forming the layer. For the metal and polysilicon layers, the thickness of a layer is easily envisaged and the resistivity of the material is known. For the diffusion layer, the depth of the diffusion regions contributes toward the effective thickness while the impurity concentration (or doping level) profile determines the resistivity.

For the MOS processes considered here, typical values of sheet resistance are given in Table 4–1.

Table 4–1 Typical sheet resistances R_s of MOS layers for 5 μm*, and Orbit 2 μm* and 1.2 μm* technologies

Layer	*R_s ohm per square*		
	5 μm	Orbit	Orbit 1.2 μm
Metal	0.03	0.04	0.04
Diffusion (or active)**	10→50	20→45	20→45
Silicide	2→4	–	–
Polysilicon	15→100	15→30	15→30
n-transistor channel	10^4 †	2×10^4 †	2×10^4 †
p-transistor channel	2.5×10^4 †	4.5×10^4 †	4.5×10^4 †

Note: In some processes a silicide layer is used in place of polysilicon.

* 5 micron (μm) technology implies minimum line width (and feature size) of 5 μm and in consequence $\lambda = 2.5$ um. Similarly, 2 μm and 1.2 μm technologies have feature sizes of 2 μm and 1.2 μm respectively.

** The figures given are for n-diffusion regions. The values for p-diffusion are 2.5 times these values.

† These values are approximations only. Resistances may be calculated from a knowledge of V_{ds} and the expressions for I_{ds} given earlier.

4.2 Sheet resistance concept applied to MOS transistors and inverters

Consider the transistor structures of Figure 4–2 and note that the diagrams distinguish the actual diffusion (active) regions from the channel regions. (*Note*: From here on, the term 'diffusion' also covers active regions in Orbit processes.) The thinox mask layout is the union of diffusion and channel regions and these regions have differing hatching patterns to stress the fact that the polysilicon and underlying silicon dioxide mask the substrate so that diffusion takes place only in the areas defined by the thinox mask which do not coincide with the polysilicon mask.

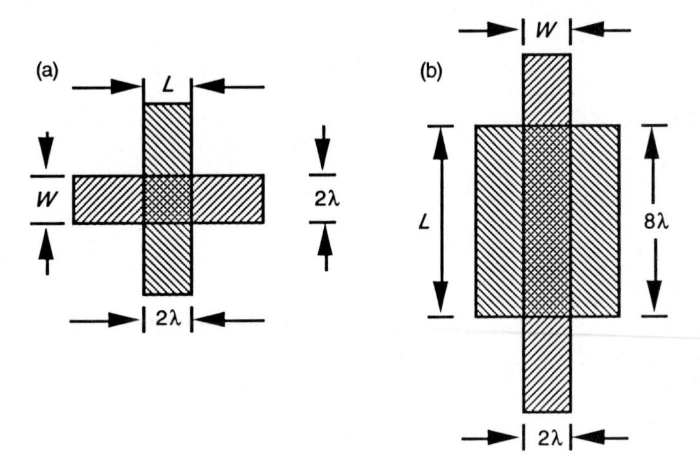

Figure 4–2 Resistance calculation for transistor channels

The simple n-type pass transistor of Figure 4–2(a) has a channel length $L = 2\lambda$ and a channel width $W = 2\lambda$. The channel is, therefore, square and channel resistance (with or without implant)

$$R = 1 \text{ square} \times R_s \frac{\text{ohm}}{\text{square}} = R_s = 10^4 \text{ ohm}*$$

The length to width ratio, denoted Z, is 1:1 in this case. The transistor structure of Figure 4–2(b) has a channel length $L = 8\lambda$ and width $W = 2\lambda$. Therefore,

$$Z = \frac{L}{W} = 4$$

Thus, channel resistance

$$R = ZR_s = 4 \times 10^4 \text{ ohm}$$

Another way of looking at this is to recognize that this channel can be regarded as four $2\lambda \times 2\lambda$ squares in series, thus giving a resistance of $4R_s$. This particular way of approaching the calculation of resistance is often useful, particularly when dealing with shapes which are not simple rectangles.

Figure 4–3 takes these considerations one step further and shows how the pull-up to pull-down ratio of an inverter is determined. In the nMOS case a simple 4:1 $Z_{p.u.}: Z_{p.d.}$ ratio obviously applies. Note, for example, that a 4:1 ratio would

*From Table 4–1.

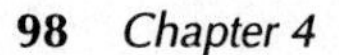

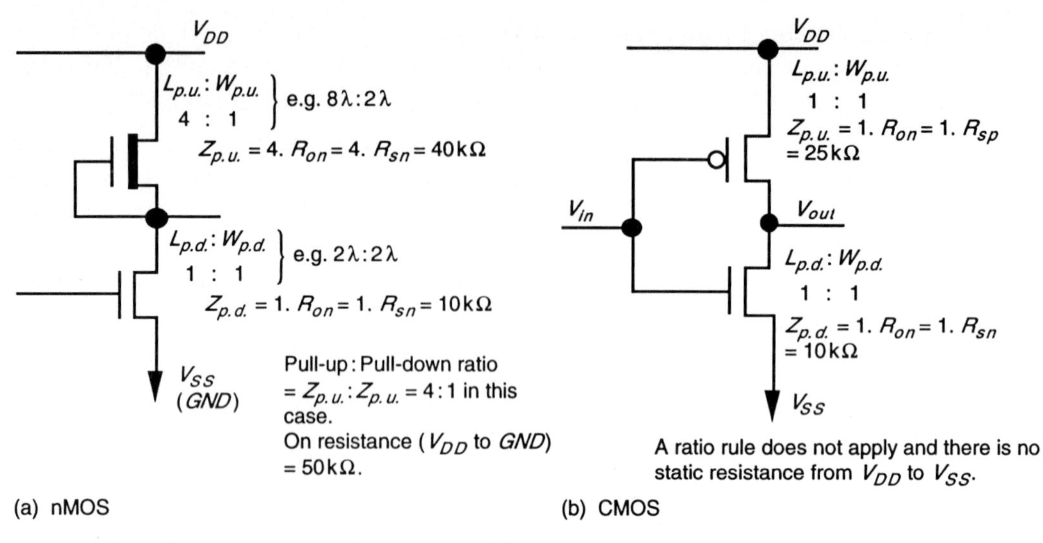

Note: R_{on} = 'on' resistance; R_{sn} = n-channel sheet resistance; R_{sp} = p-channel sheet resistance.

Figure 4–3 Inverter resistance calculation

also be achieved if the upper channel (p.u.) length $L = 4\lambda$, and width $W = 2\lambda$ with lower channel (p.d.) length $L = 2\lambda$, and width $W = 4\lambda$.

For the CMOS case, note the different value of R_s which applies for the pull-up transistor.

4.2.1 Silicides

As the line width becomes smaller, the sheet resistance contribution to RC delay increases. With the currently available polysilicon sheet resistance ranging from 15 to 100 ohm it is apparent that some of the advantages of scaling could be offset by the interconnect resistance at the gate level. Therefore the low sheet resistances of refractory silicides (2–4 ohm), which are formed by depositing metal on polysilicon and then sintering, have been investigated as an interconnecting medium.

Deposition of the metal or metal/silicon alloy prior to sintering may be done in any one of several ways:

- sputtering or evaporation;
- co-sputtering metal and silicon in the desired ratio from two independent targets;
- co-evaporation from the elements.

Although the properties of silicides make them attractive alternatives to polysilicon, there are extra processing steps which offset this advantage.

4.3 Area capacitances of layers

From the diagrams we have used to illustrate the structure of transistors, and from discussions of the fabrication processes, it will be apparent that conducting layers are separated from the substrate and each other by insulating (dielectric) layers, and thus parallel plate capacitive effects must be present and must be allowed for.

For any layer, knowing the dielectric (silicon dioxide) thickness, we can calculate area capacitance as follows:

$$C = \frac{\varepsilon_0 \varepsilon_{ins} A}{D} \text{ farads}$$

where

D = thickness of silicon dioxide
A = area of plates

(and it is assumed that ε_0, A, and D are in compatible units, for example, ε_0 in farads/cm, A in cm^2, D in cm).

$$\varepsilon_{ins} = \text{relative permittivity of } SiO_2 \doteqdot 4.0$$
$$\varepsilon_0 = 8.85 \times 10^{-14} \text{ F/cm (permittivity of free space)}$$

A normal approach is to give layer area capacitances in pF/μm^2 (where μm = micron = 10^{-6} meter $=10^{-4}$ cm). The appropriate figure may be calculated as follows:

$$C\left(\frac{\text{pF}}{\mu\text{m}^2}\right) = \frac{\varepsilon_0 \varepsilon_{ins}}{D} \frac{\text{F}}{\text{cm}^2} \times \frac{10^{12}\text{pF}}{\text{F}} \times \frac{\text{cm}^2}{10^8 \mu\text{m}^2}$$

(D in cm, ε_0 in farads/cm)

Typical values of area capacitance are set out in Table 4–2 for 5 μm technology and for Orbit 2 μm and 1.2 μm technologies.

Table 4–2 Typical area capacitance values for MOS circuits

Capacitance	*Value in pF × 10^{-4}/μm^2 (Relative values in brackets)*		
	5 μm	*2 μm*	*1.2 μm*
Gate to channel	4 (1.0)	8 (1.0)	16 (1.0)
Diffusion (active)	1 (0.25)	1.75 (0.22)	3.75 (0.23)
Polysilicon* to substrate	0.4 (0.1)	0.6 (0.075)	0.6 (0.038)
Metal 1 to substrate	0.3 (0.075)	0.33 (0.04)	0.33 (0.02)
Metal 2 to substrate	0.2 (0.05)	0.17 (0.02)	0.17 (0.01)
Metal 2 to metal 1	0.4 (0.1)	0.5 (0.06)	0.5 (0.03)
Metal 2 to polysilicon	0.3 (0.075)	0.3 (0.038)	0.3 (0.018)

Notes: Relative value = specified value/gate to channel value for that technology.
*Poly. 1 and Poly. 2 are similar (also silicides where used).

4.4 Standard unit of capacitance $\square C_g$

It is convenient to employ a standard unit of capacitance that can be given a value appropriate to the technology but can also be used in calculations without associating it with an absolute value. The unit is denoted $\square C_g$ and is defined as the gate-to-channel capacitance of a MOS transistor having $W = L =$ feature size, that is, a 'standard' or 'feature size' square as in Figure 4–2(a), for example, for lambda-based rules. (This concept, originated by VTI (USA), has been adapted here.)

$\square C_g$ may be evaluated for any MOS process. For example, for 5 µm MOS circuits:

Area/standard square = 5 µm × 5 µm = 25 µm^2 (= area of minimum size transistor)
Capacitance value (from Table 4–2) = 4×10^{-4} pF/µm^2
Thus, standard value $\square C_g = 25$ µm$^2 \times 4 \times 10^{-4}$ pF/µm^2 = .01 pF

or, for 2 µm MOS circuits (Orbit):

Area/standard square = 2 µm × 2 µm = 4 µm^2
Gate capacitance value (from Table 4–2) = 8×10^{-4} pF/µm^2
Thus, standard value $\square C_g = 4$ µm$^2 \times 8 \times 10^{-4}$ pF/µm^2 = .0032 pF

and, for 1.2 µm MOS circuits (Orbit):

Area/standard square = 1.2 µm × 1.2 µm = 1.44 µm^2
Gate capacitance value (from Table 4–2) = 16×10^{-4} pF/µm^2
Thus, standard value $\square C_g$ =1.44 µm$^2 \times 16 \times 10^{-4}$ pF/µm^2 = .0023 pF

4.5 Some area capacitance calculations

The approach will be demonstrated using λ-based geometry. The calculation of capacitance values may now be undertaken by establishing the ratio between the area of interest and the area of standard (feature size square) gate (2λ × 2λ for λ-based rules) and multiplying this ratio by the appropriate relative C value from Table 4–2. The product will give the required capacitance in $\square C_g$ units.

Consider the area defined in Figure 4–4. First, we must calculate the area relative to that of a standard gate.

$$\text{Relative area} = \frac{20\lambda \times 3\lambda}{2\lambda \times 2\lambda} = 15$$

Now:

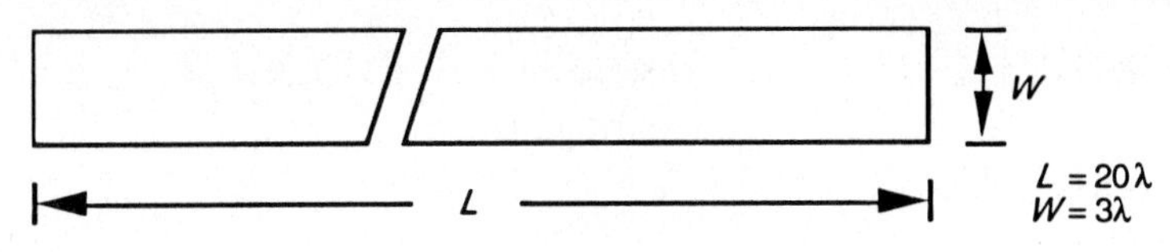

Figure 4–4 Simple area for capacitance calculation

1. Consider the area in metal 1.

   Capacitance to substrate = relative area × relative C value
   $$= 15 \times 0.075 \square C_g$$
   $$= 1.125 \square C_g$$

   That is, the defined area in metal has a capacitance to substrate 1.125 times that of a feature size square gate area.

2. Consider the same area in polysilicon.

   Capacitance to substrate $= 15 \times 0.1 \square C_g$
   $$= 1.5 \square C_g$$

3. Consider the same area in n-type diffusion.

   Capacitance to substrate $= 15 \times 0.25 \square C_g$
   $$= 3.75 \square C_g{}^*$$

Calculations of area capacitance values associated with structures occupying more than one layer, as in Figure 4–5, are equally straightforward.

Consider the metal area (less the contact region where the metal is connected to polysilicon and shielded from the substrate)

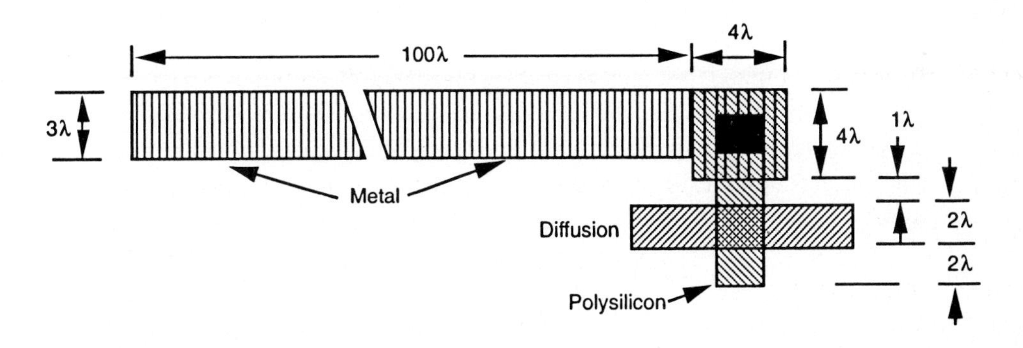

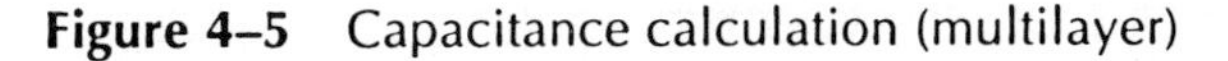

Figure 4–5 Capacitance calculation (multilayer)

* Note the relatively high capacitance values of the diffusion layer even though peripheral capacitance (see Table 4–3 in section 4.10.3) has not been allowed for. This may increase total diffusion capacitance to considerably more than the area capacitance calculated here.

$$\text{Ratio} = \frac{\text{Metal area}}{\text{Standard gate area}} = \frac{100\lambda \times 3\lambda}{4\lambda^2} = 75$$

$$\text{Metal capacitance } C_m = 75 \times 0.075 = 5.625\square C_g$$

Consider the polysilicon area (excluding the gate region)

$$\text{Polysilicon area} = 4\lambda \times 4\lambda + 3\lambda \times 2\lambda = 22\lambda^2$$

Therefore

$$\text{Polysilicon capacitance } C_p = \frac{22}{4} \times 0.1 = .55\square C_g$$

For the transistor,

$$\text{Gate capacitance } C_g = 1\square C_g$$

Therefore

$$\text{Total capacitance } C_T = C_m + C_p + C_g \doteqdot 7.20\square C_g$$

In all cases absolute values are readily evaluated by substitution of the actual value for $\square C_g$ as given in section 4.4.

It is not unusual to find metal paths of uniform 4λ width but when taking this approach in design it must be borne in mind that, compared with 3λ width paths, the capacitance will be increased by one-third.

For example, if the metal width is increased to 4λ in Figure 4–5, the capacitance C_m is increased to $7.5\square C_g$ and the capacitance of the complete structure will increase to about $9\square C_g$.

4.6 The delay unit τ

We have developed the concept of sheet resistance R_s and standard gate capacitance unit $\square C_g$. If we consider the case of one standard (feature size square) gate area capacitance being charged through one feature size square of n channel resistance (that is, through R_s for an nMOS pass transistor channel), as in Figure 4–6, we have:

$$\text{Time constant } \tau = (\,1R_s\,(\text{n channel}) \times 1\square C_g)\text{ seconds}$$

This can be evaluated for any technology and for 5 μm technology,

$$\tau = 10^4 \text{ ohm} \times 0.01 \text{ pF} = 0.1 \text{ nsec}$$

and for 2 μm (Orbit) technology,

$$\tau = 2 \times 10^4 \text{ ohm} \times 0.0032 \text{ pF} = 0.064 \text{ nsec}$$

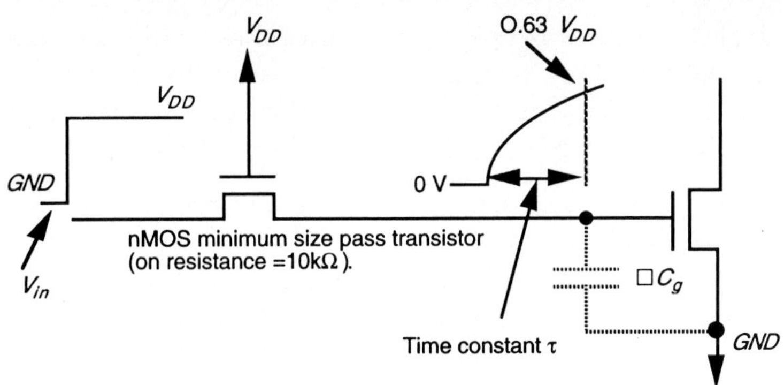

Figure 4–6 Model for derivation of τ

and for 1.2 μm (Orbit) technology,

$$\tau = 2 \times 10^4 \text{ ohm} \times 0.0023 \text{ pF} = 0.046 \text{ nsec}$$

However, in practice, circuit wiring and parasitic capacitances must be allowed for so that the figure taken for τ is often increased by a factor of two or three so that for 5 μm circuit

> τ = 0.2 to 0.3 nsec is a typical design figure used in assessing likely worst case delays.

Note that τ thus obtained is not much different from transit time τ_{sd} calculated from equation 2.2

$$\tau_{sd} = \frac{L^2}{\mu_n V_{ds}}$$

Note that V_{ds} varies as C_g charges from 0 volts to 63% of V_{DD} in period τ in Figure 4–6, so that an appropriate value for V_{ds} is the average value = 3 volts. For 5 μm technology, then,

$$\tau_{sd} = \frac{25\mu\text{m}^2 \text{ V sec}}{650 \text{ cm}^2 \text{ 3V}} \times \frac{10^9 \text{ nsec cm}^2}{10^8 \text{ }\mu\text{m}^2}$$
$$= 0.13 \text{ nsec}$$

This is very close to the theoretical time constant τ calculated above.

Since the transition point of an inverter or gate is 0.5 V_{DD}, which is close to 0.63 V_{DD}, it appears to be common practice to use transit time and time constant (as defined for the delay unit τ) interchangeably and 'stray' capacitances are usually allowed for by doubling (or more) the theoretical values calculated.

In view of this, τ is used as the fundamental time unit and all timings in a system can be assessed in relation to τ.

For 5 μm MOS technology $\tau = 0.3$ nsec is a very safe figure to use; and, for 2 μm Orbit MOS technology, $\tau = 0.2$ nsec is an equally safe figure to use; and, for 1.2 μm Orbit MOS technology, $\tau = 0.1$ nsec is also a safe figure.

4.7 Inverter delays

Consider the basic 4:1 ratio nMOS inverter. In order to achieve the 4:1 $Z_{p.u.}$ to $Z_{p.d.}$ ratio, $R_{p.u.}$ will be 4 $R_{p.d.}$ and if $R_{p.d.}$ is contributed by the minimum size transistor then, clearly, the resistance value associated with $R_{p.u.}$ is

$$R_{p.u.} = 4R_s = 40 \text{ k}\Omega$$

Meanwhile, the $R_{p.d.}$ value is $1R_s = 10$ kΩ so that the delay associated with the inverter will depend on whether it is being turned on or off.

However, if we consider a pair of *cascaded inverters*, then the delay over the pair will be constant irrespective of the sense of the logic level transition of the input to the first. This is clearly seen from Figure 4–7 and, assuming $\tau = 0.3$ nsec and making no extra allowances for wiring capacitance, we have an overall delay of $\tau + 4\tau = 5\tau$. In general terms, the delay through a pair of similar nMOS inverters is

$$T_d = (1 + Z_{p.u.}/Z_{p.d.})\tau$$

Thus, the inverter pair delay for inverters having 4:1 ratio is 5τ.

However, a single 4:1 inverter exhibits undesirable asymmetric delays since the delay in turning on is, for example, τ, while the corresponding delay in turning off is 4τ. Quite obviously, the asymmetry is worse when considering an inverter with an 8:1 ratio.

When considering CMOS inverters, the nMOS ratio rule no longer applies, but we must allow for the natural (R_s) asymmetry of the usually equal size pull-

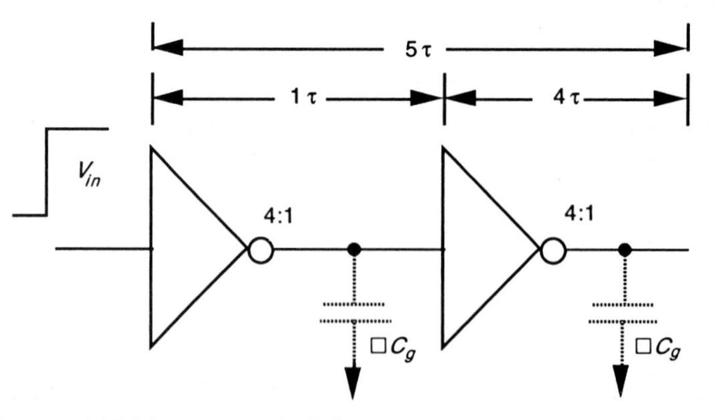

Figure 4–7 nMOS inverter pair delay

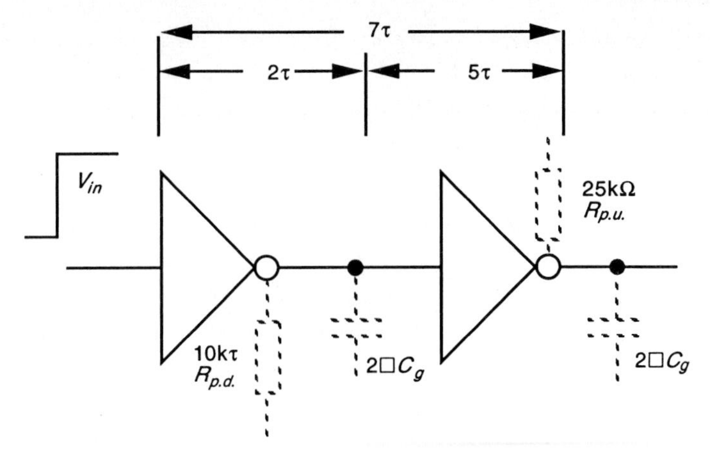

Figure 4–8 Minimum size CMOS inverter pair delay

up p-transistors and the n-type pull-down transistors. Figure 4–8 shows the theoretical delay associated with a pair of minimum size (both n- and p-transistors) lambda-based inverters. Note that the gate capacitance (= $2\square C_g$) is double that of the comparable nMOS inverter since the input to a CMOS inverter is connected to *both* transistor gates. Note also the allowance made for the differing channel resistances.

The asymmetry of resistance values can be eliminated by increasing the width of the p-device channel by a factor of two or three, but it should be noted that the gate input capacitance of the p-transistor is also increased by the same factor. This, to some extent, offsets the speed-up due to the drop in resistance, but there is a small net gain since the wiring capacitance will be the same.

4.7.1 A more formal estimation of CMOS inverter delay

A CMOS inverter, in general, either charges or discharges a capacitive load C_L and rise-time τ_r or fall-time τ_f can be estimated from the following simple analysis.

4.7.1.1 Rise-time estimation

In this analysis we assume that the p-device stays in saturation for the entire charging period of the load capacitor C_L. The circuit may then be modeled as in Figure 4–9.

The saturation current for the p-transistor is given by

$$I_{dsp} = \frac{\beta_p (V_{gs} - |V_{tp}|)^2}{2}$$

This current charges C_L and, since its magnitude is approximately constant, we have

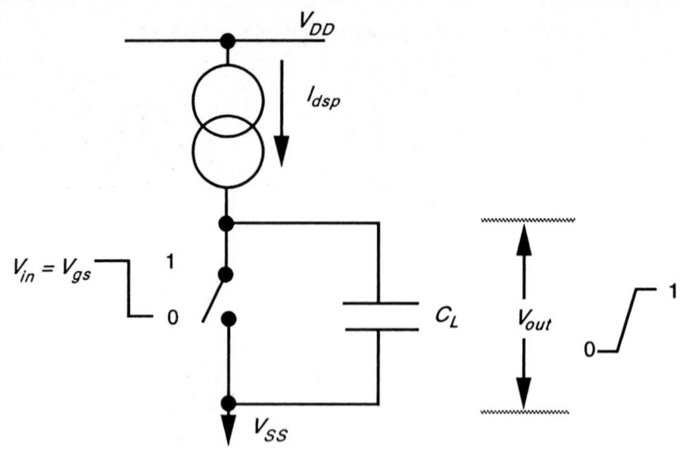

Figure 4–9 Rise-time model

$$V_{out} = \frac{I_{dsp}t}{C_L}$$

Substituting for I_{dsp} and rearranging we have

$$t = \frac{2C_L V_{out}}{\beta_p(V_{gs} - |V_{tp}|)^2}$$

We now assume that $t = \tau_r$ when $V_{out} = + V_{DD}$, so that

$$\tau_r = \frac{2V_{DD}C_L}{\beta_p(V_{DD} - |V_{tp}|)^2}$$

with $|V_{tp}| = 0.2\ V_{DD}$, then

$$\tau_r \doteqdot \frac{3C_L}{\beta_p V_{DD}}$$

This result compares reasonably well with a more detailed analysis in which the charging of C_L is divided, more correctly, into two parts: (l) saturation and (2) resistive region of the transistor.

4.7.1.2 *Fall-time estimation*

Similar reasoning can be applied to the discharge of C_L through the n-transistor. The circuit model in this case is given as Figure 4–10.

Making similar assumptions we may write for fall-time:

$$\tau_f \doteqdot \frac{3C_L}{\beta_n V_{DD}}$$

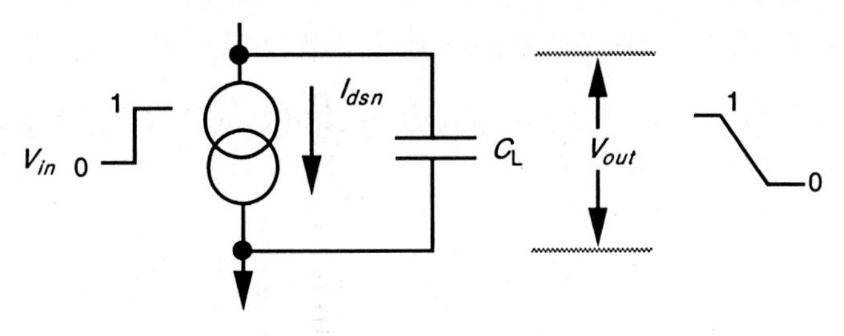

Figure 4–10 Fall-time model

4.7.1.3 Summary of CMOS rise and fall factors

Using these expressions we may deduce that:

$$\frac{\tau_r}{\tau_f} = \frac{\beta_n}{\beta_p}$$

But $\mu_n = 2.5\ \mu_p$ and hence $\beta_n \doteqdot 2.5\ \beta_p$, so that the rise-time is slower by a factor of 2.5 when using minimum size devices for both 'n' and 'p'.

In order to achieve symmetrical operation using minimum channel length, we would need to make $W_p = 2.5\ W_n$ and for minimum size lambda-based geometries this would result in the inverter having an input capacitance of $1\ \square C_g$ (n-device) + $2.5\square C_g$ (p-device) = $3.5\square C_g$ in total.

This simple model is quite adequate for most practical situations, but it should be recognized that it gives optimistic results. However, it does provide an insight into the factors which affect rise-times and fall-times as follows:

1. τ_r and τ_f are proportional to $1/V_{DD}$;
2. τ_r and τ_f are proportional to C_L;
3. $\tau_r = 2.5\ \tau_f$ for equal n- and p-transistor geometries.

4.8 Driving large capacitive loads

The problem of driving comparatively large capacitive loads arises when signals must be propagated from the chip to off chip destinations. Generally, typical off chip capacitances may be several orders higher than on chip $\square C_g$ values. For example, if the off chip load is denoted C_L then

$$C_L \geqslant 10^4 \square C_g \text{ (typically)}$$

Clearly capacitances of this order must be driven through low resistances, otherwise excessively long delays will occur.

4.8.1 Cascaded inverters as drivers

Inverters intended to drive large capacitive loads must therefore present low pull-up and pull-down resistance.

Obviously, for MOS circuits, low resistance values for $Z_{p.d.}$ and $Z_{p.u.}$ imply low $L\!:\!W$ ratios; in other words, channels must be made very wide to reduce resistance value and, in consequence, an inverter to meet this need occupies a large area. Moreover, because of the large $L\!:\!W$ ratio and since length L cannot be reduced below the minimum feature size, the gate region area $L \times W$ becomes significant and a comparatively large capacitance is presented at the input, which in turn slows down the rates of change of voltage which can take place at the input.

The remedy is to use N cascaded inverters, each one of which is larger than the preceding stage by a width factor f as shown in Figure 4–11 (showing nMOS inverters, for example).

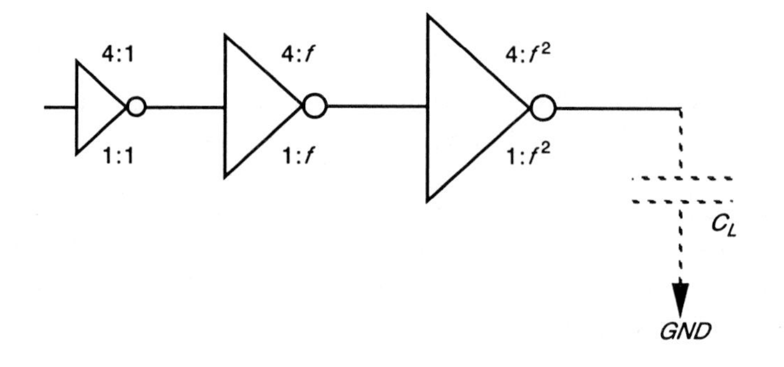

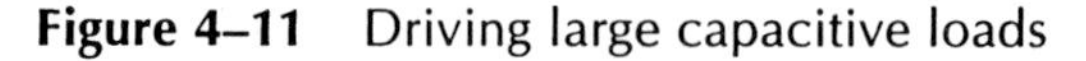
Figure 4–11 Driving large capacitive loads

Clearly, as the width factor increases, so the capacitive load presented at the inverter input increases, and the area occupied increases also. Equally clearly, the rate at which the width increases (that is, the value of f) will influence the number N of stages which must be cascaded to drive a particular value of C_L. Thus, an optimum solution must be sought as follows (this treatment is attributed to Mead and Conway).

With large f, N decreases but delay per stage increases. For 4:1 nMOS inverters

$$\left.\begin{aligned} \text{delay per stage} &= f\tau \text{ for } \Delta V_{in} \\ \text{or} &= 4f\tau \text{ for } \nabla V_{in} \end{aligned}\right\} \quad \begin{aligned} &\text{where } \Delta V_{in} \text{ indicates logic 0 to 1} \\ &\text{transition and } \nabla V_{in} \text{ indicates} \\ &\text{logic 1 to 0 transition of } V_{in} \end{aligned}$$

Therefore, total delay per nMOS pair $= 5f\tau$. A similar treatment yields delay per CMOS pair $= 7f\tau$. Now let

$$y = \frac{C_L}{\square C_g} = f^N$$

so that the choice of f and N are interdependent.

We now need to determine the value of f which will minimize the overall delay for a given value of y and from the definition of y

$$\ln(y) = N \ln(f)$$

That is

$$N = \frac{\ln(y)}{\ln(f)}$$

Thus, for N even

$$\text{total delay} = \frac{N}{2} 5 f\tau = 2.5\, Nf\tau \text{ (nMOS)}$$

$$\text{or} = \frac{N}{2} 7 f\tau = 3.5\, Nf\tau \text{ (CMOS)}$$

Thus, in all cases

$$\text{delay} \propto Nf\tau = \frac{\ln(y)}{\ln(f)} f\tau$$

It can be shown that total delay is minimized if f assumes the value e (base of natural logarithms); that is, each stage should be approximately 2.7* times wider than its predecessor. This applies to CMOS as well as nMOS inverters.

Thus, assuming that $f = e$, we have

$$\text{Number of stages } N = \ln(y)$$

and overall delay t_d

$$N \text{ even: } t_d = 2.5eN\tau \text{ (nMOS)}$$
$$\text{or } t_d = 3.5eN\tau \text{ (CMOS)}$$

$$\left.\begin{array}{l} N \text{ odd: } t_d = [2.5(N-1)+1]e\tau \text{ (nMOS)} \\ \text{or } t_d = [3.5(N-1)+2]e\tau \text{ (CMOS)} \end{array}\right\} \text{ for } \Delta V_{in}$$

or

$$\left.\begin{array}{l} t_d = [2.5(N-1)+4]e\tau \text{ (nMOS)} \\ \text{or } t_d = [3.5(N-1)+5]e\tau \text{ (CMOS)} \end{array}\right\} \text{ for } \nabla V_{in}$$

* Usually, a value of $f = 3$ is used since the curve showing delay versus f is quite flat around the minimum.

4.8.2 Super buffers

The asymmetry of the conventional inverter is clearly undesirable, and gives rise to significant delay problems when an inverter is used to drive more significant capacitive loads.

A common approach used in nMOS technology to alleviate this effect is to make use of super buffers as in Figures 4–12 and 4–13.

An inverting type is shown in Figure 4–12; considering a positive going logic transition V_{in} at the input, it will be seen that the inverter formed by T_1 and T_2 is turned on and, thus, the gate of T_3 is pulled down toward 0 volt with a small delay. Thus, T_3 is cut off while T_4 (the gate of which is also connected to V_{in}) is turned on and the output is pulled down quickly.

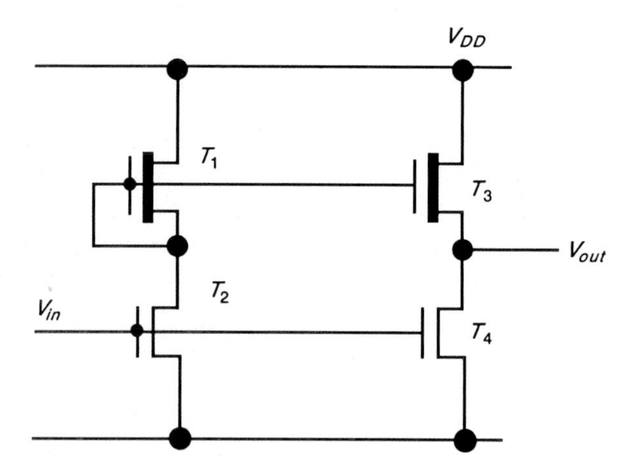

Figure 4–12 Inverting type nMOS super buffer

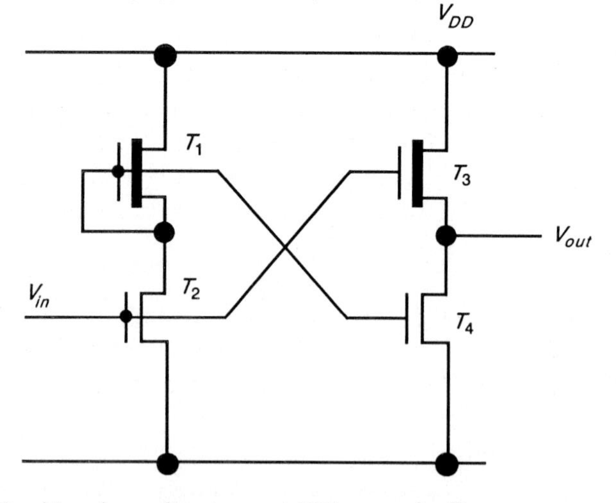

Figure 4–13 Non-inverting type nMOS super buffer

Now consider the opposite transition: when V_{in} drops to 0 volt, then the gate of T_3 is allowed to rise quickly to V_{DD}. Thus, as T_4 is also turned off by V_{in}, T_3 is made to conduct with V_{DD} on its gate, that is, with twice the average voltage that would apply if the gate was tied to the source as in the conventional nMOS inverter. Now, since $I_{ds} \alpha V_{gs}$ then doubling the effective V_{gs} will increase the current and thus reduce the delay in charging any capacitance on the output, so that more symmetrical transitions are achieved.

The corresponding non-inverting nMOS super buffer circuit is given at Figure 4–13 and, to put matters in perspective, the structures shown when realized in 5 μm technology are capable of driving loads of 2 pF with 5 nsec rise-time.

Other nMOS arrangements such as those based on the native transistor, and known as native super buffers, may be used, but such processes are not readily available to the designer and are mentioned here only briefly.

4.8.3 BiCMOS drivers

The availability of bipolar transistors in BiCMOS technology presents the possibility of using bipolar transistor drivers as the output stage of inverter and logic gate circuits. We have already seen (Chapter 2) that bipolar transistors have transconductance g_m and current/area I/A characteristics that are greatly superior to those of MOS devices. This indicates high current drive capabilities for small areas in silicon.

Bipolar transistors have an exponential dependence of the output current I_c on the input base to emitter voltage V_{be}. This means that the device can be operated with much smaller input voltage swings than MOS transistors and still switch relatively large currents. Thus, bipolar transistors have a much better switching performance, primarily as a result of the smaller input voltage swings. Only a small amount of charge must be moved during switching.

One point to consider is the possible effect of temperature T on the required input voltage V_{be}. Although V_{be} is logarithmically dependent on base width W_B, doping level N_A, electron mobility μ_n and collector current I_c it is only linearly dependent on T. This means that there is no difficulty in matching V_{be} values across a circuit, spread over an area on chip, as the temperature differences across a chip will not be sufficient to cause more than a few millivolts of difference in V_{be} between any two bipolar transistors.

The switching performance of a transistor driving a capacitive load may be visualized initially from the simple model given in Figure 4–14.

It may be shown that the time Δt necessary to change the output voltage V_{out} by an amount equal to the input voltage V_{in} is given by

$$\Delta t = \frac{C_L}{g_m}$$

where g_m is the transconductance of the bipolar transistor.

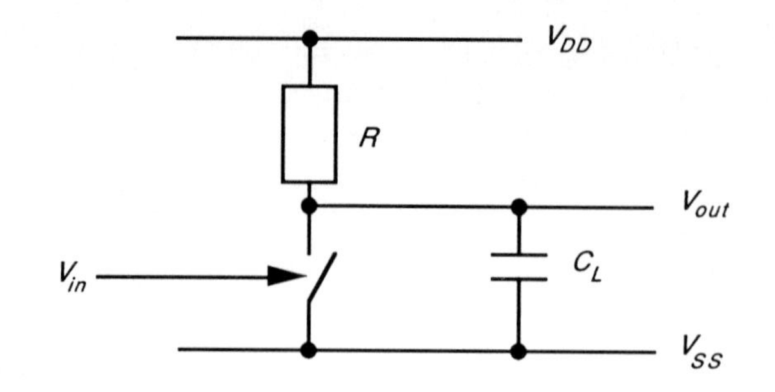

Note: The time necessary to change the output voltage by an amount that is equal to the input change is given by

$$\Delta t = C_L/g_m$$

where

g_m = device transconductance.

Figure 4–14 Driving ability of bipolar transistor

Clearly, since the bipolar transistor has a relatively high transconductance, the value of Δt is small.

A more exacting appraisal of the bipolar transistor delay reveals that it comprises two main components:

1. T_{in} — an initial time necessary to charge the base emitter junction of the bipolar (npn) transistor. Typically, for the BiCMOS transistor-based driver we are considering, T_{in} is in the region of 2ns. A similar consideration of a CMOS transistor driver in the same BiCMOS technology would reveal a figure of 1ns for T_{in}, this being the time taken to charge the input gate capacitance. As a matter of interest, a comparable figure for a GaAs driver is around 50–100 ps.
2. T_L — the time taken to charge the output load capacitance C_L and it will be noted that this time is less for the bipolar driver by a factor of h_{fe}, where h_{fe} is the bipolar transistor gain.

Although the bipolar transistor has a higher value of T_{in}, T_L is smaller because of the faster charging rate as discussed.

The combined effect of T_{in} and T_L is represented in Figure 4–15 and it will be seen that there is a critical value of load capacitance $C_{L(crit)}$ below which the BiCMOS driver is slower than a comparable CMOS driver.

A further significant parameter contributing to delay is the collector resistance R_c of a bipolar transistor. Clearly a high value for R_c will mean a long propagation delay through the transistor when charging a capacitive load. The effect can be assessed from Figure 4–16, which shows typical delay values at two values of C_L for a range of collector resistance R_c. The reason for including the buried subcollector region in the BiCMOS process is to keep R_c as low as possible.

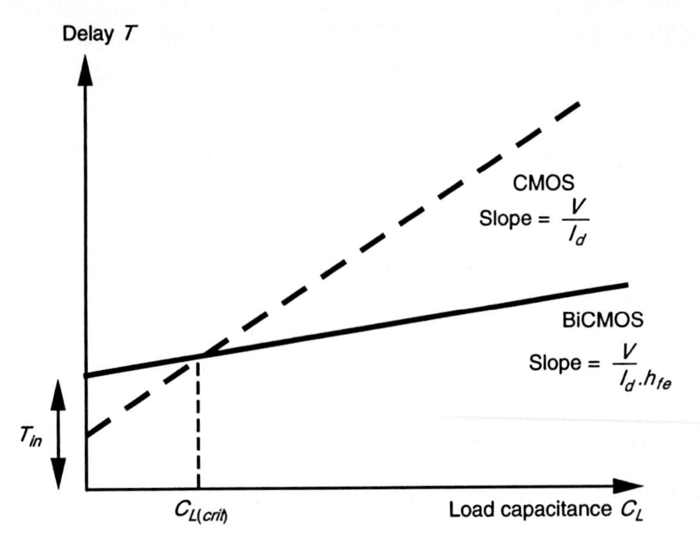

- Delay of BiCMOS inverter can be described by

  $T = T_{in} + (V/I_d)(1/h_{fe})\, C_L$

where T_{in} = time to charge up base/emitter junction

h_{fe} = transistor current gain (common emitter)

- Delay for BiCMOS inverter is reduced by a factor of h_{fe} *compared with a CMOS inverter.*

Figure 4–15 Delay estimation

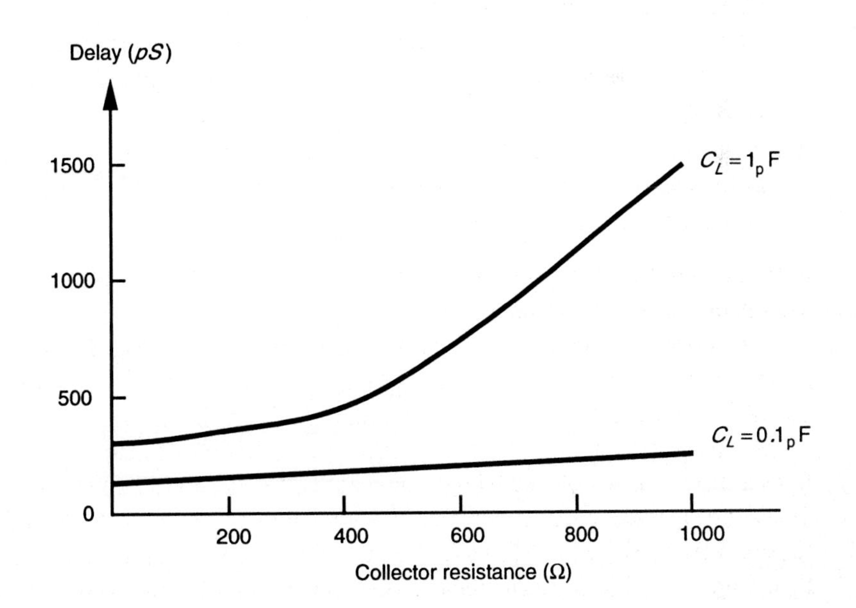

Figure 4–16 Gate delay as a function of collector resistance

BiCMOS fabrication processes produce reasonably good bipolar transistors —high g_m, high β, high h_{fe} and low R_c—without compromising or overelaborating the basic CMOS process. The availability of bipolar transistors in logic gate and driver/buffer design provides a great deal of scope and freedom for the VLSI designer.

4.9 Propagation delays

4.9.1 Cascaded pass transistors

A degree of freedom offered by MOS technology is the use of pass transistors as series or parallel switches in logic arrays. Quite frequently, therefore, logic signals must pass through a number of pass transistors in series. A chain of four such transistors is shown in Figure 4–17(a) in which all gates have been shown connected to V_{DD} (logic 1), which would be the case for a signal to be propagated to the output. The circuit thus formed may be modeled as in Figure 4–17(b) and it is then possible to evaluate the delay through the network.

The response at node V_2 with respect to time is given by

$$C\frac{dV_2}{dt} = (I_1 - I_2) = \frac{[(V_1 - V_2) - (V_2 - V_3)]}{R}$$

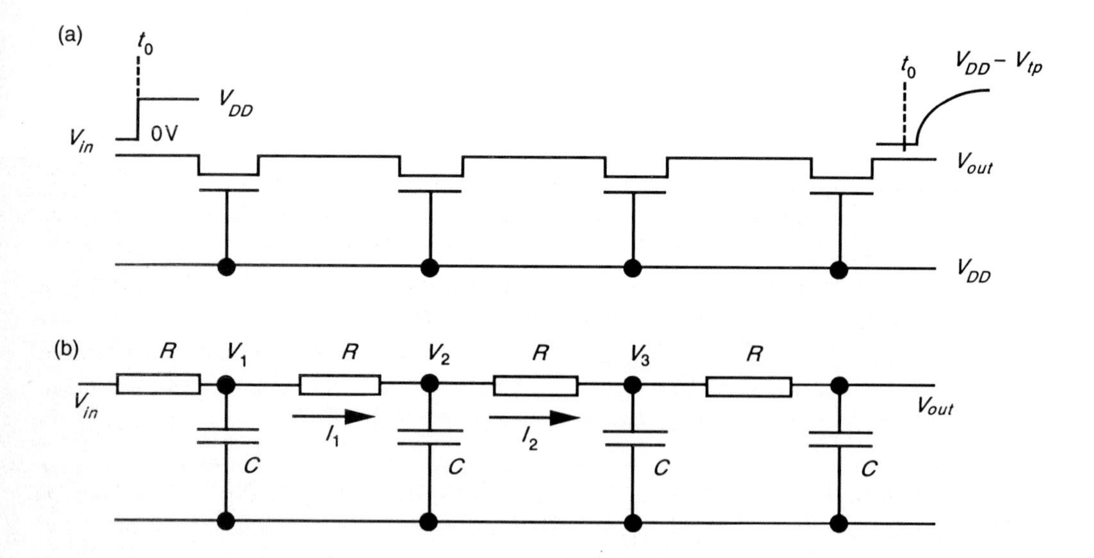

Figure 4–17 Propagation delays in pass transistor chain

In the limit as the number of sections in such a network becomes large, this expression reduces to

$$RC\frac{dV}{dt} = \frac{d^2V}{dx^2}$$

where

R = resistance per unit length
C = capacitance per unit length
x = distance along network from input.

The propagation time τ_p for a signal to propagate a distance x is such that

$$\tau_p \alpha x^2$$

The analysis can be simplified if all *Rs* and *Cs* are lumped together, then

$$R_{total} = nrR_s$$
$$C_{total} = nc\square C_g$$

where r gives the relative resistance per section in terms of R_s and c gives the relative capacitance per section in terms of $\square C_g$.

Then, it may be shown that overall delay τ_d for n sections is given by

$$\tau_d = n^2 rc(\tau)$$

Thus, the overall delay increases rapidly as n increases and in practice no more than *four* pass transistors should be normally connected in series. However, this number can be exceeded if a buffer is inserted between each group of four pass transistors *or* if relatively long time delays are acceptable.

4.9.2 Design of long polysilicon wires

Long polysilicon wires also contribute distributed series R and C as was the case for cascaded pass transistors and, in consequence, signal propagation is slowed down. This would also be the case for wires in diffusion where the value of C may be quite high, and for this reason the designer is discouraged from running signals in diffusion except over very short distances.

For long polysilicon runs, the use of buffers is recommended. In general, the use of buffers to drive long polysilicon runs has two desirable effects. First, the signal propagation is speeded up and, second, there is a reduction in sensitivity to noise.

The reason why noise may be a problem with slowly rising signals may be deduced by considering Figure 4–18. In the diagram the slow rise-time of the signal at the input of the inverter (to which the signal emerging from the long

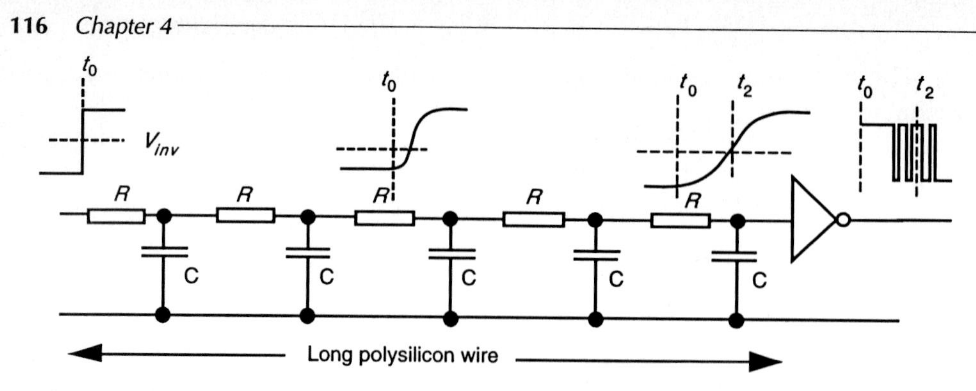

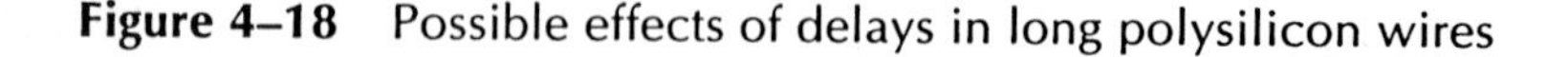

Figure 4–18 Possible effects of delays in long polysilicon wires

polysilicon line is connected) means that the input voltage spends a relatively long time in the vicinity of V_{inv} so that small disturbances due to noise will switch the inverter state between '0' and '1' as shown at the output point.

Thus it is essential that long polysilicon wires be driven by suitable buffers to guard against the effects of noise and to speed up the rise-time of propagated signal edges.

4.10 Wiring capacitances

In section 4.5 we considered the area capacitances associated with the layers to substrate and from gate to channel. However, there are other significant sources of capacitance which contribute to the overall wiring capacitance. Three such sources are discussed below.

4.10.1 Fringing fields

Capacitance due to fringing field effects can be a major component of the overall capacitance of interconnect wires. For fine line metallization, the value of fringing field capacitance (C_{ff}) can be of the same order as that of the area capacitance. Thus, C_{ff} should be taken into account if accurate prediction of performance is needed.

$$C_{ff} = \varepsilon_{SiO_2}\varepsilon_0 l\left[\frac{\pi}{\ln\left\{1+\frac{2d}{t}\left(1+\sqrt{\left(1+\frac{t}{d}\right)}\right)\right\}}-\frac{t}{4d}\right]$$

where

l = wire length
t = thickness of wire
d = wire to substrate separation

Then, total wire capacitance

$$C_w = C_{area} + C_{ff}$$

4.10.2 Interlayer capacitances

Quite obviously the parallel plate effects are present between one layer and another. For example, some thought on the matter will confirm the fact that, for a given area, metal to polysilicon capacitance must be higher than metal to substrate. The reason for not taking such effects into account for simple calculations is that the effects occur only where layers cross or when one layer underlies another, and in consequence interlayer capacitance is highly dependent on layout. However, for regular structures it is readily calculated and contributes significantly to the accuracy of circuit modeling and delay calculation.

4.10.3 Peripheral capacitance

The source and drain n-diffusion regions (n-active regions for Orbit processes) form junctions with the p-substrate or p-well at well-defined and uniform depths; similarly for p-diffusion (p-active) regions in n-substrates or n-wells. For diffusion regions, each diode thus formed has associated with it a peripheral (side-wall) capacitance in picofarads per unit length which, in total, can be considerably greater than the area capacitance of the diffusion region to substrate; the smaller the source or drain area, the greater becomes the relative value of the peripheral capacitance.

For Orbit processes, the n-active and p-active regions are formed by impurity implant at the surface of the silicon and thus, having negligible depth, they have negligible peripheral capacitance.

However, for n-and p-regions formed by a diffusion process, the peripheral capacitance is important and becomes particularly so as we shrink the device dimensions.

In order to calculate the total diffusion capacitance we must add the contributions of area and peripheral components

$$C_{total} = C_{area} + C_{periph}$$

Typical values follow in Table 4–3. For further considerations on capacitive effects the reader is referred to Arpad Barna, *VHSIC — Technologies and Tradeoffs*, Wiley, 1981.

Table 4–3 Typical values for diffusion capacitances

Diffusion capacitance	*Typical values*		
	5 μm	*2 μm*	*1.2 μm*
Area C (*C_{area}*) (as in Table 4–2)	1.0×10^{-4} pF/μm^2	1.75×10^{-4} pF/μm^2	3.75×10^{-4} pF/μm^2
Periphery (*C_{periph}*)	8.0×10^{-4} pF/μm	negligible*	negligible*

* Assuming implanted regions of negligible depth.

4.11 Choice of layers

Frequently, in designing an arrangement to meet given specifications, there are several possible ways in which the requirements may be met, including the choice between the layers on which to route certain data and control signals. However, there are certain commonsense constraints which should be considered:

- V_{DD} and V_{SS} *(GND)* should be distributed on metal layers wherever possible and should not depart from metal except for 'duck unders', preferably on the diffusion layer when this is absolutely essential. A consideration of R_s values will reveal the reason for this.
- Long lengths of polysilicon should be used only after careful consideration because of the relatively high R_s value of the polysilicon layer. Polysilicon is unsuitable for routing V_{DD} or V_{SS} other than for very small distances.
- With these restrictions in mind, it is generally the case that the resistances associated with transistors are much higher than any reasonable wiring resistance, so that there is no real danger of any problem due to voltage divider effects between wiring and transistor resistances.
- Capacitive effects must also be carefully considered, particularly where fast signal lines are required and particularly in relation to signals on wiring having relatively high values of R_s. Diffusion (or active) areas have relatively high values of capacitance to substrate and are harder to drive in consequence. Charge sharing may also cause problems in certain circuits or architectures and must be carefully considered. Over small equipotential regions, the signal on a wire can be treated as being identical at all points. Within each region the delay associated with signal propagation is small in comparison with gate delays and with signal delays in systems connected by the wires.

Thus the wires in a MOS system can be modeled as simple capacitors. This concept leads to the establishment of electrical rules (guidelines) for communication paths (wires) as given in Table 4–4.

The factors set out in Tables 4–4 and 4–5 help to put matters in perspective.

Table 4–4 Electrical rules

Layer	*Maximum length of communication wire*		
	lambda-based (5 μm)	*μm-based (2 μm)*	*μm-based (1.2 μm)*
Metal	chip wide	chip wide	chip wide
Silicide	2,000λ	NA	NA
Polysilicon	200λ	400 μm	250 μm
Diffusion (active)	20λ*	100 μm	60 μm

* Taking account of peripheral and area capacitances. NA = not applicable.

Table 4–5 Choice of layers

Layer	*R*	*C*	*Comments*
Metal	Low	Low	Good current capability without large voltage drop ... use for power distribution and global signals.
Silicide	Low	Moderate	Modest RC product. Reasonably long wires are possible. Silicide is used in place of polysilicon in some nMOS processes.
Polysilicon	High	Moderate	RC product is moderate; high IR drop.
Diffusion (active)	Moderate	High	Moderate IR drop but high C. Hence hard to drive.

4.12 Observations

This chapter has completed our examination of the factors determining the characteristics and performance of MOS circuits in silicon. Useful concepts have been introduced and tables of typical parameter values have been set out to allow ready estimation of the performance of simple designs. Methods of dealing with larger capacitive loads, for example 'off chip' loads, have also been discussed.

All the basic information for carrying out and evaluating simple design work is now in place and will be put into practice following a discussion on scaling effects.

4.13 Tutorial exercises

1. A particular layer of MOS circuit has a resistivity $\rho = 1$ ohm cm. A section of this layer is 55 μm long and 5 μm wide and has a thickness of 1 μm. Calculate the resistance from one end of this section to the other (along the length). Use the concept of sheet resistance R_s. What is the value of R_s?
2. A particular section of a layout (as in Figure 4–19) includes a 3λ wide metal path which crosses a 2λ wide polysilicon path at right angles. Assuming that the layers are separated by a 0.5 μm thick layer of silicon dioxide, find the capacitance between the two layers.

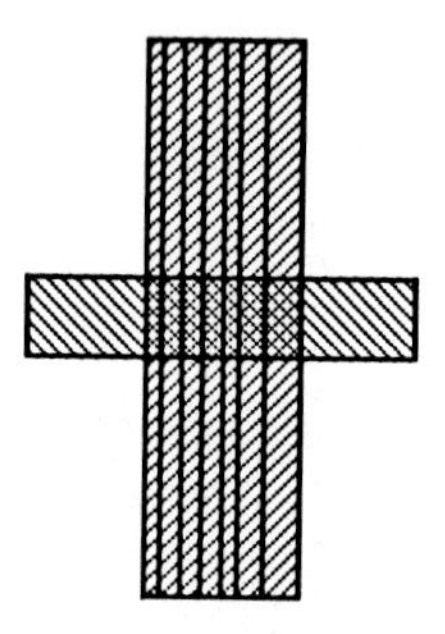

Figure 4–19 Layout detail for Question 2

The polysilicon layer in turn crosses a 4λ wide diffusion region at right angles to form a transistor. Using the tables provided in the text, find the gate to channel capacitance. Compare it with the metal to polysilicon capacitance already calculated.

Assume $\lambda = 2.5$ μm in all cases.

3. Two nMOS inverters are cascaded to drive a capacitive load $C_L = 16\square C_g$ as shown in Figure 4–20. Calculate the pair delay (V_{in} to V_{out}) in terms of τ for the inverter geometry indicated in the figure. What are the ratios of each inverter?

   If strays and wiring are allowed for, it would be reasonable to increase the capacitance to ground across the output of each inverter by $4\square C_g$. What is the pair delay allowing for strays?

   Assume a suitable value for τ and evaluate this pair delay.
4. An off chip capacitance load of 5pF is to be driven from (a) CMOS and (b) nMOS inverters. Set out suitable arrangements giving appropriate channel $L{:}W$ ratios and dimensions. Calculate the number of inverter stages required, and the delay exhibited by the overall arrangement driving the 5pF load.

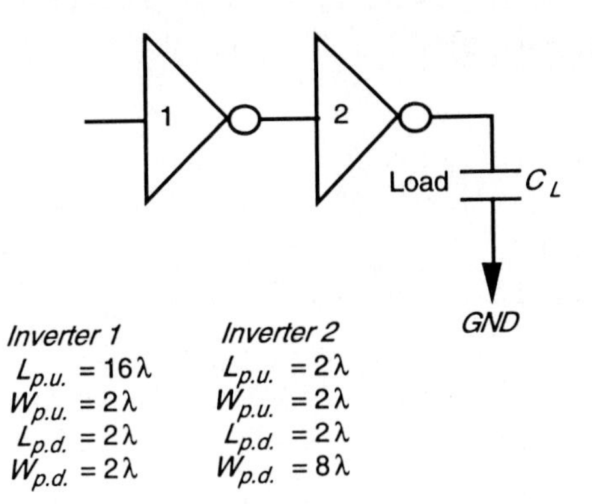

Figure 4–20 Circuit for Question 3

5. *A worked example*: Using the parameters given in this chapter calculate the C_{in} and C_{out} values of capacitance for the structure represented in Figure 4–21.

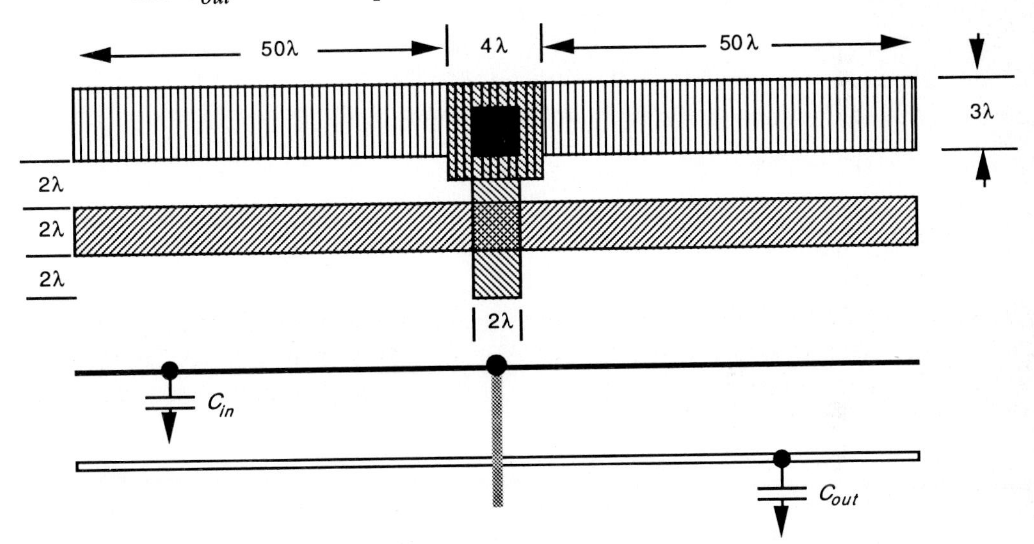

Figure 4–21 Structure for Question 5

Solution: The input capacitance C_{in} is made up of three components — metal bus capacitance C_m, polysilicon capacitance C_p, and the gate capacitance C_g. Thus

$$C_{in} = C_m + C_p + C_g$$

$$C_m = [2 \times (50 \times 3)\lambda^2 \times 6.25\ \mu\text{m}^2/\lambda^2]\{0.3 \times 10^{-4}\ \text{pF}/\mu\text{m}^2\}$$
$$= .05625\ \text{pF}$$

$$C_p = [(4 \times 4 + 2 \times 2 + 2 \times 1)\lambda^2 \times 6.25\ \mu\text{m}^2/\lambda^2]\ \{0.4 \times 10^{-4}\ \text{pF}/\mu\text{m}^2\}$$
$$= .0055\ \text{pF}$$

$$C_g = 1\square C_g = .01 \text{ pF}$$

Thus

$$C_{in} = .05625 + .0055 + .01 = .07175 \text{ pF } (=> 7\,\square C_g)$$

Now, the output capacitance C_{out} is contributed by the diffusion area C_{da} and peripheral C_{dp} capacitances so that (assuming the transistor is off) we have

$$C_{out} = C_{da} + C_{dp}$$

$$= [(51 \times 2)\lambda^2 \times 6.25 \ \mu\text{m}^2/\lambda^2] \times 1 \times 10^{-4} \text{ pF}/\mu\text{m}^2 + [2 \times (51 + 2)\lambda \times 2.5 \ \mu\text{m}/\lambda] \times 8 \times 10^{-4} \text{ pF}/\mu\text{m}$$

$$= .06375 + .212 = .27575 \text{ pF (note significance of } C_{dp})$$

5 Scaling of MOS circuits

Little things are pretty.

Proverb

Good things come in small packages.

Proverb

Objectives

VLSI fabrication technology is still in the process of evolution which is leading to smaller line widths and feature size and to higher packing density of circuitry on a chip.

The scaling down of feature size generally leads to improved performance and it is important therefore to understand the effects of scaling. There are also future limits to scaling down which may well be reached in the next decade.

Although this chapter may be seen by some to interrupt the flow of the text toward actual VLSI design, the authors considered this an appropriate topic following the previous chapters dealing with basic parameters and characteristics which, of course, are all affected by scaling.

Microelectronic technology may be characterized in terms of several indicators, or figures of merit. Commonly, the following are used:

- Minimum feature size
- Number of gates on one chip
- Power dissipation
- Maximum operational frequency
- Die size
- Production cost.

Many of these figures of merit can be improved by shrinking the dimensions of transistors, interconnections and the separation between features, and by adjusting the doping levels and supply voltages. Accordingly, over the past decade, much effort has been directed toward the upgrading of process technology and the resultant scaling down of devices and feature size.

In the design processes postulated by Mead and Conway and used for most examples in this text, it has been the practice to dimension all layouts in terms of λ. A value may then be allocated to λ, prior to manufacture, which is in line with the capabilities of the silicon foundry or is determined by current technology and/or meets the specifications which have been set out for the circuit. One benefit of this approach lies in the fact that the design rules have been formulated in such a way as to allow *limited* direct scaling of the dimensions of circuits, so that today's design is not automatically outdated when line widths are reduced (i.e. the value allocated to λ is reduced) by advances in tomorrow's technology.

Scaling is therefore an important factor, and it is essential for the designer to understand the implementation and the effects of scaling. In writing this chapter, the authors gratefully acknowledge the useful contributions made by Dr A. Osserain and Dr B. Hochet, both of the Swiss Federal Institute of Technology, Lausanne, Switzerland.

This chapter discusses scaling and its effect on performance and indicates some problems and ultimate limitations.

5.1 Scaling models and scaling factors

The most commonly used models are the constant electric field scaling model and the constant voltage scaling model. They both present a simplified view, taking only first degree effects into consideration, but are easily understood and well suited to educational needs. Recently, a combined voltage and dimension scaling model has been presented (Bergmann, 1991).

In this chapter, the application of each of the three models will be illustrated. To assist in visualization, it is useful to refer to Figure 5–1 which indicates the

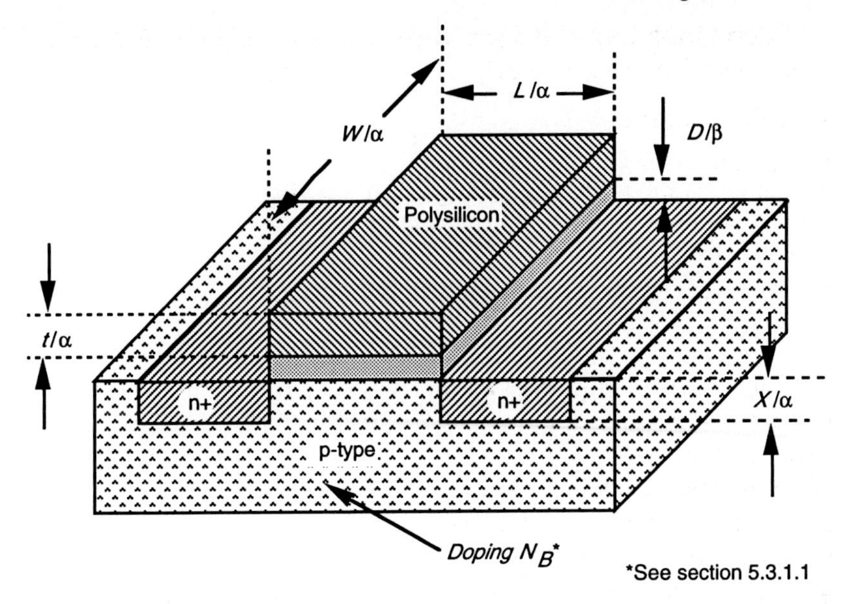

Figure 5–1 Scaled nMOS transistor (pMOS similar)

device dimensions and substrate doping level which are associated with the scaling of a transistor.

In order to accommodate the three models, two scaling factors — $1/\alpha$ and $1/\beta$ — are used. $1/\beta$ is chosen as the scaling factor for supply voltage V_{DD} and gate oxide thickness D, and $1/\alpha$ is used for all other linear dimensions, both vertical and horizontal to the chip surface. For the constant field model and the constant voltage model, $\beta = \alpha$ and $\beta = 1$ respectively are applied.

5.2 Scaling factors for device parameters

In this section, simple derivations and calculations reveal the effects of scaling.

5.2.1 Gate area A_g

$$A_g = L.W.$$

where L and W are the channel length and width respectively. Both are scaled by $1/\alpha$.

Thus A_g is scaled by $1/\alpha^2$

5.2.2 Gate capacitance per unit area C_0 or C_{ox}

$$C_0 = \frac{\varepsilon_{ox}}{D}$$

where ε_{ox} is the permittivity of the gate oxide (thinox) [= $\varepsilon_{ins.}\varepsilon_0$] and D is the gate oxide thickness which is scaled by $1/\beta$

$$\text{Thus } C_0 \text{ is scaled by } \frac{1}{1/\beta} = \beta$$

5.2.3 Gate capacitance C_g

$$C_g = C_0\, L.W.$$

$$\text{Thus } C_g \text{ is scaled by } \beta\frac{1}{\alpha^2} = \frac{\beta}{\alpha^2}$$

5.2.4 Parasitic capacitance C_x

$$C_x \text{ is proportional to } \frac{A_x}{d}$$

where d is the depletion width around source or drain which is scaled by $1/\alpha$, and A_x is the area of the depletion region around source or drain which is scaled by $1/\alpha^2$.

$$\text{Thus } C_x \text{ is scaled by } \frac{1}{\alpha^2}\cdot\frac{1}{1/\alpha} = \frac{1}{\alpha}$$

5.2.5 Carrier density in channel Q_{on}

$$Q_{on} = C_0 . V_{gs}$$

where Q_{on} is the average charge per unit area in the channel in the 'on' state. Note that C_0 is scaled by β and V_{gs} is scaled by $1/\beta$.

Thus Q_{on} is scaled by 1

5.2.6 Channel resistance R_{on}

$$R_{on} = \frac{L}{W}\frac{1}{Q_{on}\mu}$$

where μ is the carrier mobility in the channel and is assumed constant.

Thus R_{on} is scaled by $\frac{1}{\alpha}\frac{1}{1/\alpha}1 = 1$

5.2.7 Gate delay T_d

T_d is proportional to $R_{on}.C_g$

Thus T_d is scaled by $\frac{1.\beta}{\alpha^2} = \frac{\beta}{\alpha^2}$

5.2.8 Maximum operating frequency f_0

$$f_0 = \frac{W}{L}\frac{\mu C_0 V_{DD}}{C_g}$$

or, f_0 is inversely proportional to delay T_d.

Thus f_0 is scaled by $\frac{1}{\beta/\alpha^2} = \frac{\alpha^2}{\beta}$

5.2.9 Saturation current I_{dss}

$$I_{dss} = \frac{C_0\mu}{2}\frac{W}{L}(V_{gs} - V_t)^2$$

noting that both V_{gs} and V_t are scaled by $1/\beta$, we have

I_{dss} is scaled by $\beta\,(1/\beta)^2 = 1/\beta$

5.2.10 Current density J

$$J = \frac{I_{dss}}{A}$$

where A is the cross-sectional area of the channel in the 'on' state which is scaled by $1/\alpha^2$

So, J is scaled by $\frac{1/\beta}{1/\alpha^2} = \frac{\alpha^2}{\beta}$

5.2.11 Switching energy per gate E_g

$$E_g = \frac{1C_g}{2}(V_{DD})^2$$

So, E_g is scaled by $\frac{\beta}{\alpha^2} \cdot \frac{1}{\beta^2} = \frac{1}{\alpha^2 \beta}$

5.2.12 Power dissipation per gate P_g

P_g comprises two components such that

$$P_g = P_{gs} + P_{gd}$$

where the static component

$$P_{gs} = \frac{(V_{DD})^2}{R_{on}}$$

and the dynamic component

$$P_{gd} = E_g \, f_0$$

It will be seen that both P_{gs} and P_{gd} are scaled by $1/\beta^2$

So, P_g is scaled by $1/\beta^2$

5.2.13 Power dissipation per unit area P_a

$$P_a = \frac{P_g}{A_g}$$

So, P_a is scaled by $\frac{1/\beta^2}{1/\alpha^2} = \alpha^2/\beta^2$

5.2.14 Power-speed product P_T

$$P_T = P_g . T_d$$

So, P_T is scaled by $\frac{1}{\beta^2} \cdot \frac{\beta}{\alpha^2} = \frac{1}{\alpha^2 \beta}$

5.2.15 Summary of scaling effects

It is useful to summarize the scaling effects in a convenient form. Table 5–1 sets out scaling effect for various key parameters of MOS FET devices and for the three scaling models mentioned earlier.

Table 5–1 Scaling effects

Parameters		*Combined V and D*	*Constant E*	*Constant V*
V_{DD}	Supply voltage	$1/\beta$	$1/\alpha$	1
L	Channel length	$1/\alpha$	$1/\alpha$	$1/\alpha$
W	Channel width	$1/\alpha$	$1/\alpha$	$1/\alpha$
D	Gate oxide thickness	$1/\beta$	$1/\alpha$	1
A_g	Gate area	$1/\alpha^2$	$1/\alpha^2$	$1/\alpha^2$
C_0 (or C_{ox})	Gate C per unit area	β	α	1
C_g	Gate capacitance	β/α^2	$1/\alpha$	$1/\alpha^2$
C_x	Parasitic capacitance	$1/\alpha$	$1/\alpha$	$1/\alpha$
Q_{on}	Carrier density	1	1	1
R_{on}	Channel resistance	1	1	1
I_{dss}	Saturation current	$1/\beta$	$1/\alpha$	1
A_c	Conductor X-section area	$1/\alpha^2$	$1/\alpha^2$	$1/\alpha^2$
J	Current density	α^2/β	α	α^2
V_g	Logic 1 level	$1/\beta$	$1/\alpha$	1
E_g	Switching energy	$1/\alpha^2.\beta$	$1/\alpha^3$	$1/\alpha^2$
P_g	Power dispn per gate	$1/\beta^2$	$1/\alpha^2$	1
N	Gates per unit area	α^2	α^2	α^2
P_a	Power dispn per unit area	α^2/β^2	1	α^2
T_d	Gate delay	β/α^2	$1/\alpha$	$1/\alpha^2$
f_0	Max. operating frequency	α^2/β	α	α^2
P_T	Power-speed product	$1/\alpha^2.\beta$	$1/\alpha^3$	$1/\alpha^2$

Constant E: $\beta = \alpha$; Constant V: $\beta = 1$

5.3 Some discussion on and limitations of scaling

Although scaling down does have many desirable effects, some of the associated effects may cause problems which eventually become severe enough to prevent further miniaturization.

5.3.1 Substrate doping

So far, in discussing the various effects, we have neglected the built-in (junction) potential V_B, which in turn depends on the substrate doping level, and this is acceptable so long as V_B is small compared with V_{DD}. However, when this no longer holds, then the effects of V_B must be included.

Furthermore, substrate doping impinges on many of the characteristics of transistors fabricated on it. Thus further discussion is warranted.

5.3.1.1 *Substrate doping scaling factors*

As the channel length of a MOS transistor is reduced, the depletion region widths must also be scaled down to prevent the source and drain depletion regions from meeting. Depletion region width d for the junctions is given by

$$d = \sqrt{\frac{2\varepsilon_{si}\varepsilon_0 V}{qN_B}}$$

where

ε_{si} = relative permittivity of silicon ($\doteqdot 12$)
ε_0 = permittivity of free space (= 8.85×10^{-14} F/cm)
V = effective voltage across the junction = $V_a + V_B$
q = electron charge
N_B = doping level of substrate
V_a (maximum value = V_{DD}) = applied voltage
V_B = built-in (junction) potential

and

$$V_B = \frac{kT}{q}\ln\left(\frac{N_B N_D}{n_i^2}\right)$$

where N_D is the source or drain doping, and n_i is the intrinsic carrier concentration in silicon.

In, say, 5 μm technology, V_B is in the region of 500 mV whilst applied voltage V_a (= V_{DD}) is commonly 5 V so that V_B may be neglected for scaling considerations. Under these circumstances,

$$d = \sqrt{\frac{2\varepsilon_{si}\varepsilon_0 V_{DD}}{qN_B}}$$

If V_{DD} is scaled by $1/\beta$ and d by $1/\alpha$, then N_B can be scaled by α^2/β (Bergmann, 1991).

For some more recent technologies, N_B is increased to reduce d so that V_B is also enlarged. (For example, if $N_B = 10^{15}$ cm^{-3} and $N_D = 10^{20}$ cm^{-3} then V_B = 0.88 V). At the same time, V_{DD} is also scaled down, and is thus no longer large compared with V_B so that V_B must be taken account of in scaling.

Thus, for the combined voltage and dimension scaling model applied to a transistor for which we have a known V_a, we may write

$$V_a = mV_B$$

where m is a real number, so that

$$V = V_a + V_B = mV_B + V_B$$

Now if we scale V_a by $1/\beta$ we have

$$V_s = \frac{mV_B}{\beta} + V_B \text{ so that scaling factor} = \frac{\beta + m}{\beta(m+1)}$$

where V_s is the effective scaled voltage across the depletion region. Consequently, N_B should be scaled by

$$\frac{\alpha^2(\beta+m)}{\beta(m+1)}$$

so that d scales by $1/\alpha$.

This model not only expresses the effects of the relationship between V_a and V_B, but also shows their relation to the scaling factor β. Where m is large and β is small, the scaling factor for N_B reverts to α^2/β, but in other cases this model becomes significant.

5.3.1.2 *Depletion width*

In the previous discussion, N_B is increased to reduce the depletion width, but this also increases the threshold voltage V_t which is against the required trends for scaling down.

In [Hoen] N_B must be kept below $1.3 \times 10^{19}\,\text{cm}^{-3}$. At higher values of N_B, the maximum electric field which can be applied to the gate oxide is insufficient to invert the substrate so that no channel can be formed.

However, the technology of deep channel implantation increases N only near the source and drain to substrate junctions. Thus, N_B can be maintained at a satisfactory level in the channel region and this problem is thus reduced. Nonetheless, depletion width d and built-in potential V_B will impose limitations on scaling.

It can be shown (Grove, 1967) that

$$E_{max} = \frac{2V}{d}$$

where E_{max} is the maximum electric field induced in the one-sided step junction.

When N_B is increased by α and if $V_a = 0$, then V_B is increased by $\ln \alpha$ and d is decreased by

$$\sqrt{\frac{\ln\alpha}{\alpha}}$$

Therefore, the electric field E across the depletion region is increased by $\sqrt{\alpha/\ln\alpha}$ and will thus reach the critical level E_{crit} with increasing N_B.

Figure 5–2(a) shows the depletion width d as a function of substrate concentration N_B and supply voltage V_{DD}. The dashed line indicates the maximum depletion width for $E_{max} = E_{crit}$. Substituting into the equation for d, we have

$$d = \sqrt{\frac{2\varepsilon_{si}\varepsilon_0}{qN_B}\left(\frac{E_{crit} \cdot d}{2}\right)}$$

whence

$$d = \frac{\varepsilon_{si}\varepsilon_0 (E_{crit})}{qN_B}$$

The area of Figure 5–2(a) above the dashed line is the region where the increased electric field will induce breakdown. Thus, the point at which the dashed line and the $V_a = 0$ line intersect indicates the maximum allowable substrate doping level, which is about $N_B = 3 \times 10^{17}\text{cm}^{-3}$ (for $N_D = 3 \times 10^{20}\text{cm}^{-3}$). At higher values of N_B junction tunneling will occur. Therefore allowable values for d fall below the dashed line and above the $V_a = 0$ line.

Figure 5–2(b) shows the maximum electric field in the depletion layer versus N_B. Any applied voltage greater than $V_a = 0$ will cause breakdown to occur at lower values of N_B.

In the foregoing discussions, the effects of N_D have been assumed to be negligible.

5.3.2 Limits of miniaturization

The minimum size of a transistor is determined by both process technology and the physics of the device itself. The reduction of device geometry currently depends mainly on alignment accuracy and on the resolution of photolithographic technology; the limit on feature size is now at 0.3 μm, but the increasing availability of direct write E-beam technology will allow this limit to be further reduced.

The size of a transistor is usually defined in terms of its channel length L. As the channel length is scaled down, the edge of the depletion region around the source comes closer to that around the drain. In order to prevent punch-through and maintain transistor action, it can be shown that the channel length L must be at least $2d$. Therefore, L is in turn determined by the substrate concentration N_B and supply voltage V_{DD} (which determines V_a).

Applying the conclusions from the previous section, we may estimate the minimum possible channel length as 0.14 μm. The minimum transit time for an electron to travel from source to drain can also be calculated. From (Sze, 1985),

$$v_{drift} = \mu E$$

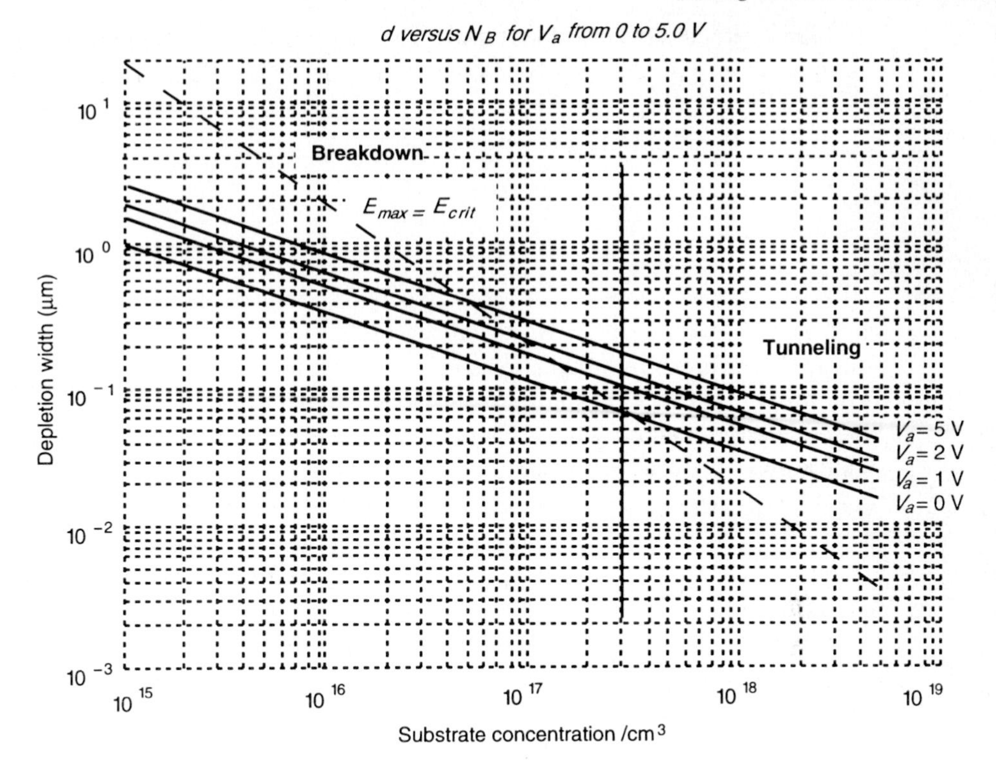

Figure 5–2(a) Substrate concentration/cm³

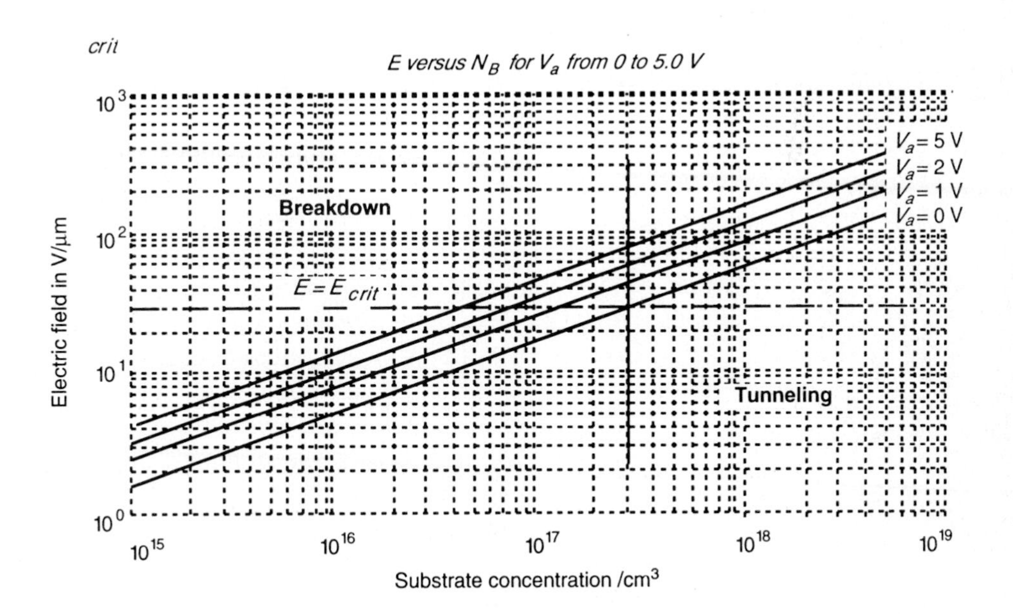

Figure 5–2(b) Depletion width d and electric field E versus substrate doping N_B

where v_{drift} is the carrier drift velocity, and

$$L = 2d$$

so that transit time τ is given by

$$\tau = \frac{L}{v_{drift}} = \frac{2d}{\mu E}$$

The maximum carrier drift velocity is approximately equal to v_{sat} where the saturation velocity $v_{sat} = 1 \times 10^7$cm/sec (Sze, 1985), regardless of the supply voltage. Therefore the minimum transit time may be assumed to occur for a minimum size transistor when V_a is approximately 0 V. Transit times may be assessed from Figure 5–3. Note that Figure 5–3(a) assumes a transistor of size $L = 2d$ with zero space between source and drain depletion regions.

5.3.3 Limits of interconnect and contact resistance

Since the width, thickness and spacing of interconnects are each scaled by $1/\alpha$, cross-section areas must scale by $1/\alpha^2$. Thus, for short distance interconnections the conductor length is also scaled by $1/\alpha$, so that resistance is increased by α. For constant field scaling, current I is also scaled by $1/\alpha$ so that IR drop remains constant as a device is scaled, and thus represents a higher proportion of the supply voltage V_{DD} which is also scaled by $1/\alpha$. Thus driving capability and noise margins are degraded.

With decreasing device dimensions, we are also seeing further increases in the levels of integration and consequent increases in die size. This lengthens the interconnections from one side of the chip to the other and, therefore, both resistance and capacitance of the interconnects are increased, producing much larger time constant values. Thus the effects of increased propagation delays, signal decay, and clock skew will decrease maximum achievable operating frequency, even though the smaller transistors produce gates with less delay.

One solution to this problem has been to make use of multilayer interconnections with thicker, wider conductors and thicker separating layers. This will reduce both R and C and also reduce die size. Other measures include the use of cascaded drivers and repeaters to reduce the effects of long interconnects.

A further option is to use optical interconnection techniques where a very high level of integration is required for high speed circuits. In order to use such techniques, optical fibers, laser diodes, receivers, and amplifiers must be included in the integrated circuit. Performance will vary with the materials used, but rough estimations can be made for comparison with metal interconnects. To start our considerations, a model may be set out as in Figure 5–4.

The propagation delay T_p along a single aluminum interconnect can be calculated from the following approximate equation (Sakurai, 1983):

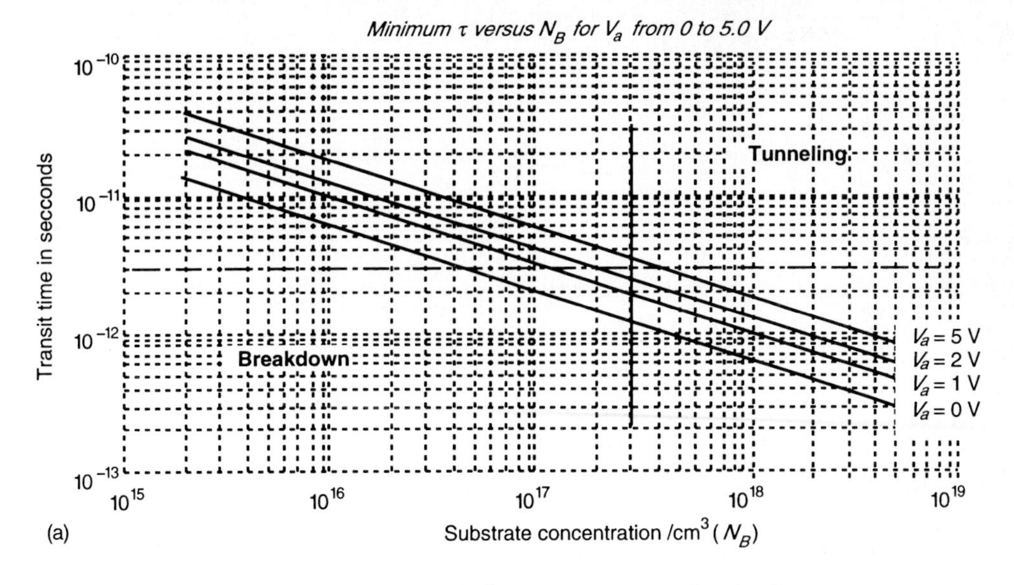

Figure 5–3(a) Transit time τ versus substrate concentration/cm^3

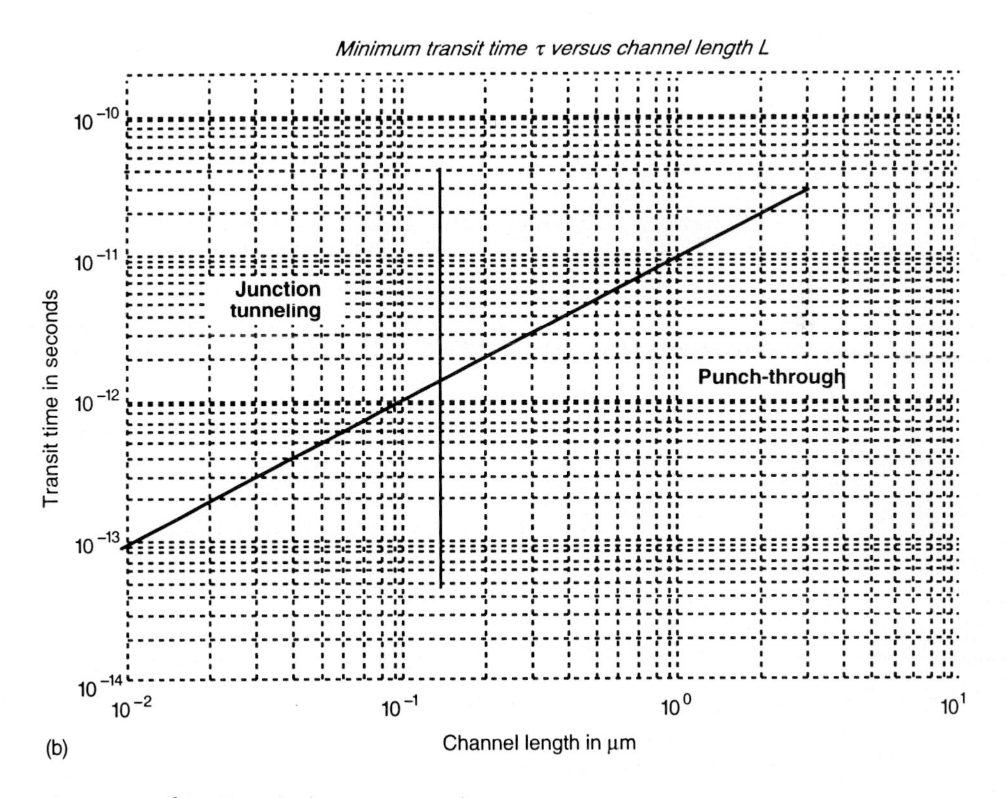

Figure 5–3(b) Transit time τ versus L

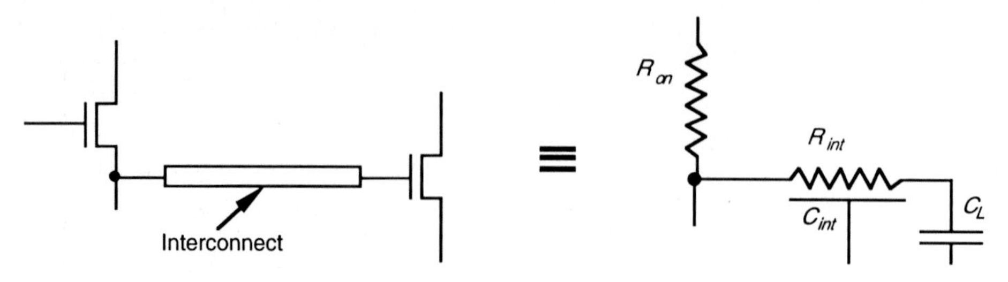

Figure 5–4 Model of metal interconnect

$$T_p = R_{int}\, C_{int} + 2.3\,(R_{on}\, C_{int} + R_{on}\, C_L + R_{int}\, C_L)$$

whence

$$T_p \doteqdot (2.3\, R_{on} + R_{int})C_{int}$$

Now

$$R_{int} = \rho \frac{L}{HW}$$

$$C_{int} = \varepsilon_{ox}\,[1.15.W/t_{ox} + 2.28\,(H/t_{ox})^{0.222}]L$$

where

R_{on} is the ON resistance of the transistor
R_{int} is the resistance of the interconnect
C_{int} is the capacitance of the interconnect
t_{ox} is the thickness of the dielectric oxide
ρ is the resistivity of the interconnect
L, W, H are the length, width and height (thickness) of the interconnect

$$\varepsilon_{ox} = 3.4515 \times 10^{-5}\ \text{pF}/\mu\text{m} \text{ — the permittivity of } SiO_2$$

If we use a value of $\rho = 3\ \mu\Omega$cm for aluminum (Bakoglu and Meindl, 1985), and if we choose $t_{ox} = 0.8\ \mu$m for thick oxide with interconnect $L = 1$ cm, $W = 3\ \mu$m and $H = 1\ \mu$m, we then have the propagation delay T_p given by

$$T_p = (2.3 \times 5\ k\Omega + 0.1\ k\Omega)2.5\ \text{pF} = 29\ \text{nsec}$$

Optical fibers can be used to replace metal interconnects in critical applications, and Figure 5–5 shows this in schematic form. R_{int} and C_{int} may be assumed to be zero, and the time needed for the output driver to transfer a logic state is given by

$$T_p = 2.3\, R_{on}\, C_L + t_{laser} + t_{int} + t_{rec}$$

where

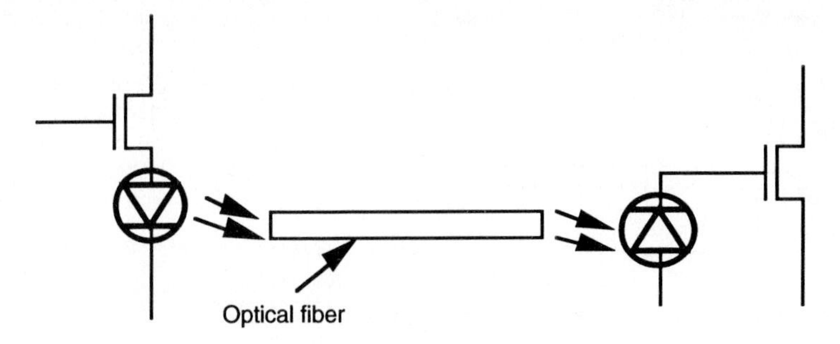

Figure 5–5 Electro-optical interconnection

C_L is the input capacitance of the laser diode
t_{laser} is the delay time through the laser diode
t_{int} is the propagation delay along the optical fiber interconnnect
t_{rec} is the receiver delay time

and

$$t_{int} = \frac{nL}{c}$$

where

n is the retractive index for the optic fiber material
L is the length of the fiber
c is the free space speed of light ($c = 3 \times 10^8$ m/sec).

Since laser diodes and receivers can work at frequencies above 10 GHz, each of them presents a relatively short delay — typically around 100 psec. The capacitance of a discrete laser diode is about 1 pF (Hutcheson, 1987) and the refractive index of commonly used material for fiber optics is between 1.5 and 2.0.

Evaluating the propagation delay we have

$$T_p = 2.3 \times 5 \times 10^3 \times 1 \times 10^{-12} + 1 \times 10^{-10} + \frac{2 \times 10^{-4}}{3 \times 10^8} + 1 \times 10^{-10} = 11.7 \text{ nsec}$$

Delay time versus line length and width may be assessed from Figure 5–6. It is obvious that the longer the interconnect, the more speed advantage arises from the use of fiber optics. In considering delay time versus line width, it may be shown that R_{on} is the dominant factor for aluminum whilst R_{int} contributes the major component for poly.

The performance of laser diodes and receivers can be improved if they are formed as part of an integrated circuit. GaAs is a material which allows this integration since it can accommodate both electronic components and optical interconnections in the one chip.

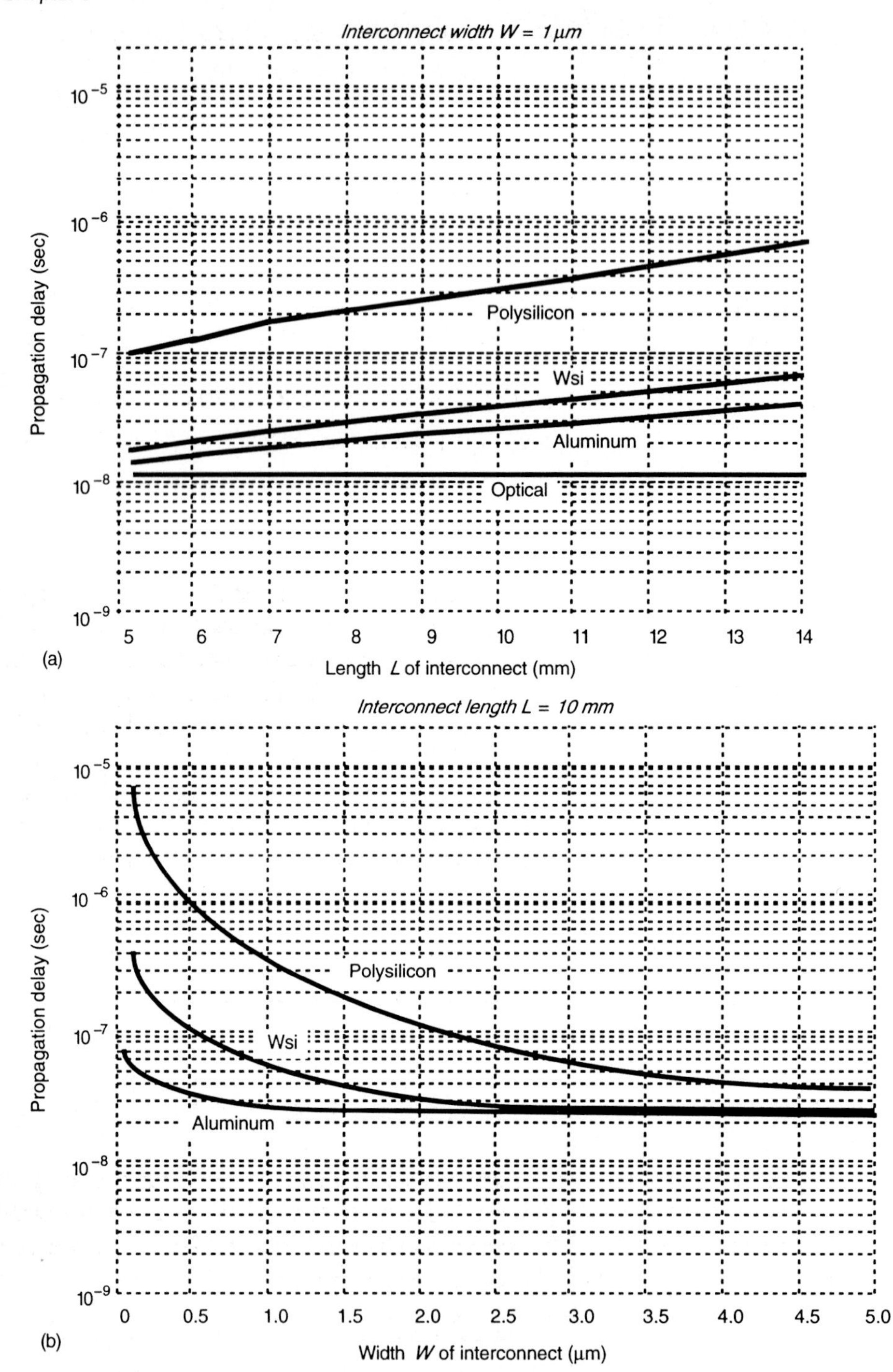

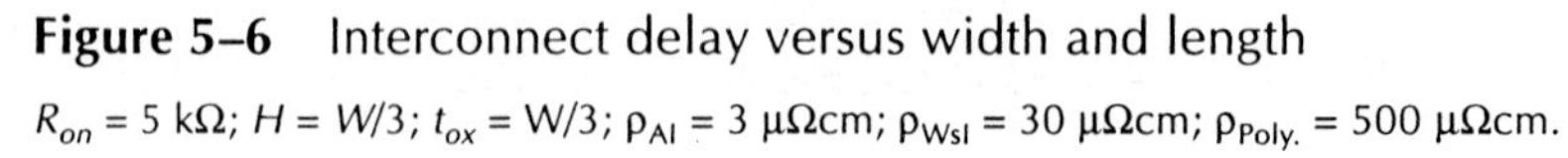

Figure 5–6 Interconnect delay versus width and length

R_{on} = 5 kΩ; $H = W/3$; $t_{ox} = W/3$; ρ_{Al} = 3 μΩcm; ρ_{WsI} = 30 μΩcm; $\rho_{Poly.}$ = 500 μΩcm.

5.4 Limits due to subthreshold currents

One of the major concerns in the scaling of devices is the effect on subthreshold current I_{sub} which is directly proportional to $exp(V_{gs} - V_t)q/kT$.

When a transistor is in the off state, then the value of $V_{gs} - V_t$ is negative and should be as large as possible to minimize I_{sub}. As voltages are scaled down, the ratio of $V_{gs} - V_t$ to kT will reduce so that subthreshold current increases quite dramatically. For this reason it may be desirable to scale both V_{gs} and V_t together with V_{DD} by factor 1/b rather than 1/a, since a is generally greater than b. However, this increases electric field strengths and thus lowers breakdown voltages.

The maximum electric field across a depletion region is given by (Grove, 1967):

$$E_{max} = \frac{2(V_a + V_B)}{d}$$

This applies to a one-sided step junction.

As discussed previously, $(V_a + V_B)$ is scaled by $(\beta + m)/\beta\ (m + 1)$ and d is scaled by $1/\alpha$. Therefore, E_{max} is scaled by $\alpha\,(\beta + m)/\beta\ (m + 1)$. Again, if α is greater than β, then more electric field stress will be applied across depletion regions of scaled-down transistors.

At the same time, the junction breakdown voltage BV must be considered. BV is given by (Grove, 1967):

$$BV = \frac{\varepsilon_{si}\varepsilon_0 (E_{crit})^2}{2qN_B}$$

It will be seen that BV is thus scaled by $\beta(m + 1)/\alpha^2\,(\beta + m)$ and will decrease. Extra care is therefore required in estimating the breakdown voltage for scaled devices. It should be noted that electric fields are greater and BV is greater at the corners of diffusion regions underlying or abutting silicon dioxide.

5.5 Limits on logic levels and supply voltage due to noise

Major advantages in the scaling of devices are smaller gate delay time, that is, higher operating frequencies and lower power dissipation. However, the decreased inter-feature spacing and greater switching speeds inevitably result in noise problems. Noise may also be amplified and is thus a major concern.

The mean square current fluctuation in the channel is given by

$$(i^2) = 4kTR_n g_m \Delta f$$

where R_n is the equivalent noise resistance at the input and Δf is the bandwidth.

F. M. Klaassen and J. Prins (1966) have investigated the thermal noise in a MOS transistor over a range of substrate doping levels N_B from 10^{14} to 10^{17} cm^{-3}. When a transistor works in saturation, g_m is no longer proportional to the gate voltage V_g, and can be expressed as

$$g_m \doteqdot BV_p$$

where

$$B = \frac{\mu W C_{ox}}{L}$$

and V_p is the pinch off voltage given by

$$V_p = V'_g - \frac{1}{2}\left(\frac{a}{C_{ox}}\right)^2 \left[\frac{(1 + 4V'_g C_{ox})^{1/2}}{a} - 1\right]$$

where

$$V'_g = V_g - V_t + V_B$$
$$a = (2\varepsilon_{si} q N_B)^{1/2}$$

V_B is the junction (built-in) potential.

Then, the equivalent noise resistance R_n is given by

$$R_n = \left(\frac{1V'_g}{2V'_p} + \frac{1}{6}\right) g_m^{-1}$$

where

$$V'_p = V_p + V_B$$

Since V_p is a monotonically decreasing function of the gate oxide thickness t_{ox} and substrate doping N_B, R_n is also a monotonically decreasing function of the same parameters.

Consequently, the main factor of the thermal noise $R_n g_m$ is given by

$$R_n g_m = \frac{1}{2}\left(\frac{V_g - V_t + V_B}{V_p + V_B}\right) + \frac{1}{6}$$

This indicates that $R_n g_m$ is strongly and directly dependent on t_{ox} and N_B and also, to a lesser extent, on V_g. Experimental results, as in Figure 5–7, support this theory (Klaassen and Prins, 1966).

Considering current technology in which t_{ox} is scaled, the effect of N_B is smaller (Sah, Wu and Hielscher, 1966), and V'_p and V'_g are replaced by V_p and V_g respectively. Thus, the expression for $R_n g_m$ becomes

$$R_n g_m = \frac{1}{2}\left[\frac{V_g}{V_g - \frac{1}{2}\left(\frac{a}{C_{ox}}\right)^2\left(\frac{1+4V_g'C_{ox}}{a}\right)^{1/2} - 1}\right] + \frac{1}{6}$$

When constant field scaling is applied, V_g is scaled by $1/\alpha$, C_{ox} and N_B are scaled by α. Consequently, $R_n\, g_m$ is only slightly reduced owing to the increased value of C_{ox}. Thus, the ratio of logic level to thermal noise is degraded by almost the same factor.

Flicker noise has been the subject of many studies since A. L. McWhorter introduced his model in 1956. The noise was observed and explained as the result of fluctuations of carriers trapped in the channel by surface states.

As a conclusion of the investigations by F. M. Klaassen, the change in the number of trapped carriers dn_t due to the change in the number of induced free carriers d_n presents a current fluctuation Δi at the output, such that

$$\Delta i^2 \approx \frac{q\mu s I V_d}{L^2 f}$$

where

$s = dn_t / d_n$ — the surface state efficiency
I = the DC drain current
f = the frequency
V_d = the applied drain voltage.

Usually the output noise is represented by an equivalent noise voltage source ΔV at the input (Klaassen, 1971), such that

$$\Delta V^2 = \frac{qs(V_d - 0.5V_d)}{C_g f}$$

When the transistor operates in saturation, then $V_d \doteqdot V_g$, so that

$$\Delta V^2 = \frac{1}{2}\frac{qsV_g}{C_g f}$$

where

V_g = the effective applied gate voltage
C_g = the gate capacitance.

Since s is a process dependent factor, the flicker noise is scaled by one for constant field scaling or by α^2 / β^2 for the combined scaling model.

In addition to noise sources already considered, other noise inputs are due to mutual inductive and mutual capacitive coupling, and these alone could impose practical limitations on the lowest usable operating voltages.

Considering the cross-talk between two parallel signal lines on a chip, the coupling model presented (Watts, 1989) shows that capacitive noise is proportional to $C.dV/dt$, where C is the inter-line capacitance, and dV/dt is approximately equal to V_a/t_r, where t_r is the rise time of the coupled signal. The inductive noise is related to LdI/dt, where L is the mutual inductance and $dI/dt \approx I_{sat}/t_r$. Therefore, cross-talk noise increases as operating frequency increases and t_r is reduced.

There are also other noise sources due to external influences, such as radio frequency signals, voltage spikes or voltage drops on power lines or ground connections, unterminated signal lines and lines with non-uniform impedance characteristics.

A typical peak to peak noise of at least 100 mV may be observed on the power and ground lines of even well-designed multilayer PC boards (Long & Butner, 1989).

Scaling down exacerbates the effect of both internally and externally generated noise and this degrades both the production yield and the reliability of high density chip layouts.

In order to assess the effects of noise on the probability of failure P_F in a circuit, we may consider a situation where, as in Figure 5–7, the minimum signal to noise ratio *(SNR)* for satisfactory operation is assumed to be four, and the mean noise is assumed to be zero with a standard deviation $\Sigma = 100$ mV. P_F may be estimated by using the Gaussian distribution

$$Q(x) = \frac{1}{\sqrt{2\pi}} \int_x^{\infty} \left(\frac{e^{-\mu^2/2}}{2} \right) du$$

The P_F values for a range of supply rail voltages and for the conditions specified are derived by integrating the appropriate area at one end of the Gaussian curve and are shown in Figure 5–8.

5.7 Limits due to current density

High purity aluminum seems the most attractive, and is thus the most widely used, material for forming interconnections in VLSI chips. However, the scaling down of dimensions also increases the current density in interconnects by the same factor if constant field scaling is applied. When the current density in aluminum approaches 10^6 Amps/cm^2 (10 mA/μm^2), the interconnects are likely to be burned off owing to metal migration. Thus, allowable current densities are set well below this limit and figures of J = 1 to 2 mA/μm^2 are commonly used.

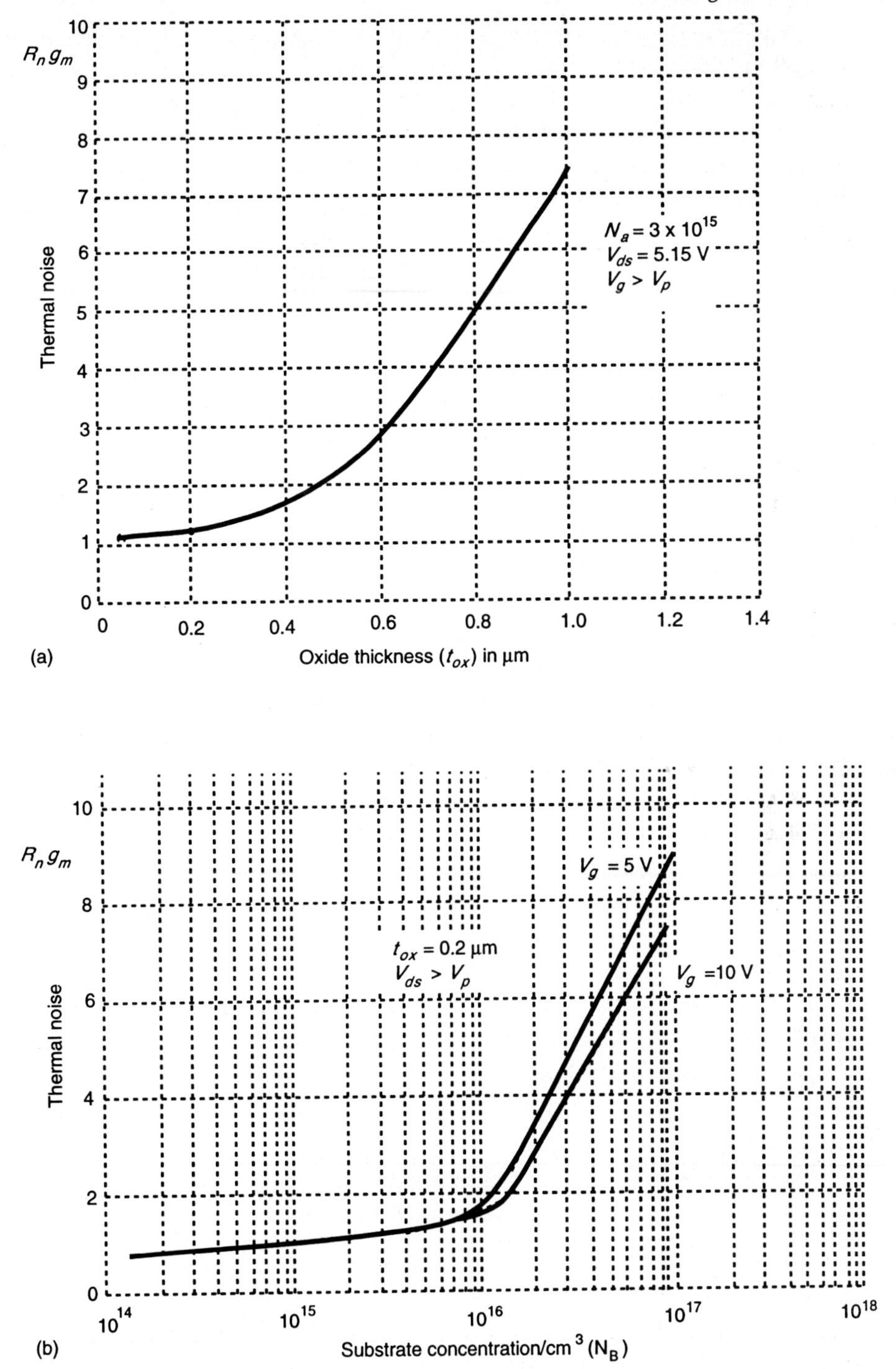

Figure 5–7 Thermal noise versus oxide thickness (a) and substrate doping (b)

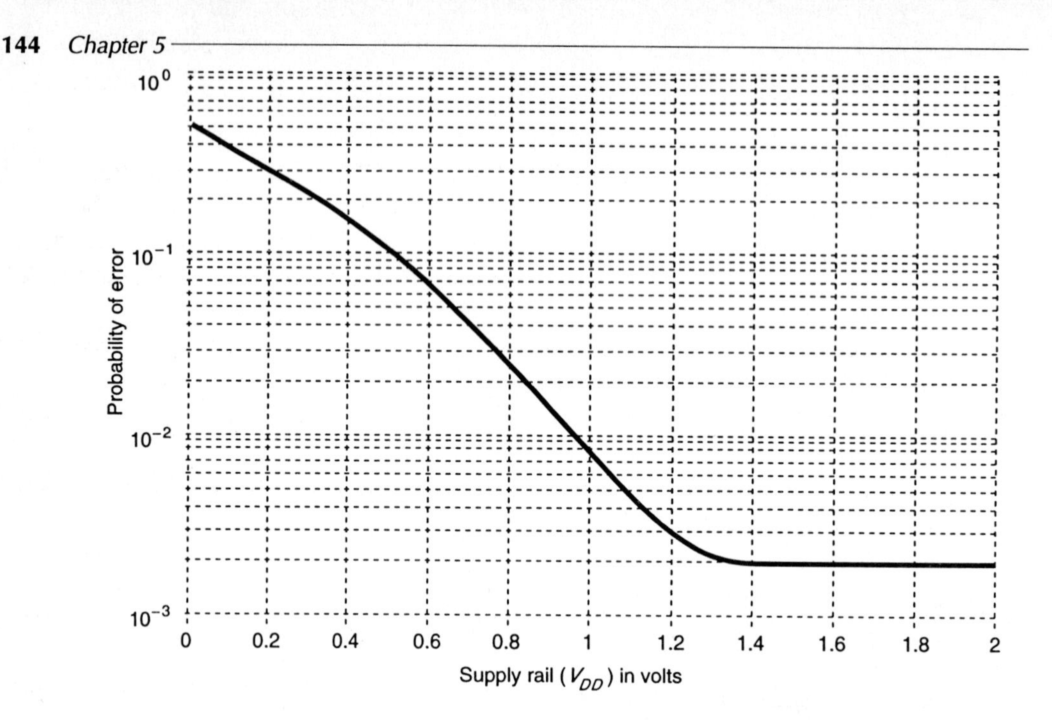

Figure 5–8 Probability of error versus supply voltage

5.8 Observations

Scaling has not only been developed theoretically, but has also been widely applied in fabrication facilities as equipment improves. This has provided a direct and simple way of making smaller, faster chips. Current designs are often scaled down by 10% to 20% in linear dimensions as a way of providing better performance whilst faster smaller chip designs are completed.

The contributions made by scaling are significant. In particular, power dissipations are reduced, switching speeds are increased and chip size is reduced, which in turn improves production costs. However, the continual scaling down of dimensions is now pushing the technology toward both technological and, ultimately, real physical limits.

Recent history has demonstrated that a new generation of microelectronic products emerges about every four years and that the device dimensions are scaled down by about 20% for each new generation. If this pace is maintained, then the ultimate physical limits on device size for silicon chips will be reached within the next decade. For further progress in the direction of high speed circuits, one must look to the development of alternative materials such as gallium arsenide (GaAs) and to super conductors.

5.9 References

Bakoglu, H. B., and Meindl, J. D. (1985, May) 'Optimal interconnection circuits for VLSI', *IEEE Transactions on Electronic Devices*, Vol. ED-32, No. 5.

Bergmann, N. W. (1991) 'A combined voltage and dimension scaling model for VLSI circuits', Proceedings of Microelectronics Conference, Melbourne, 24–25 June 1991, 101–4.

Grove, A. S. (1967) *Physics and Technology of Semiconductor Devices*, John Wiley and Sons Inc., New York.

Hutcheson, L. D. (1987) *Integrated Optical Circuits and Components: Design and Applications,* Marcel Dekker, New York.

Klaassen, F. M., and Prins, J. (1966) 'Thermal noise in MOS Transistors', *Philips Research Report*, Vol. 22.

Klaassen, F. M. (1971) 'Characterisation of Low 1/f Noise in MOS Transistors', *IEEE Transactions on Electronic Devices,* Vol. ED-18, No. 10, 887–91.

Long, S. I., and Butner, S.E. (1989) *Gallium Arsenide Digital Integrated Circuit Design*, McGraw-Hill, USA.

Meindl, J. D. (1986) 'Interconnection limits on ultra large scale integration', *Proceedings VLSI '85*, North Holland, 13–19.

Sah, C. T., Wu, S. Y., and Hielscher, F. H. (1966) 'The effects of fixed bulk charge on the thermal noise in metal-oxide-semiconductor transistors', *IEEE Transactions on Electronic Devices,* Vol. ED-13, 410–14.

Sakurai, T. (1983, August) 'Approximation of wiring delay in MOSFET LSI', *IEEE Journal of Solid State Circuits*, Vol. SC-18, 418–26.

Sze, S. M. (1985) *Semiconductor Devices: Physics and Technology*, Bell Telephone Laboratories, USA.

Watts, R. K. (1989) *Submicron Integrated Circuits*, John Wiley and Sons Inc., New York.

6 Subsystem design and layout

Logic is simply the architecture of human reason.

Evelyn Waugh

Tall oaks from little acorns grow.

David Everett

Objectives

Having now covered the basic MOS and BiCMOS technologies, the behavior of components formed by MOS layers, the basic units which help to characterize behavior, and a set of design rules, we are in a position to undertake the design of some of the subsystems (leaf-cells) from which larger systems are composed.

The most basic leaf-cells are the common logic gate arrangements and these are dealt with in nMOS, CMOS and BiCMOS forms. In the case of CMOS gates there are several ways in which the logic may be configured, most of which are dealt with in this chapter.

The concepts of structured design, which leads to system designs of high 'regularity', are introduced through examples. A highly regular design is to be sought for VLSI systems since high regularity implies the detailed design of relatively few leaf-cells which are then replicated many times and interconnected to form the system.

Other commonly applied concepts, such as two-phase clocks and buses for the interconnection paths between leaf-cells and subsystems, are also introduced and illustrated. Important aspects of power distribution on chip and associated limitations are discussed.

The chapter also includes an introduction to common subsystem designs such as multiplexers and shift registers.

6.1 Some architectural issues

In all design processes, a logical and systematic approach is essential. This is particularly so in the case of the design of a VLSI system which could otherwise take so long as to render the whole system obsolete before it is off the drawing board. Take, for example, the case of a relatively straightforward MSI logic circuit comprising, say, 500 transistors. A reasonable time to allocate to the design and proving of such a circuit could be some two engineer-months. Consider now the design of a 500,000 transistor VLSI system. Even if a linear relationship exists between complexity and design time, the required design time would be 2000 engineer-months or 170 engineer-years. In fact, design time tends to rise exponentially with increased complexity. Obviously, then, we must adopt design methods which allow the handling of complexity in reasonable periods of time and with reasonable amounts of labor.

Certainly we are not about to tackle 500,000 transistor designs in this text, but some sensible concepts applied even at the subsystem (leaf-cell) level can be most worthwhile and can also be directly compatible with larger system design requirements. Guidelines may be set out as follows:

1. Define the requirements (properly and carefully).
2. Partition the overall architecture into appropriate subsystems.
3. Consider communication paths carefully in order to develop sensible interrelationships between subsystems.
4. Draw a floor plan of how the system is to map onto the silicon (and alternate between 2, 3 and 4 as necessary).
5. Aim for regular structures so that design is largely a matter of replication.
6. Draw suitable (stick or symbolic) diagrams of the leaf-cells of the subsystems.
7. Convert each cell to a layout.
8. Carefully and thoroughly carry out a design rule check on each cell.
9. Simulate the performance of each cell/subsystem.

The whole design process will be greatly assisted if considerable care is taken with:

1. the *partitioning* of the system so that there are clean and clear subsystems with a minimum interdependence and complexity of interconnection between them;

2. the *design simplification* within subsystems so that architectures are adopted which allow the exploitation of a cellular design concept. This allows the system to be composed of relatively few standard cells which are replicated to form highly regular structures.

In designing digital systems in MOS technology there are two basic ways of building logic circuits, which will now be discussed.

6.2 Switch logic

Switch logic is based on the 'pass transistor' or on transmission gates. This approach is fast for small arrays and takes no static current from the supply rails. Thus, power dissipation of such arrays is small since current only flows on switching.

Switch (pass transistor) logic is similar to logic arrays based on relay contacts in that the path through each switch is isolated from the signal activating the switch. In consequence, the designer has a considerable amount of freedom in implementing architectural features compared with bipolar logic-based designs.

A number of texts on switching theory, some dating from the 1950s and 1960s, have sections on relay/switch logic and the reader is referred to such material for generating ideas for implementation in MOS switch logic. An example is Marcus, *Switching Circuits for Engineers*, Prentice Hall, 1962; 3rd edn, 1975.

Basic *And* and *Or* connections are set out in Figure 6–1, but many combinations of switches are possible.

6.2.1 Pass transistors and transmission gates

Switches and switch logic may be formed from simple n- or p-pass transistors or from transmission gates (complementary switches) comprising an n-pass and a p-pass transistor in parallel as shown in Figure 6–2. The reason for adopting the apparent complexity of the transmission gate, rather than using a simple n-switch or p-switch in most CMOS applications, is to eliminate the undesirable threshold voltage effects which give rise to the loss of logic levels in pass transistors as indicated in Figure 6–2. No such degradation occurs with the transmission gate, but more area is occupied and complementary signals are needed to drive it. 'On' resistance, however, is lower than that of the simple pass transistor switches.

When using nMOS switch logic, there is one restriction which must always be observed: *no* pass transistor gate input may be driven through *one or more pass transistors* (see Figure 6–2). As shown, logic levels propagated through pass transistors are degraded by threshold voltage effects. Since the signal out of pass transistor T_1 does not reach a full logic 1, but rather a voltage one transistor

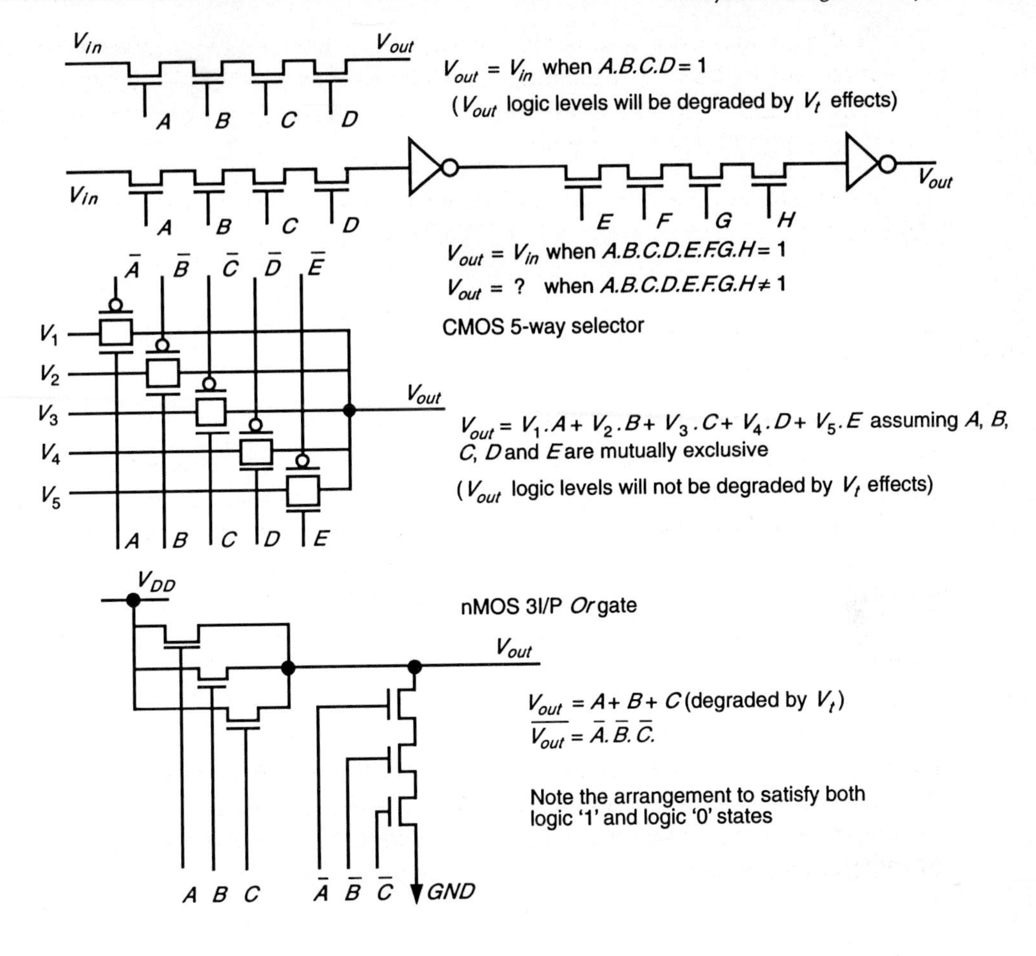

Figure 6–1 Some switch logic arrangements

threshold below a true logic 1, this degraded voltage would not permit the output of T_2 to reach an acceptable logic 1 level.

6.3 Gate (restoring) logic

Gate logic is based on the general arrangement typified by the inverter circuits (the inverter being the simplest gate).

Both *Nand* and *Nor* and, with CMOS, *And* and *Or* gate arrangements are available. Inverters are also employed to complement and restore logic levels that have been degraded (e.g. because they have passed through pass transistors).

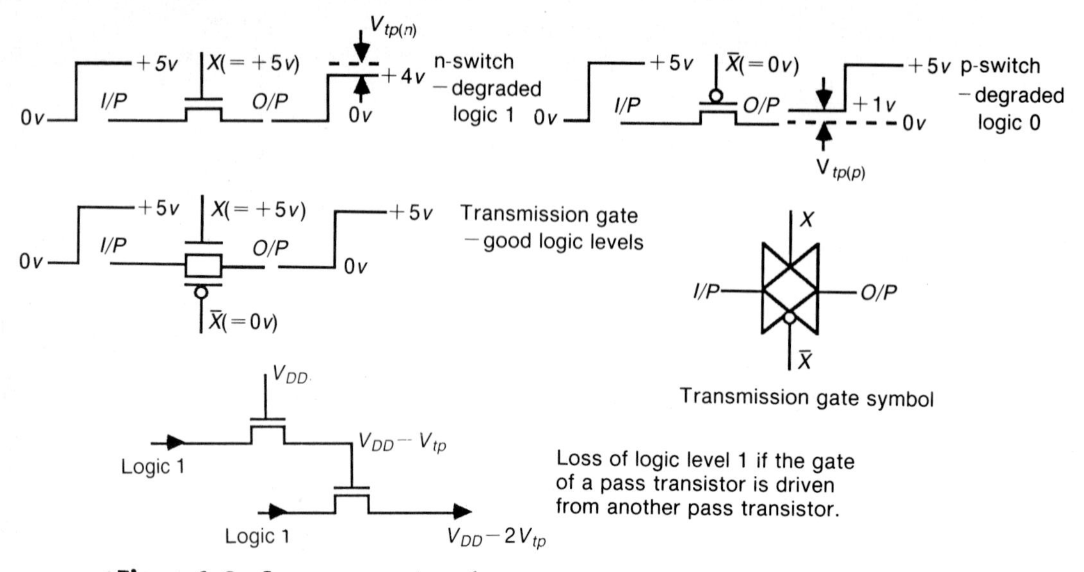

Figure 6–2 Some properties of pass transistors and transmission gates

6.3.1 The inverter

Some of the most commonly used inverter circuit diagrams — the inverter symbol, and the corresponding stick and symbolic diagrams — should be familiar by now. An assortment is reproduced here in Figure 6–3. Note that it is often useful to indicate the nMOS inverter $Z_{p.u.}/Z_{p.d.}$ ratio and/or the channel length to width ratio for each MOS transistor as shown.

In achieving the desired pull-up to pull-down ratio, several possibilities emerge, two of which are illustrated in Figures 6–4 and 6–5 for an 8:1 nMOS inverter. Note the effect that the different approaches have on power dissipation P_d and on the area occupied by the inverter. Also note the resistance and capacitance values. The CMOS inverter carries no static current and thus has no power dissipation unless switching. The reader must not, however, imagine that CMOS circuits have no dissipation problems. The switching dissipation for fast CMOS logic circuits will be considerable.

6.3.2 Two-input nMOS, CMOS and BiCMOS *Nand* gates

Two-input *Nand* gate arrangements are given in Figure 6–6. The nMOS (and pseudo-nMOS) $L{:}W$ ratios should be carefully noted since they must be chosen to achieve the desired overall $Z_{p.u.}/Z_{p.d.}$ ratio (where $Z_{p.d.}$ is contributed in this case by *both* input transistors in series).

In order to arrive at the required $L{:}W$ ratios for an nMOS (or pseudo-nMOS)

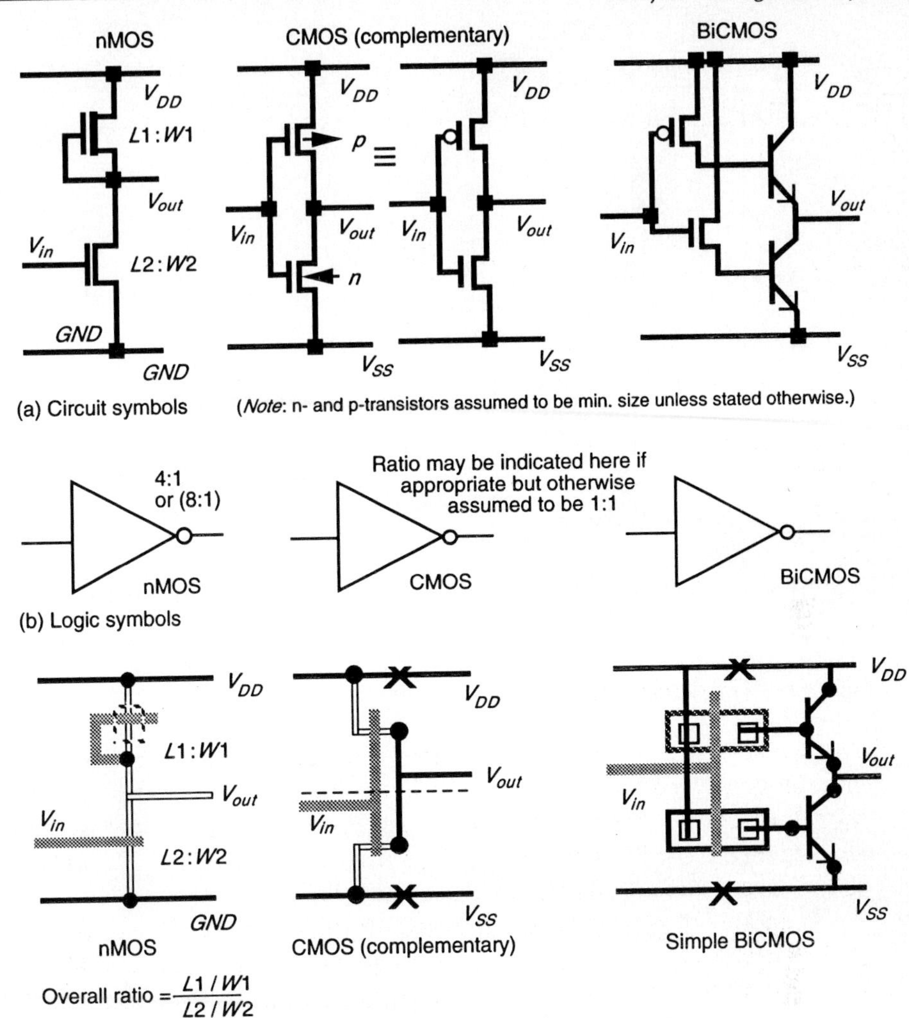

Figure 6–3 nMOS, CMOS and BiCMOS inverters

Nand gate with *n* inputs, it is only necessary to consider the very simple circuit model of the gate in the condition when all *n* pull-down transistors are conducting as in Figure 6–7.

The critical factor here is that the output voltage V_{out} must be near enough to ground to turn off any following inverter-like stages, that is

$$V_{out} \leqslant V_t = 0.2V_{DD}$$

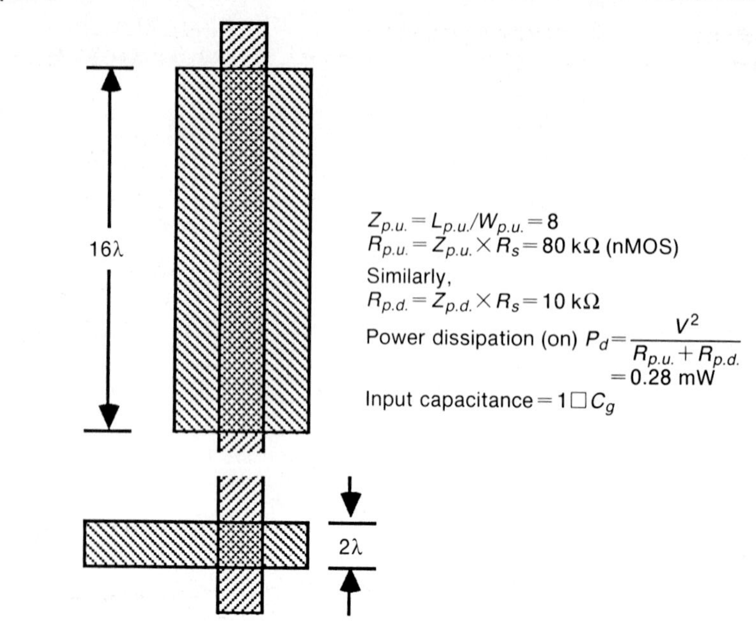

Figure 6–4 8:1 nMOS inverter (minimum size p.d.)

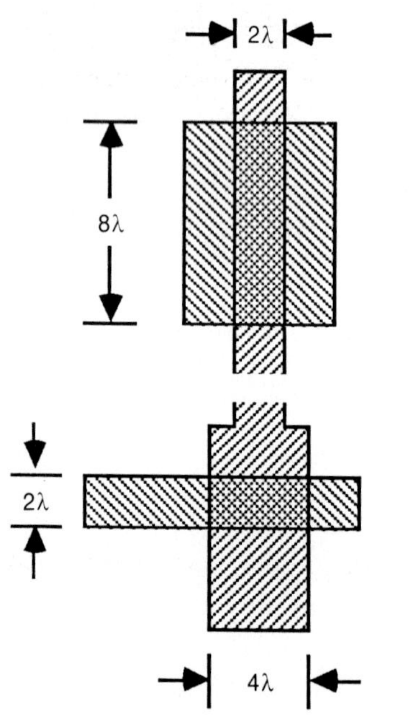

$Z_{p.u.} = L_{p.u.}/W_{p.u.} = 4$
$R_{p.u.} = Z_{p.u.} \times R_s = 40\ k\Omega$ (nMOS)
Similarly,
$R_{p.d.} = Z_{p.d.} \times R_s = 5\ k\Omega$

Power dissipation (on) $P_d = \frac{V^2}{R_{p.u.} + R_{p.d.}} = 0.56$ mW

Input capacitance $= 2\square C_g$

Note: A 4:1 inverter is formed if the p.d. width is halved.

Figure 6–5 An alternative 8:1 nMOS inverter

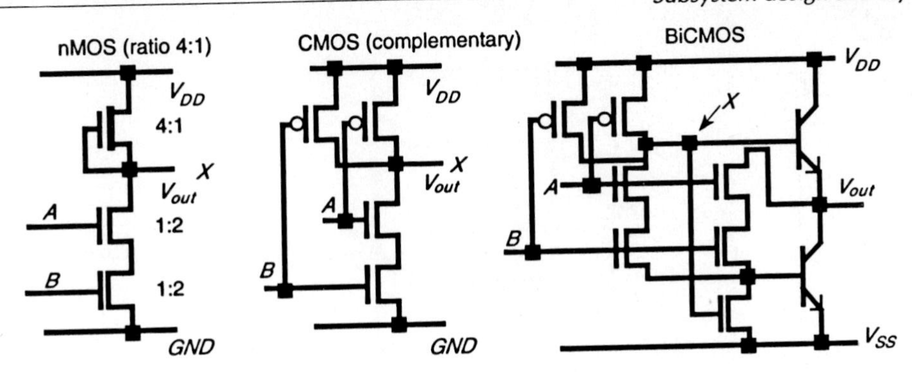

(a) Circuit diagrams *Note*: n- and p- transistors assumed to be minimum size unless stated otherwise.

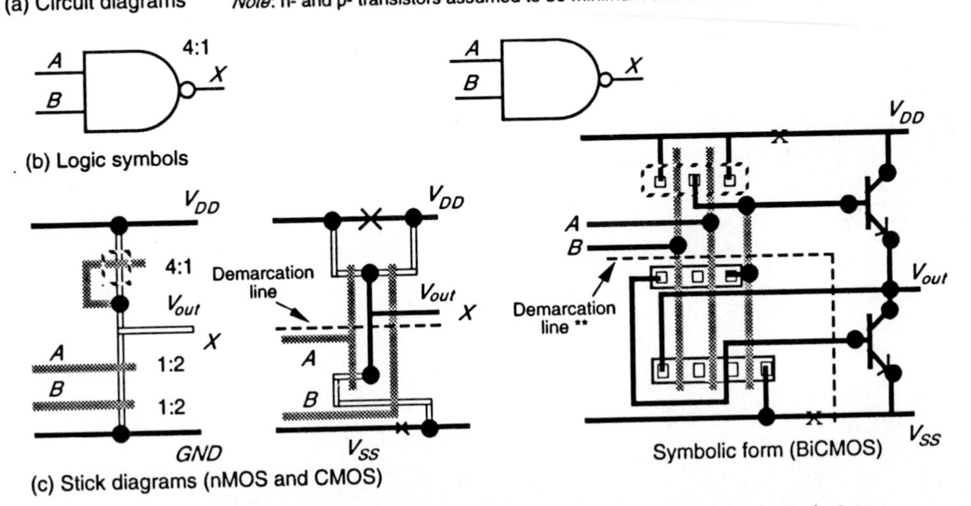

(b) Logic symbols

(c) Stick diagrams (nMOS and CMOS)

Note: The natural 2.5:1 asymmetry of the CMOS inverter is improved to 1.25:1 (or better) owing to the two n-type pull-down transistors in series for the two I/P *Nand*.

** Demarcation line (edge of n-well) may be shown if required.

Figures 6–6(a)–(c) nMOS, CMOS and BiCMOS 2-input *Nand* gates

Thus

$$\frac{V_{DD} \times nZ_{p.d.}}{nZ_{p.d.} + Z_{p.u.}} \leqslant 0.2V_{DD}$$

where $Z_{p.d.}$ applies for any one pull-down transistor. The boundary condition then is

$$\frac{nZ_{p.d.}}{nZ_{p.d.} + Z_{p.u.}} = 0.2$$

$$\text{whence nMOS } \textit{Nand} \text{ ratio} = \frac{nZ_{p.d.}}{nZ_{p.d.}} = \frac{4}{1}$$

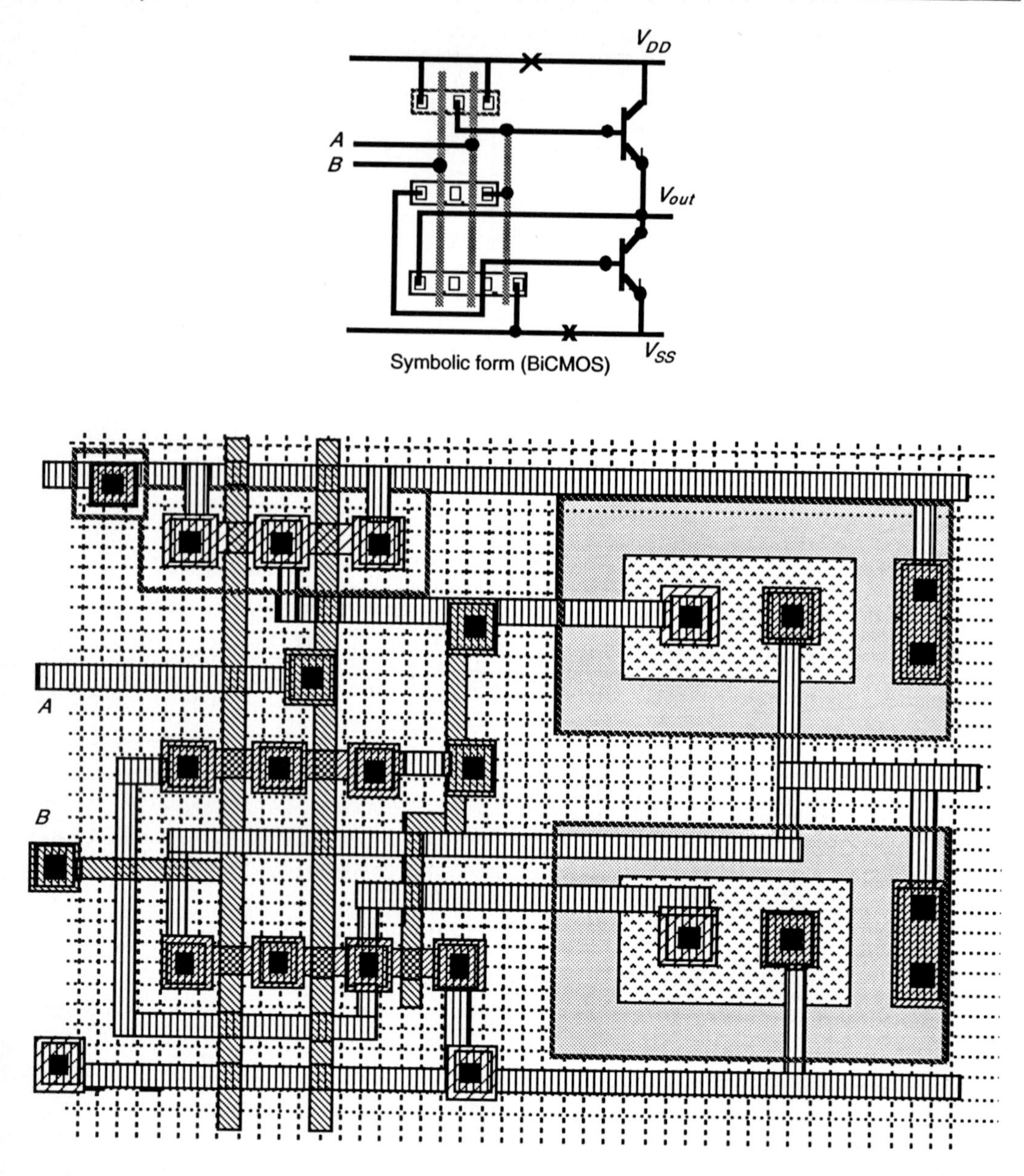

Figure 6–6(d) A BiCMOS two-input *Nand* gate

that is, the ratio between $Z_{p.u.}$ and the sum of all the pull-down $Z_{p.d.}$s must be 4:1 (as for the nMOS inverter).

This ratio must be adjusted appropriately if input signals are derived through pass transistors.

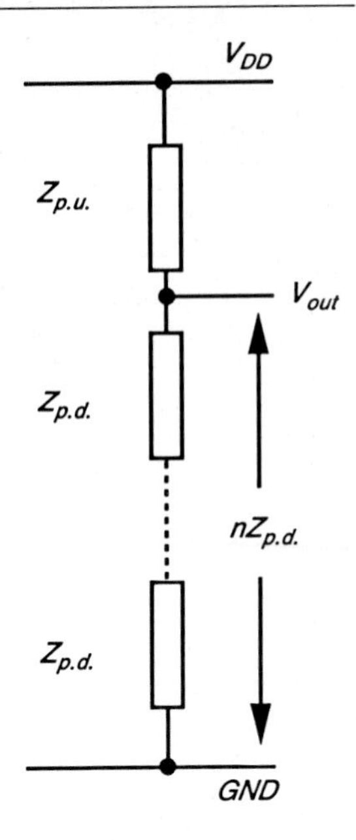

Figure 6–7 nMOS *Nand* ratio determination

Further consideration of the nMOS *Nand* gate geometry reveals two significant factors:

1. nMOS *Nand* gate *area requirements* are considerably greater than those of a corresponding nMOS inverter, since not only must pull-down transistors be added in series to provide the desired number of inputs, but, as inputs are added, so must there be a corresponding adjustment of the length of the pull-up transistor channel to maintain the required overall ratio.
2. nMOS *Nand* gate *delays* are also increased in direct proportion to the number of inputs added. If each pull-down transistor is kept to minimum size ($2\lambda \times 2\lambda$), then each will present 1 $\square C_g$ at its input, but if there are n such inputs, then the length and *resistance* of the pull-up transistor must be increased by a factor of n to keep the correct ratio. Thus, delays associated with the nMOS *Nand* are

$$\tau_{Nand} = n\tau_{inv}$$

where n is the number of inputs and τ_{inv} is the corresponding nMOS inverter delay. (The alternative approach of keeping $Z_{p.u.}$ constant and widening the pull-down channels has the same effect, since in this case C_g for each pull-down transistor will be increased to $n\square Cg$.)

Furthermore, the rise time of the nMOS *Nand* output is dependent on the actual input(s) on which the ∇ transition takes place.

In consequence of these properties, the nMOS *Nand* gate is used *only* where absolutely necessary and the number of inputs is restricted.

The CMOS *Nand* gate has no such restrictions but, bearing in mind the remarks on asymmetry (Figure 6–6), it is necessary to allow for extended fall-times on capacitive loads owing to the number of n-transistors in series forming the pull-down. Some adjustment of transistor geometry may be necessary for this reason and to keep the transfer characteristic symmetrical about $V_{DD}/2$.

The BiCMOS gate shown is a practical version and is thus more complex than the simple intuitive version. However, it has considerable load-driving capabilities and is most useful where a large fan-out is required or where there is some other form of high capacitance load on the output. A typical mask layout for this gate, using Orbit™ 2 μm design rules, is given in monochrome form in Figure 6–6(d) and in color as Color plate 8(a).

The relatively easy conversion from symbolic form to mask layout for the BiCMOS 2-input *Nor* gate is illustrated by Figure 6–6(d) and Color plate 8(a).

6.3.3 Two-input nMOS, CMOS and BiCMOS *Nor* gates

Two-input *Nor gate* arrangements are given in Figure 6–8; note here that the nMOS (or pseudo nMOS) form of *Nor* gate can be expanded to accommodate any reasonable number of inputs (e.g. see Color plate 9) and, in those technologies, is preferred to the *Nand* gate when there is a choice (which is usually the case if logical expressions are suitably manipulated).

Since both 'legs' of the two-input nMOS *Nor* gate provide a path to ground from the pull-up transistor, the ratios must be such that any one conducting pull-down leg will give the appropriate inverter-like transfer characteristic. Thus, each leg has the same ratio as would be the case for an nMOS inverter. This applies irrespective of the number of inputs accommodated.

The area occupied by the nMOS (or pseudo-nMOS) *Nor* gate is reasonable since the pull-up transistor dimensions are unaffected by the number of inputs accommodated. In consequence, the *Nor* gate is as fast as the corresponding inverter and is the preferred inverter-based nMOS (or pseudo-nMOS) logic gate when a choice is possible.

Obviously, the ratio between $Z_{p.u.}$ and $Z_{p.d.}$ of any one leg must be appropriate to the source from which that input is driven for nMOS designs (namely 4:1 driven from another inverter-based circuit or 8:1 if driven via one or more pass transistors) but will be uniformly 3:1 for a pseudo-nMOS design where any series switching is by transmission gate.

The CMOS *Nor* gate (see Color plate 9) consists of a pull-up p-transistor-based structure, which implements the logic 1 conditions and a complementary n-transistor arrangement to implement the logic 0 conditions at the output. In the case of the *Nor* gate, the p-structure consists of transistors in series, one for each input, while the n pull-down arrangement has as many transistors in parallel as there are inputs to the *Nor* gate. Thus, the already predominant resistance of the p-

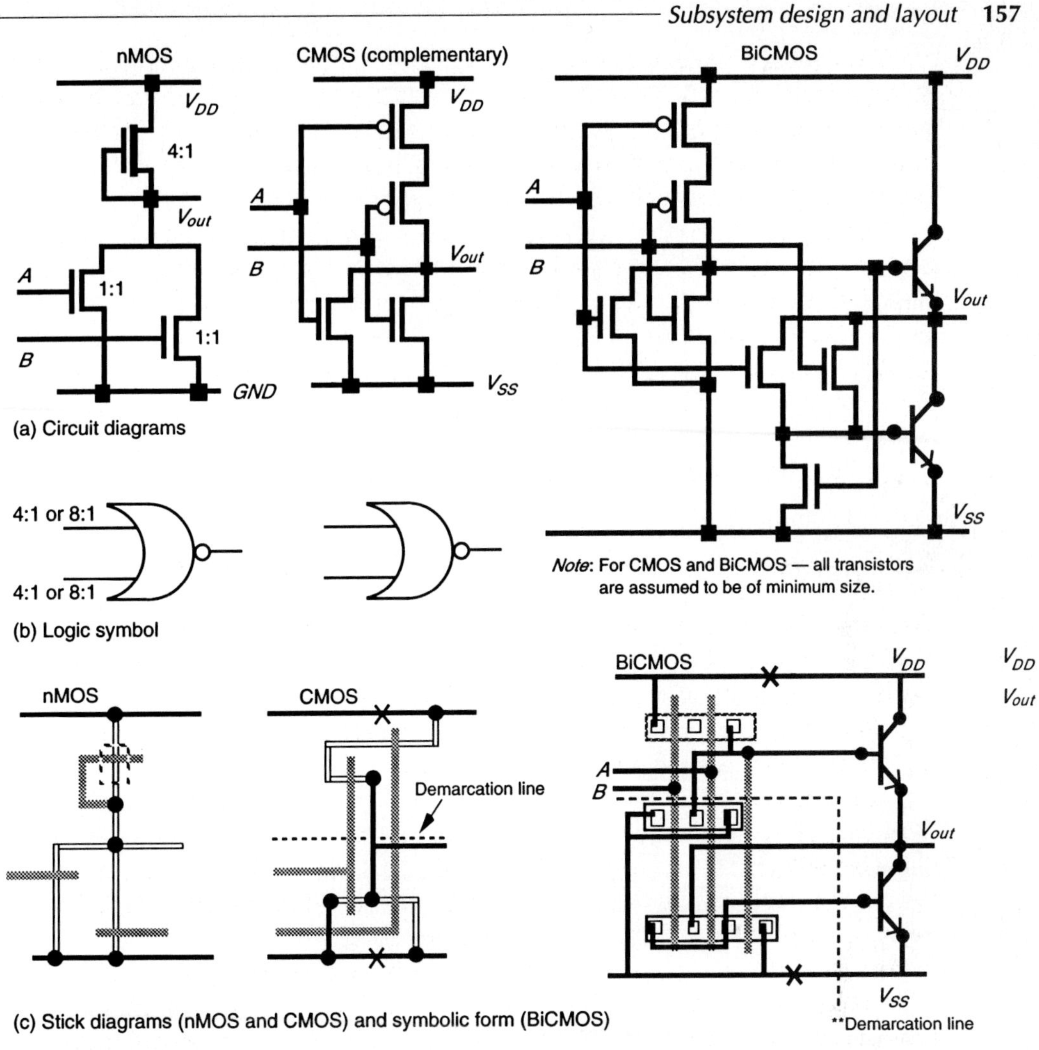

Figures 6–8(a)–(c) nMOS, CMOS and BiCMOS two-input *Nor* gate

devices is aggravated in its effect by the number connected in series. Rise-time and fall-time asymmetry on capacitive loads is thus increased and there will also be a shift in the transfer (V_{in} vs V_{out}) characteristic which will reduce noise immunity. For these reasons, CMOS (complementary logic) *Nor* gates with more than two inputs may require adjustment of the p- and/or n- transistor geometries (*L:W* ratios).

The CMOS *Nand* gate, on the other hand, benefits from the connection of p-transistors in parallel, but once again the geometries may require thought when several inputs are required.

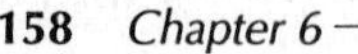

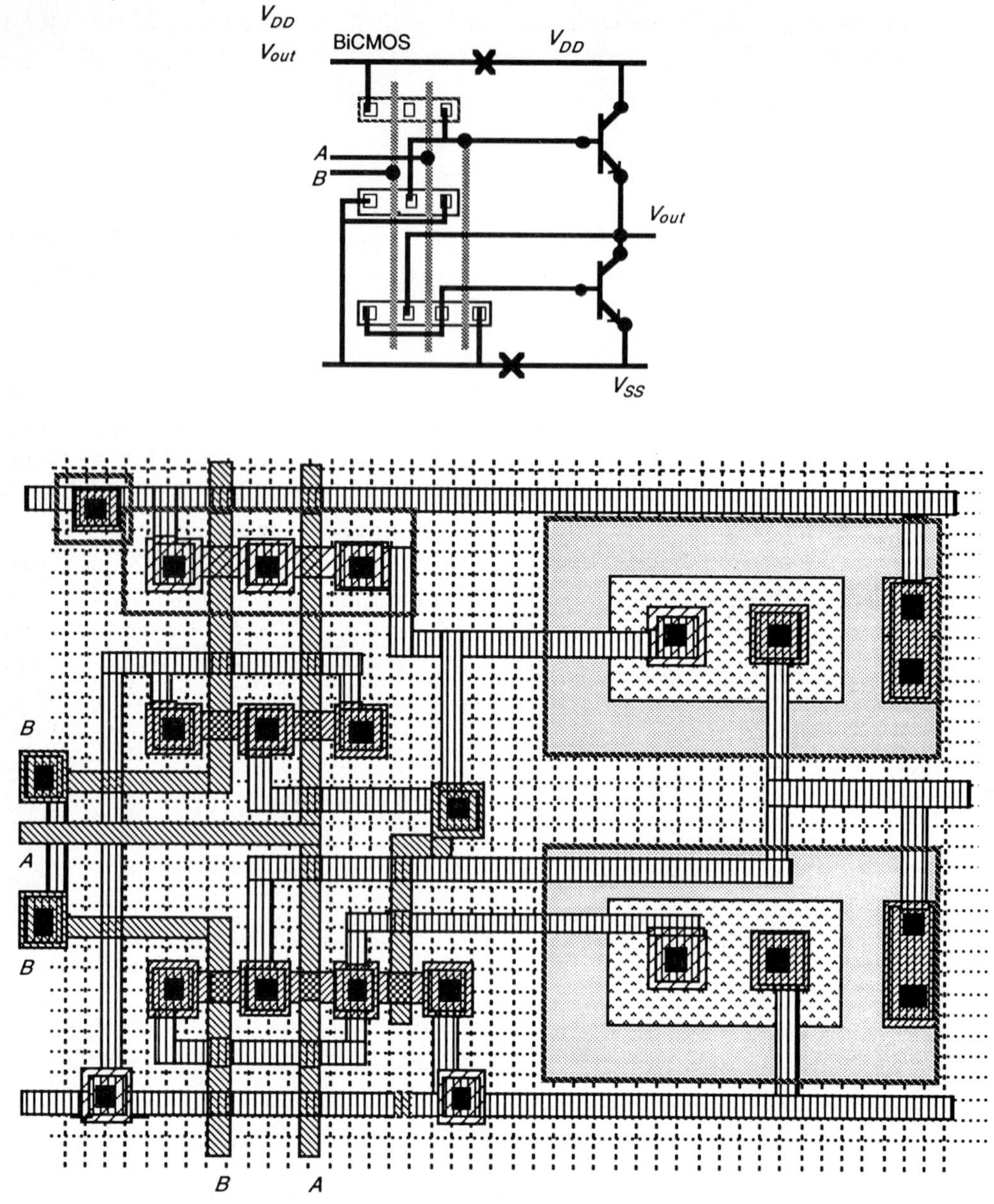

Figure 6–8(d) A BiCMOS two-input *Nor* gate

The BiCMOS *Nor* gate shown is a practical version and, as for the BiCMOS *Nand,* is more complex than a simple intuitive version. However, it also has considerable capacitive load-driving capabilities and is most useful where a large fan-out is required or where there is some other form of high capacitance load on the output such as in the I/O region of a chip.

The relatively easy conversion from symbolic form to mask layout is illustrated for the BiCMOS two-input *Nor* gate in Figure 6–8(d) and Color plate 8(b). The mask layout has been drawn using the Orbit™ 2 μm BiCMOS rule set.

6.3.4 Other forms of CMOS logic

The availability of both n- and p-transistors makes it possible for the CMOS designer to explore and exploit various alternatives to inverter-based CMOS logic.

6.3.4.1 Pseudo-nMOS logic

Clearly, if we replace the depletion mode pull-up transistor of the standard nMOS circuits with a p-transistor with gate connected to V_{SS}, we have a structure similar to the nMOS equivalent. This approach to logic design is illustrated by the three-input *Nand* gate in Figure 6–9. The circuit arrangements look and behave much like nMOS circuits and appropriate ratio rules must be applied.

In order to determine the required ratio, we consider the arrangement of Figure 6–10 in which a pseudo-nMOS inverter is being driven by another similar inverter, and we consider the conditions necessary to produce an output voltage of V_{inv} for an identical input voltage. As for the nMOS analysis, we consider the conditions for which $V_{inv} = V_{DD}/2$.

At this point the n-device is in saturation (i.e. $0 < V_{gsn} - V_{tn} < V_{dsn}$) and the p-device is operating in the resistive region (i.e. $0 < V_{dsp} < V_{gsp} - V_{tp}$). Equating currents of the n-transistor and the p-transistor, and by suitable rearrangement of

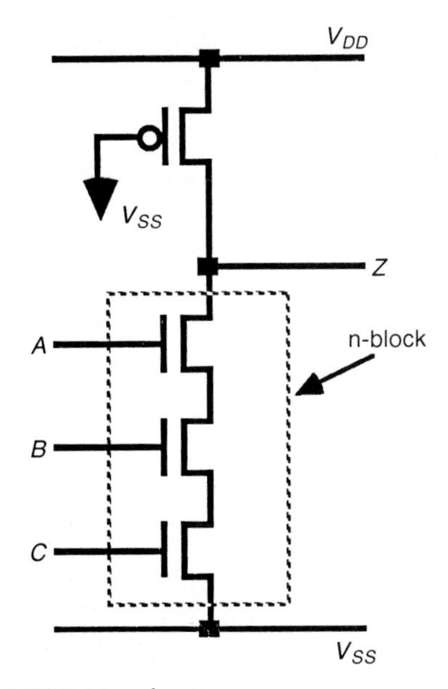

Figure 6–9 pseudo-nMOS *Nand* gate

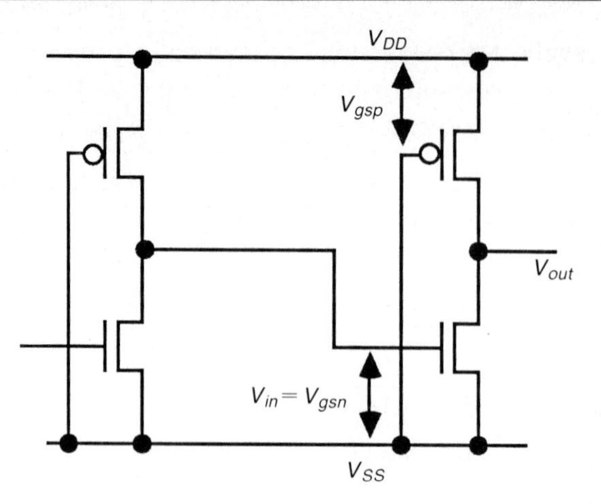

Figure 6–10 Pseudo-nMOS inverter when driven from a similar inverter

the resultant expression, we obtain

$$V_{inv} = V_{tn} + \frac{(2\mu_p / \mu_n)^{1/2} [(-V_{DD} - V_{tn}) V_{dsp} - V_{dsp}{}^2]^{1/2}}{(Z_{p.u.} / Z_{p.d.})^{1/2}}$$

where

$$Z_{p.u.} = L_p / W_p$$

and

$$Z_{p.d.} = L_n / W_n$$

With

$$V_{inv} = 0.5\ V_{DD}$$

$$V_{tn} = |V_{tp}| = 0.2\ V_{DD}$$

$$V_{DD} = 5\text{V}$$

$$\mu_n = 2.5\ \mu_p$$

we obtain

$$\frac{Z_{p.u.}}{Z_{p.d.}} = \frac{3}{1}$$

A transfer characteristic, V_{out} vs V_{in}, can be drawn and, as for the nMOS case, the characteristic will shift with changes of $Z_{p.u.}/Z_{p.d.}$ ratio.

Two points require comment:

1. Since the channel sheet resistance of the p-pull-up is about 2.5 times that of the n-pull-down, and allowing for the ratio of 3:1, the pseudo-nMOS inverter presents a resistance between V_{DD} and V_{SS} which is, say, 85 kΩ compared with 50 kΩ for a comparable 4:1 nMOS device. Thus, power dissipation is reduced to about 60% of that associated with the comparable nMOS device.
2. Owing to the higher pull-up resistance, the inverter pair delay is larger by a factor of 8.5:5 than the 4:1 minimum size nMOS inverter.

6.3.4.2 Dynamic CMOS logic

The actual logic (see Figure 6–11(a) for the schematic arrangement) is implemented in the inherently faster nMOS logic (the n-block); a p-transistor is used for the non-time-critical precharging of the output line 'Z' so that the output capacitance is charged to V_{DD} during the off period of the clock signal ϕ. During this same period the inputs are applied to the n-block and the state of the logic is then evaluated during the on period of the clock when the bottom n-transistor is turned on. Note the following:

1. Charge sharing may be a problem unless the inputs are constrained not to change during the on period of the clock.
2. *Single phase dynamic logic structures cannot be cascaded* since, owing to circuit delays, an incorrect input to the next stage may be present when evaluation begins, so that its output is inadvertently discharged and the wrong output results.

One remedy is to employ a four-phase clock in which the actual signals used are the derived clocks ϕ_{12}, ϕ_{23}, ϕ_{34}, and ϕ_{41}, as illustrated in Figure 6–11(b).

The basic circuit of Figure 6–11(a) is modified by the inclusion of a transmission gate as in Figure 6–11(c), the function of which is to sample the output during the 'evaluate' period and to hold the output state while the next stage logic evaluates. For this strategy to work, the next stage must operate on overlapping but later clock signals. Clearly, since there are four different derived clock signals which are used in sequential pairs (e.g. ϕ_{12} and ϕ_{23} in Figure 6–11(c)), there are four different gate clocking configurations. These configurations are usually identified by a type number which reflects the last of the clock periods activating the gate. For example, the gate shown would be identified as 'type 3' since the output Z is precharged during ϕ_2 and is evaluated during ϕ_3 (the transmission gate is clocked by ϕ_{23}). In order to avoid erroneous evaluations, the gates must be connected in allowable sequences as set out in Table 6–1.

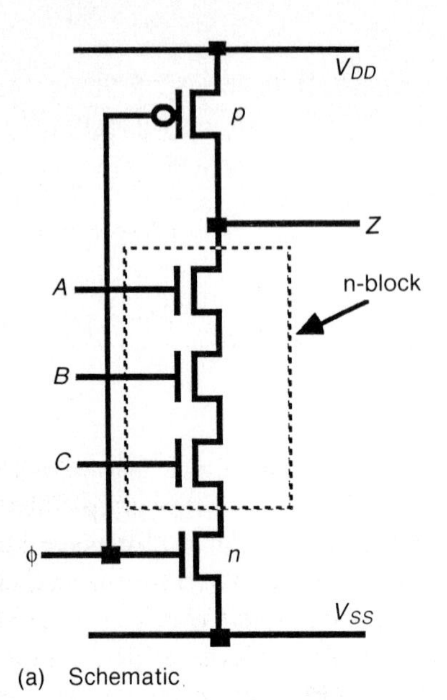

(a) Schematic

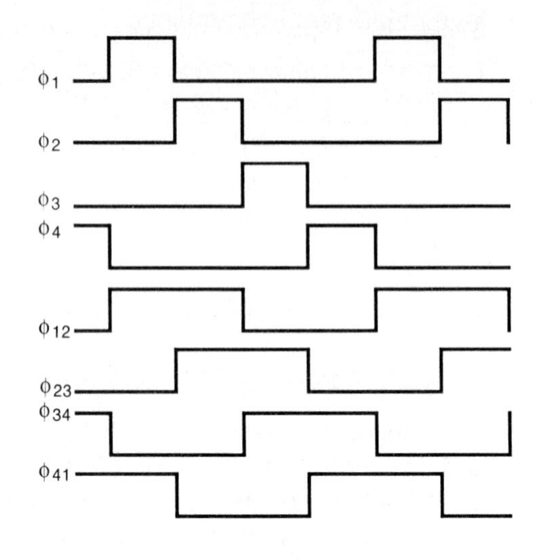

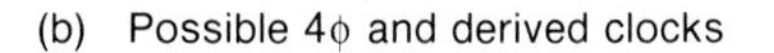
(b) Possible 4ϕ and derived clocks

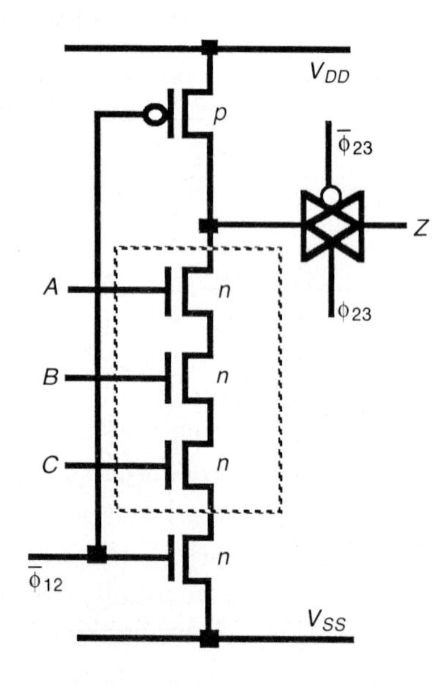

(c) Type 3 arrangement

Figure 6–11 Dynamic CMOS logic three-input *Nand* gate

Table 6–1 Dynamic logic types and sequences

Gate type	*Evaluate clock*	*Transmission gate clock*	*Allowable next types*
Type 1	$\bar{\phi}_{34}$	ϕ_{41}	Types 2 or 3
Type 2	$\bar{\phi}_{41}$	ϕ_{12}	Types 3 or 4
Type 3	$\bar{\phi}_{12}$	ϕ_{23}	Types 4 or 1
Type 4	$\bar{\phi}_{23}$	ϕ_{34}	Types 1 or 2

6.3.4.3 Clocked CMOS (C²MOS) logic

The general arrangement may be made clearer by Figure 6–12. The logic is implemented in both n- and p-transistors in the form of a pull-up p-block and a complementary n-block pull-down structure (Figure 6–12(a)), as for the inverter-based CMOS logic discussed earlier. However, the logic in this case is evaluated (connected to the output) only during the on period of the clock. As might be expected, a clocked inverter circuit forms part of this family of logic as shown in Figure 6–12(b). Owing to the extra transistors in series with the output, slower rise-times and fall-times can be expected.

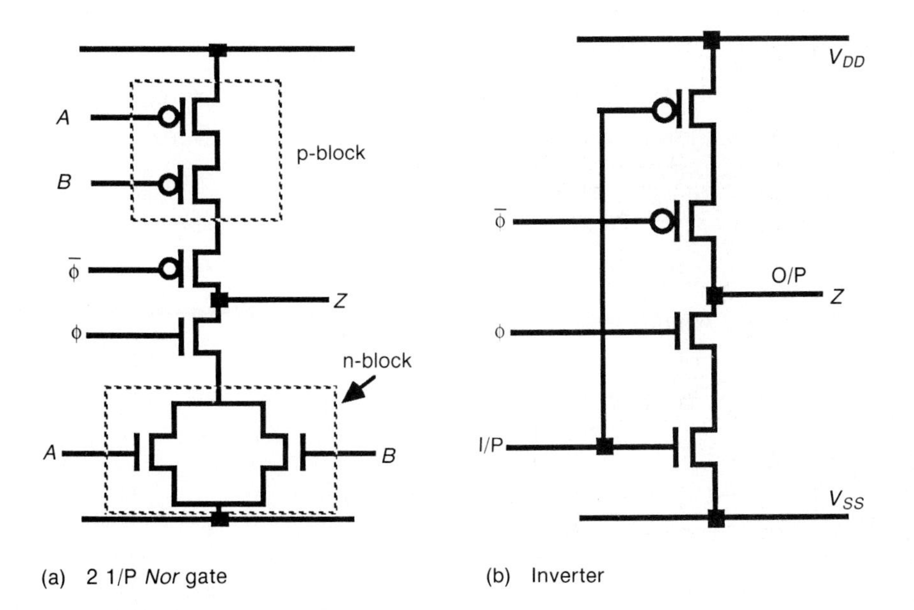

Figure 6–12 Clocked CMOS (C²MOS) logic

6.3.4.4 CMOS domino logic

An extension to the dynamic CMOS logic discussed earlier is set out in Figure 6–13. This modified arrangement allows for the cascading of logic structures

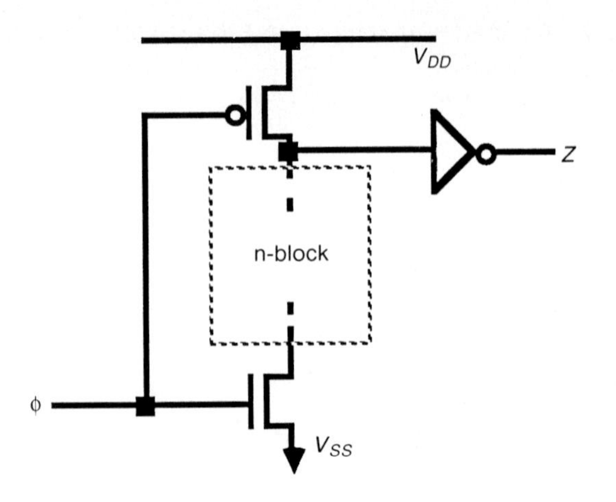

Figure 6–13 CMOS domino logic

using only a single phase clock. This requires a static CMOS buffer in each logic gate.

The following remarks will help to place this type of logic in the scheme of things:

1. Such logic structures can have smaller areas than conventional CMOS logic.
2. Parasitic capacitances are smaller so that higher operating speeds are possible.
3. Operation is free of glitches since each gate can make only one '1' to '0' transition.

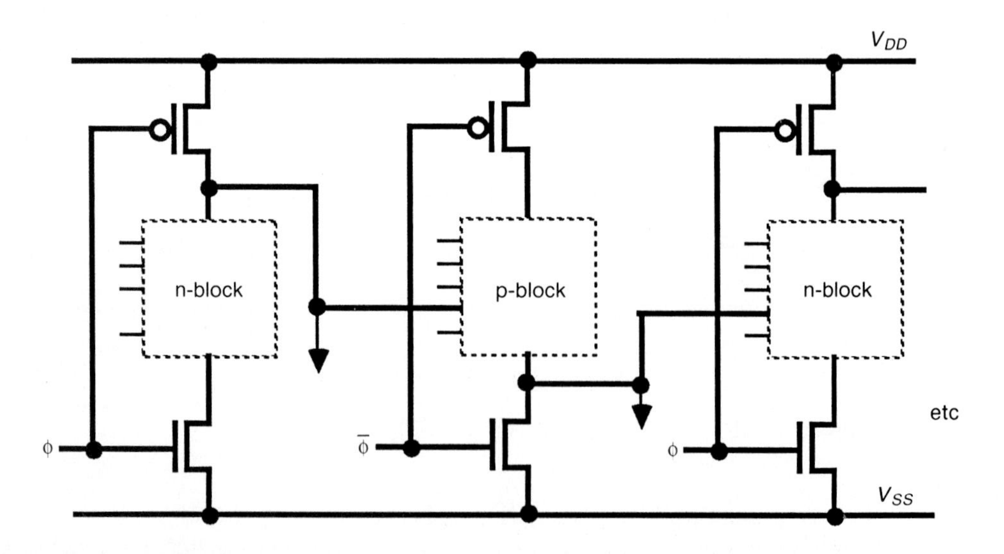

Figure 6–14 n-p CMOS logic

4. Only non-inverting structures are possible because of the presen inverting buffer.
5. Charge distribution may be a problem and must be considered.

6.3.4.5 n-p CMOS logic

This is another variation of basic dynamic logic arrangement, in which the actual logic blocks are alternately 'n' and 'p' in a cascaded structure as in Figure 6–14. The precharge and evaluate transistors are fed from the clock ϕ and clockbar $\bar{\phi}$ alternately, and clearly the functions of the top and bottom transistors also alternate, between precharge and evaluate.

Other forms of CMOS logic are also possible, but this text does not attempt to give an exhaustive treatment.

6.4 Examples of structured design (combinational logic)

The best way to illustrate the nature of and approach to structured design is to work through some examples.

6.4.1 A parity generator

A circuit is to be designed to indicate the parity of a binary number or word. The requirement is indicated in Figure 6–15 for an $(n + 1)$-bit input.

Since the number of bits is undefined, we must find a general solution on a cascadable bit-wise basis so that n can have any value. A suitably regular structure is set out in Figure 6–16. From this, we may recognize a standard or basic one-bit cell from which an n-bit parity generator may be formed. Such a cell is shown in Figure 6–17.

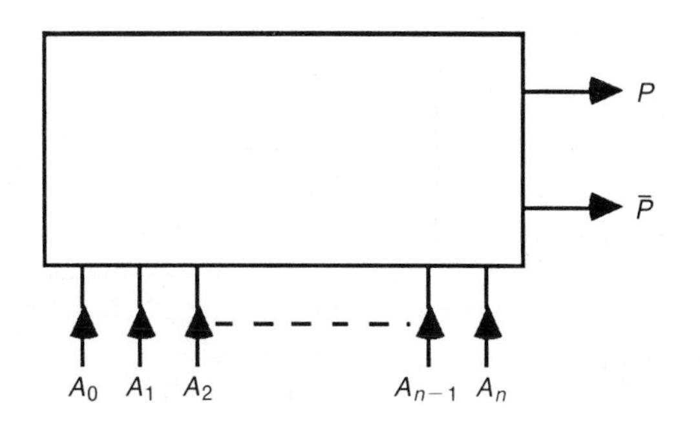

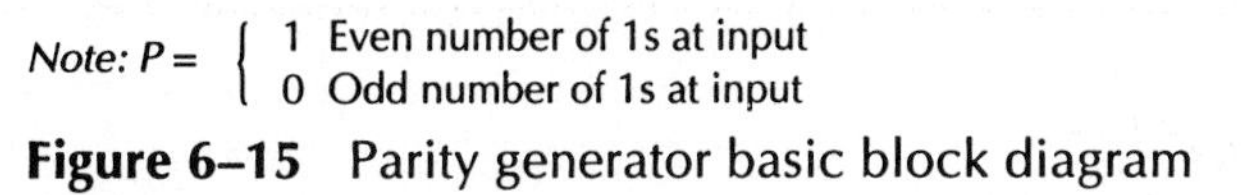

Figure 6–15 Parity generator basic block diagram

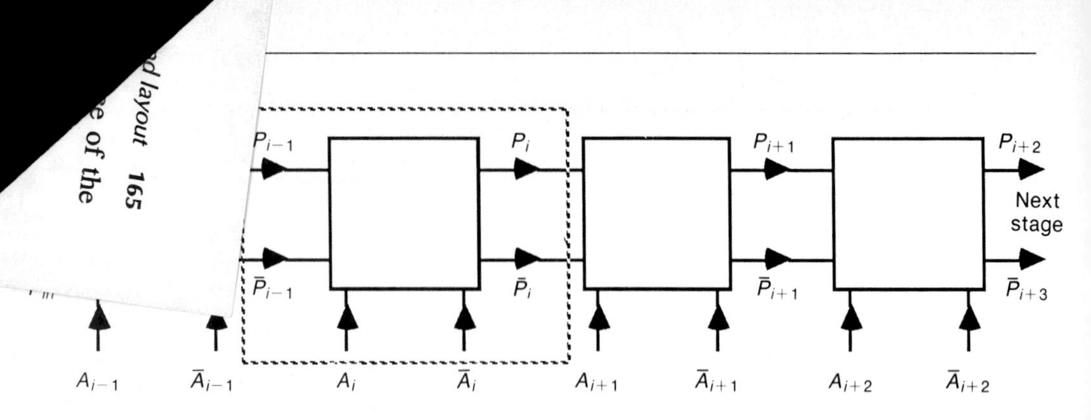

Note: Parity requirements are set at the left-most cell where $P_{in} = 1$ sets even and $P_{in} = 0$ sets odd parity.

Figure 6–16 Parity generator — structured design approach

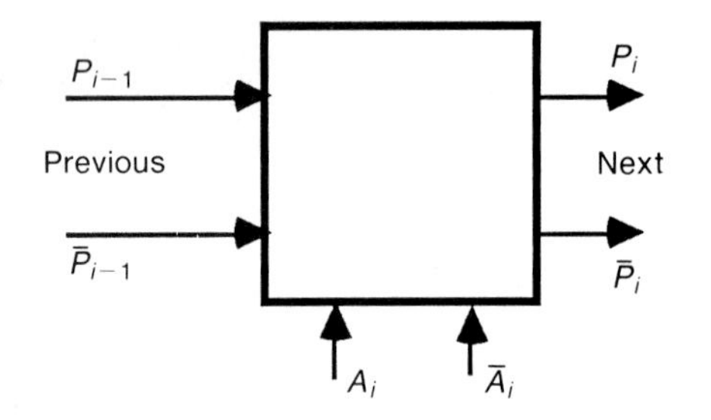

Figure 6–17 Parity generator — basic one-bit cell

It will be seen that parity information is passed from one cell to the next and is modified or not by a cell, depending on the state of the input lines A_i and $\bar{A}_i$.

A little reflection will readily reveal that the requirements are

$$A_i = 1 \text{ parity is changed, } P_i = \bar{P}_{i-1}$$
$$A_i = 0 \text{ parity is unchanged, } P_i = P_{i-1}$$

A suitable arrangement for such a cell is given in stick diagram form in Figure 6–18(a) (nMOS) and 6–18(b) (CMOS). The circuit implements the function

$$P_i = \bar{P}_{i-1} \cdot A_i + P_{i-1} \cdot \bar{A}_i$$

Note that a cell boundary may be chosen in each case so that cells may be cascaded at will.

When converting stick diagrams to layouts, care must be taken that the boundary is set so that *no design rule violations occur when cells are butted together.* Obviously, the boundary must also be chosen so that wastage of area is avoided and, where possible, so that *design rule errors are not present when a cell is checked in isolation,* although this may not always be possible.

Also, note that inlet and corresponding outlet points should match up both in *layer* and *position,* so that direct interconnection between cells is achieved when cells are butted.

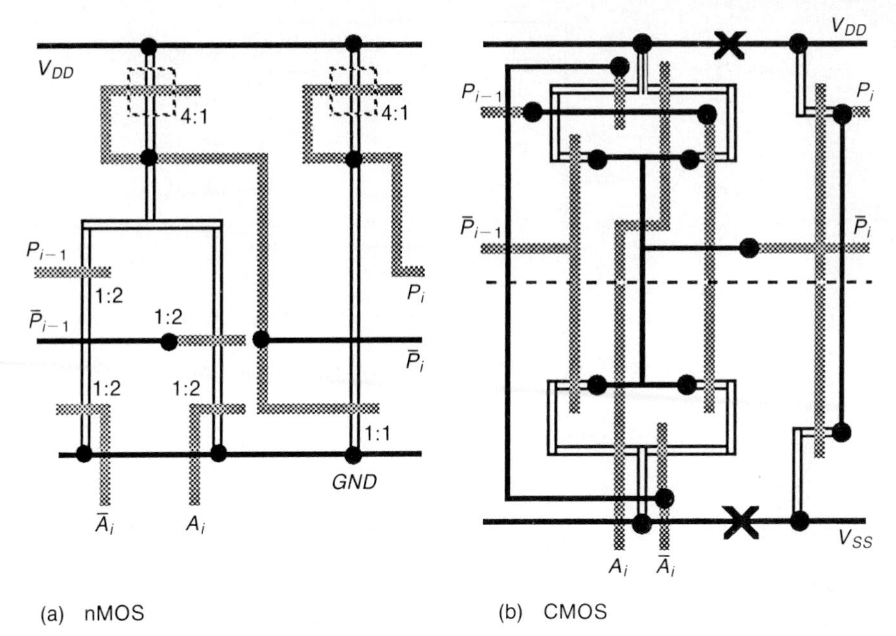

(a) nMOS (b) CMOS

Figure 6–18 Stick diagrams (parity generator)

6.4.2 Bus arbitration logic for n-line bus

(This example and its solutions are similar to an example accredited to Professor John Newkirk in VTI course material.)

The functional requirements of this circuit are given by Figure 6–19 and associated truth table. If the highest priority line A_n is Hi (Logic 1), then output line A_n^p will be Hi and all other output lines Lo (Logic 0), irrespective of the state of the other input lines $A_1 --- A_{n-1}$. Similarly, A_{n-1}^p will be Hi only when A_{n-1} is Hi and A_n is Lo; again the state of all input lines of lower priority $(A_1 ---- A_{n-2})$ will have no effect and all other output lines will be Lo.

This requirement can be expressed algebraically as follows:

$$A_n^p = A_n$$

$$A_{n-1}^p = \overline{A}_n . A_{n-1} \qquad [\overline{A}_{n-1}^p = A_n + \overline{A}_{n-1}]$$

$$A_{n-2}^p = \overline{A}_n . \overline{A}_{n-1} . A_{n-2} \qquad [\overline{A}_{n-2}^p = A_n + A_{n-1} + \overline{A}_{n-2}]$$

$$\vdots \qquad \vdots$$

$$A_n^p = \overline{A}_n . \overline{A}_{n-1} . \overline{A}_{n-2} ---------- \overline{A}_3 . \overline{A}_2 . A_1 \qquad \text{(etc.)}$$

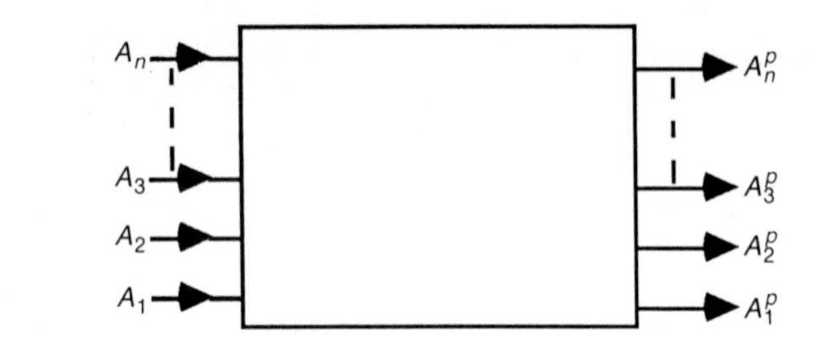

Truth table

A_n	.....	A_3	A_2	A_1	A_n^p	.....	A_3^p	A_2^p	A_1^p
0	.....	0	0	0	0	.....	0	0	0
0		0	0	1	0		0	0	1
0		0	1	X	0		0	1	0
0		1	X	X	0		1	0	0
.		.	.	.	.				.
.		.	.	.	.				.
.		.	.	.	.				.
.		.	.	.	.				.
1	.....	X	X	X	1	.....	0	0	0

*X = Don't care

Figure 6–19 Bus arbitration logic and truth table

A direct but unstructured implementation of these expressions may be readily envisaged and a suitable arrangement of switch (pass transistor) logic is given in Figure 6–20.

This implementation seems the obvious one, but it does suffer from the fact that as the input line under consideration moves down in significance so the complexity of the logic grows. For example, we have shown only the top three lines in Figure 6–20, but it will be seen that:

A_n requires one diffusion path and no switches
A_{n-1} requires two diffusion paths and two switches
A_{n-2} requires three diffusion paths and four switches

and so on.

This is not a regular structure and is not well suited for VLSI implementation. Therefore, we must take a cellular approach by setting out the requirements in alternative fashion as in Figure 6–21.

A regular structure having been arrived at, the requirements for each cell may be expressed as follows:

$$A_i^p \begin{cases} = g_{i+1} \text{ if } A_i = 1 \\ \text{or } 0 \text{ otherwise} \end{cases}$$

$$g_i \begin{cases} = 0 \quad \text{if } A_i = 1 \\ \text{or } g_{i+1} \text{ otherwise} \end{cases}$$

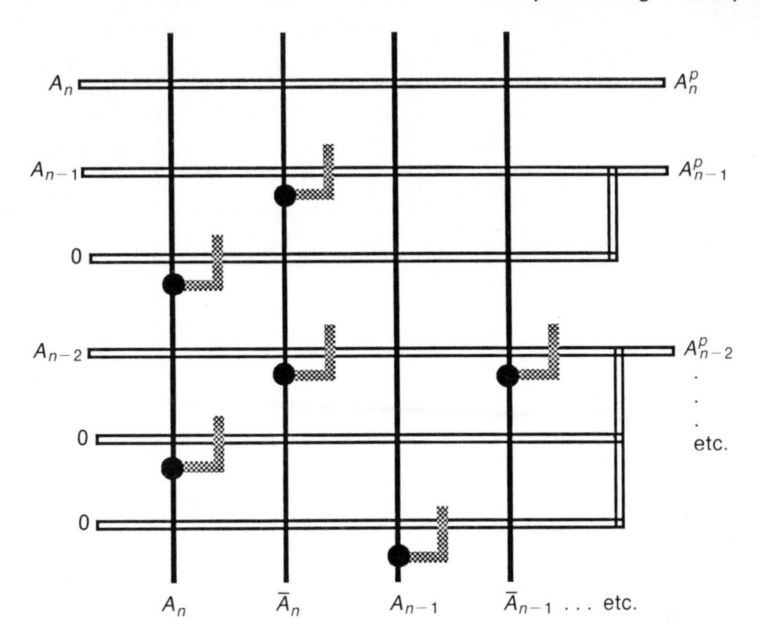

Figure 6–20 Stick diagram — bus arbitration logic

These requirements may be met by the circuit of Figure 6–22, but care must be taken not to cascade more than four cells without buffering the grant line.

The art of arriving at conveniently expressed relationships which allow a sructured design is one which must be cultivated and it is often helped by adopting an 'if, then, else (or otherwise)' approach. The solution to the problem under consideration can be formulated after expressing the need of each cell in words:

If $A_i = 1$ then $A_i^p = g_{i+1}$,
else $A_i^p = 0$ (if $A_i = 0$)
If $A_i = 0$ then $g_i = g_{i+1}$
else $g_i = 0$ (if $A_i = 1$)

} both A_i^p and g_i can be derived from g_{i+1}

From which we could deduce

$$A_i^p = A_i . g_{i+1}$$

$$g_i = \overline{A}_i . g_{i+1}$$

However, there is a danger with expressions of the conventional Boolean type — a tendency to ignore the fact that MOS *switch* logic is such that not only must the logic 1 condition be satisfied, but it is also necessary to deliberately *satisfy* the logic 0 conditions. The TTL logic designer is used to working with logic circuits in which the output must be logic 0 if the logic 1 output conditions are not satisfied. However, some MOS switch-based logic circuits have the property

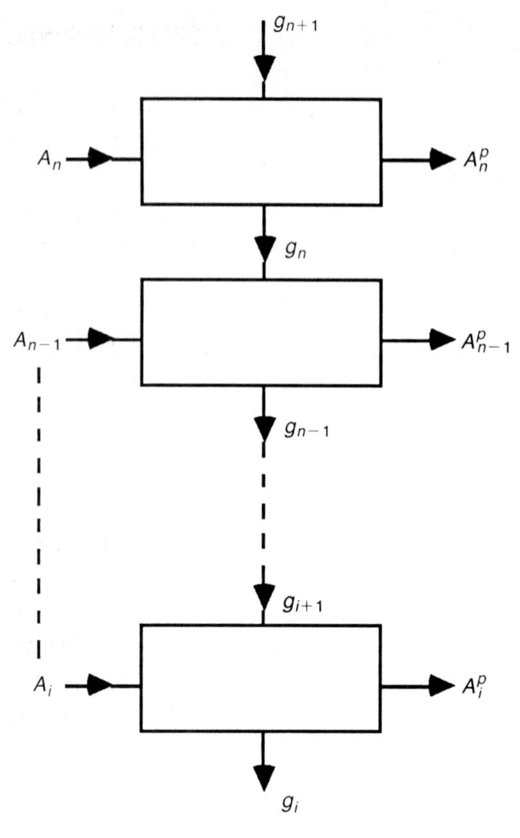

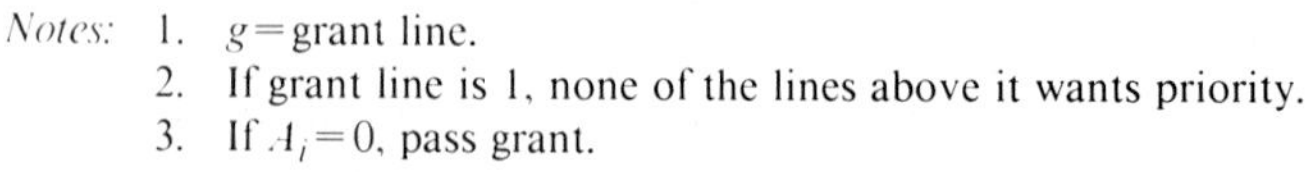
Notes: 1. g = grant line.
2. If grant line is 1, none of the lines above it wants priority.
3. If $A_i = 0$, pass grant.

Figure 6–21 Bus arbitration logic — structured design

that if the logic 1 output conditions are not met, then the output can be indeterminate or, if some storage capacitance is present (for example, input capacitance C_g of an inverter), then the output can remain at logic 1 even after the conditions which caused it no longer exist. Thus, it is necessary to deliberately implement the 'else' conditions. We must, therefore, write expressions for both the logic 1 and logic 0 conditions of the output lines, thus

$$A_i^p = A_i . g_{i+1}; \ \overline{A}_i^p = \overline{A}_i + \overline{g}_{i+1}$$

$$g_i = \overline{A}_i g_{i+1}; \ \overline{g}_i = A_i + \overline{g}_{i+1}$$

which is the circuit realized in Figure 6–22. This circuit is suitable for implementation in nMOS or in CMOS technology. Although in the CMOS case there is the possibility of replacing the n-type pass transistors by transmission gates, there is no advantage

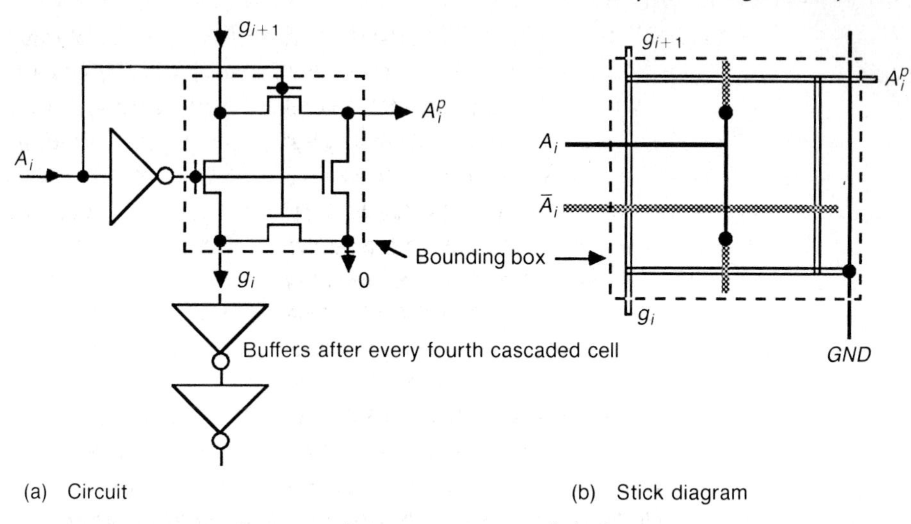

Figure 6–22 Bus arbitration logic — structured design

to be gained from this as the degrading of logic 1 level is counteracted by the presence of buffers after every fourth cell. There is clearly an area advantage in using simple n-type pass transistors and the only difference, therefore, between an nMOS and a CMOS design will be the type of buffer (inverter) stages.

6.4.3 Multiplexers (data selectors)

Multiplexers are widely used and have many applications. They are also commonly available in a number of standard configurations in TTL and other logic families. In order to arrive at a standard cell for multiplexers, we will consider a commonly used circuit, the four-way multiplexer.

The requirements and general arrangement of a four-way multiplexer are set out in Figure 6–23, from which we may write

$$Z = I_0.\overline{S}_1.\overline{S}_0 + I_1.\overline{S}_1.S_0 + I_2.S_1.\overline{S}_0 + I_3.S_1.S_0$$

where S_1 and S_0 are the selector inputs. Note that in this case we do not need to be concerned about undefined ouput conditions since, if S_1 and S_0 have defined logic states, output Z must always be connected to one of I_0 to I_3.

Thus, a direct n-switch logic implementation follows which is given as Figure 6–24(a) in stick diagram form with a standard-cell-based mask layout following as Figure 6–25 and in Color plate 10.

A transmission-gate-based CMOS stick diagram is given in Figure 6–24(b). A mask layout of this figure appears as Color plate 11 and it can be seen that all n-transistors are placed below the demarcation line and close to the V_{SS} rail to allow ready configuration of the p-well and V_{SS} contacts. The p-transistors are

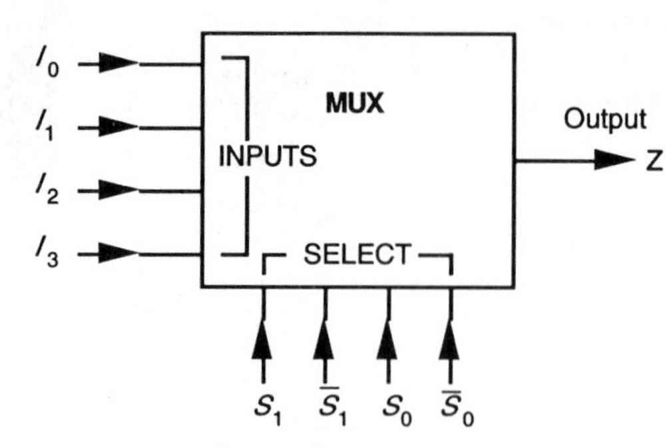

Truth table

S_1	S_0	Z
0	0	I_0
0	1	I_1
1	0	I_2
1	1	I_3

Figure 6–23 Selector logic circuit

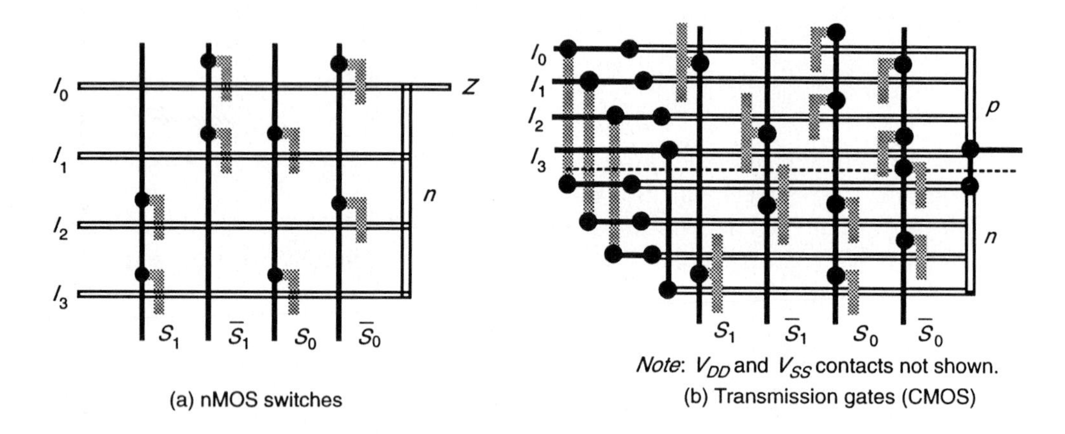

Figure 6–24 Switch logic implementations of a four-way multiplexer

similarly placed above the notional demarcation line and close to V_{DD}. Note that logic 1 levels will not be degraded by this arrangement as those in the nMOS version are.

Now, if we can establish standard cells from which a four-way multiplexer can be composed, then we will also cover the case of the two-way multiplexer. Such a cell will, by inference, also be suitable for constructing an 8-way or a 16-way multiplexer.

For the nMOS case a standard cell is illustrated in Figure 6–25. The standard cell in this case measured $7\lambda \times 11\lambda$ and is shown in the dotted outline. Note that two versions of the cell are needed to complete the network, one version with a pass transistor as shown and the other version without. If computer-aided design tools are used, the two versions may be designed as one cell suitably parameterized to include or exclude the pass transistor. Note that in Figure 6–25 the dimensions do not include the end connection to Z.

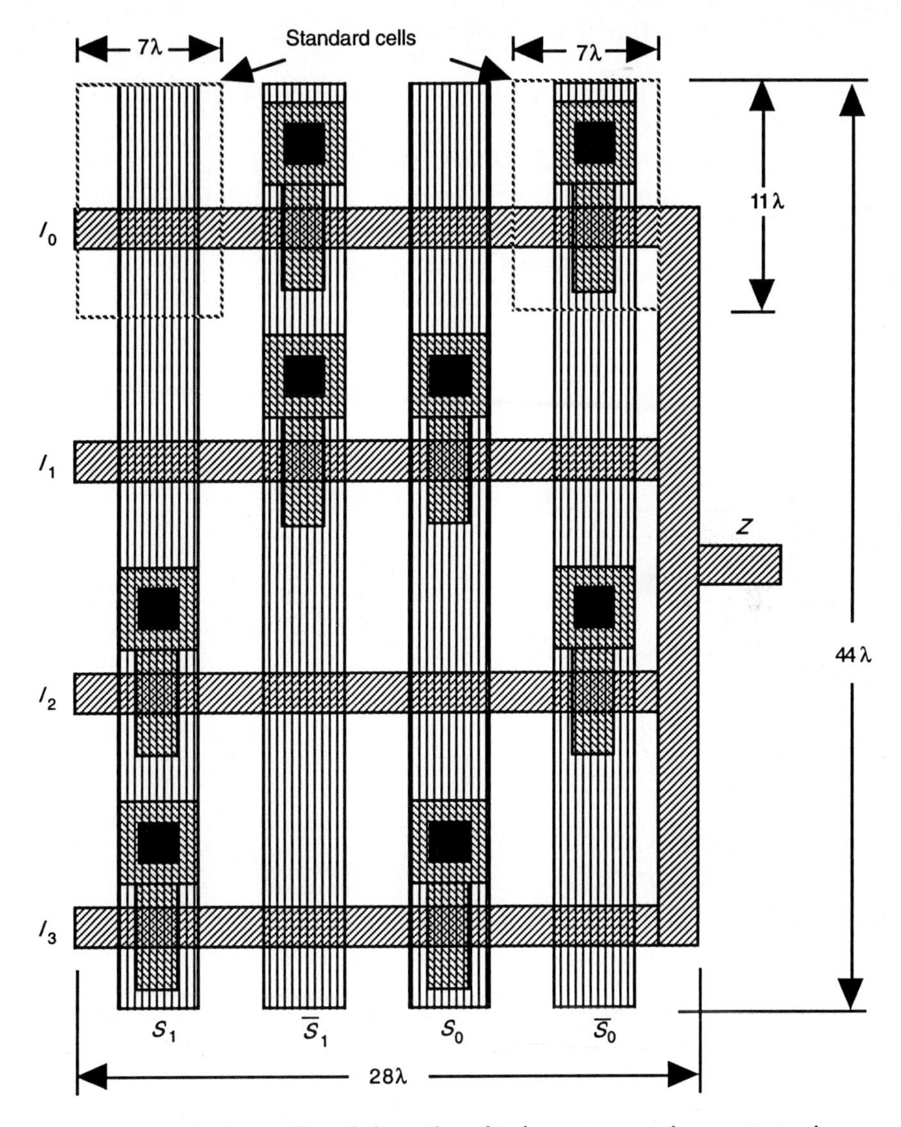

Figure 6–25 Four-way n-switch based multiplexer (MUX) layout (see also Color plate 10)

Note also that this layout places the metal select lines over the top of pass transistors. This practice is acceptable in this situation where a transistor gate is actually driven from and connected to the particular metal line which runs across it. This method of economizing in area must be used with caution when locating transistors under metal layers to which they are not connected, and may not be acceptable when the underlying transistors are used as storage points to hold a charge and retain a logic level.

6.4.4 A general logic function block

An arrangement to generate any function of two variables (A, B) is readily formed from any form of four-way multiplexer.

The general approach is indicated in Figure 6–26. It will be seen that the required function is generated by driving the multiplexer select inputs from the required two variables A and B and by 'programming' the inputs I_0–I_3 appropriately with 0s and 1s, as indicated in the figure. Larger multiplexers may be similarly employed to generate any function of up to four variables (16-way multiplexer).

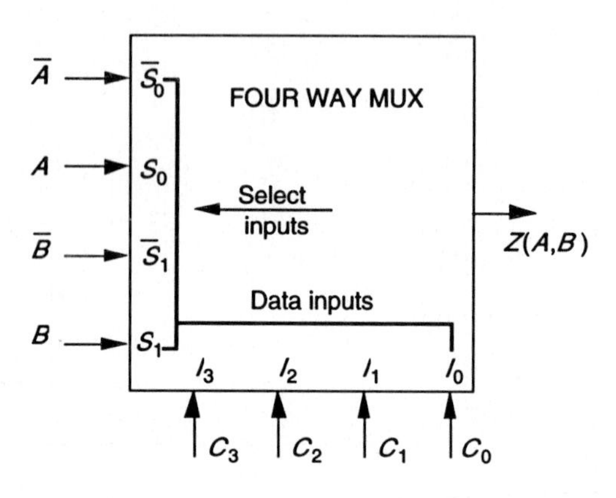

INPUT PROGRAMMING				FUNCTION $Z(A,B)$	
C_3	C_2	C_1	C_0		
0	0	0	0	0	$Z = 0$
0	0	0	1	$\bar{A}.\bar{B}$; $\overline{A+B}$	*Nor*
0	0	1	0	$A.\bar{B}$	
0	0	1	1	$\bar{B}$	Not B
0	1	0	0	$\bar{A}.B$	
0	1	0	1	$\bar{A}$	Not A
0	1	1	0	$A.\bar{B}+\bar{A}.B$	*Exclusive-Or*
0	1	1	1	$\bar{A}+\bar{B}$; $\overline{AB}$	*Nand*
1	0	0	0	$A.B$	*And*
1	0	0	1	$\bar{A}.\bar{B}+A.B$	Comparator
1	0	1	0	A	$O/P = A$
1	0	1	1	$A+\bar{B}$	
1	1	0	0	B	$O/P = B$
1	1	0	1	$\bar{A}+B$	
1	1	1	0	$A+B$	*Or*
1	1	1	1	1	$Z = 1$

Figure 6–26 General logic function block (two variables)

6.4.5 A four-line Gray code to binary code converter

As a further exercise, which employs a very widely used logic arrangement (the *Exclusive-Or* gate), consider the requirement for code conversion from Gray to binary as set out in Table 6–2.

By inspecting (or mapping from) Table 6–2, it will be seen that the following expressions relate the two codes:

$$\left.\begin{aligned} A_0 &= \overline{G}_0.A_1 + G_0.\overline{A}_1 \\ A_1 &= \overline{G}_1.A_2 + G_1.\overline{A}_2 \\ A_2 &= \overline{G}_2.A_3 + G_2.\overline{A}_3 \end{aligned}\right\} \text{ \textit{Exclusive-Or} operations}$$

$$A_3 = G_3$$

A suitable arrangement is set out in Figure 6–27, and the only detailed design required is that of a two input *Exclusive-Or* gate. Many arrangements are possible to implement this operation, but let us consider an *Exclusive-Or* gate made up of standard logic gates, as in Figure 6–28.

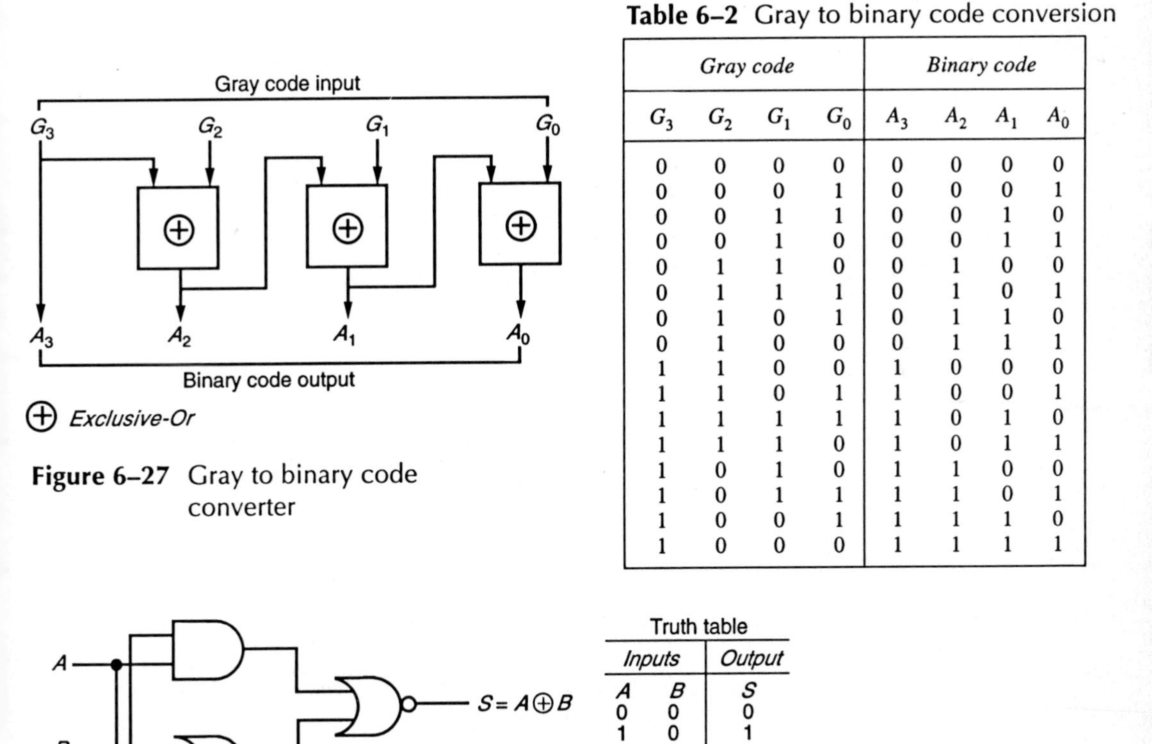

Figure 6–27 Gray to binary code converter

Table 6–2 Gray to binary code conversion

Gray code				*Binary code*			
G_3	G_2	G_1	G_0	A_3	A_2	A_1	A_0
0	0	0	0	0	0	0	0
0	0	0	1	0	0	0	1
0	0	1	1	0	0	1	0
0	0	1	0	0	0	1	1
0	1	1	0	0	1	0	0
0	1	1	1	0	1	0	1
0	1	0	1	0	1	1	0
0	1	0	0	0	1	1	1
1	1	0	0	1	0	0	0
1	1	0	1	1	0	0	1
1	1	1	1	1	0	1	0
1	1	1	0	1	0	1	1
1	0	1	0	1	1	0	0
1	0	1	1	1	1	0	1
1	0	0	1	1	1	1	0
1	0	0	0	1	1	1	1

Truth table

Inputs		*Output*
A	*B*	*S*
0	0	0
1	0	1
0	1	1
1	1	0

Figure 6–28 One possible arrangement for an *Exclusive-Or* gate

A mask layout for this arrangement is presented in Figure 6–29; note that the p-well and p^+ mask outlines are included in the layout together with the p-well and substrate contacts. Simulation of the design yields the results given in Figure 6–30. The simulator used was the ISD* program PROBE™ and the circuit extractor NET™. Likely circuit delays can be seen quite plainly.

6.4.6 The programmable logic array (PLA)

This arrangement provides a general, structured and regular way of mapping multiple output combinational logic expressions onto silicon. See Appendix C.

6.5 Some clocked sequential circuits

6.5.1 Two-phase clocking

The clocked circuits to be considered here will be based on a two-phase non-overlapping clock signal as defined by Figure 6–31.

A two-phase clock offers a great deal of freedom in sequential circuit design if the clock period and the duration of the signals ϕ_1 and ϕ_2 are correctly chosen. If this is the case, data is allowed to become stable before any further transfer takes place and there is no chance of race conditions occurring.

Clocked circuitry is considerably easier to design than the corresponding asynchronous sequential circuitry. It does, however, usually pay the penalty of being slower. However, at this stage of learning VLSI design we will concentrate on two-phase clocked sequential circuits alone and thus simplify design procedures. When studying Figure 6–31, it is necessary to recognize the fact that ϕ_1 and ϕ_2 do not need to be symmetrical as shown. For a given clock period, each clock phase period and its associated underlap period can be varied if the need arises in optimizing a design.

A number of techniques are used to generate the two clock phases. One popular method is illustrated in Figure 6–32 and it will be seen that the output frequency is one-quarter of that of the input clock.

A very simple arrangement using combinational logic and generating a two-phase clock at the frequency of a single-phase input clock is set out in Figure 6–33(a). The input clock signal *C* is used to provide a delayed version of itself (*CD*) by passing it through an even number of inverters. The delay thus produced determines the underlap period for the two-phase clock. Waveforms are as shown in Figure 6–33(b). The phase 1 signal ϕ_1 (PH1) is generated by *Anding C* with *CD* whilst the phase 2 signal ϕ_2 (PH2) is produced by *Noring C* with *CD* (that is, *Anding C′* with *CD′*). Clearly, the minimum underlap period will be that generated

---

* ISD refers to Integrated Silicon Design Pty Ltd of Adelaide, South Australia. The use of their design tools is acknowledged with thanks and appreciation.

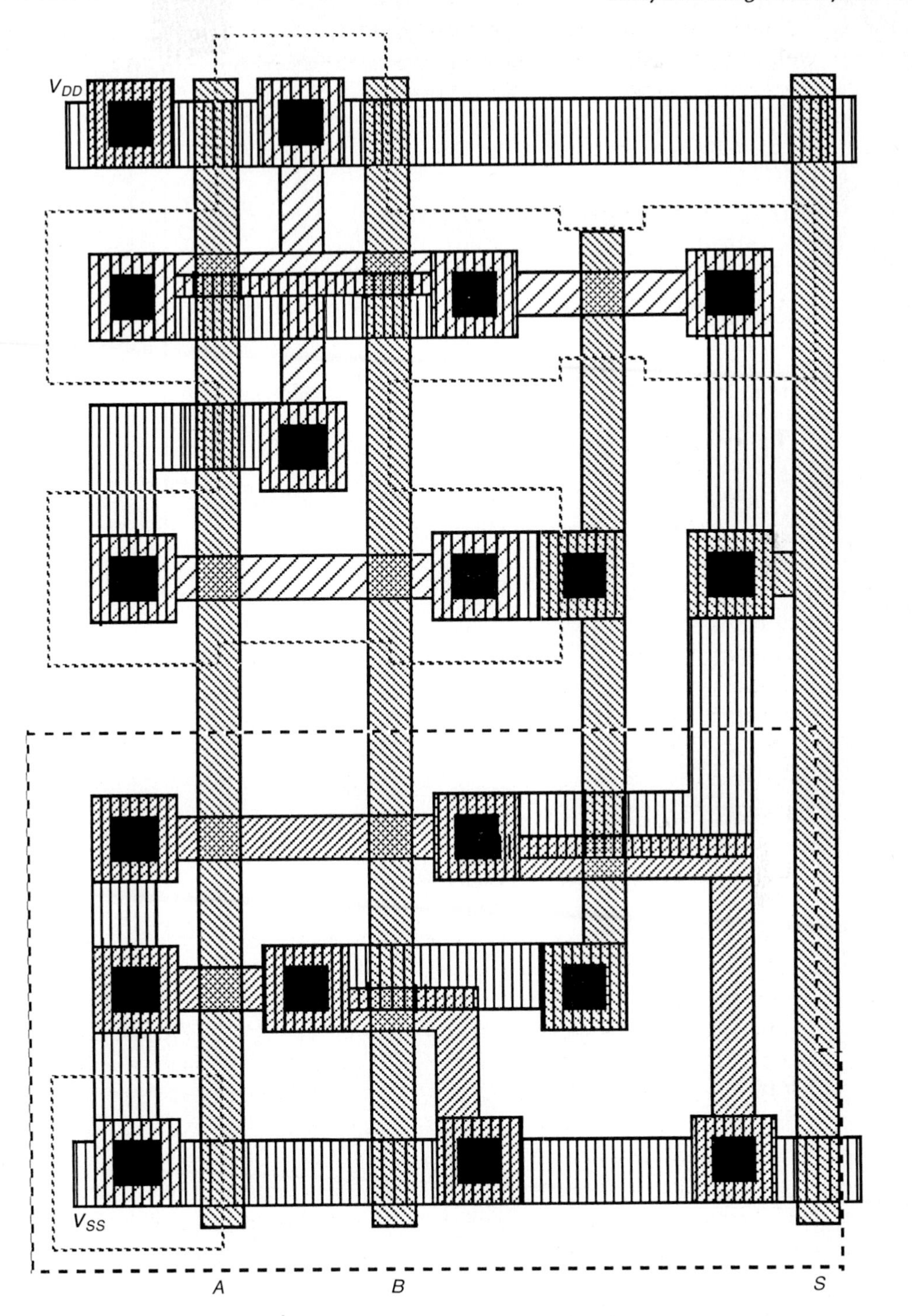

Figure 6–29 Mask layout for *Exclusive-Or* gate of Figure 6–28

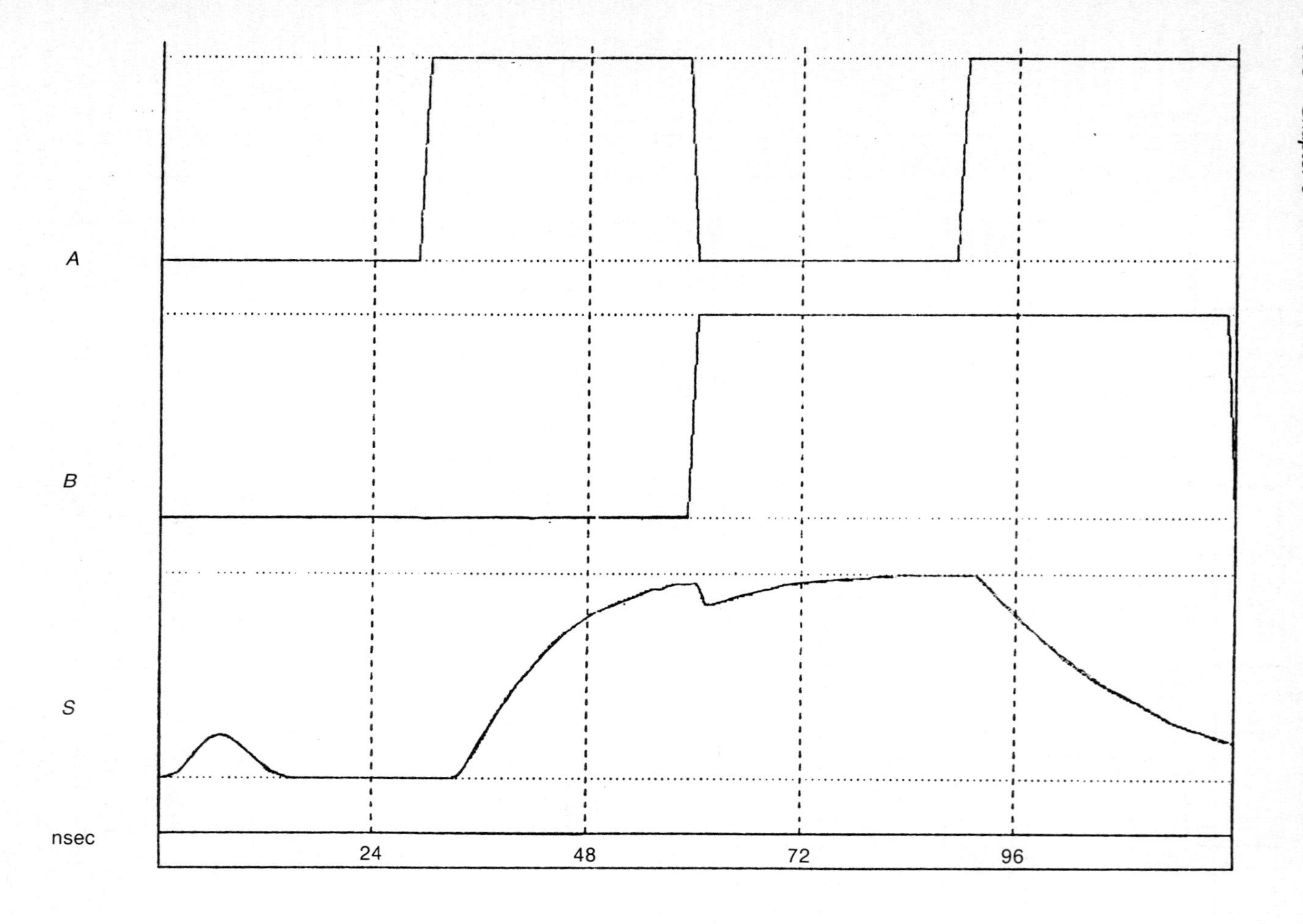

Figure 6–30 Simulation results for *Exclusive-Or* gate

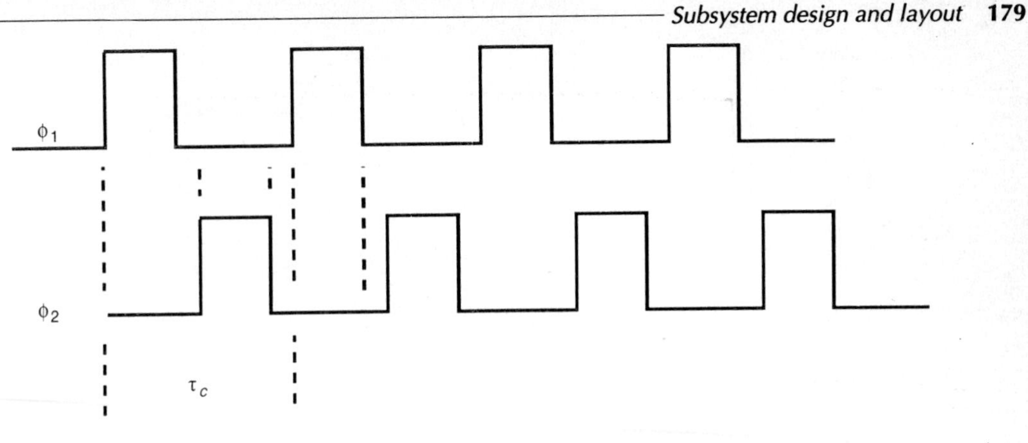

Notes: 1. τ_c = clock period
2. $\phi_1(t) . \phi_2(t) = 0$; *all* t

Figure 6–31 Two-phase clocking

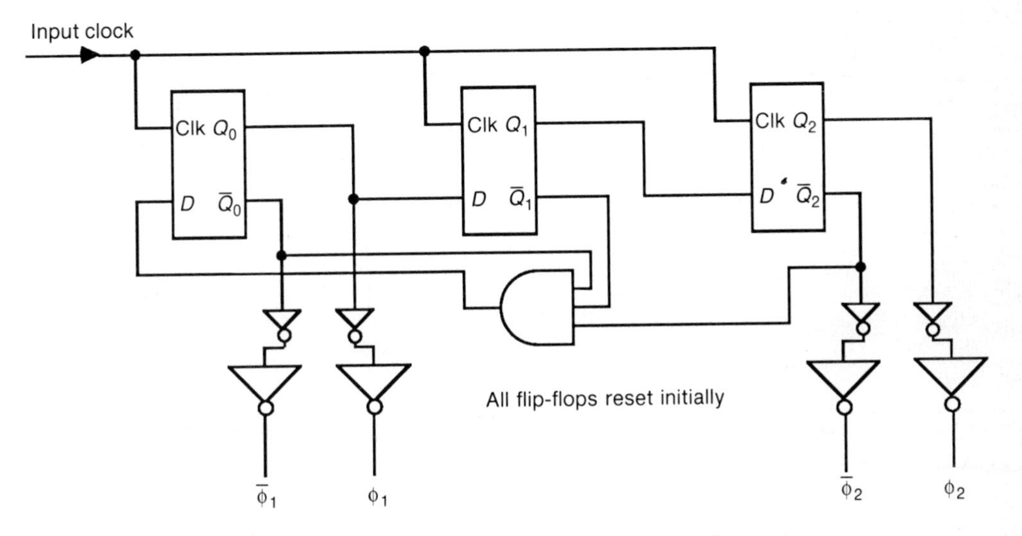

Figure 6–32 Two-phase clock generator using D flip-flops

by the delay through two inverters and this is also the increment by which the delay may be increased by adding further inverter pairs.

Since clock lines often feed many stages and are associated with long bus lines, they often present quite considerable capacitance to the clock line drivers. Here then is a case where a bipolar capability can be used to advantage to drive the high capacitance load. This approach is demonstrated in Figure 6–34, which uses bipolar-based output stages and also produces the complements of the two phases since complementary clocks are almost invariably required. Simulation waveforms are given in Figure 6–34(b) and a possible mask layout is presented as Color plate 12.

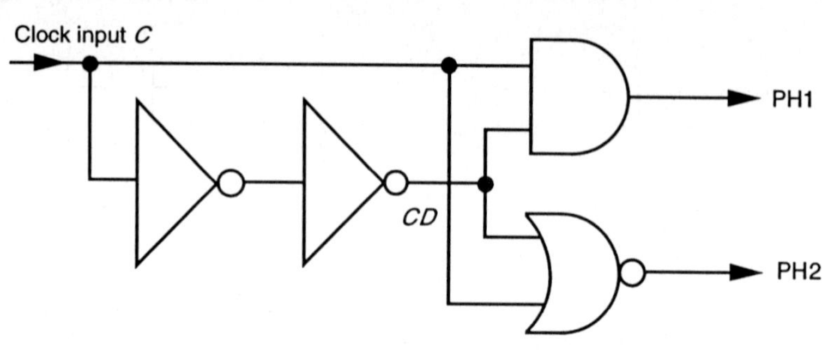

Figure 6–33(a) Simple two-phase clock generator circuit — basic form

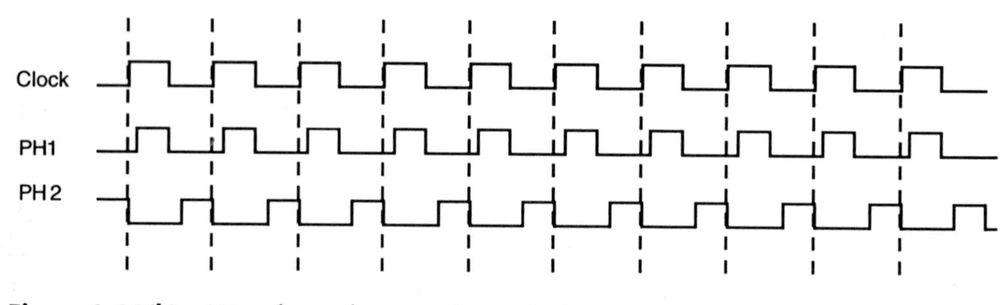

Figure 6–33(b) Waveforms for two-phase clock generator

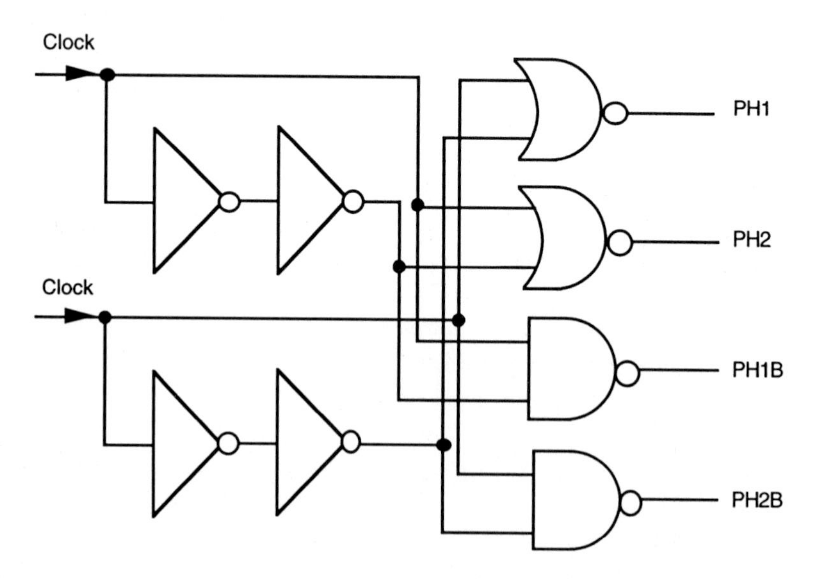

Figure 6–34(a) Two-phase clock generator (with complementary outputs) for BiCMOS logic implementation

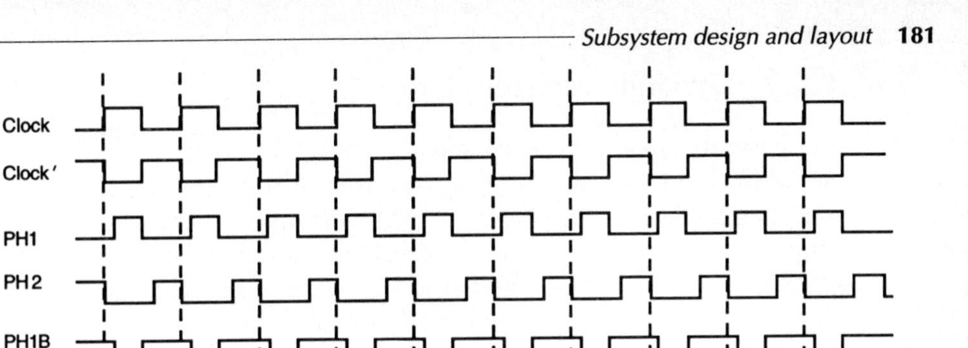

Figure 6–34(b) Waveforms for circuit of Figure 6–34(a)

6.5.2 Charge storage

A necessary feature of sequential circuits is a facility to remember or take account of previous conditions. An obvious area of application of such a facility is in memory elements, registers, finite state machines, etc.

MOS technology takes advantage of the excellent insulating properties of silicon dioxide layers on integrated circuits to store charges in capacitors, including the gate-channel capacitance of transistors. Such storage is known as *dynamic storage* since, in a reasonably short time, stored charges will leak away and will have to be refreshed if data/conditions are to be retained.

Considering charges on the gate capacitance, the leaking away of the charge is mainly due to leakage currents I_s across the channel to substrate reversed biased diode. At room temperature and for typical 5 μm dimensions and typical voltages, this current is in the order of 0.1 nA, and so an approximate idea of holding time can be obtained from the simple circuit model of Figure 6–35 which considers 1□Cg initially charged to 5 volts.

This simple model indicates storage times of up to, say, 0.25 msec to discharge from V_{DD} to V_{inv} (= 0.5 V_{DD}), but it should be noted that current I_s doubles for every 10°C rise in temperature so that the storage time is halved.

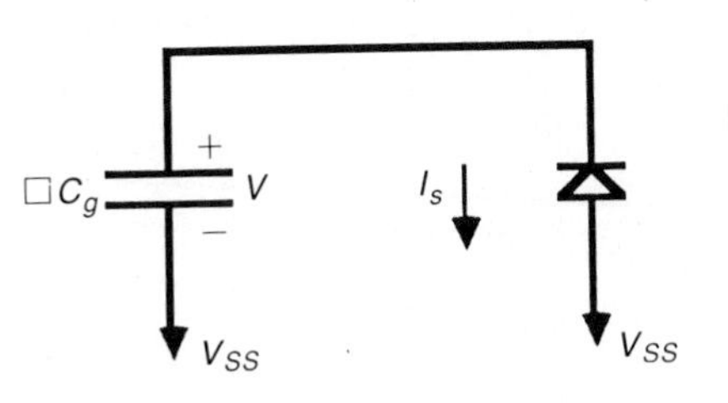

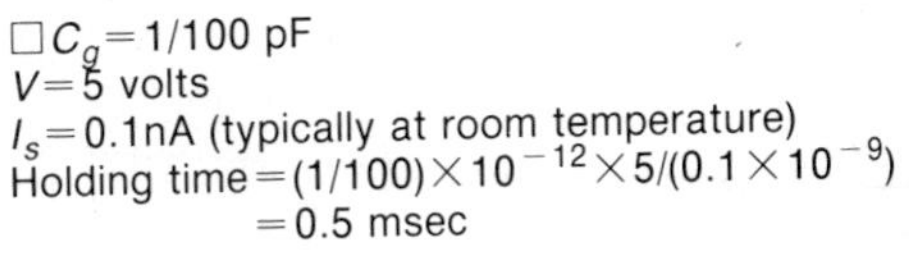

Figure 6–35 Simple stored charge model

6.5.3 Dynamic register element

The basic dynamic register element is shown in Figure 6–36 in mixed stick/circuit notation and may be seen to consist of three transistors for nMOS and four for CMOS per stored bit in complemented form. The element's operation is simple to appreciate. $(V_{in})_t$ is clocked in by ϕ_1 (or ϕ_2) of the clock and charges the gate capacitance C_g of the inverter to V_{in}. If subscript t is taken to represent the time during which ϕ_1 (say) is at logic 1 and subscript $t + 1$ is taken to indicate the period during which ϕ_1 is at logic 0, then the available output will be $(\overline{V}_{in})_{t+1}$ which will be maintained by the stored charge on the gate until C_g discharges or until the next ϕ_1 signal occurs.

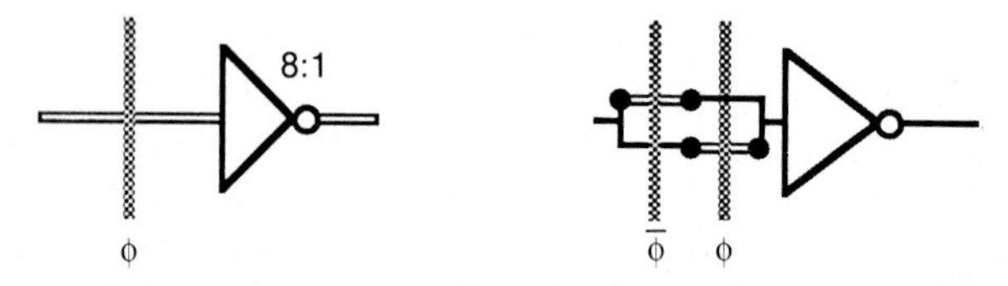

(a) nMOS pass transistor switched (b) CMOS transmission gate switched

Figure 6–36 Basic inverting dynamic storage cells

If uncomplemented storage is essential, the basic element is modified as indicated in Figure 6–37 and will be seen to consist of six transistors for nMOS and eight for CMOS. Data clocked in on ϕ_1 is stored on C_{g1} and the corresponding output appears at the output of inverter 1. On ϕ_2 this value is clocked into and stored by C_{g2} and the output of inverter 2 then presents the 'true' form of the stored bit. Note that data read in on ϕ_1 is not available at the output until sometime following the next positive edge of the clock signal ϕ_2.

Dynamic storage elements and the corresponding register arrays are used in situations where signals are updated frequently (i.e. at < 0.25 msec intervals).

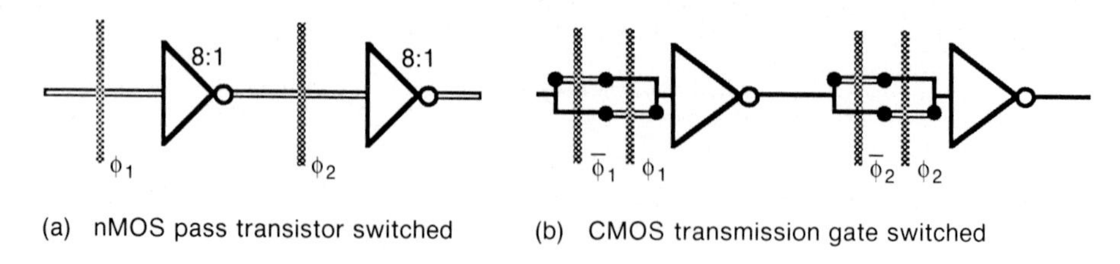

(a) nMOS pass transistor switched (b) CMOS transmission gate switched

Figure 6–37 Non-inverting dynamic storage cells

6.5.4 A dynamic shift register

Cascading the basic elements of Figure 6–37 gives a serial shift register arrangement which may be extended to n bits. A four-bit serial right shift nMOS register is illustrated in Figure 6–38(a). Data bits are shifted in when $\phi_1.LD$ is present, one bit being entered on each ϕ_1 signal (provided that LD is logic 1). Each bit is stored in C_{g1} as it is entered, and then transferred complemented into C_{g2} during the next ϕ_2. Thus, after a ϕ_1 followed by ϕ_2 signal, the stored bit is present at the output of inverter 2. On the next ϕ_1, the next input bit is stored in C_{g1} and simultaneously the first bit stored is passed on to inverter pair 3 and 4 by being stored in $C_{g3,}$ and so on. It will be seen that bits are thus clocked to the right along the shift register on each ϕ_1 followed by ϕ_2 sequence. Once four bits are stored, the data is available in parallel form at the outputs of inverters 2, 4, 6 and 8, and is also available in serial form from the output of inverter 8 when $\phi_1.RD$ is high as further clock sequences are received (where RD is the serial read control signal). The operation of the CMOS version (Figure 6–38(b)) is similar, transmission gates replacing inter-stage pass transistors and C_{in1} replacing C_{g1}, etc., as the storage capacitance.

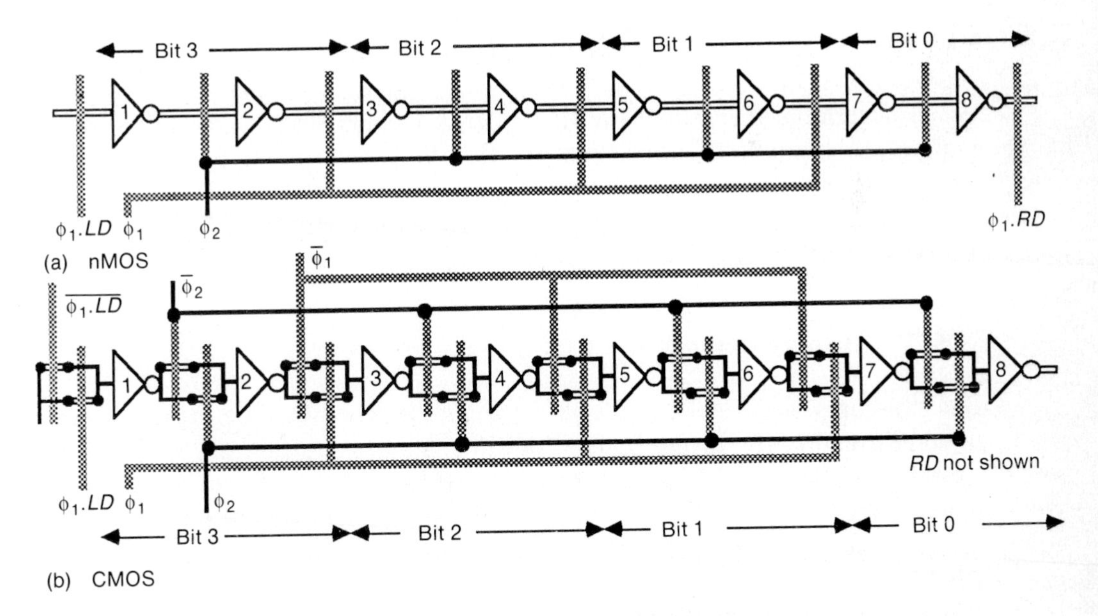

Figure 6–38 Four-bit dynamic shift registers (nMOS and CMOS)

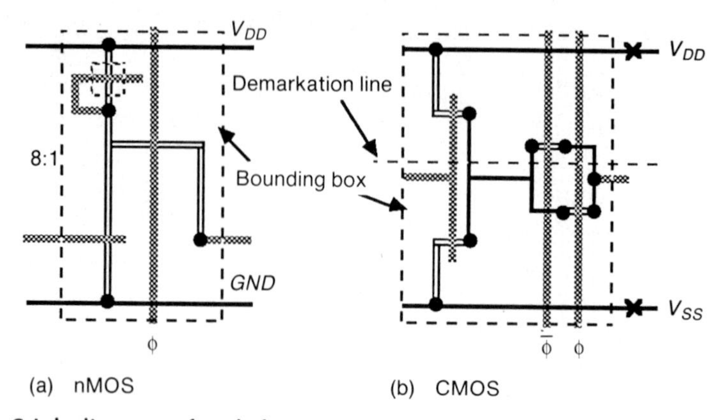

Figure 6–39 Stick diagrams for shift register cells

Many variations of this basic arrangement are possible, but in general they are all based on the basic cell consisting of an inverter and a pass transistor or a transmission gate. Suitable standard cells are shown in stick diagram form in Figure 6–39 with the corresponding mask layouts in Figure 6–40. Note that two nMOS layouts are given (using butting and buried contacts respectively) and one possible CMOS layout is suggested (see also Color plate 7).

6.6 Other system considerations

When designing at leaf-cell level, it is easy to lose sight of overall system requirements and restrictions. In particular, the use of buses to interconnect subsystems and circuits must always be most carefully considered; such matters and the current-carrying capacity of aluminum wiring used for V_{DD} and GND or V_{SS} rails are often overlooked completely.

6.6.1 Bipolar drivers for bus lines

Bus structures carrying data or control signals are generally long and connected to and through a significant number of circuits and subsystems. Thus, the bus capacitances are appreciable and thought must be given to the manner in which any bus line is to be driven. Otherwise, the propagation of signals may be a slow process. Clearly, the capacitive load-driving properties of bipolar transistors in a BiCMOS process make bipolar drivers an attractive proposition for bus lines. However, this must be approached with some caution as the speed of bipolar drivers is only fully realized with bus lines for which there is only one source of drive, for example, as in the case of clock line drivers. Bipolar drivers are not so suitable where one or other of several sources drives a common bus since under

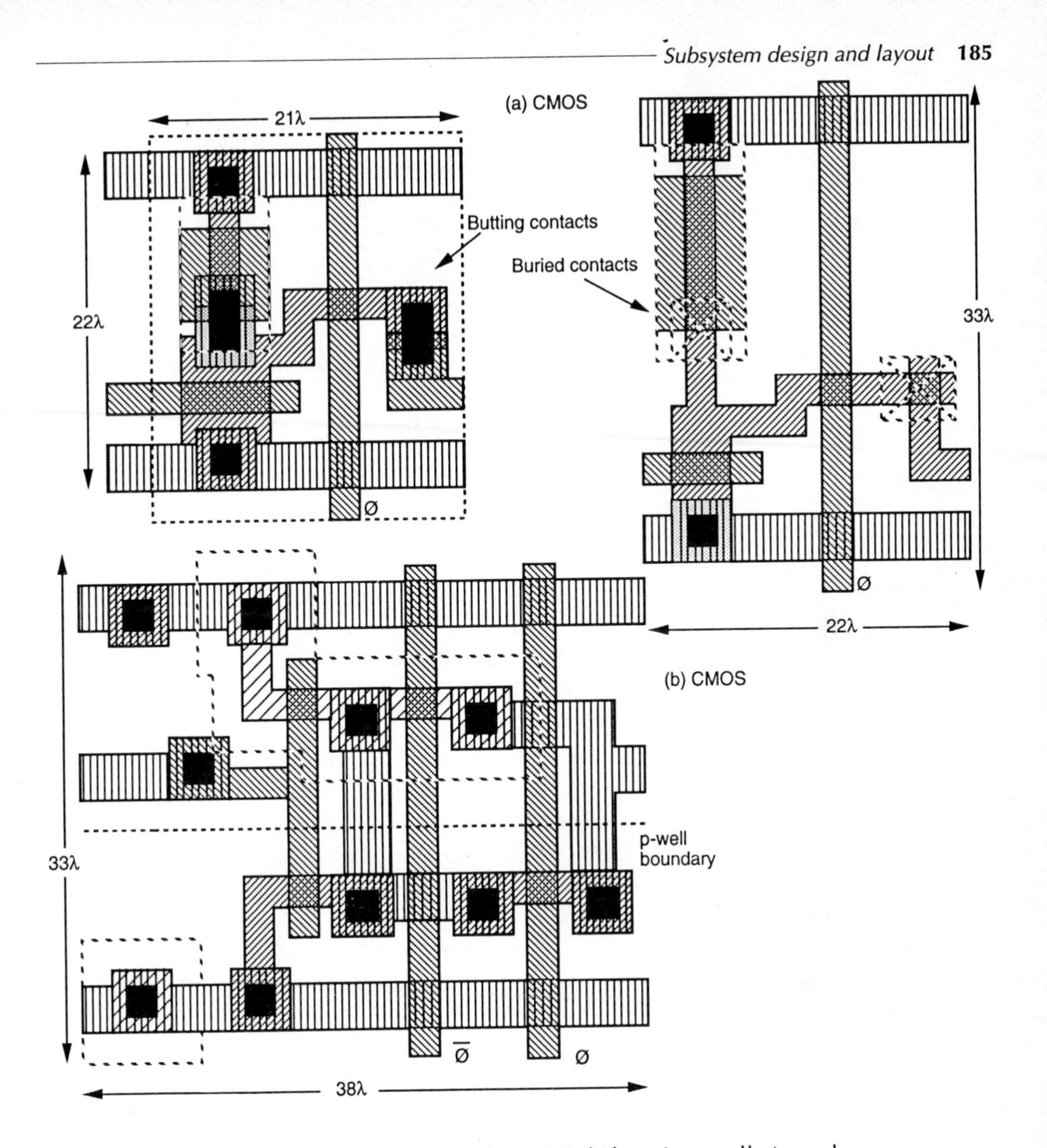

Figure 6–40 Mask layouts for nMOS and CMOS shift register cells (see also Color plate 7)

those circumstances a series switch must be inserted between each source of drive and the bus. The series resistance of such a switch to a large extent negates the speed advantage. In such cases MOS transistor drivers are often used and the following basic approaches may be considered.

6.6.2 Basic arrangements for bus lines

There are three classes of bus — passive, active, and precharged. A *passive* bus rail is a floating rail to which signals may be connected from drivers through series switches, for example, pass transistors, to propagate along the bus and from which signals may be taken, also through pass transistors (see Figure 6–41).

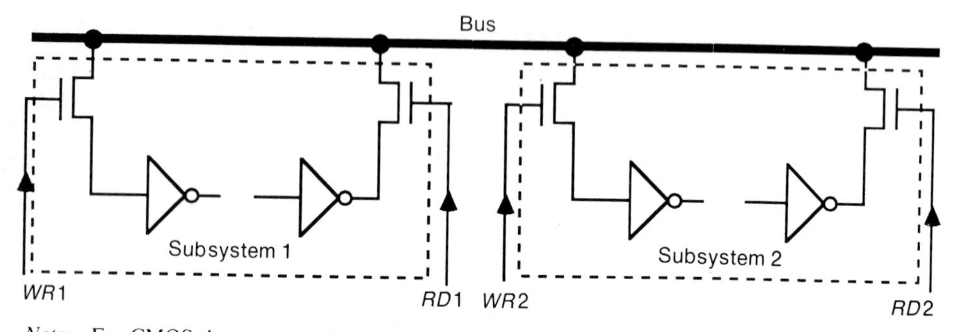

Note: For CMOS the pass transistors could become transmission gates.

Figure 6–41 Passive bus — nMOS or CMOS

A form of *active* bus is to treat the bus rail as a wired *Nor* connection which has a common pull-up $R_{p.u.}$ and n-type pull-down transistors or series n-type transistor logic pull-downs where there are circuits which must be selected to drive the bus. Signals are taken off the bus in a similar manner and the general arrangement is given as Figure 6–42. This arrangement is not suited to complementary CMOS logic-based designs since it is based on pull-down logic only.

The passive bus suffers from ratio problems in that, for any reasonable area restrictions on the bus driver circuits, the bus will be slow to respond, particularly for the ΔV (logic 0 to 1) transitions, because of the relatively high value pull-up resistance of the drivers and the associated series pass transistor or transmission gate.

The active bus is better in that more time is available for the bus to charge to V_{DD}, since $R_{p.u.}$ is always connected to the bus and there are no series pass transistors between $R_{p.u.}$ and the bus. However, there are still ratio problems which limit the speed of the bus if reasonable area is to be occupied.

6.6.3 The precharged bus concept

The *precharged* bus approach limits the effects of bus capacitance in that a single pull-up transistor which is turned on only during ϕ_2 (say) provides for the bus to charge during the ϕ_2 on period; the size of this transistor can be made relatively

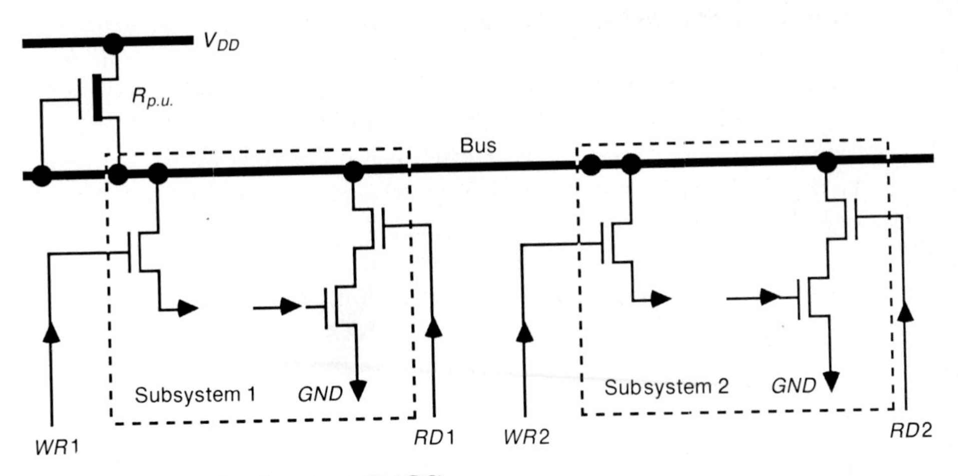

Figure 6–42 Active bus (not CMOS)

large (i.e. a low L:W ratio) and, therefore, have a low resistance. There are no ratio problems between it and the bus drivers since they are never turned on at the same time. The bus drivers merely pull down (or not) the precharged bus by discharging C_{Bus}. The arrangement is given at Figure 6–43 and, in effect, a ratioless precharged wired *Nor* circuit is formed by the bus system. However, care must be taken in nMOS systems when using logic 1 levels from the bus since the bus never reaches V_{DD}, due to threshold voltage effects in the precharging transistor.

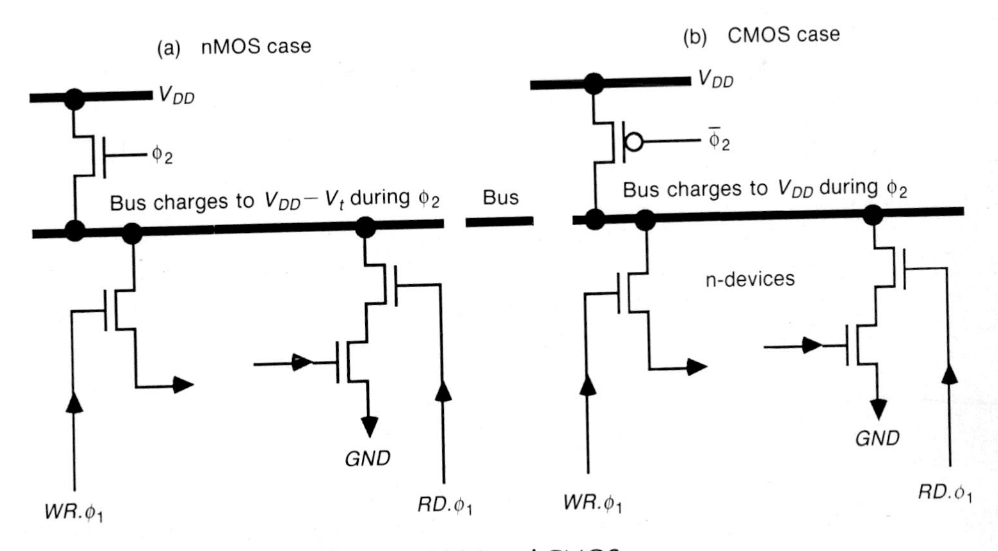

Figure 6–43 Precharged bus — nMOS and CMOS

Cross-talk and delay factors are also of significance in bus design. For example, many signals on chip may be propagated for some considerable distance (in chip dimension terms) along metal buses. Now metal buses even of minimum width are relatively wide (e.g. $3\lambda = 7.5$ μm in 5 μm technology), and thus have significant area capacitance to substrate (almost 2.5×10^{-4} pF per μm length for the example). This does not give rise to serious delay-line effects since the metal exhibits a low resistance (approximately 0.01 Ω/μm for the example) but a metal bus of any length presents a significant capacitive load to the driver. For example, a bus 400λ (1000 μm) long will present a total $C = 0.25$ pF.

Since metal also has appreciable thickness — typically 1 μm — the edge of a long bus represents a significant area. For the 400λ long bus considered above the area of each edge will be 1000 μm^2. This may give rise to cross-talk noise between two or more buses which run side by side for any appreciable length. This problem is not as serious in silicon chip designs as in GaAs technology, for example, owing to the relatively low dielectric constant (approx. = 4) for the silicon dioxide which will form the dielectric between the edges of two parallel buses.

Bus structures are widely used and will be further discussed in following sections of this text.

6.6.4 Power dissipation for CMOS and BiCMOS circuits

For pseudo-nMOS type circuitry, current and power are readily determined in a manner similar to nMOS. However, for complementary inverter-based circuits we may proceed by first recognizing that the very short current pulses which flow when circuits of this type are switching between states are generally negligible in comparison with charge and discharge currents of circuit capacitances. Then we may see that overall dissipation is composed of two terms:

1. P_l the dissipation due to the leakage current I_l through an 'off' transistor. Consequently, for n transistors, we have

$$P_l = n.I_l V_{DD}$$

where $I_l = 0.1$ nA, typically at room temperature.

2. P_s is the dissipation due to energy supplied to charge and discharge the capacitances associated with each switching circuit. Assuming that the output capacitance of a stage can be combined with the input capacitance(s) of the stage(s) it is driving and then represented collectively as C_L, then, for n identical circuits switched by a square wave at frequency f it may be shown that

$$P_s = C_L V_{DD}{}^2 f$$

The total power dissipation P_T is thus

$$P_T = P_1 + P_s$$

from which the average current may be deduced.

Power dissipation for bipolar devices can be simply modeled by

$$P = V_{cc} \times I_c$$

where V_{cc} is the supply voltage and I_c is the current through the device.

It may be seen that BiCMOS switching devices will exhibit a constant value for power dissipation, not frequency-dependent like CMOS.

6.6.5 Current limitations for V_{DD} and GND (V_{SS}) rails

A problem often ignored is that of metal migration for high current densities in metal conductors. If the current density exceeds a threshold value then one finds that metal atoms begin to move in the direction of the current. For aluminum conductors this threshold value is

$$J_{th} \doteqdot 1 \text{ to } 2 \text{ mA}/\mu\text{m}^2$$

The danger points occur where there is a narrowing or constriction in the conductor. At these points the current density is at its highest and metal is transported from the constricted regions which, in consequence, become even more constricted and eventually may blow like a fuse. The actual mechanism of atomic transport of metal in a thin film carrying relatively high currents is well understood, but the science of predicting the location and the time of such occurrences is not well developed.

By way of example, we may consider the question of how many nMOS 8:1 inverters (as in a dynamic shift register) can be driven by a minimum size conductor assuming lambda-based rules and 5 μm technology. From the design rules, the metal is 3λ wide, which corresponds to 7.5 μm. The thickness of the conductor is about 1 μm as shown in Figure 6–44.

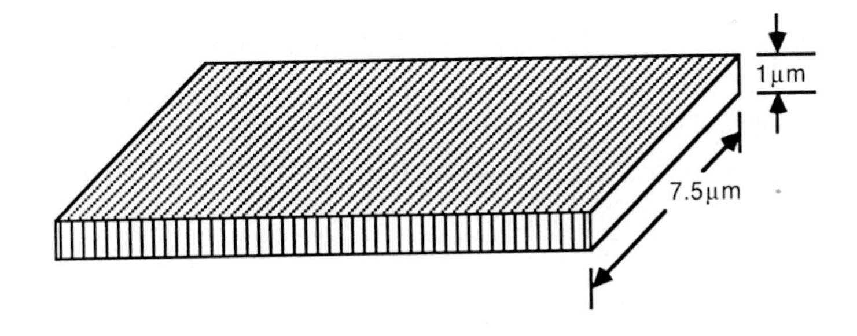

Figure 6–44 A minimum size metal path or wire

For 8:1 inverter (e.g. 8:1 p.u. and 1:1 p.d.)

$$R = (8 + 1) \times 10^4\ \Omega$$
$$= 90\ \text{k}\Omega$$

Therefore

$$\text{Current } I = \frac{5}{90} = 0.06 \text{ mA per inverter.}$$

Now, with a wire cross-section of 7.5 μm^2, the current density limitation $J_{th} = 1$ mA/μm^2 implies that a current of 7.5 mA can be supplied. Thus about 125 inverters can be driven.

One approach that may be pursued to allow some increase in the current density above the specified critical limit is to take advantage of the 'relaxation effect' that occurs in the metal when electron flow occurs in short pulses rather than at a steady state level.

However, the important factor here is that a standard (minimum) width metal conductor can only support a subsystem of quite modest size. Thus, in a design of any complexity we must ensure that this fact is not overlooked and power rail distribution becomes an important and often complex issue.

6.6.6 Further aspects of V_{DD} and V_{SS} rail distribution

Ideally, the power distibution rails (power distribution buses) for a chip should provide a constant and equal voltage supply to each and every device on the chip. Rails should also be able to supply the current required by every device. Clearly, these ideals are not achievable in practice and issues which determine the limitations are:

1. metal migration imposed current density restrictions — as already discussed in the preceding section;
2. the *IR* drop due to rail series resistance;
3. the series inductance of the rails.

The IR drops are readily calculated, provided that the currents in any bus section can be estimated since the metal bus cross-sectional area and length for that section are known.

For a parent bus supplying current to other uniformly distributed short bus branches along the length L of the parent bus, then the current at any distance x from the source is given by

$$I_x = I_L\left(\frac{1 - x}{L}\right)$$

where

I_L = the total load current supplied by the parent bus
x = the distance from the source.

The voltage drop at, say, the far end of the bus can be estimated from

$$\Delta V = \rho . \frac{I_L}{A}\left(\text{integral from 0 to } L \text{ of} \left(1 - \frac{x}{L}\right) dx.\right)$$

$$= \rho . I_L . \frac{L}{2A}$$

where

ρ = resistivity of metal
A = cross-sectional area of metal bus.

However, the bus structure is not usually as regular as envisaged here so that estimation of the voltage drop at any point is not as simple a matter as implied above if accurate determination is required.

The transmission line nature of any wiring introduces the possibility of voltage transients due to its self-inductance L_0. The transient changes in voltage due to the presence of self-inductance can be modeled by

$$\Delta V = L_0 \frac{dI}{dt}$$

where

dI/dt is the rate of change of line current.

Regarding the bus/oxide/substrate structure as a microstrip, the inductance L_0 is given by

$$L_0 = Z_0 \sqrt{\frac{\varepsilon_{eff}}{c}}$$

where

Z_0 is the characteristic impedance
c is the velocity of light
ε_{eff} is the effective dielectric constant.

In general terms, line impedance Z_0 is given by

$$Z_0 = \left(\frac{L}{C}\right)^{1/2}$$

where

L and C are the values per unit length of the bus.

Clearly, transient voltages induced in either the V_{DD} or the GND (V_{SS}) rail may lead to noise margin problems for inverters and gates.

IR drops generally can give rise to deterioration in the noise margins. This can be visualized with the aid of Figure 6–45.

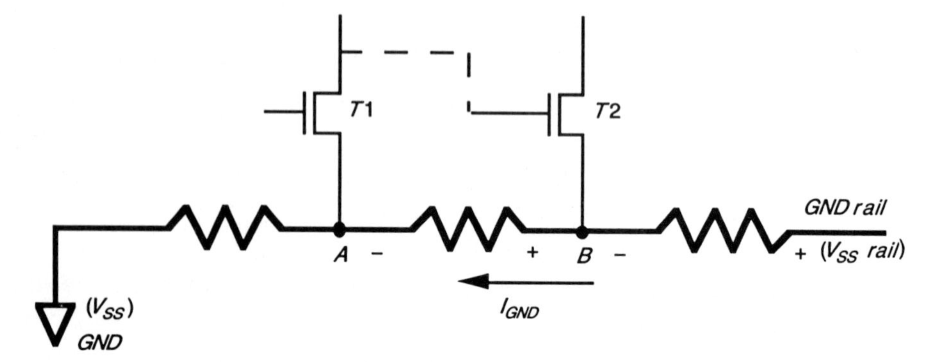

Figure 6–45 Ground (V_{SS}) rail noise

It may be seen that if *T1* is switched on, then any transient at point *A* and/or DC voltage induced in the *GND* (V_{SS}) rail from point *B* to *GND* will affect the noise margins for the next stage, *T2*.

6.7 Observations

The material presented up to this point has provided a basic toolkit and has introduced techniques with which to approach the task of VLSI system design. In this context it is most important to learn by doing and the tutorial exercises included in the text are designed for that purpose. Instructors may wish to introduce further exercises from which students will undoubtedly benefit greatly. The best (perhaps the only) way to learn to design VLSI systems is to do it.

The next three chapters tackle the design of a four-bit data path system as part of a four-bit microprocessor.

6.8 Tutorial exercises

1. (a) Construct a color-coded stick diagram to represent the design of the following integrated nMOS and CMOS structures and indicate pull-up/pull-down ratios in each case:

(i) three-input *Nand* gate;

(ii) three-input *Nor* gate;

(iii) 8:1 multiplexer circuit incorporating an enable control line;

(iv) a dual-serial shift register capable of holding and shifting (right) two 4-bit words;

(v) a selectively loadable dynamic register to hold one four-bit word (parallel).

(b) For question 1(a)(iv) draw the corresponding transistor circuit diagrams.

2. Construct a stick diagram for an nMOS or CMOS parity generator as in Figure 6–46. The required response is such that $Z = 1$ if there is an even number (including zero) of 1s on the inputs and $Z = 0$ if there is an odd number. (Use simple color coding for stick diagrams.)

   Configure your design in a modular expandable fashion so that the inputs could be increased to five or more quite readily, using the basic cell suggested in Figure 6–47.

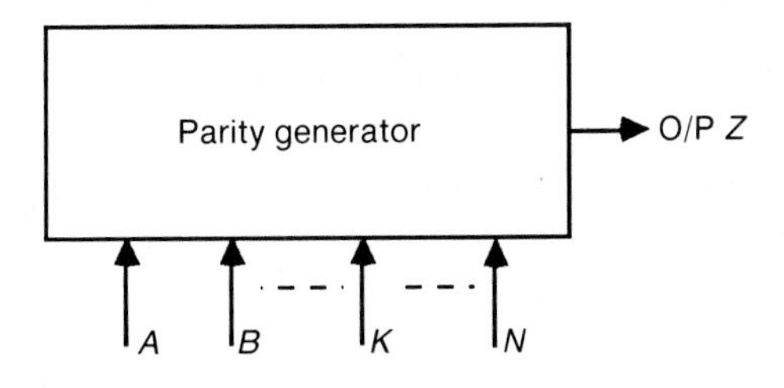

Figure 6–46 Parity generator outline

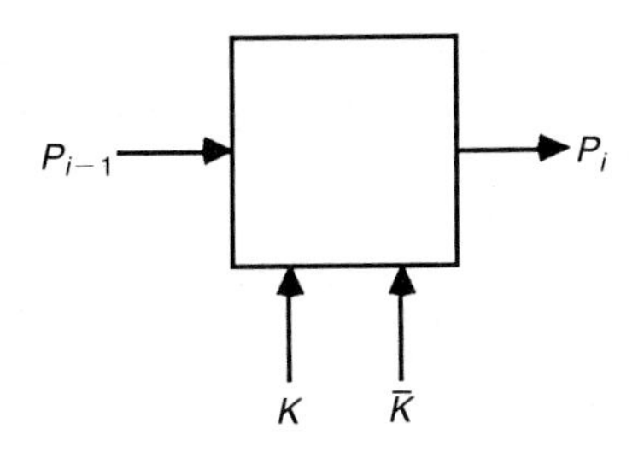

Figure 6–47 Parity generator cell

3. (a) Construct a color-coded stick diagram to represent the design of an integrated nMOS structure to decode the three input lines E_0, E_1, and E_2 into eight output lines $Z_0, Z_1 \ldots, Z_7$, in accordance with the following truth table.

Truth table

E_2	E_1	E_0	Z_0	Z_1	Z_2	Z_3	Z_4	Z_5	Z_6	Z_7
0	0	0	0	1	1	1	1	1	1	1
0	0	1	1	0	1	1	1	1	1	1
0	1	0	1	1	0	1	1	1	1	1
0	1	1	1	1	1	0	1	1	1	1
1	0	0	1	1	1	1	0	1	1	1
1	0	1	1	1	1	1	1	0	1	1
1	1	0	1	1	1	1	1	1	0	1
1	1	1	1	1	1	1	1	1	1	0

(b) Discuss the expandability or otherwise of your structure and the ease with which it would translate to CMOS.

4. A priority encoder is a combinational circuit in which each input is assigned a priority with respect to the other inputs, and the output code generated at any time is that associated with the highest priority input then present.

Construct a color-coded stick diagram to implement such a structure as in the following table with Figure 6–48.

Truth table

E_2	E_1	E_0	P_1	P_0
0	0	0	0	0
0	0	1	1	1
0	1	0	1	0
0	1	1	1	0
1	0	0	0	1
1	0	1	0	1
1	1	0	0	1
1	1	1	0	1

5. Referring to section 6.4.6 and to the development of an *Exclusive-Or* gate, design an alternative form of two I/P *Exclusive-Or* using transmission gates and inverters only. Your design should include a stick diagram and a mask layout.

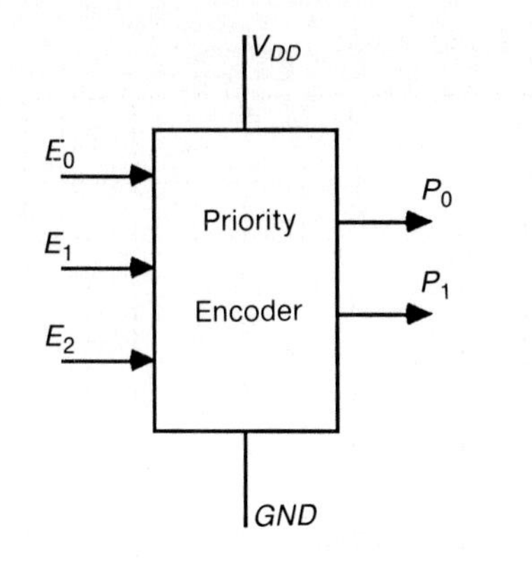

Truth table

E_2	E_1	E_0	P_1	P_0
0	0	0	0	0
0	0	1	1	1
0	1	0	1	0
0	1	1	1	0
1	0	0	0	1
1	0	1	0	1
1	1	0	0	1
1	1	1	0	1

Figure 6–48 Priority encoder

7 Subsystem design processes

One of the pleasantest things in the world is going on a journey.
Sir John Harrington

The longest journey starts with a single step.
Mao Zedong

Objectives

This chapter and the following two carry through the design of a digital system using a top-down approach. The complete system environment is that of a 4-bit microprocessor which is readily envisaged as an interconnection of four major architectural blocks — ALU, Control Unit, I/O Unit and Memory.

The design developed in this text is that of the ALU or data path, which itself divides readily into four subsystems. This chapter concentrates our attention on the design of one of the ALU subsystems — the Shifter.

The whole design process clearly illustrates the step-by-step nature of structured design and the inherently regular nature of properly conceived subsystem architecture. A general design process is developed and set out in this chapter.

7.1 Some general considerations

The first question to ask about any design methodology is the time-honored 'What's in it for me? Is it going to be worthwhile investing the time to learn?'. To answer the second part first, remarkably little time is needed to learn the rudiments of VLSI design. This is largely thanks to the Mead and Conway methodology which originally brought VLSI design within the scope of the ordinary electronics engineer. In fact, the average undergraduate student of electrical or electronic engineering can acquire a basic level of competence in VLSI design for an investment of about 40 hours of lectures spread over one or more academic terms or semesters. Similarly, a 10-day full-time continuing education course can quite readily bring practicing professional engineers or computer scientists up to a similar standard. A basic level of competence is taken as the ability to apply the design methodology and make use of design tools and procedures to the point where a chip design of several hundred transistors (or higher for regular structures) can be tackled.

The first part of the question — 'What's in it for me?' — may be quite simply answered as: Providing better ways of tackling some problems, providing a way of designing and realizing systems that are too large, too complex, or just not suited to 'off-the-shelf' components and providing an appreciation and understanding of IC technology.

'Better' may include:

1. *Lower unit cost* compared with other approaches to the same requirement. Quantity plays a part here but even small quantities, if realized through cooperative ventures such as the multiproject chip (MPC) or multiproduct wafer (MPW), can be fabricated for as little as $200 (MPC) or $500 (MPW) per square millimetre of silicon, including the bonding and packaging of five or six chips per customer.
2. *Higher reliability.* High levels of system integration usually greatly reduce interconnections — a weak spot in any system.
3. *Lower power dissipation, lower weight, and lower volume* compared with most other approaches to a given system.
4. *Better performance* — particularly in terms of speed power product.
5. *Enhanced repeatability.* There are fewer processes to control if the whole system or a very large part of it is realized on a single chip.
6. *The possibility of reduced design/development periods* (particularly for more complex systems) if suitable design procedures and design aids are available.

7.1.1 Some problems

Some of the problems associated with VLSI design are:

1. How to design large complex systems in a reasonable time and with reasonable effort. This is a problem shared with other approaches to system design.
2. The nature of architectures best suited to take full advantage of VLSI and the technology.
3. The testability of large/complex systems once implemented in silicon.

Problems 1 and 3 are greatly reduced if two aspects of standard practice are accepted:

- Approach the design in a top-down manner and with adequate computer-aided tools to do the job. Partition the system sensibly, aiming for simple interconnection between subsystems and high regularity within subsystems. Generate and then verify each section of the design.
- Design testability into the system from the outset and be prepared to devote a significant proportion (e.g. up to 30%) of the total chip area to test and diagnostic facilities.

These problems are the subject of considerable research and development activity at this time.

In tackling the design of a system, we must bear in mind that topological properties are generally far more significant than the logical operations being performed. It may be said that it is better to duplicate (or triplicate, etc.) rather than communicate. This is indeed the case, and it is an approach which seems wrong to more traditional designers. In fact, even in relatively straightforward designs, as much as 40–50% of the chip may be taken up with interconnections, and it is true to say that interconnections generally pose the most acute problems in the design of large systems. Communications must therefore be given the highest priority early in the design process and a *communications strategy* should be evolved and adhered to throughout that process.

Accordingly, the architecture should be carefully chosen to allow the design objectives to be realized *and* to allow high regularity in realization.

7.2 An illustration of design processes

- Structured design begins with the concept of hierarchy.
- It is possible to divide any complex function into less complex subfunctions. These may be subdivided further into even simpler subfunctions and so on — the bottom level being commonly referred to as 'leaf-cells'.
- This process is known as top-down design.

- As a system's complexity increases, its organization changes as different factors become relevant to its creation.
- Coupling can be used as a measure of how much submodules interact. Clever systems partitioning aims at reducing implicit complexity by minimizing the amount of interaction between subparts; thus independence of design becomes a reality.
- It is crucial that components interacting with high frequency be physically proximate, since one may pay severe penalties for long, high-bandwidth interconnects.
- Concurrency should be exploited — it is desirable that all gates on the chip do useful work most of the time.
- Because technology changes so fast, the adaptation to a new process must occur in a short time. Thus a technology-independent description becomes important.

In representing a design, several approaches may be used at different stages of the design process; for example:

- conventional circuit symbols;
- logic symbols;
- stick diagrams;
- any mixture of logic symbols and stick diagrams that is convenient at a particular stage;
- mask layouts;
- architectural block diagrams;
- floor plans.

We will illustrate various representations during the course of the following design exercise to illustrate design processes.

7.2.1 The general arrangement of a 4-bit arithmetic processor

The 4-bit microprocessor has been chosen as a design example because it is particularly suitable for illustrating the design and interconnection of common architectural blocks.

Figure 7–1 sets out the basic architecture of most, if not all, microprocessors. At this stage we will consider the design of the data path only, but matters relevant to other blocks will follow in later chapters.

The data path has been separated out in Figure 7–2 and it will be seen that the structure comprises a unit which processes data applied at one port and presents its output at a second port. Alternatively, the two data ports may be combined as

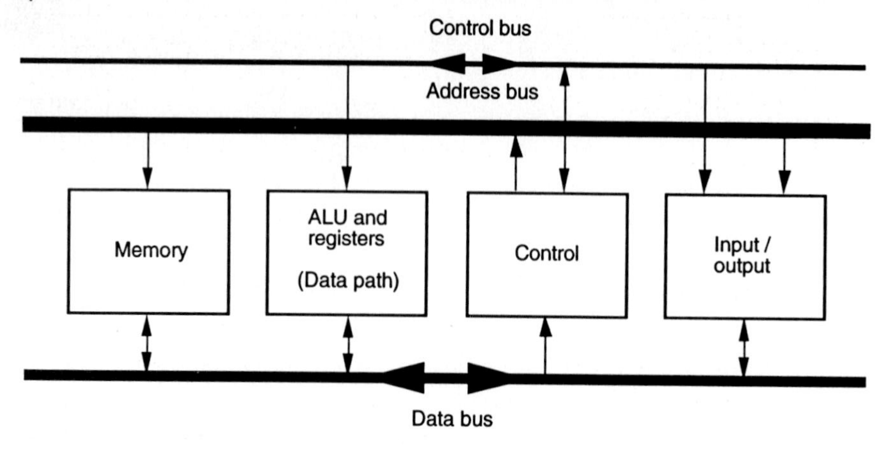

Figure 7–1 Basic digital processor structure

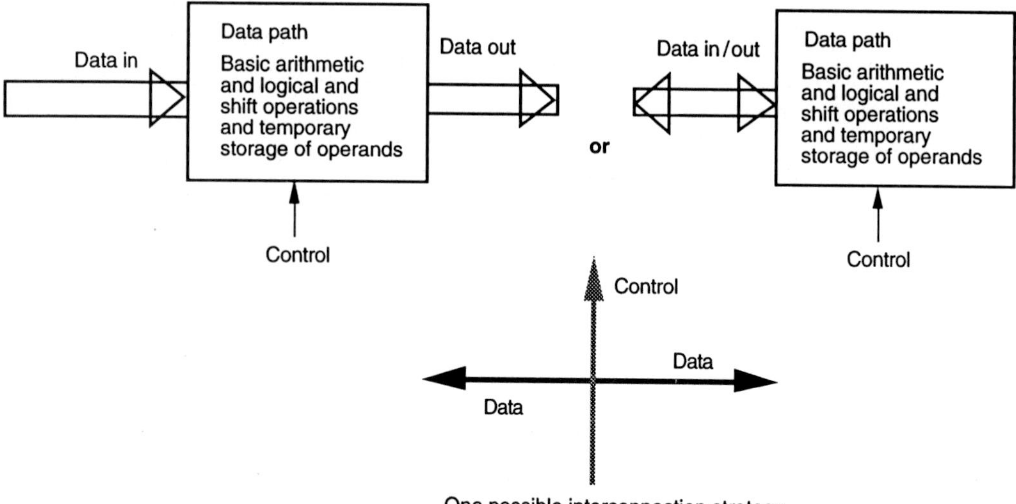

Figure 7–2 Communications strategy for data path

a single bidirectional port if storage facilities exist in the data path. Control over the functions to be performed is effected by control signals as indicated.

At this early stage it is essential to evolve an interconnections strategy (as shown) to which we will then adhere.

Now we will decompose the data path into a block diagram showing the main subunits. In doing this it is useful to anticipate a possible *floor plan* to show the planned relative disposition of the subunits on the chip and thus on the mask layouts. A block diagram is presented in Figure 7–3.

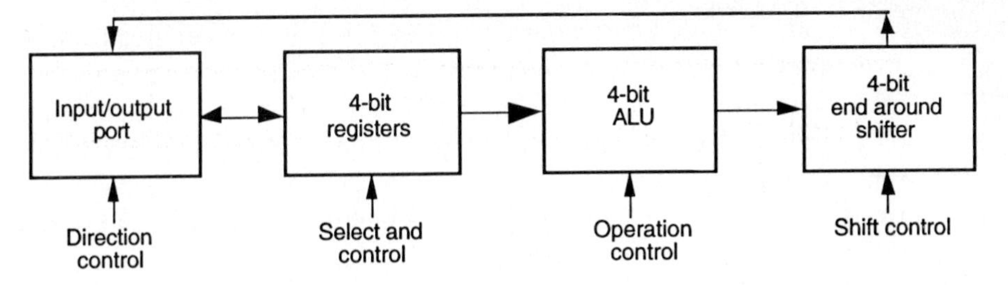

Figure 7–3 Subunits and basic interconnections for data path

A further decision must then be made about the nature of the bus architecture linking the subunits. The choices in this case range from one-bus, two-bus or three-bus architecture. Some of the possibilities are shown in Figure 7–4.

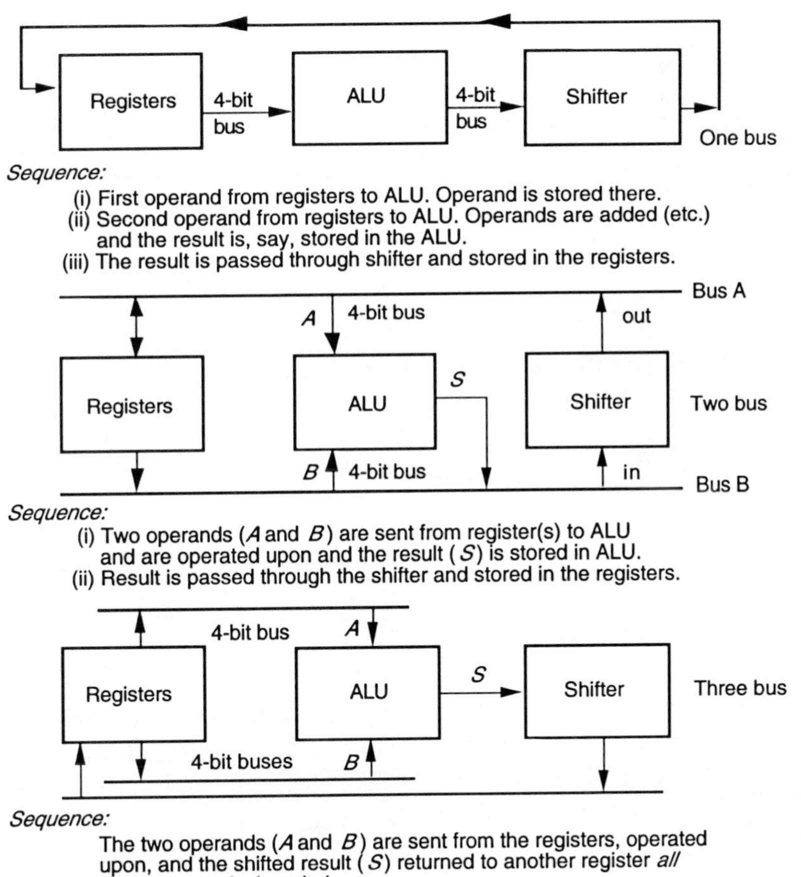

Figure 7–4 Basic bus architectures

In pursuing this particular design exercise, it was decided to implement the structure with a two-bus architecture. In our planning we can now extend on our interconnections strategy by planning for power rails and notionally making some basic allocation of layers on which the various signal paths will be predominantly run. These additional features are illustrated in Figure 7–5, together with a tentative floor plan of the proposed design which includes some form of interface (I/O) to the parent system data bus (see Figure 7–1).

The proposed processor will be seen to comprise a register array in which 4-bit numbers can be stored, either from an input/output port or from the output of the ALU via a shifter. Numbers from the register array can be fed in pairs to the ALU to be added (or subtracted, etc.) and the result can be shifted or not, before being returned to the register array or possibly out through the I/O port. Obviously, data connections between the I/O port, ALU, and shifter must be in the form of 4-bit buses. Simultaneously, we must recognize that each of the blocks must be suitably connected to control lines so that its function may be defined for any of a range of possible operations.

The required arrangement has been turned into a very tentative floor plan, as in Figure 7–5, which indicates a possible relative disposition of the blocks and also indicates an acceptable and sensible interconnection strategy indicated by the lines showing the preferred direction of data flow and control signal distribution. At this stage of learning, floor plans will be very tentative since we will not as yet be able to accurately assess the area requirements, say for a 4-bit register or a 4-bit adder.

Overall interconnection strategy having been determined, stick diagrams for the circuits comprising sections of the various blocks may be developed, conforming to the required strategy. An interactive process of modification may well then take place between the various stages as the design progresses. During the design process, and in particular when defining the interconnection strategy and designing the stick diagrams, care must be taken in allocating the layers to the various data or control paths. We must remember that:

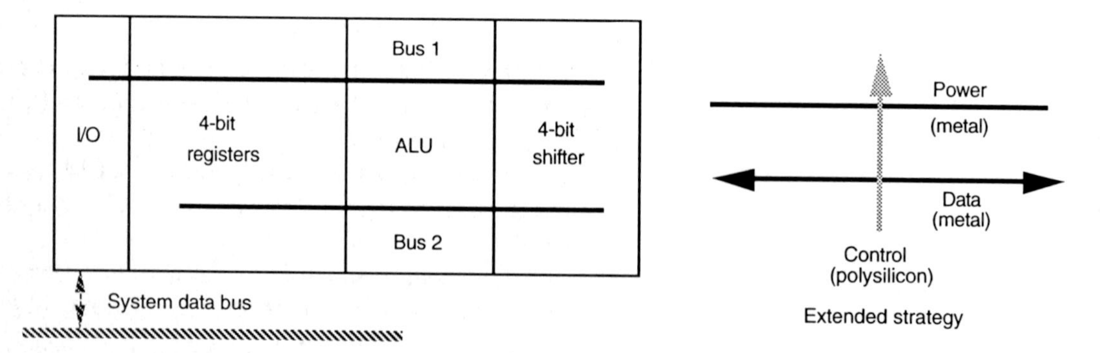

Figure 7–5 Tentative floor plan for 4-bit data path

1. Metal can cross polysilicon or diffusion without any significant effect (with some reservations to be discussed later).
2. Wherever polysilicon crosses diffusion a transistor will be formed. This includes the second polysilicon layer for processes that have two.
3. Wherever lines touch on the same level an interconnection is formed.
4. Simple contacts can be used to join diffusion or polysilicon to metal.
5. To join diffusion and polysilicon we must use either a buried contact or a butting contact (in which case all three layers are joined together at the contact) or two contacts, diffusion to metal then metal to polysilicon.
6. In some processes, a second metal layer is available. This can cross over any other layers and is conveniently employed for power rails.
7. First and second metal layers may be joined using a *via*.
8. Each layer has particular electrical properties which must be taken into account.
9. For CMOS layouts, p- and n-diffusion wires must not directly join each other, nor may they cross either a p-well or an n-well boundary.

With these factors in mind, we may now adopt suitable tactics to meet the strategic requirements when we approach the design of each subunit in turn.

7.2.2 The design of a 4-bit shifter

Any general purpose *n*-bit shifter should be able to shift incoming data by up to n – 1 places in a right-shift or left-shift direction. If we now further specify that all shifts should be on an 'end-around' basis, so that any bit shifted out at one end of a data word will be shifted in at the other end of the word, then the problem of right shift or left shift is greatly eased. In fact, a moment's consideration will reveal, for a 4-bit word, that a 1-bit shift right is equivalent to a 3-bit shift left and a 2-bit shift right is equivalent to a 2-bit shift left, etc. Thus we can achieve a capability to shift left or right by zero, one, two, or three places by designing a circuit which will shift right only (say) by zero, one, two, or three places.

The nature of the shifter having been decided on, its implementation must then be considered. Obviously, the first circuit which comes to mind is that of the shift register in Figures 6–38, 6–39, and 6–40. Data could be loaded from the output of the ALU and shifting effected; then the outputs of each stage of the shift register would provide the required parallel output to be returned to the register array (or elsewhere in the general case).

However, there is danger in accepting the obvious without question. Many designers, used to the constraints of TTL, MSI, and SSI logic, would be conditioned to think in terms of such standard arrangements. When designing VLSI systems, it pays to set out exactly what is required to assess the best approach. In this case, the shifter must have:

- input from a four-line parallel data bus;
- four output lines for the shifted data;
- means of transferring input data to output lines with any shift from zero to three bits inclusive.

In looking for a way of meeting these requirements, we should also attempt to take best advantage of the technology; for example, the availability of the switch-like MOS pass transistor and transmission gate.

We must also observe the strategy decided on earlier for the direction of data and control signal flow, and the approach adopted should make this feasible. Remember that the overall strategy in this case is for data to flow horizontally and control signals vertically.

A solution which meets these requirements emerges from the days of switch and relay contact based switching networks — the *crossbar switch.* Consider a direct MOS switch implementation of a 4×4 crossbar switch, as in Figure 7–6.

The arrangement is quite general and may be readily expanded to accommodate n-bit inputs/outputs. In fact, this arrangement is an overkill in that any input line can be connected to any or all output lines — if all switches are closed, then all inputs are connected to all outputs in one glorious short circuit. Furthermore, 16 control signals ($sw_{00} - sw_{15}$), one for each transistor switch, must be provided to drive the crossbar switch, and such complexity is highly undesirable. An adaptation of this arrangement recognizes the fact that we can couple the switch gates together

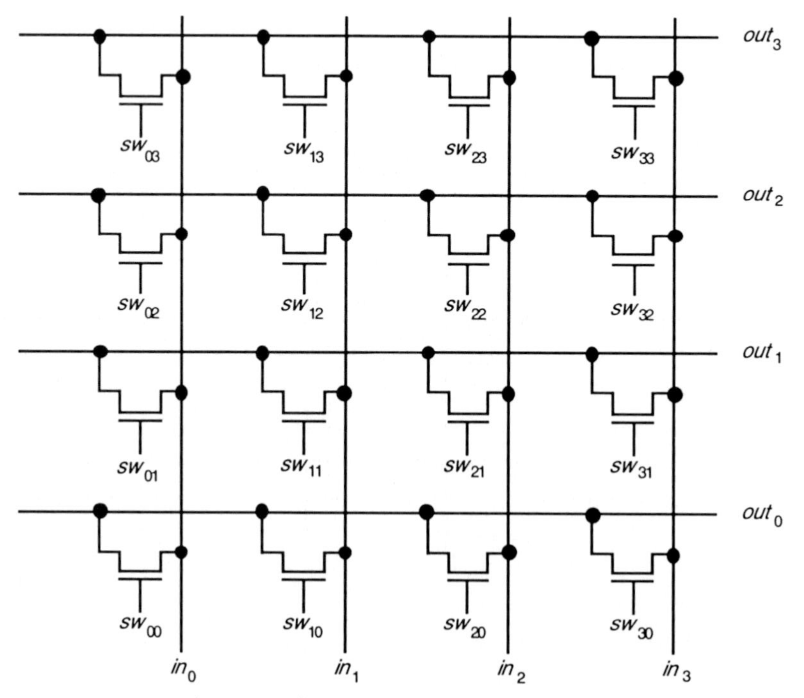

Figure 7–6 4×4 crossbar switch

in groups of four (in this case) and also form four separate groups corresponding to shifts of zero, one, two and three bits. The arrangement is readily adapted so that the in-lines also run horizontally (to conform to the required strategy).

The resulting arrangement is known as a *barrel shifter* and a 4 × 4-bit barrel shifter circuit diagram is given in Figure 7–7. The interbus switches have their gate inputs connected in a staircase fashion in groups of four and there are now four shift control inputs which must be mutually exclusive in the active state. CMOS transmission gates may be used in place of the simple pass transistor switches if appropriate.

The structure of the barrel shifter is clearly one of high regularity and generality and it may be readily represented in stick diagram form. One possible implementation, using simple n-type switches, is given in Figure 7–8.

The stick diagram clearly conveys regular topology and allows the choice of a standard cell from which complete barrel shifters of any size may be formed by replication of the standard cell. It should be noted that standard cell boundaries must be carefully chosen to allow for butting together side by side and top to bottom to retain the overall topology. The mask layout for standard cell number 2 (arbitrary choice) of Figure 7–8 may then be set out as in Figure 7–9. Once the standard cell dimensions have been determined, then any $n \times n$ barrel shifter may be configured and its outline, or bounding box, arrived at by summing up the dimensions of the replicated standard cell. The use of simple n-type switches in

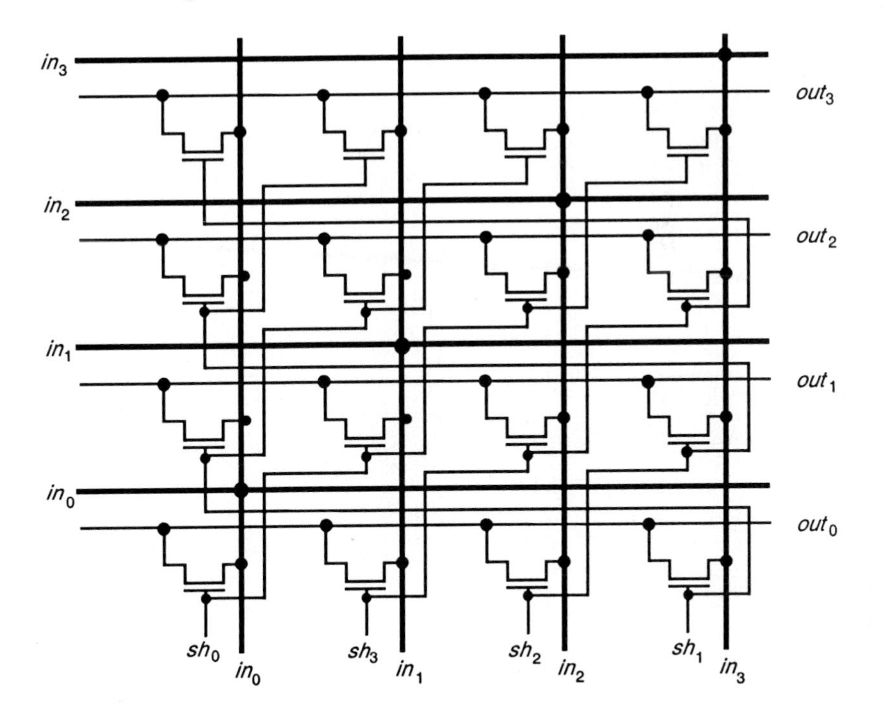

Figure 7–7 4 × 4 barrel shifter

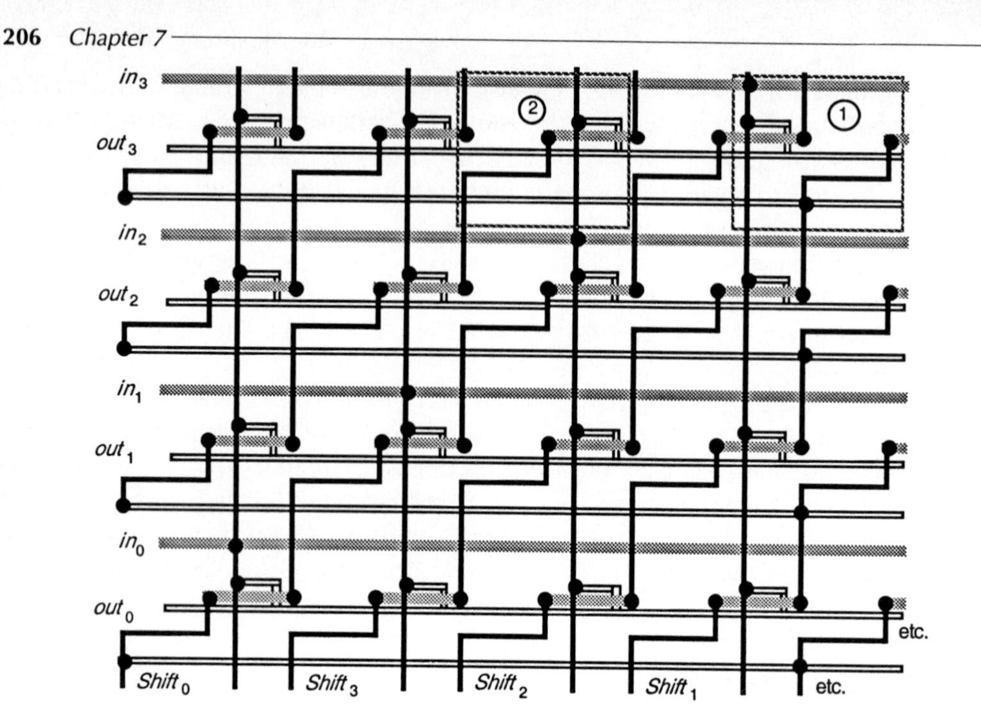

Figure 7–8 One possible stick diagram for a 4 × 4 barrel shifter

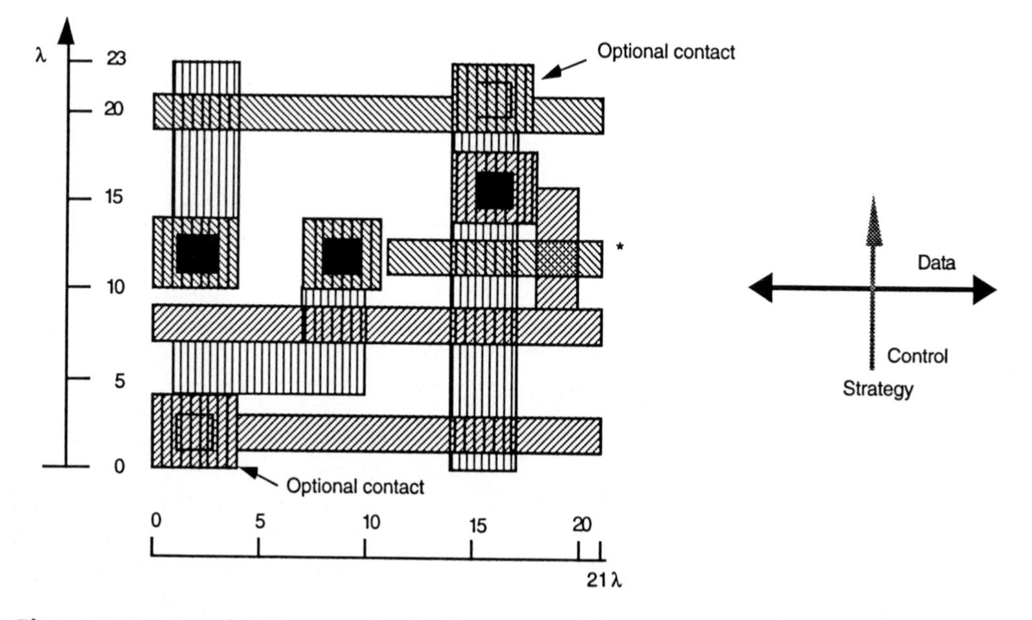

Figure 7–9 Barrel shifter standard cell 2 — mask layout

* If this particular cell is checked for design rule errors in isolation, then an error will be generated owing to insufficient extension of polysilicon over thinox where shown. This error will *not* be present when cells are butted together. This effect is caused by the particular choice of cell boundaries and care must be taken when making such choices.

a CMOS environment might be questioned. Although there will be a degrading of logic 1 levels through n-type switches, this generally does not matter if the shifter is followed by restoring circuitry such as inverters or gate logic. Furthermore, as there will only ever be one n-type switch in series between an input and the corresponding output line, the arrangement is fast.

The minimum size *bounding box* outline for the 4 × 4-way barrel shifter is given in Figure 7–10. The figure also indicates all inlet and outlet points around the periphery together with the layer on which each is located. This allows ready placing of the shifter within the floor plan (Figure 7–5) and its interconnection with the other subsystems forming the datapath. It also emphasizes the fact that, as in this case, many subsystems need external links to complete their architecture. In this case, the links shown on the right of the bounding box must be made and must be allowed for in interconnections and overall dimensions. This form of representation also allows the subsystem geometric characterization to be that of the bounding box alone for composing higher levels of the system hierarchy.

7.3 Observations

At this stage it is convenient to examine the way we have approached the design of a system and of a particular subsystem in detail. The steps involved may be set out as follows:

1. Set out a specification together with an architectural block diagram.
2. Suitably partition the architecture into subsystems which are, as far as possible, self-contained and which give as simple interconnection requirements as possible.
3. Set out a tentative floor plan showing the proposed physical disposition of subsystems on the chip.
4. Determine interconnection strategy.
5. Revise 2, 3 and 4 interactively as necessary.
6. Choose layers on which to run buses and the main control signals.
7. Take each subsystem in turn and conceive a regular architecture to conform to the strategy set out in 4. Set out circuit and/or logic diagrams as appropriate. Remember that MOS switch-based logic is such that both the logic 1 and logic 0 conditions of an output must be deliberately satisfied (not as in TTL logic, where if logic 1 conditions are satisfied then logic 0 conditions follow automatically).
8. Develop stick diagrams adopting suitable tactics to observe the overall strategy (4) and choice of layers (6). Determine suitable *standard cell(s)* from which the subsystem may be formed.

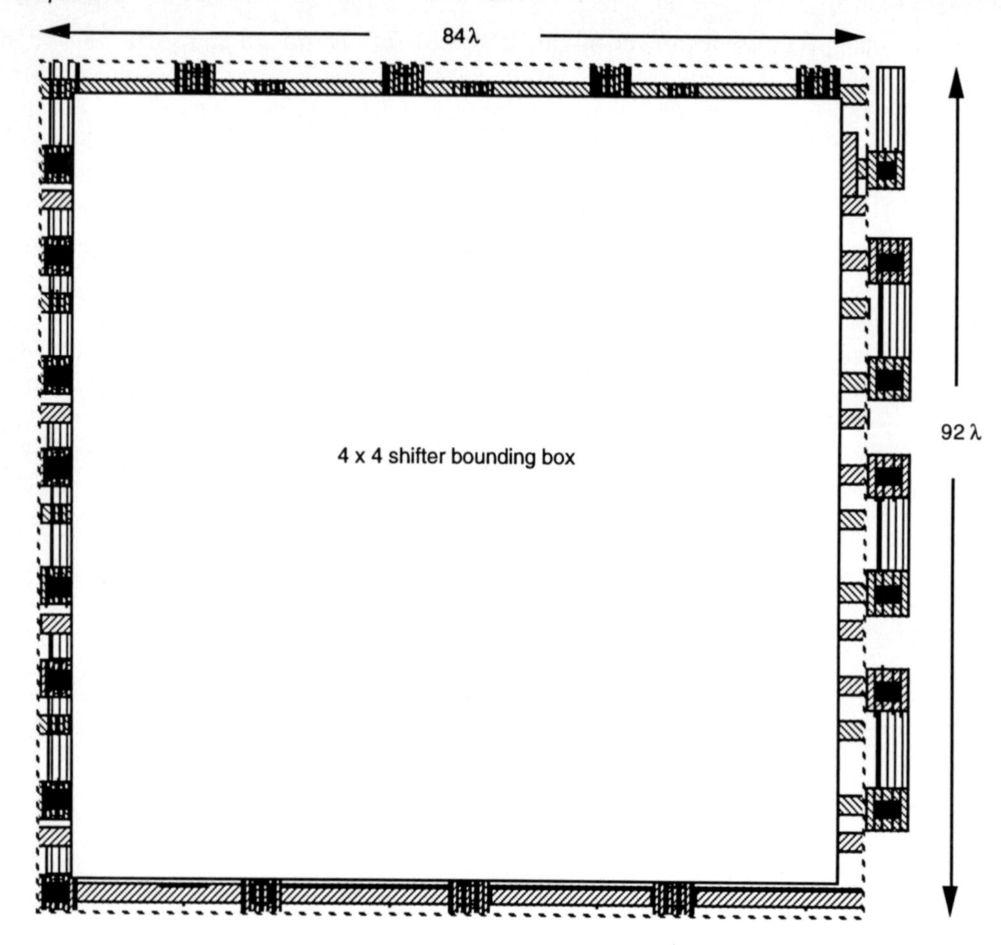

Figure 7–10 Bounding box for 4 × 4 barrel shifter

9. Produce mask layouts for the standard cell(s), making sure that cells can be butted together, side by side and top to bottom, without design rule violation or waste of space. Determine overall dimension of the standard cell(s).
10. Cascade and replicate standard cells as necessary to complete the desired subsystem. This may now be characterized in *bounding box* form with positions and layers of inlets and outlets. External links etc. *must* be allowed for.

7.4 Tutorial exercises

1. (a) Set out the mask layout for a 4-way multiplexer using transmission gate switches.

   (b) Determine the overall dimensions (in terms of lambda) for your design.

   (c) Compare the attributes of this design with those of the n-type pass transistor version given in Chapter 6.

2. Using a block diagram (symbolic) form of representation for the 4-way multiplexer, draw up an interconnection diagram showing four such multiplexers configured as a 4-bit shift left/right shifter subsystem. The shifter should meet the same overall logical requirements as the barrel shifter designed in this chapter. You should carefully specify the control signals needed for this shifter design.

3. Estimate the area that will be occupied by your design of question 2, assuming the use of the multiplexer design of question 1. Compare this with the barrel shifter design developed in this chapter.

8 Illustration of the design process — computational elements

Progress is a comfortable disease.

e. e. cummings

Beneath this slab
John Brown is stowed
He watched the ad(d)s
And not the road.

Ogden Nash

Objectives

In this chapter we progress to the arithmetic subsystem of the 4-bit data path. Once again, high regularity should be the aim of the designer. If the subsystems are regular and therefore composed of relatively few actual leaf-cell circuits, then the designer can concentrate on the main problem of VLSI design — the routing of interconnections and communication paths. It is hoped that this fact is beginning to emerge as this design progresses. Properly conceived communications — both at leaf-cell and at system levels — are the key to good design.

The arithmetic circuitry required here is relatively simple but does lead into a further consideration of adder circuitry and a fairly comprehensive survey of arrangements for multiplication.

8.1 Some observations on the design process

The design of the shifter, as the first subsystem of the proposed 4-bit data path, has illustrated some important features:

1. First and foremost, try to put requirements into words (an *if, then, else* approach often helps you do this) so that the most appropriate architecture or logic can be evolved.
2. If a standard cell (or cells) can be arrived at, then the actual detailed design work is confined to relatively small areas of simple circuitry (leaf-cells). Such cells can usually have their performance simulated with relative ease so that an idea of the performance of the complete subsystem may be deduced.
3. If generality as well as regularity is achieved then, for example, any size of shifter can be built up by simple replication and butting together of the standard cell(s).
4. Design is largely a matter of the topology of communications rather than detailed logic circuit design.
5. Once standard cell layouts are designed, overall area calculations can be precisely made (*not* forgetting to allow for any necessary links or other external terminations). Thus, accurate floor plan areas may be allocated.
6. VLSI design methodology for MOS circuits is not hard to learn.
7. The design rules are simple and straightforward in application.
8. A structured and orderly approach to system design is highly beneficial and becomes essential for large systems.

8.2 Regularity

So far we have used regularity as a qualitative parameter. Regularity should be as high as possible to minimize the design effort required for any system. The level of any particular design as far as this aspect is concerned may be measured by quantifying regularity as follows:

$$\text{Regularity} = \frac{\text{Total number of transisitors on the chip}}{\text{Number of transistor circuits that must be designed in detail}}$$

The denominator of this expression will obviously be greatly reduced if the whole chip, or large parts of it, can be fabricated from a few standard cells, each of which is relatively simple in structure.

For the 4 × 4-bit barrel shifter just designed, the regularity factor is given by

$$\text{Regularity} = \frac{16}{1} = 16$$

However, an 8×8-bit shifter, for example, would require no more detailed design and would have a regularity factor of 64.

Good system design can achieve regularity factors of 50 or 100 or more, and inherently regular structures, such as memories, achieve very high figures indeed.

8.3 Design of an ALU subsystem

Having designed the shifter, we may now turn our attention to another subsystem of the 4-bit data path (as in Figure 8–1). A convenient and appropriate choice is the ALU.

The heart of the ALU is a 4-bit adder circuit and it is this which we will actually design, indicating later how it may be readily adapted to subtract and perform logical operations. Obviously, a 4-bit adder must take the sum of two 4-bit numbers, and it will be seen we have assumed that all 4-bit quantities are presented in parallel form and that the shifter circuit has been designed to accept and shift a 4-bit parallel sum from the ALU.

Let us now specify that the sum is to be *stored* in parallel at the output of the adder from where it may be fed through the shifter and back to the register array. Thus, a single 4-bit data bus is needed from the adder to the shifter and another 4-bit bus is required from the shifted output back to the register array (since the shifter is merely a switch array with no storage capability). As far as the input to the adder is concerned, the two 4-bit parallel numbers to be added are to be presented in parallel on two 4-bit buses. We can also decide on some of the basic aspects of system timing at this stage and will assume clock phase ϕ_1 as being the phase during which signals are fed along buses to the adder input and during

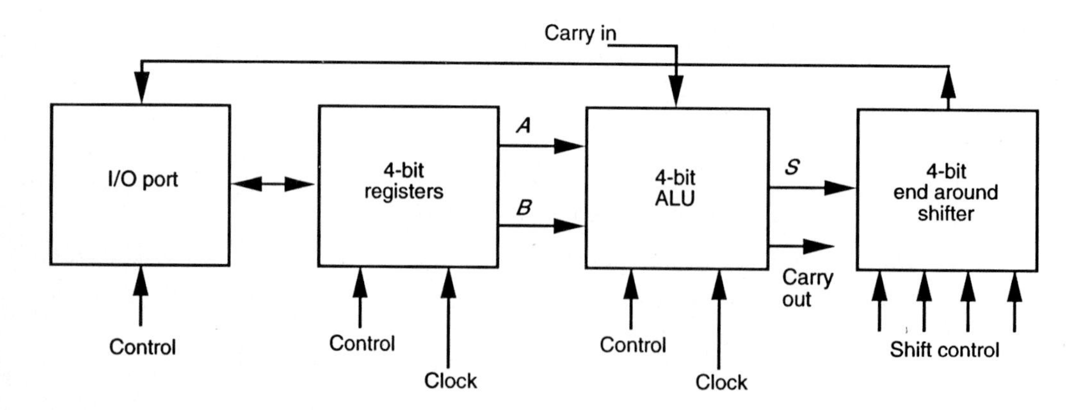

Figure 8–1 4-bit data path for processor (block diagram)

which their sum is stored at the adder output. Thus clock signals are required by the ALU as shown. The shifter is unclocked but must be connected to four shift control lines. It is also necessary to provide a 'carry out' signal from the adder and, in the general case, to provide for a possible 'carry in' signal, as indicated in Figure 8–1.

8.3.1 Design of a 4-bit adder

In order to derive the requirements for an *n*-bit adder, let us first consider the addition of two binary numbers A + B as follows:

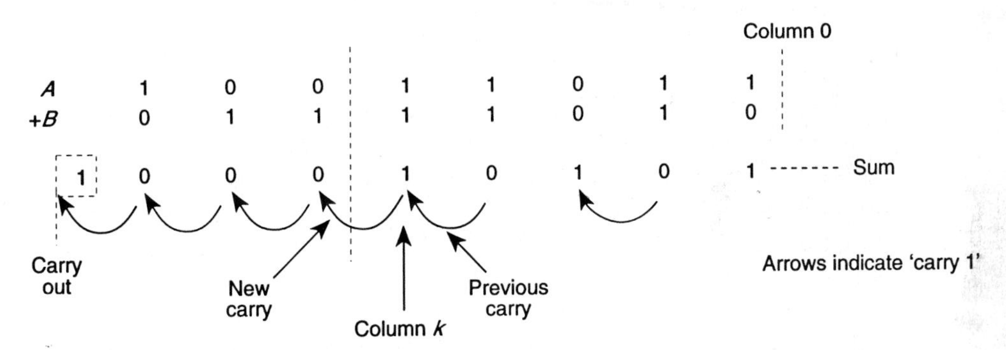

It will be seen that for any column *k* there will be *three* inputs — the corresponding bits of the input numbers, A_k and B_k, and the 'previous carry' — *carry in* (C_{k-1}). It will also be seen that there are two outputs, the *sum* (S_k) and a *new carry* (C_k).

We may thus set out a truth table for the *k* column of any adder, as in Table 8–1.

Table 8–1 Truth table for binary adder

Inputs			*Outputs*	
A_k	B_k	*Previous carry* C_{k-1}	*Sum* S_k	*New carry* C_k
0	0	0	0	0
0	1	0	1	0
1	0	0	1	0
1	1	0	0	1
0	0	1	1	0
0	1	1	0	1
1	0	1	0	1
1	1	1	1	1

Conventionally, and assuming that we are not implementing a 'carry look ahead' facility, we may write *standard adder equations,* which fully describe the entries in Table 8–1.

One form of these equations is:

Sum $$S_k = H_k \overline{C}_{k-1} + \overline{H}_k C_{k-1}$$

New carry $$C_k = A_k B_k + H_k C_{k-1}$$

where

Half sum $$H_k = \overline{A}_k B_k + A_k \overline{B}_k$$

Previous carry is indicated as C_{k-1} and $0 \leqslant k \leqslant n-1$ for n-bit numbers.

These equations may be directly implemented as *And-Or* functions or, most economically, S_k and H_k can be directly implemented with *Exclusive-Or* gates. However, for VLSI custom implementation there are none of the standard logic packages which are the delight of the TTL logic designer. It may be advantageous, then, to restate the requirements in another way.

8.3.1.1 Adder element requirements

Table 8–1 reveals that the *adder requirements* may be stated thus:

If $A_k = B_k$ then $S_k = C_{k-1}$
else $S_k = \overline{C}_{k-1}$

and for the carry C_k

If $A_k = B_k$ then $C_k = A_k = B_k$ *
else $C_k = C_{k-1}$

* This relationship could also have been stated as:

Carry $C_k = 1$ when $A_k = B_k = 1$

or

$C_k = 0$ when $A_k = B_k = 0$

8.3.1.2 A standard adder element

A 1-bit adder element may now be represented as in Figure 8–2. Note that any number of such elements may be cascaded to form any size of adder and that the element is quite general.

Note also that this standard adder element may itself be composed from a number of replicated subcells. Regularity and generality must be aimed at in all levels of the architecture.

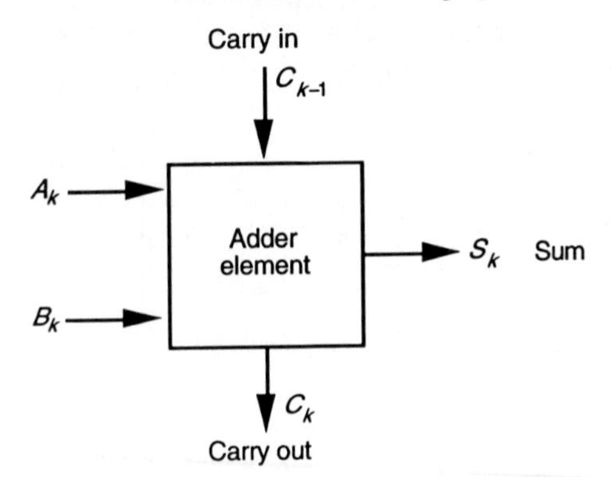

n such elements would be cascaded to form an *n*-bit adder.

Figure 8–2 Adder element

Using multiplexers is an implementation of the logic circuitry for the adder element that is easy to follow, resulting directly from the way in which the requirements are stated in words (see section 8.3.1.1). This approach is illustrated in Figures 8–3 and 8–4 and it may be seen that the words used to describe the adder logic are directly implemented by the various paths through the multiplexers.

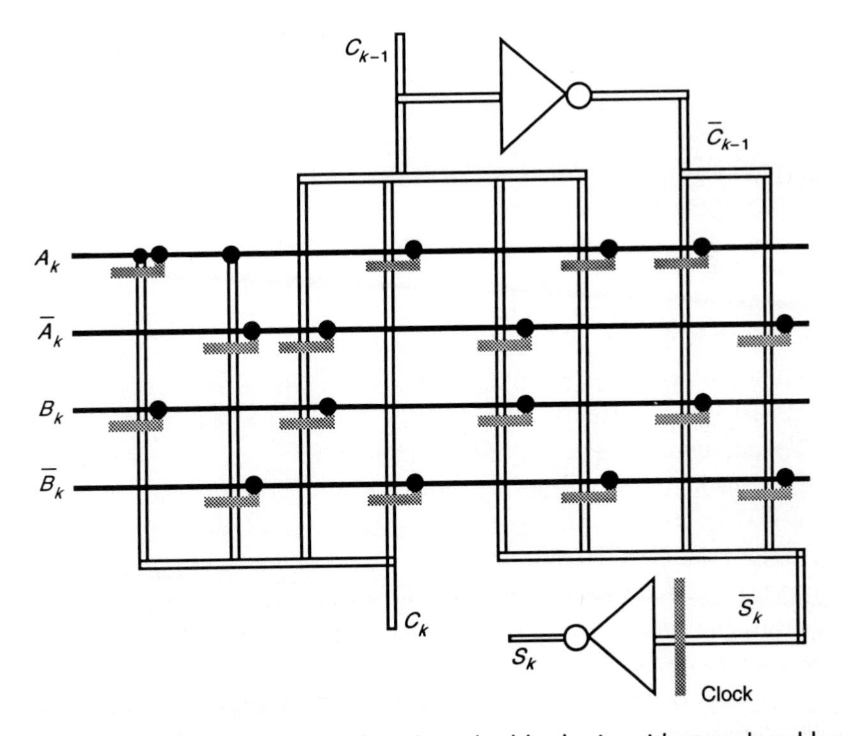

Figure 8–3 Multiplexer (n-switches)-based adder logic with stored and buffered sum output

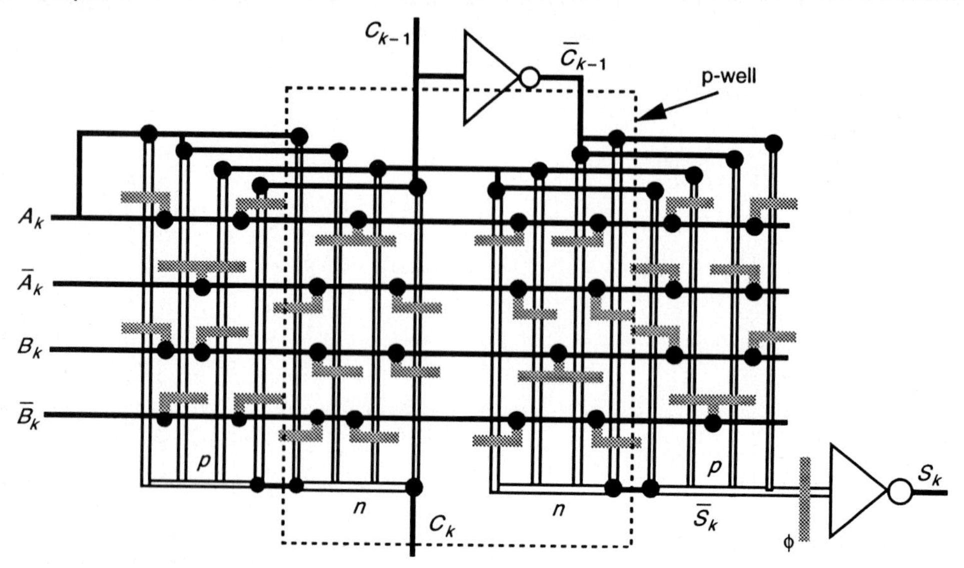

Figure 8–4 CMOS version of adder logic

In these figures the multiplexers form C_k and $\overline{S}_k$ *(not* S_k*)* to allow single inverter storage or buffering of S_k if this is needed. In fact, one design actually implemented in silicon (see Figure 10–9) uses an nMOS multiplexer-based version of the adder. The basic logic requirements of this adder element, or bit-slice, are thus readily met in nMOS or CMOS technology. However, some practical factors must now be taken into account.

In order to form an n-bit adder, n adder elements must be cascaded with *carry out* of one element connecting to *carry in* of the next more significant element. Thus, the carry chain as a whole will consist of many pass transistors in series. This will give a very slow response and the carry line must therefore be suitably buffered between adder elements. (Remember, no more than four pass transistors in series — see section 4.9.) Also, we have assumed that both complements, $\overline{A}_k$ and $\overline{B}_k$, of the incoming bits are available. This may not be the case. Furthermore, signals A_k and B_k are to be derived from buses interconnecting the register with the ALU and may thus be taken off the bus through pass transistors. If this is the case, then these signals could not be used directly to drive the pass transistors of the multiplexers (see section 6.2.1). Finally, we must allow for storing the sum at the output of the adder, as discussed earlier in this section.

More practical general arrangements are shown in Figure 8–5. It will be seen that the adder element now contains all necessary buffering (at the expense of increased area). Seven inverter stages are required, deployed as follows (from top to bottom of Figure 8–5):

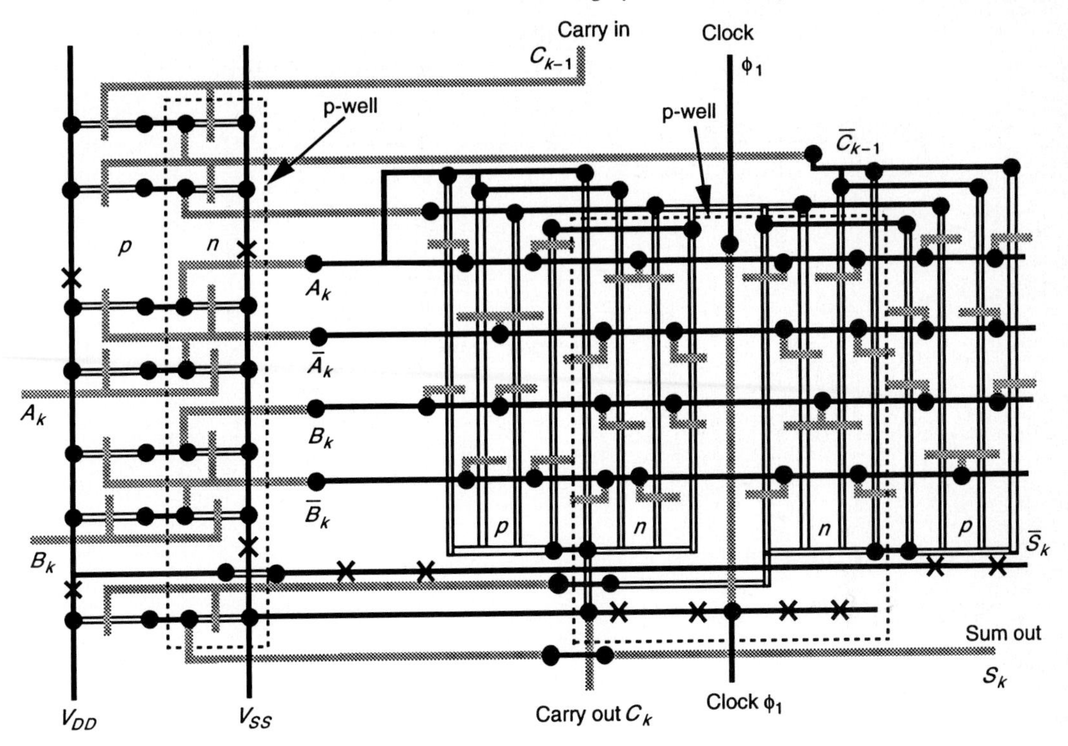

Figure 8–5 CMOS adder element

- Two inverters to form $\overline{C}_{k-1}$ and C_{k-1} (buffered)
- Two inverters to form $\overline{A}_k$ and A_k (buffered)
- Two inverters to form $\overline{B}_k$ and B_k (buffered)
- One inverter to act as a dynamic store for S_k.

Note that only one inverter is required to store the sum digit S_k *provided* that $\overline{S}_k$ rather than $\overline{S}_k$ is formed by the multiplexer. The observant reader will have already noted that the logic is configured to form the complement $\overline{S}_k$.

8.3.1.3 *Standard cells required to be designed for the adder element*

The stick diagram of Figure 8–5 shows that the adder consists of three parts:

1. the multiplexers (nMOS or CMOS);
2. the inverter circuits (4:1 and 8:1 ratio nMOS or CMOS);
3. the communication paths.

The first choice is that of technology — for example, nMOS or CMOS — and this in turn decides the detailed nature of the multiplexer and inverters. Both technologies lend themselves well to a replicated standard cell approach, only two standard cells being required for the complete adder element. The first cell, given in Figure 8–6, is the very simple cell from which multiplexers are formed. Note that the one cell design may appear with a transistor (includes poly. and contact) or without a transistor (omits contact cut and/or poly.). The second cell required is an inverter.

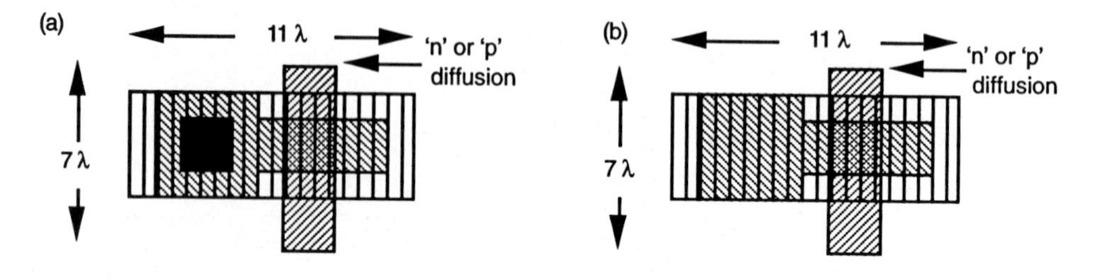

Figures 8–6 Multiplexer cell with or without cut

For nMOS, two versions of an inverter are needed — one for an 8:1 ratio and a second version for a 4:1 ratio. However, only one standard inverter cell design is actually needed with a choice of widths for the pull-down channel as shown in Figures 8–7 and 8–8. For completeness, a butting contact based inverter design is shown as Figure 8–7 since these contacts were used at one time by a number of nMOS fabricators.

The use of a 'standard' nMOS inverter with choice of width for the pull-down channel is a common practice. However, note that the narrow channel for the 4 :1 configuration in Figure 8–7 has been placed so that its edges are on *whole λ boundaries,* not half λ boundaries as would be the case if narrowing had been carried out symmetrically. *Always* design mask layouts having edges on whole λ boundaries. Some design rule checking software and some fabrication processes might not accept half λ edges.

More commonly, buried contacts would be used to join diffusion and poly. layers in nMOS fabrication and suitable buried contact based inverter designs are given in Figure 8–8.

In this case the vertical dimension is larger than that of Figure 8–7, but there are occasions where the lack of any metal regions in the center of the inverter is a positive advantage. For the layout shown, two metal bus lines could be run through the cell and across the inverter from side to side. This might prove a considerable advantage in saving space in certain layouts, such as register or memory arrays where data buses must run through each storage element. This could not be done when using a compact butting contact design because of the need to maintain 3λ metal to metal separation from the rails and the metal layer 'cover' on the butting contact.

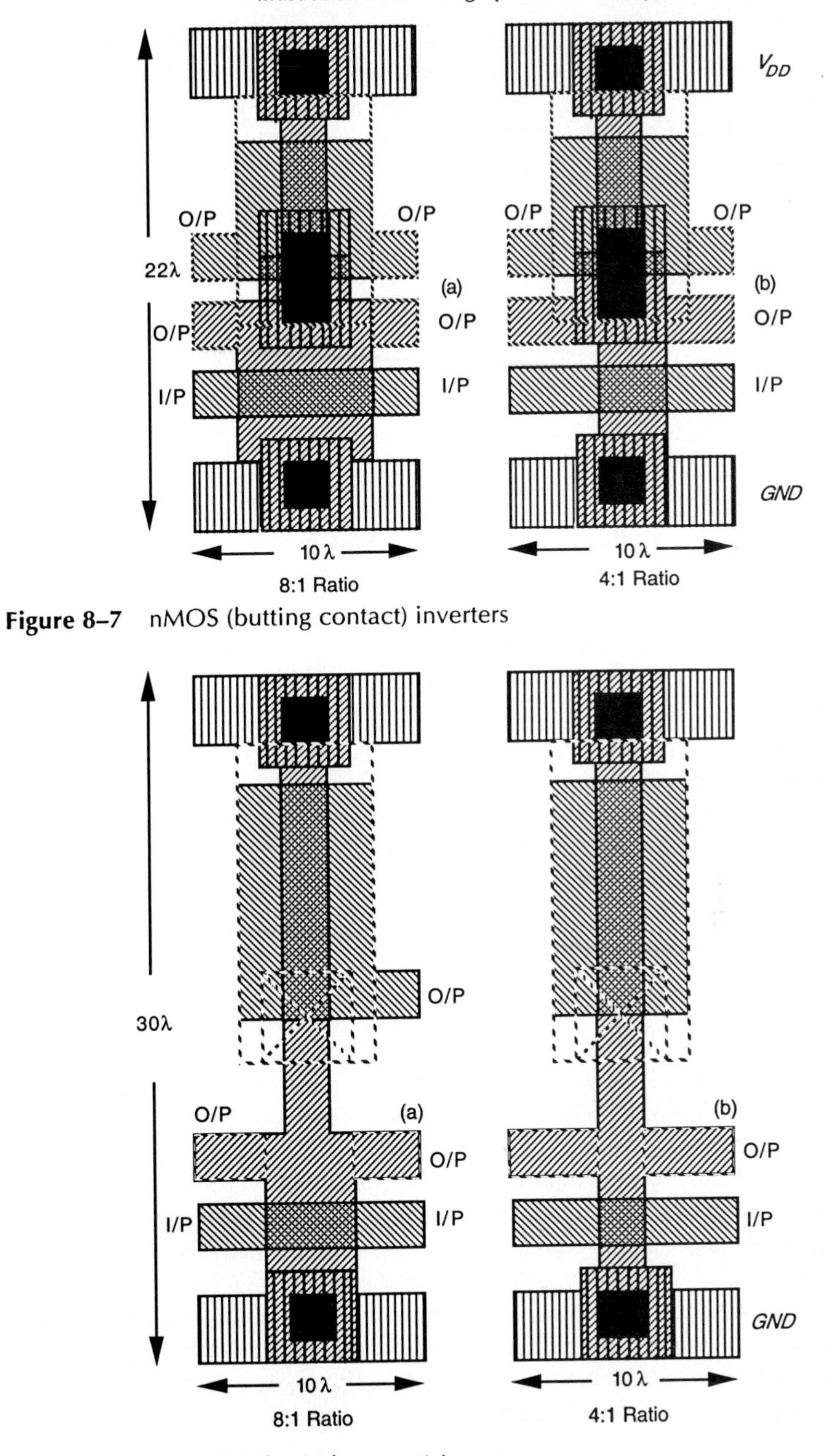

Figure 8–7 nMOS (butting contact) inverters

Figure 8–8 nMOS (buried contact) inverters

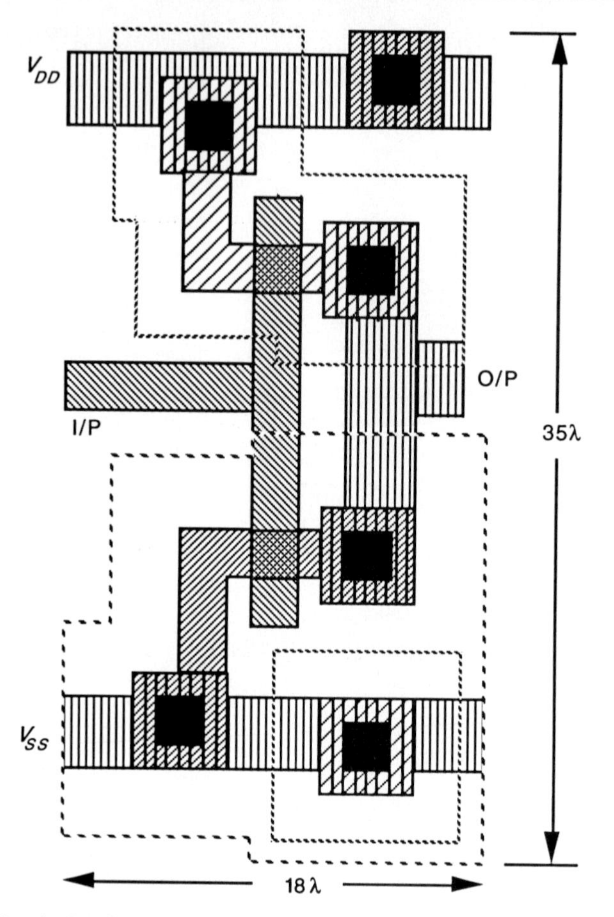

Figure 8–9 A CMOS inverter design

For CMOS-based designs, as set out in Figure 8–5, we normally need a complementary CMOS inverter. A possible mask layout is shown in Figure 8–9.

8.3.1.4 Adder element bounding box

We may now assess the area requirements for, say, the CMOS adder element as in Figure 8–5. First estimate the bounding box for the multiplexer area of the adder. Each standard multiplexer cell (Figure 8–6) is $7\lambda \times 11\lambda$ and there are 16 such elements side by side 'horizontally' and four stacked 'vertically'. We must also allow at least an additional 6λ width for the metal to metal spacings required by the clock bus passing through the center. In the vertical direction we must allow spacings for the interconnections between the tops of the multiplexers (an estimated additional 30λ) and a further 10λ for the connection out from $\overline{S}_k$ and C_k at the bottom. Thus, the bounding box must be at least $16 \times 7\lambda + 6\lambda = 118\lambda$ 'wide' and $4 \times 11\lambda + 30\lambda + 10\lambda = 84\lambda$ in 'height', as shown in Figure 8–10.

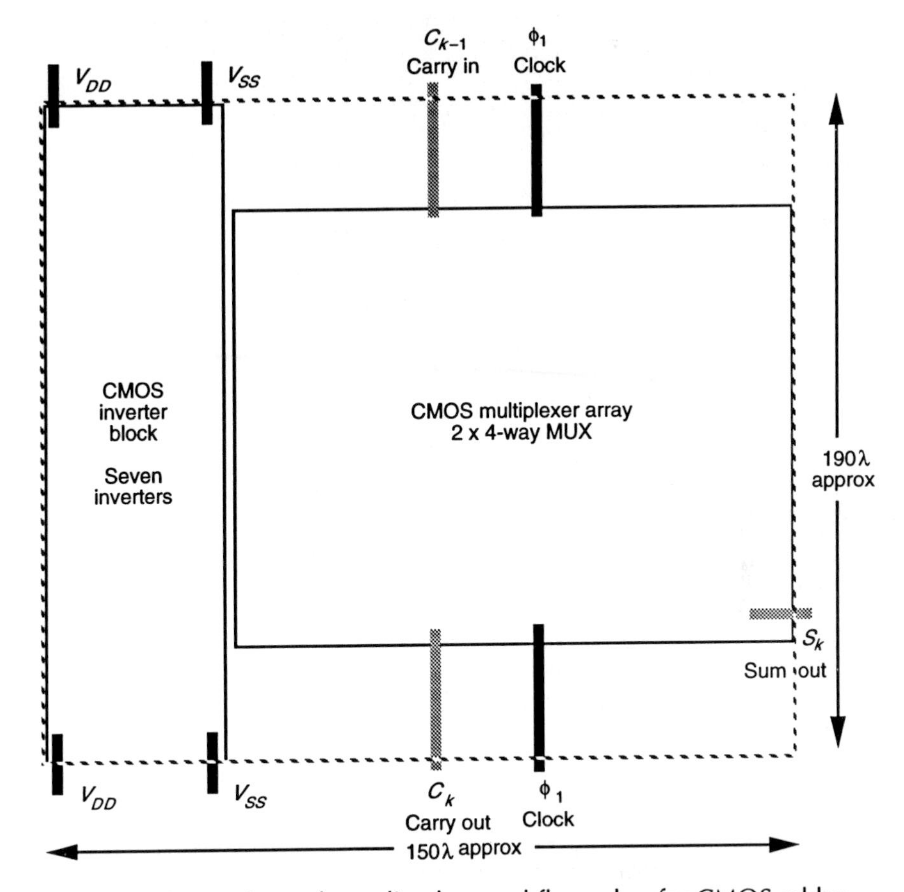

Figure 8–10 Approximate bounding box and floor plan for CMOS adder element

To complete an assessment of the approximate area to be occupied by the CMOS adder element we need to allow for the seven inverters shown in Figure 8–5. We have already determined a bounding box outline for a suitable CMOS inverter circuit (see Figure 8–9) and it will be seen that each inverter occupies a rectangle measuring 18λ 'wide' and 35λ 'high'. Thus, seven inverters alone placed side by side will occupy an area of $126\lambda \times 35\lambda$ and, allowing, say, an additional 50% width for space between each for connections, we have an overall area requirement of about $190\lambda \times 35\lambda$ for the inverters. Thus, the overall bounding box (or *floor plan outline*) for a complete adder element will be approximately that given as the overall outline in Figure 8–10. Note that vertical distribution of power is required by this layout, but the direction of global power distribution may be reviewed as the design of the complete processor — floor plan as in Figure 7–5 — progresses. Details of inlet/outlet points on the inverter block and overall adder element bounding boxes will be worked out as part of the next tutorial exercise.

The 4-bit adder is then formed by cascading four adder elements as indicated in Figure 8–11(a) and an initial assessment of the minimum floor plan area requirement follows from the 4-bit adder bounding box of Figure 8–11(b). This is the second subsystem of the floor plan of Figure 7–5, the first being the barrel shifter of Figure 7–10.

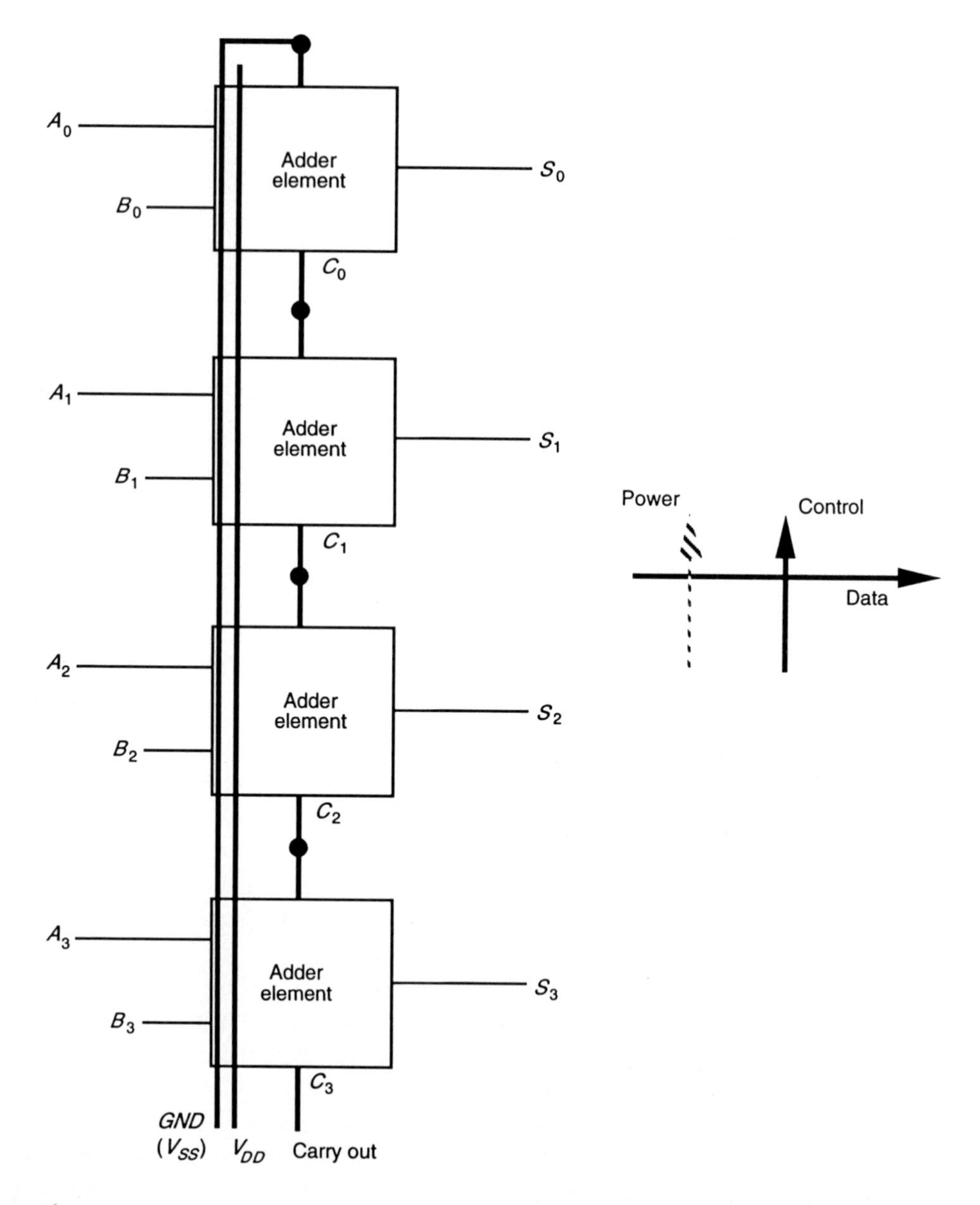

Figure 8–11(a) 4-bit adder

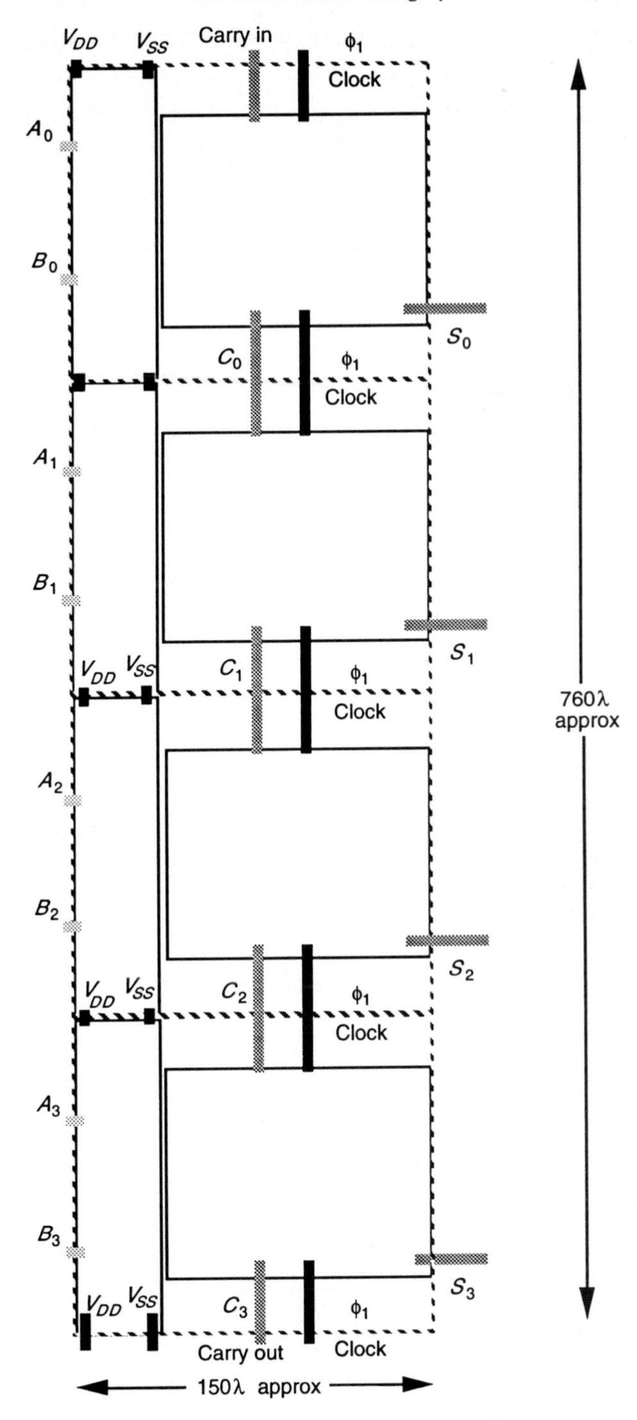

Figure 8–11(b) 4-bit adder outline

8.3.2 Implementing ALU functions with an adder

An arithmetic and logical operations unit (ALU) must, obviously, be able to *add* two binary numbers ($A + B$), and must also be able to *subtract* ($A - B$).

From the point of view of logical operations it is essential to be able to *And* two binary words ($A.B$). It is also desirable to *Or* ($A + B$) and perhaps also detect *Equality*, and of course we also need an *Exclusive-Or* function.

Subtraction by an adder is an easy operation provided that the binary numbers A and B are presented in *twos complement* form. In this case, to find the difference $A - B$ it is only necessary to complement B (exchange 1 for 0 and vice versa for all bits of B), add 1 to the number thus obtained, and then *add* this quantity to A using the standard addition process discussed earlier. The output of the adder will then be the required difference in twos complement form. Note that the complement facility necessary for subtraction can also serve to form the *logical complement* (which is indeed exchanging 0 for 1 and vice versa).

It is highly desirable to keep the architecture of the ALU as simple as possible, and it would be nice if the adder could be made to perform logical operations as readily as it performs subtraction. In order to examine this possibility, consider the standard adder equation set out in section 8.3.1 and reproduced here:

Sum $\quad S_k = \overline{H}_k C_{k-1} + H_k \overline{C}_{k-1}$

New carry $\quad C_k = A_k B_k + H_k C_{k-1}$

where Half sum $\quad H_k = \overline{A}_k B_k + A_k \overline{B}_k$

Consider, first, the *Sum* output if C_{k-1} is held at logical 0, then

$$S_k = H_k.1 + \overline{H}_k.0 = H_k$$

that is $\quad S_k = H_k = \overline{A}_k B_k + A_k \overline{B}_k$ — An *Exclusive-Or* operation

Now, hold C_{k-1} at logical 1, then

$$S_k = H_k.0 + \overline{H}_k.1 = \overline{H}_k$$

that is

$S_k = \overline{H}_k = \overline{A}_k \overline{B}_k + A_k B_k$ — An *Exclusive-Nor* (*Equality*) operation

Next, consider the *carry* output of each element, first if C_{k-1} is held at logical 0. Then

$C_k = A_k.B_k + H_k.0 = A_k.B_k$ — an *And* operation

Now, if C_{k-1} is held at logical 1, then

$$C_k = A_k.B_k + H_k.1 = A_k.B_k + \overline{A}_k.B_k + A_k.\overline{B}_k$$

Therefore

$C_k = A_{k.} + B_k$ — an *Or* operation

Thus it may be seen that suitable switching of the carry line between adder elements will give the ALU logical functions. A possible arrangement of the adder elements for both arithmetic and logical functions is suggested in Figure 8–12.

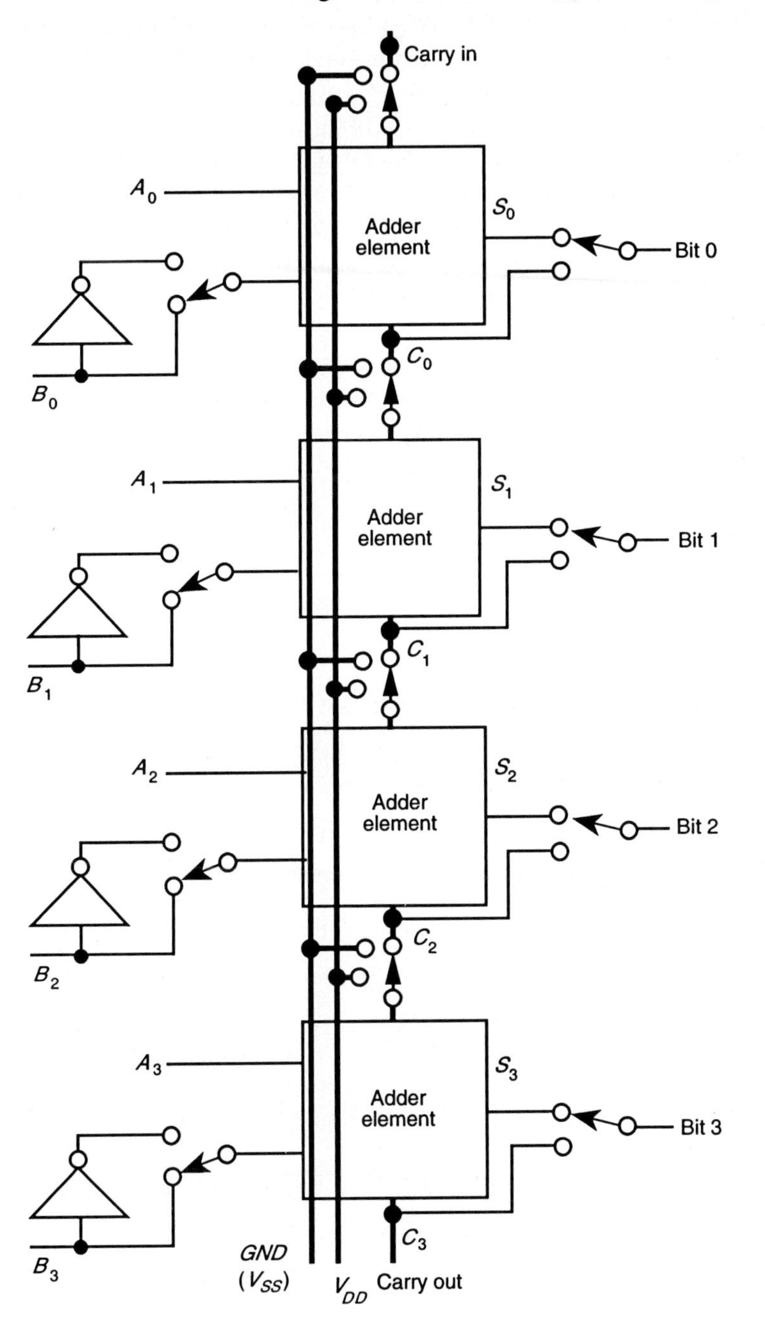

Figure 8–12 4-bit ALU

8.4 A further consideration of adders

A further consideration of aspects of adder circuitry is desirable since adders are the basic elements of all arithmetic processes. Also, so far, we have taken a very simple and direct approach to implementing the adder equations and have not considered refinement or optimization of performance.

In order to broaden the scope of our discussion, let us first consider some of the commonly used alternative forms of the adder equations introduced in Section 8.3.1 and repeated here for convenience.

Sum $S_k = \overline{H}_k C_{k-1} + H_k \overline{C}_{k-1}$

New carry $C_k = A_k B_k + H_k C_{k-1}$

where

Half sum $H_k = \overline{A}_k B_k + A_k \overline{B}_k$

The expressions may also make use of lowercase letters. New carry may also be expressed in terms of the previous carry c_{k-1} with a *propagate* signal p_k and *generate* signal g_k, where

$$p_k (= H_k) = a_k \oplus b_k \text{ and } g_k = a_k.b_k$$

Then we may write,

new carry $c_k = p_k.c_{k-1} + g_k$

or $c_k = (a_k + b_k)\, c_{k-1} + a_k.b_k$

and sum $s_k = a_k \oplus b_k \oplus c_{k-1}$

The sum may also be expressed in terms of the carry in c_{k-1} and carry out signals c_k together with the input bits a_k and b_k as follows:

$$s_k = \overline{c}_k.(a_k + b_k + c_{k-1}.) + a_k.b_k.c_{k-1}$$

Such manipulations lead, for example, to the complementary CMOS logic circuit in Figure 8–13.

However, an alternative and perhaps more direct realization, which leads to the concept of a carry chain, is set out in Figure 8–14. This in turn, when considering carry circuits alone, leads to a popular arrangement known as the *Manchester carry-chain*.

8.4.1 The Manchester carry-chain

Instead of the carry passing through a complete transmission gate as in Figure 8–14, the carry path is precharged by the clock signal and the carry path may then be gated by a single n-type pass transistor as shown in Figure 8–15.

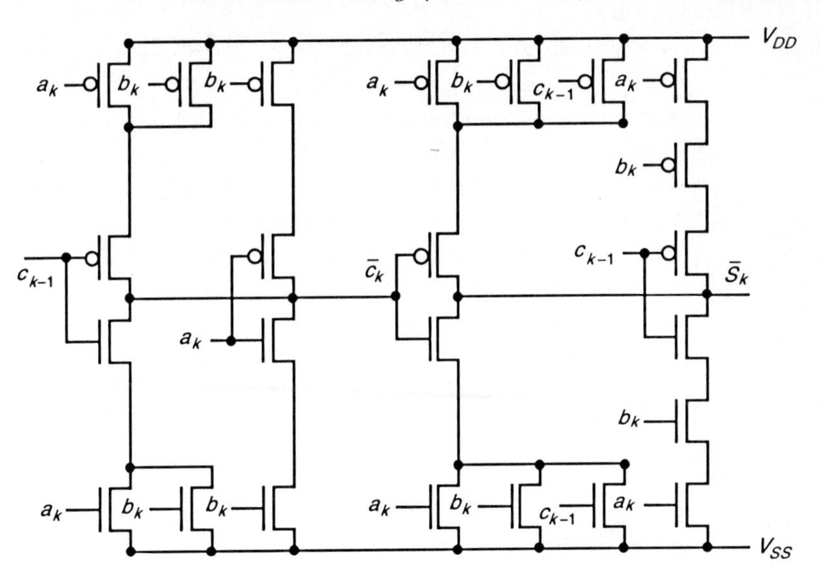

Figure 8–13 One possible (symmetrical) adder cell arrangement

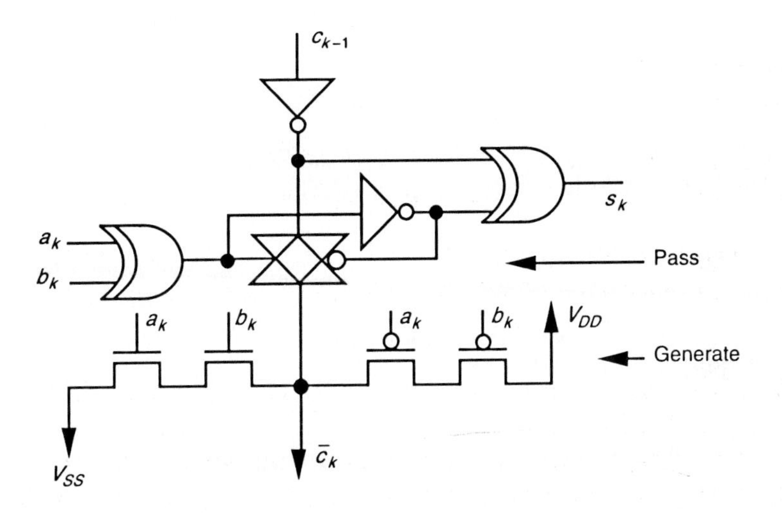

Figure 8–14 An adder element based on the pass/generate concept

Although individual Manchester carry cells are fast, care must be taken when cascading them since this effectively connects pass transistors in series. We have already seen that the delay goes up as the square of n where n is the number connected in series. Obeying the rules set out earlier to cover this situation, we must buffer after every four carry chain cells as shown in Figure 8–16.

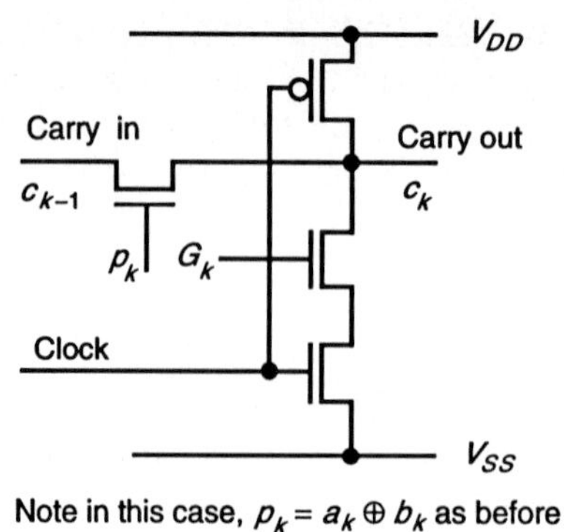

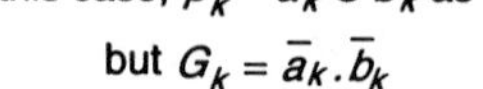

Figure 8–15 Manchester carry-chain element

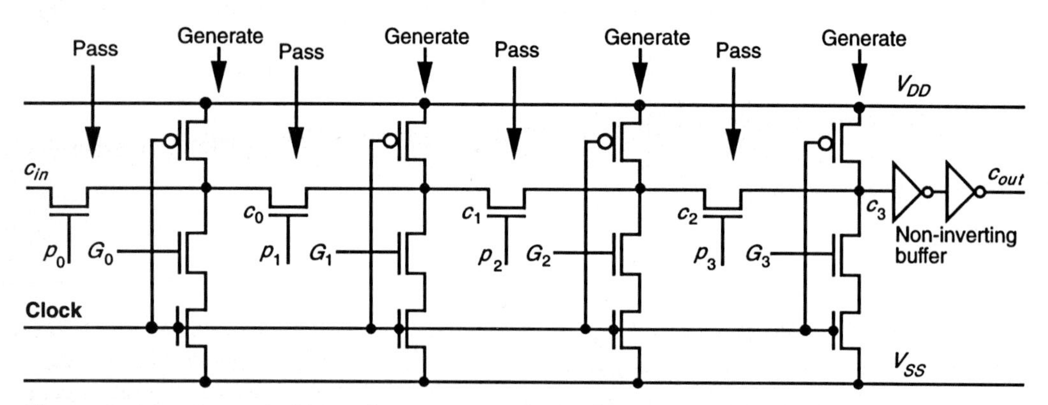

Figure 8–16 Cascaded Manchester carry-chain elements with buffering

In BiCMOS technology it is possible to implement this arrangement and achieve speed improvement by a factor of two over the CMOS arrangement. However, this approach functions with lower input voltage swings to achieve the full speed advantage (Hotta et al., 1986).

8.4.2 Adder enhancement techniques*

In the case of small adders ($n < 8$-bits), it is generally advantageous to adopt the relatively simple hardware of the ripple through carry. Thus, the carry completion time is clearly directly proportional to n. On the other hand, large adders (up to say $n = 64$ or even $n = 128$-bits) cannot afford to wait for the long completion time of a large ripple through carry line. Thus special techniques must be adopted

* This section is based on material provided by Dr B. Hochet of the Swiss Federal Institute of Technology (EPFL), Lausanne, Switzerland. The authors gratefully acknowledge this contribution.

to improve addition time. This improvement is possible only through some increase in complexity and, in consequence, at the expense of increased area in silicon. The next subsections discuss three techniques for effecting faster carry generation and each approach is characterized by a different area/performance ratio.

8.4.2.1 Carry select adders

For this arrangement — also referred to as a conditional sum adder — the adder is divided into blocks. Each block is composed of two adders, one with a logical 0 *carry in* and the other with a logical 1 *carry in*. The *sum* and *carry out* generated are then selected by the actual *carry in* which comes from the *carry out* output of the previous block as shown in Figure 8–17.

8.4.2.1.1 Optimization of the carry select adder

Let us consider an n-bit ripple carry adder. The computation time T is given by:

$$T = k_1 n$$

where k_1 is the delay through one adder cell.

If we now divide the adder into blocks, each with two parallel paths, then the completion time T becomes

$$T = k_1 . \frac{n}{2} + k_2$$

where k_2 is the time needed by the multiplexer of the next block to select the actual output carry. A decision now has to be made on the size, in bits, of each adder block and clearly this could be 1-bit, in which case the number of multiplexers is a maximum, or two or more bits resulting in fewer multiplexers. If there are many multiplexers, then the ripple through effects occur in the multiplexer chain rather than in the carry chain through the blocks. Consequently, an optimum value must be sought for the block size.

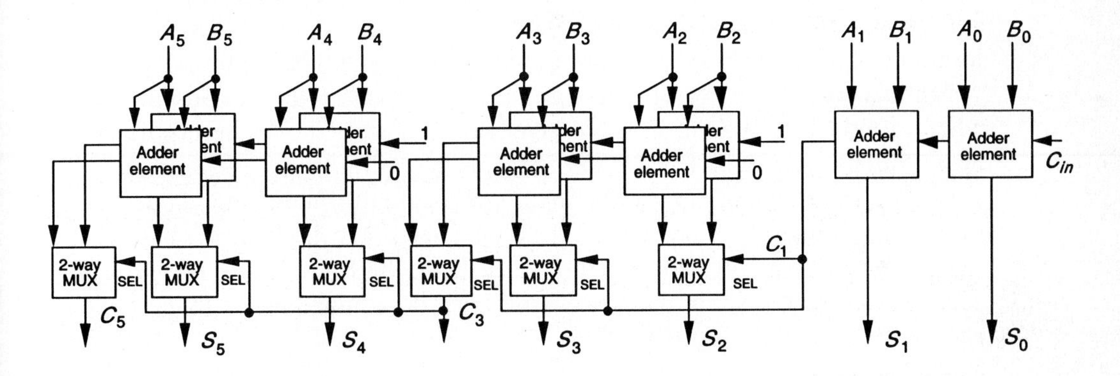

Figure 8–17 Carry select adder structure (6-bit)

Suppose the n-bit adder is divided into M blocks, and that each block contains P adder cells in series, and considering the arrangement of Figure 8–17, we may see that the completion time T for the overall carry output signal is composed of two parts:

- the propagation delay through the first block
- the propagation delay through the multiplexers

so that,

$$T = P\,k_1\,. + (M-1)k_2$$

noting that $n = M.P$, the minimum value for T is reached when

$$M = \sqrt{(n.k_1 / k_2}$$

As a further improvement, each succeeding block may be extended by one or more stages to account for the delay in the multiplexer. For instance, if the delay in the multiplexer is equal to the cell delay, then the size P of the succeeding block should be increased by one. On the other hand, if the multiplexer delay is twice that for the cell delay, then each block may have two more adder cells than the previous one; that is, P can be increased by two from one block to the next. The actual optimum increase in P from one block to its successor depends on the ratio between k_1 and k_2. However, care must be taken to properly allow for the multiplexer delay which will also depend on the number of inputs, that is, on P, increasing as P increases.

It should also be noted that the adder blocks do not have to be ripple carry adders but may use any of the available enhancement techniques, such as carry look-ahead or carry skip techniques. In such cases, the optimization requirements may be different from those discussed here.

8.4.2.2 *Carry skip adders*

When computing an addition with a ripple through adder, the completion time will sometimes be small since the carries, generated at several positions, are formed simultaneously as shown (e.g. with three carries) in Figure 8–18.

In this case, the carry propagation may be likened to the domino principle, where, if one falls, then each successive stage is knocked over in turn up to the next point at which a different carry is formed.

In this example, assuming the input *carry in* = 1, three simultaneous carry propagation chain reactions occur. It may be seen that the longest chain is the second one, which takes seven cell delays (from the fourth bit to the 11th bit). Thus, the addition time for these two numbers is determined by the longest chain, and in this case will be given by

$$T = 7.k + k'$$

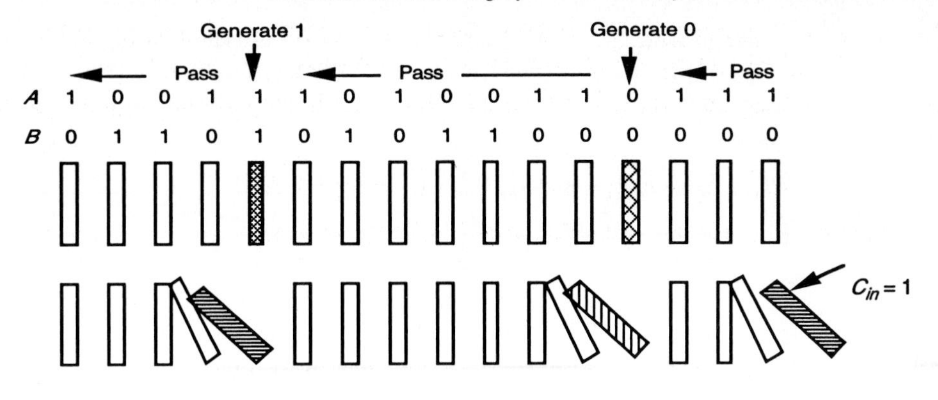

Figure 8–18 Diagrammatic representation of carry skip adder

where k is the cell delay and k' is the time needed to compute the 11th bit sum using the carry in to the 11th bit.

If, for a ripple carry adder, the input bits A_i and B_i are different for all bit positions, then the input carry is propagated at all bit positions and never generated. The addition is thus only completed after the carry has propagated along the entire adder. In this case, the computation delay must be nk, and although it may be less than this quite frequently, the worst case must be assumed in all cases when using the adder in, say, high speed or real-time or other time-critical applications.

Carry skip adders take advantage of both the generation and the propagation of the carry signal. They may be divided into blocks where, for each block, a special circuit is used to detect the condition when A and B bits differ in all bit positions in the block (that is $p_i = 1$ for all 'i' in the block). The output signal from such a circuit is called the *block propagation signal.* If the *block propagation signal* = 1, then the carry signal entering the block can bypass it and be transmitted through a multiplexer to the next block. Figure 8–19 sets out the schematic structure of a 24-bit carry skip adder, subdivided into four blocks and based on this approach.

8.4.2.2.1 Optimization of the carry skip adder

Once again there will be factors which determine the optimum block size for this arrangement and in this case we assume equal size blocks. Let k_1 denote the time needed by the carry signal to propagate through the adder cell, and k_2 the time needed for a carry to skip over a block. Further, let us divide the n-bit carry skip adder into M blocks — each block containing P adder cells. Since, as was the case for the ripple carry adder, the actual computing time depends on the configuration of the input numbers, the completion time may well be small but may also reach the worst case. We must thus evaluate and optimize the worst case conditions as depicted in Figure 8–20.

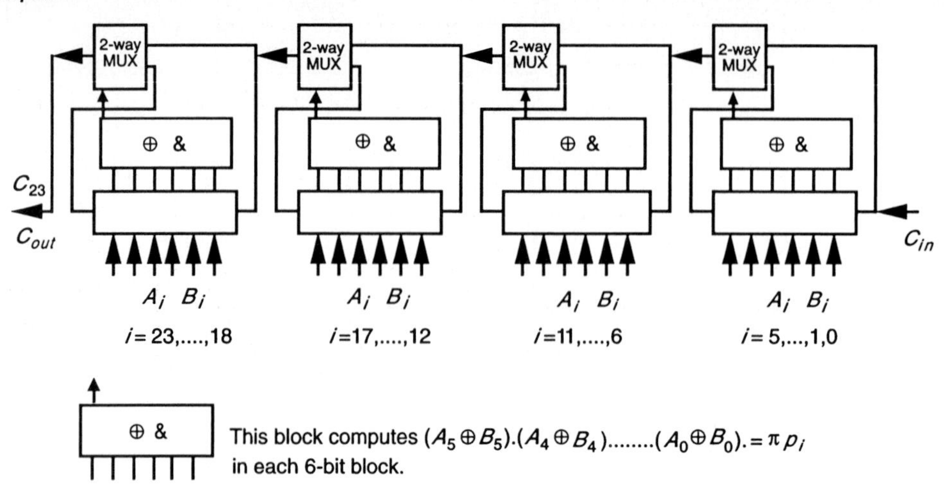

Figure 8–19 Structure of a 24-bit (for example) carry skip adder

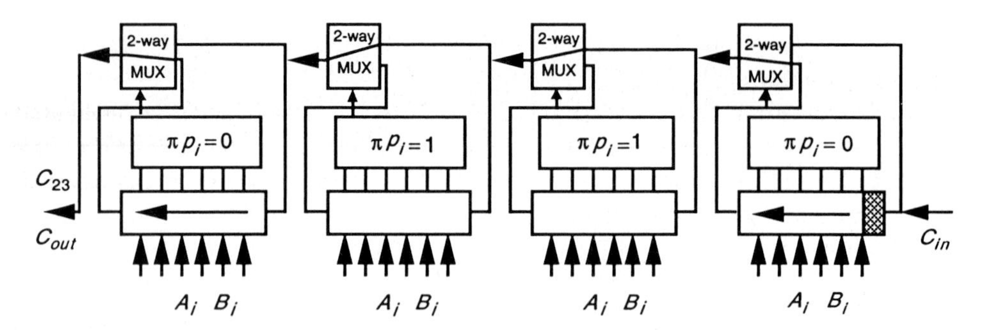

Figure 8–20 Worst case carry propagation for carry skip adder

The total (worst case) propagation delay time T is given by

$$T = 2(P-1).k_1 + (M-2)\,k_2$$

where

$$P = n/M$$

The minimum value of T is reached when

$$M = \sqrt{(2n.k_1/k_2}$$

As for the carry select adders, a further improvement may be achieved if the adder is divided into blocks of differing sizes (Guyot et al., 1987).

Finally, Figure 8–21 shows a possible arrangement for a block, complete with its multiplexer and block propagation signal generating circuit. This particular realization leads to good regularity and thus to a high density layout in silicon.

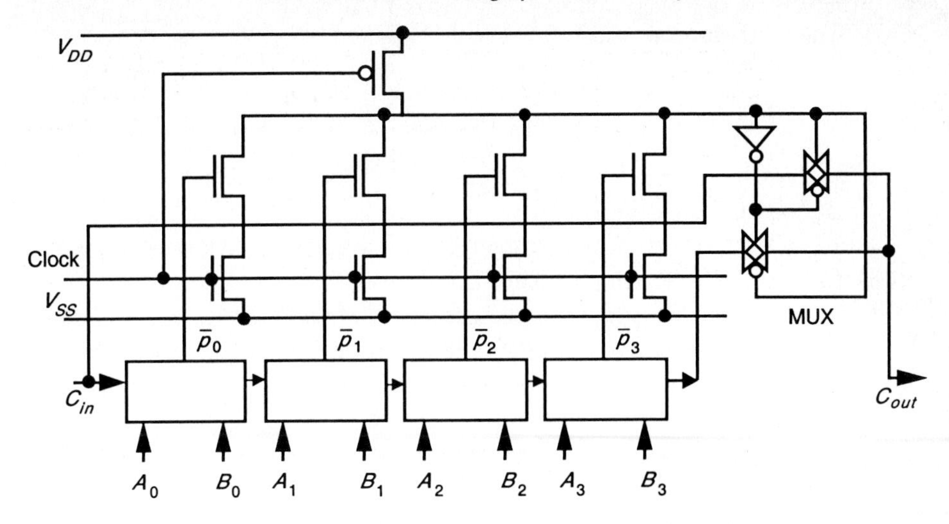

Figure 8–21 Possible implementation of the block propagation concept

8.4.2.3 Carry look-ahead (CLA) adders

We have considered some other methods of improving adder throughput time and may now turn to algebra to seek a general solution to this problem. This is to be found in rearranging the expressions for the adder (given in section 8.3.1), in particular the expression for carry

$$C_k = A_k B_k + H_k . C_{k-1}$$

noting that $H_k = A_k B_k + A_k B_k$ the expression can be rearranged into the form

$$C_k = A_k . B_k + (A_k + B_k) . C_{k-1}$$

Thus for C_0 we may write

$$C_0 = A_0 . B_0 + (A_0 + B_0) C_{in}$$

which allows for an input carry; and, therefore

$$C_1 = A_1 . B_1 + (A_1 + B_1)\ C_0$$

may then be written as

$$C_1 = A_1 . B_1 . + (A_1 . + B_1) . A_0 . B_0 . + (A_1 . + B_1 .) . (A_0 + B_0) . C_{in}$$

and, similarly

$$C_2 . = A_2 . B_2 . + (A_2 + B_2) . A_1 . B_1 . + (A_2 + B_2) . (A_1 + B_1) . A_0 . B_0 . + \\ (A_2 + B_2) . (A_1 + B_1) . (A_0 + B_0) . C_{in}$$

The next stage would be

$$C_3 = A_3.B_3 + (A_3 + B_3).\ A_2.B_2 + (A_3 + B_3).(A_2 + B_2).A_1.B_1 + (A_3 + B_3).(A_2 + B_2).(A_1 + B_1).A_0.B_0 + (A_3 + B_3).(A_2 + B_2).(A_1 + B_1).(A_0 + B_0).C_{in}$$

and so on for further stages.

If there is no input carry, then C_{in} becomes 0 and the last term in each expression for carry will be eliminated.

Although these expressions become very lengthy as the bit significance increases, each expression is only three logic levels deep, so the delay in forming the carry is constant irrespective of bit position. However, the logic does rapidly become over-cumbersome and also presents problems in 'fan-out' and 'fan-in' requirements on the gates used. A compromise, usually adopted, is a combination of 'carry look-ahead' and 'ripple through' as indicated in Figure 8–22. The 3-bit groups shown were arbitrarily chosen to illustrate the approach.

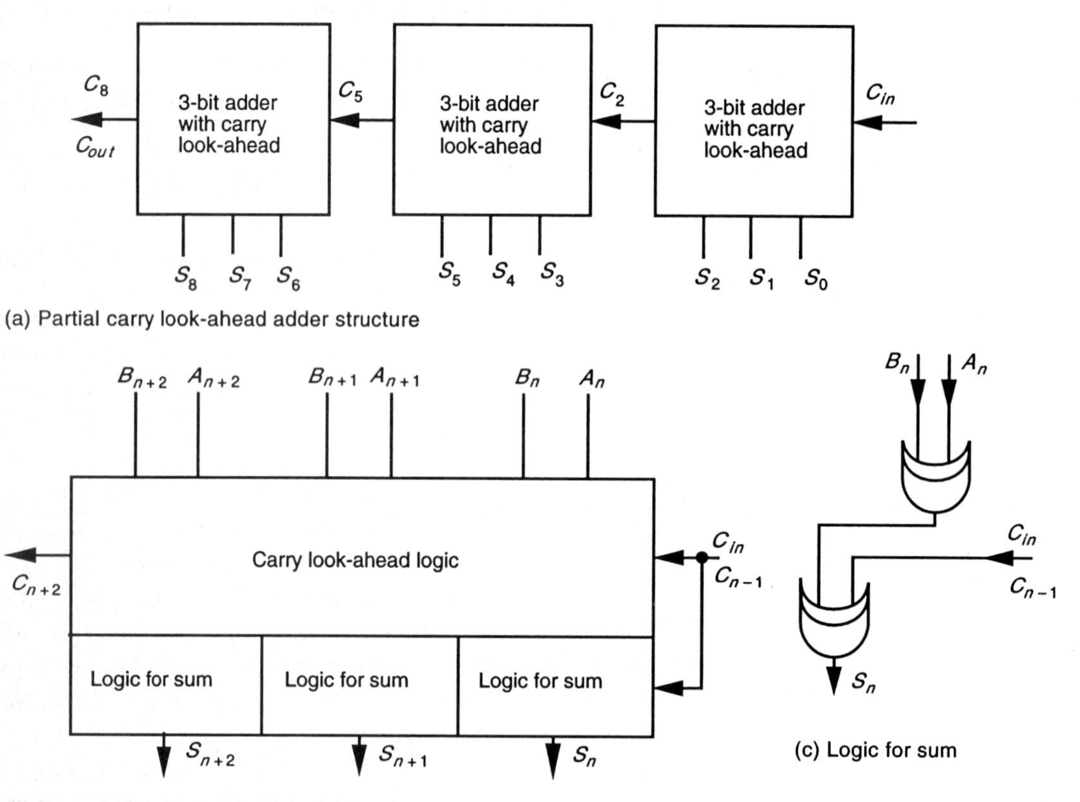

Figure 8–22 Carry look-ahead and ripple through compromise

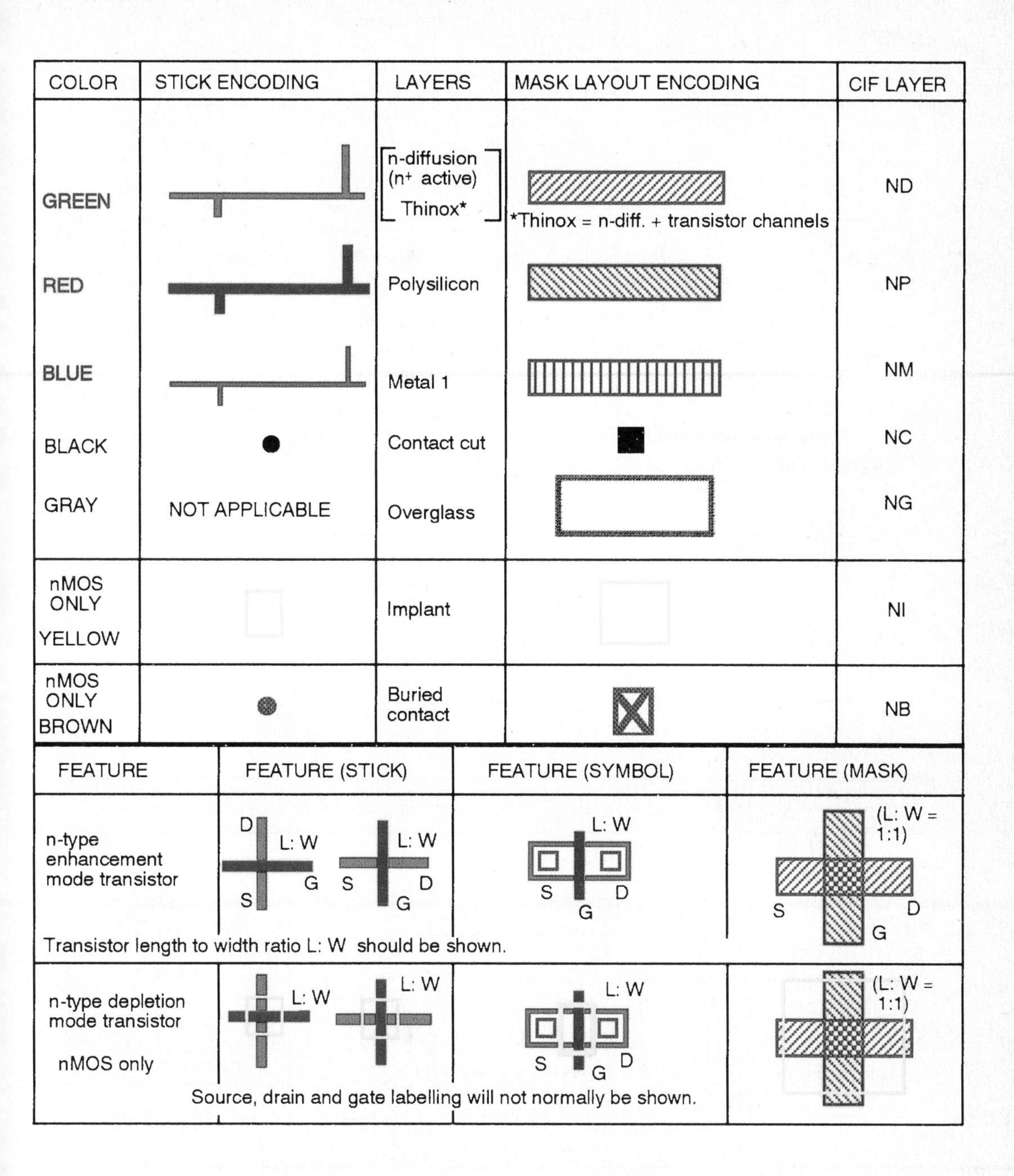

Color plate 1(a) Encodings for a simple single metal nMOS process. (See Figure 3–1(a) for nMOS monochrome encoding details.)

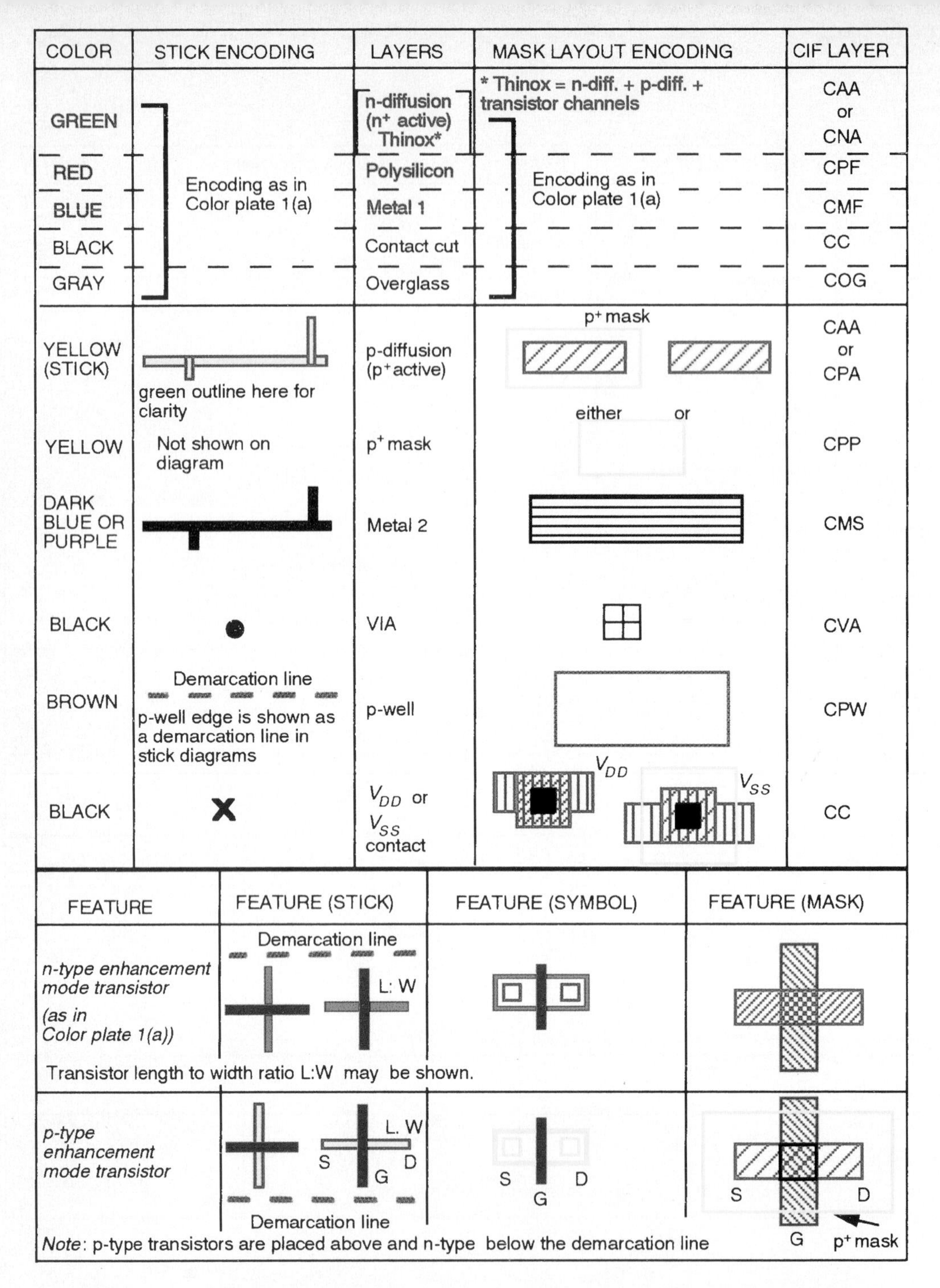

Color plate 1(b) Color encodings for a double metal CMOS p-well process. The same well encoding and demarcation line is used for an n-wll process. For a p-well process, the n features are in the well. For an n-well process, the p features are in the well. (See Figure 3–1(b) for CMOS monochrome encoding details.)

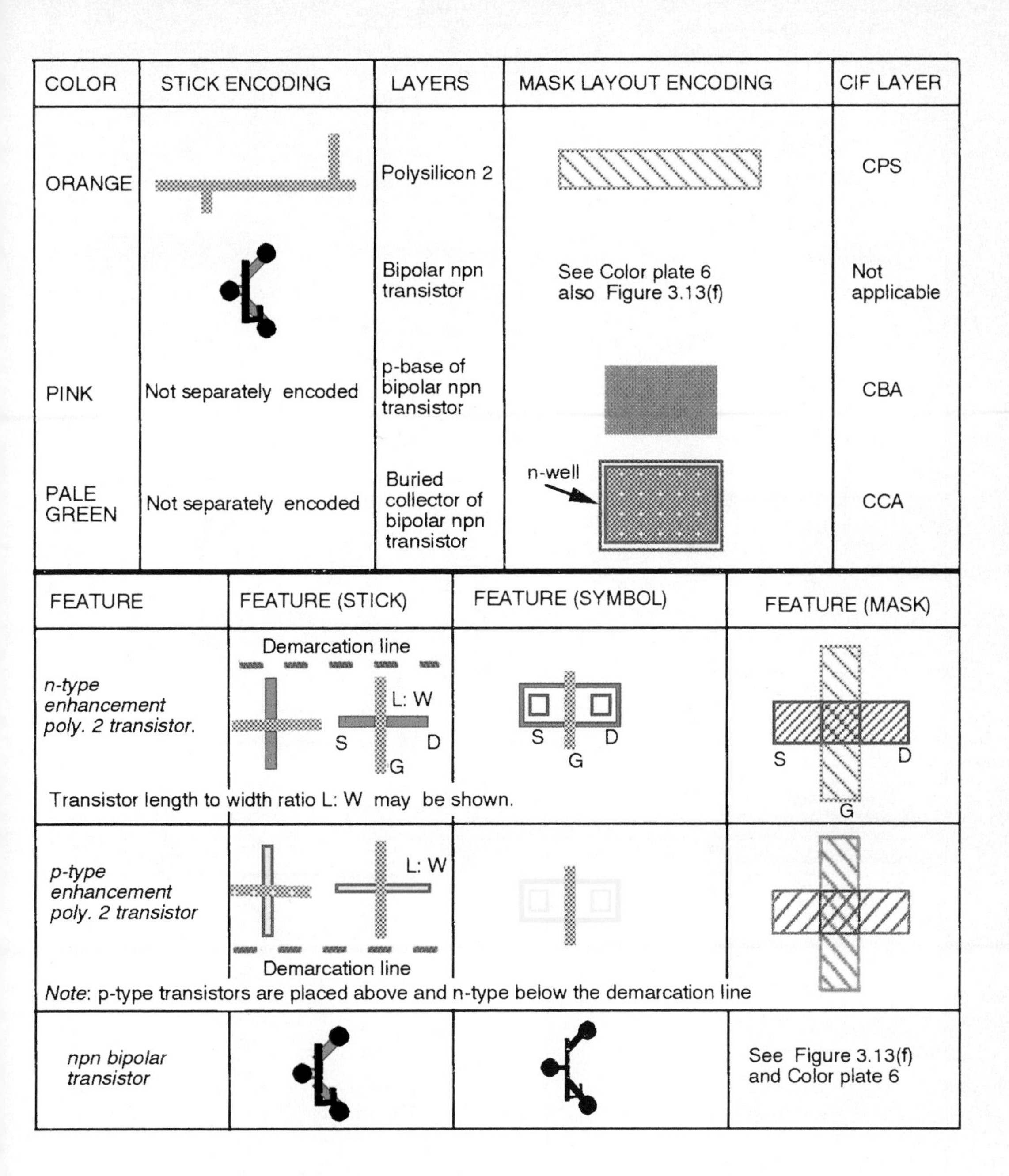

COLOR	STICK ENCODING	LAYERS	MASK LAYOUT ENCODING	CIF LAYER
ORANGE		Polysilicon 2		CPS
		Bipolar npn transistor	See Color plate 6 also Figure 3.13(f)	Not applicable
PINK	Not separately encoded	p-base of bipolar npn transistor		CBA
PALE GREEN	Not separately encoded	Buried collector of bipolar npn transistor	n-well	CCA

FEATURE	FEATURE (STICK)	FEATURE (SYMBOL)	FEATURE (MASK)
n-type enhancement poly. 2 transistor.	Demarcation line; L: W; S; D; G	S; D; G	S; D; G
Transistor length to width ratio L: W may be shown.			
p-type enhancement poly. 2 transistor	L: W; Demarcation line		
Note: p-type transistors are placed above and n-type below the demarcation line			
npn bipolar transistor			See Figure 3.13(f) and Color plate 6

Color plate 1(c) Additional encodings for a double metal double poly. BiCMOS n-well process. The same well encoding and demarcation line as in Figure 3–1(b) is used for an n-well process. For a p-well process, the n features are in the well. (See Color plate 6 for additional BiCMOS color encoding details and see Figure 3–1(c) for monochrome encoding details.)

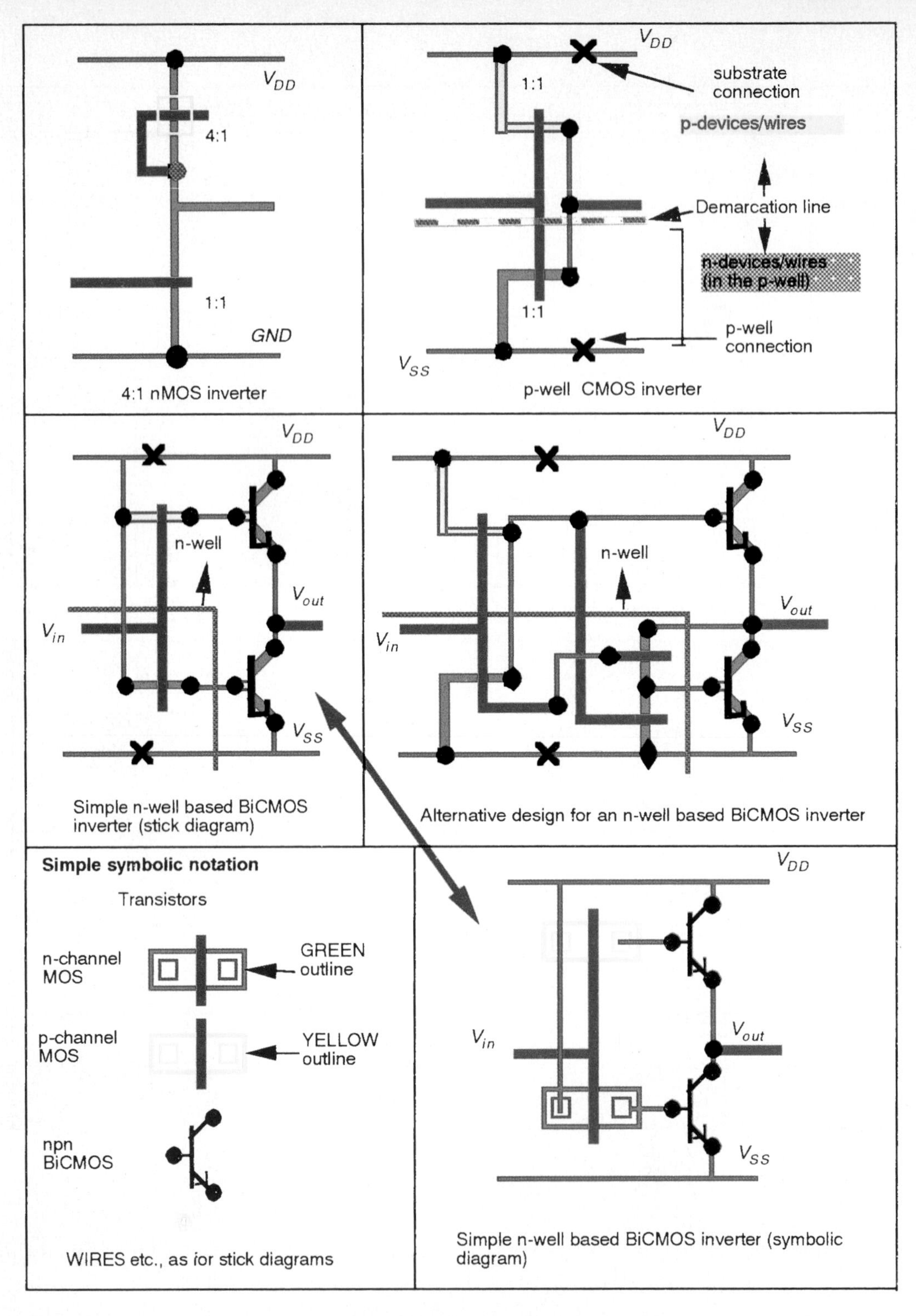

Color plate 1(d) Color stick diagram examples. (See Figure 3–1(d) Monochrome stick diagrams and simple symbolic encoding.)

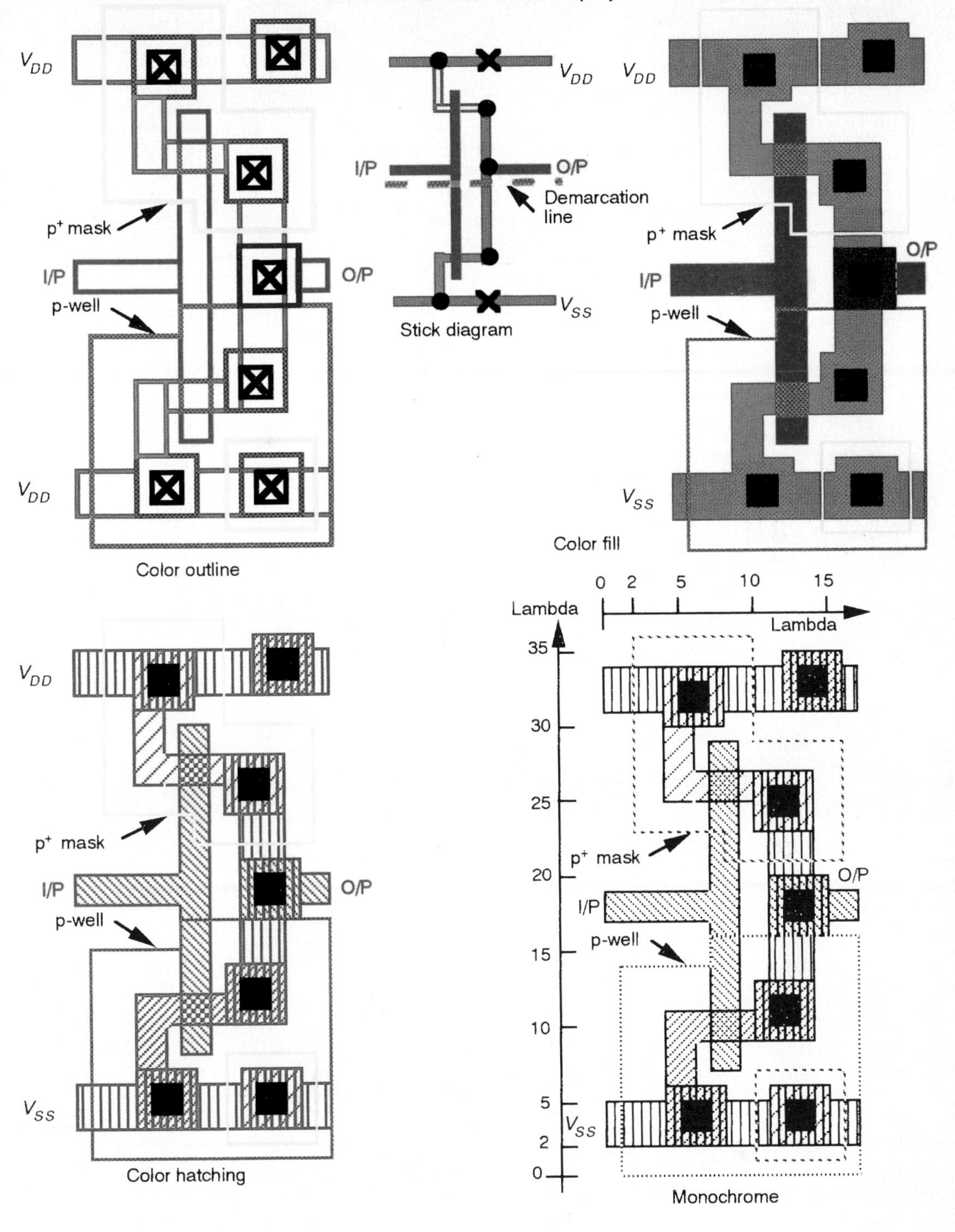

Color plate 2 Example layout encodings

Design rules for wires (interconnects) (ORBIT 2 μm CMOS)

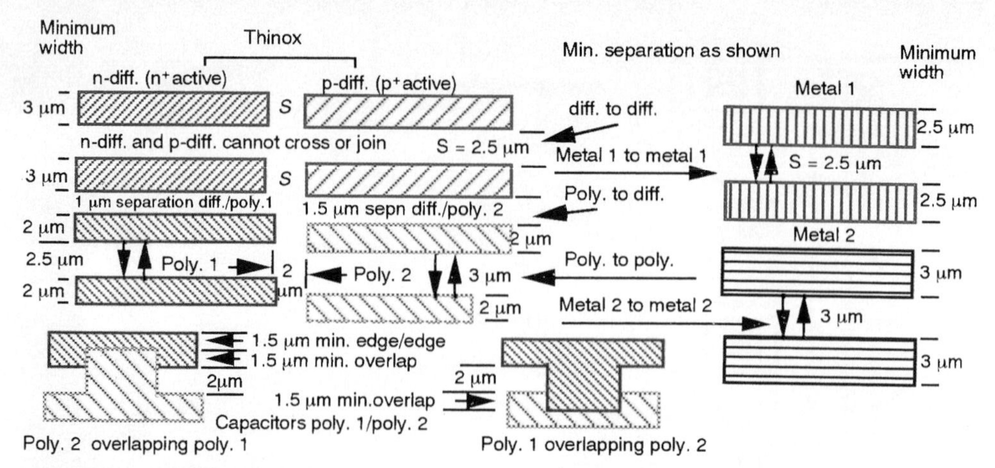

Otherwise poly. 2 must not be coincident with poly. 1

Note: Where no separation is specified, wires may overlap or cross (e.g. metal may cross any layer). For p-well CMOS, n-diff. wires can only exist inside and p-diff. wires outside p-well. For n-well CMOS, p-diff. wires can only exist inside and n-diff. wires outside n-well.

Avoid coincident edges where metal 1 and metal 2 runs follow the same path for > 25 μm length (underlap metal 1 edges by 0.8 μm).

Transistor related design rules (ORBIT 2 μm CMOS)

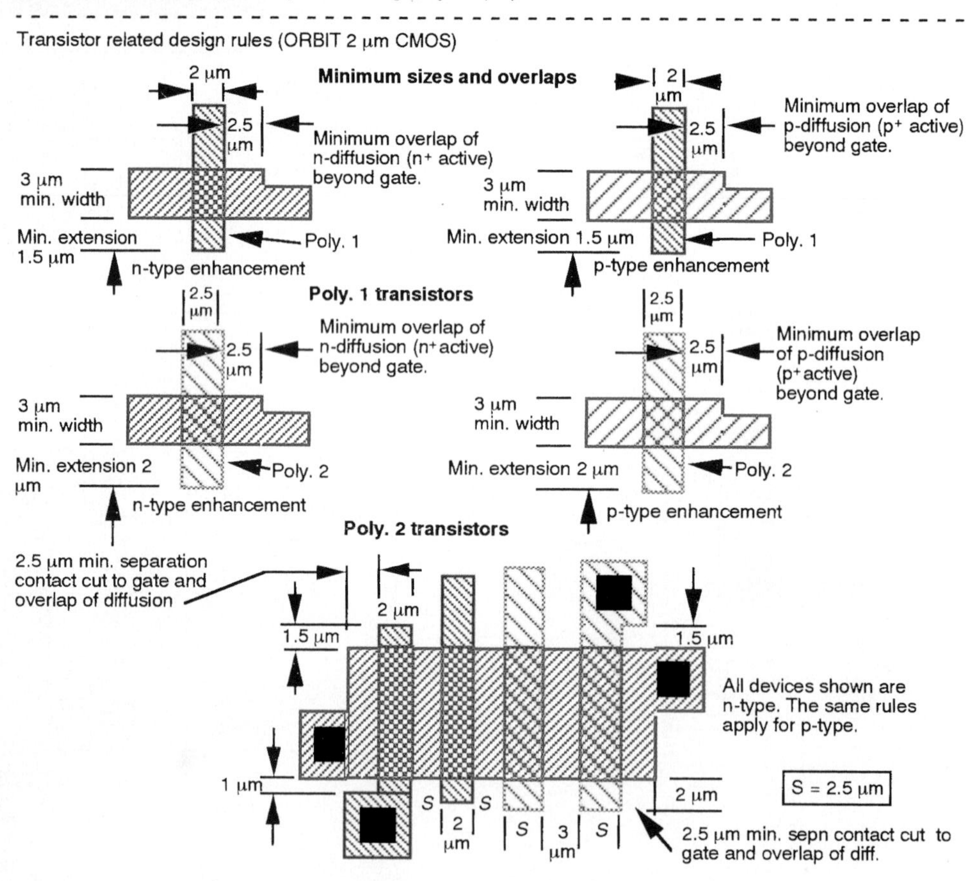

Color plate 3 ORBIT™ 2 μm design rules (a) (b)

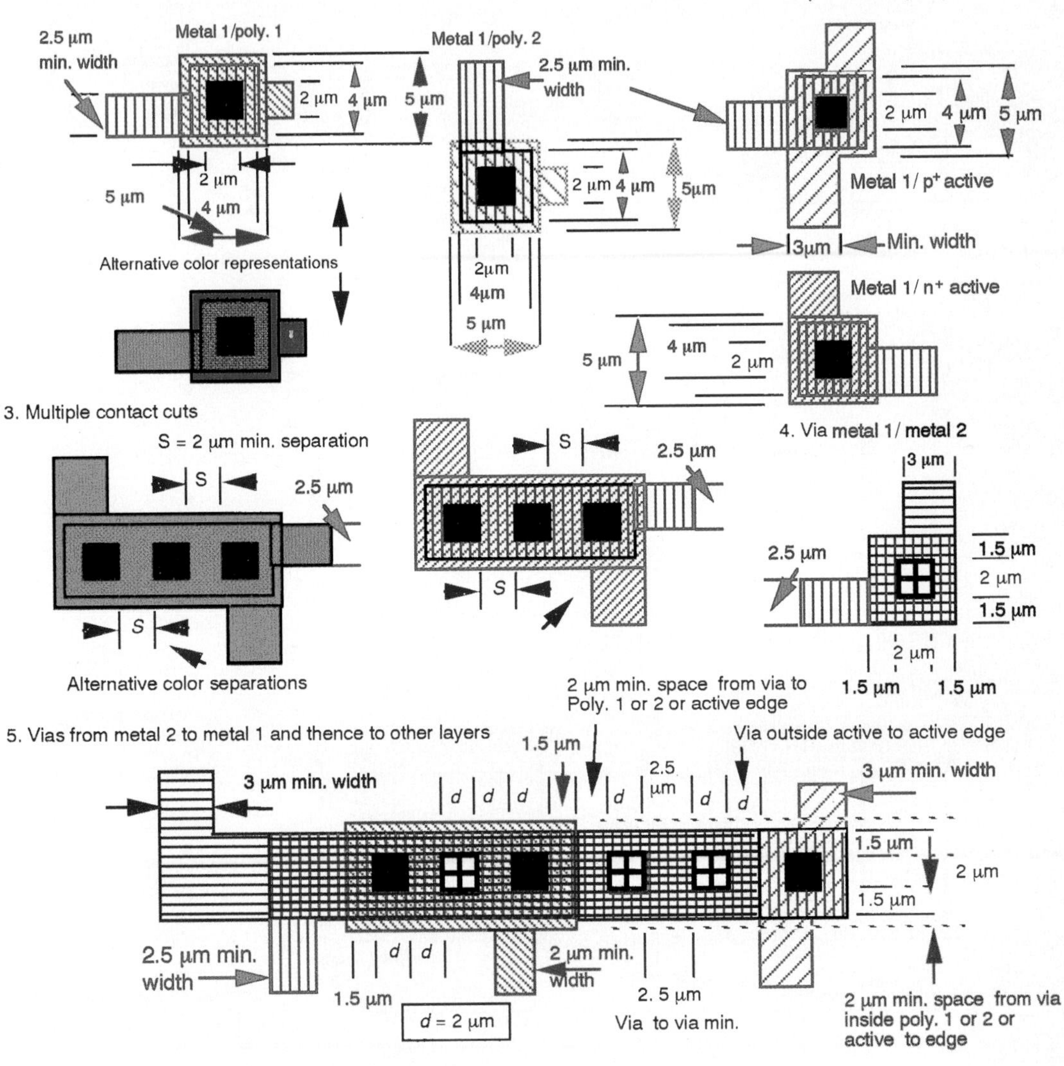

Note that vias must not be placed over contacts.

Color plate 4 ORBIT™ 2 μm design rules (c)

Rules for n-well and V_{DD} and V_{SS} contacts (ORBIT 2 μm CMOS process)

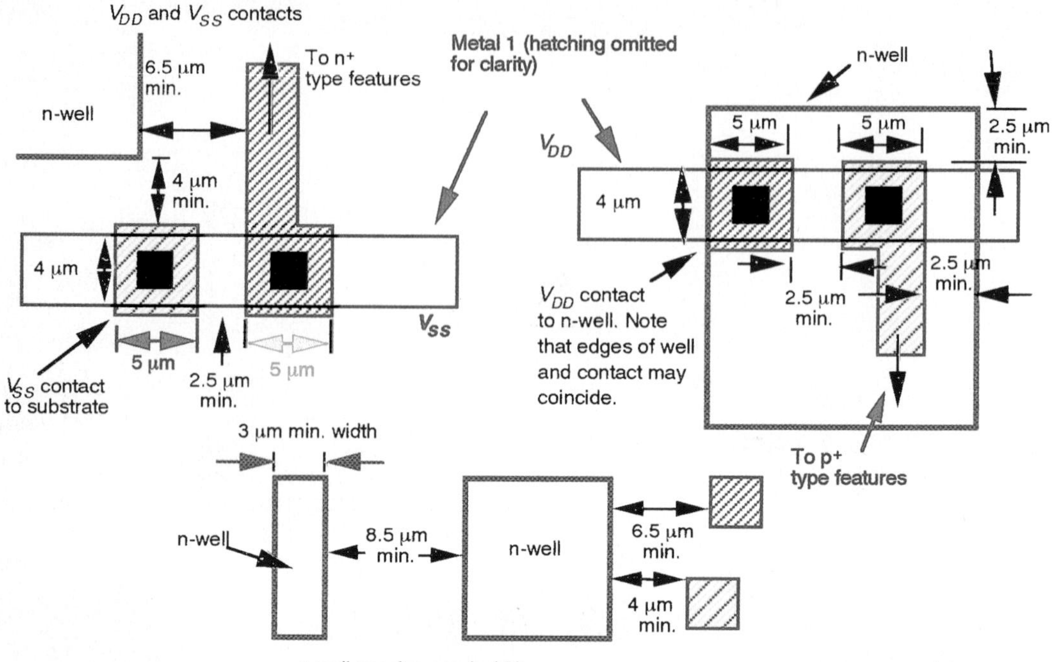

Rules for pad and overglass geometry (ORBIT 2 μm CMOS)

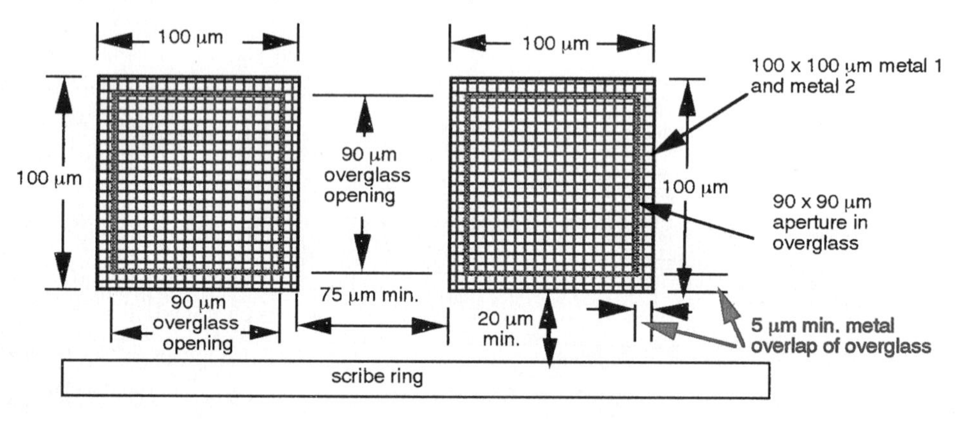

Other rules and encodings:

Via overlap of pad 2 μm
Pad to active separation 20 μm minimum
Color encoding for overglass mask . . . gray.

Color plate 5 ORBIT™ 2 μm design rules (d) (e)

Special rules for BiCMOS transistors (ORBIT 2 μm CMOS)

Note: For clarity, layers have not been drawn transparent. Note that BCCD underlies the entire area and the p-base underlies all within its boundary.

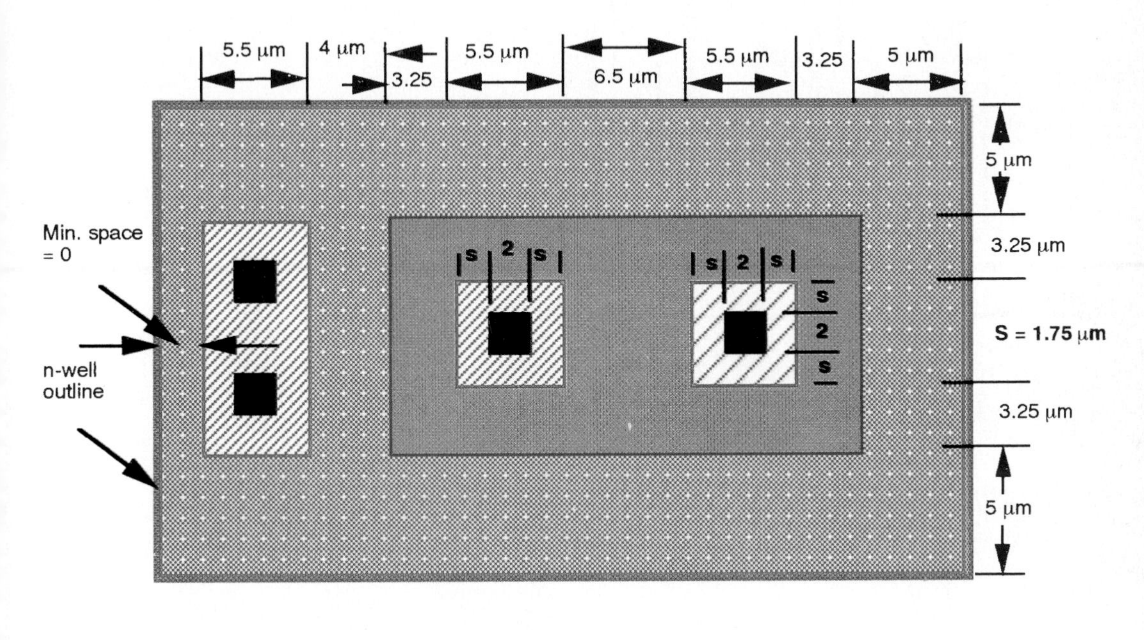

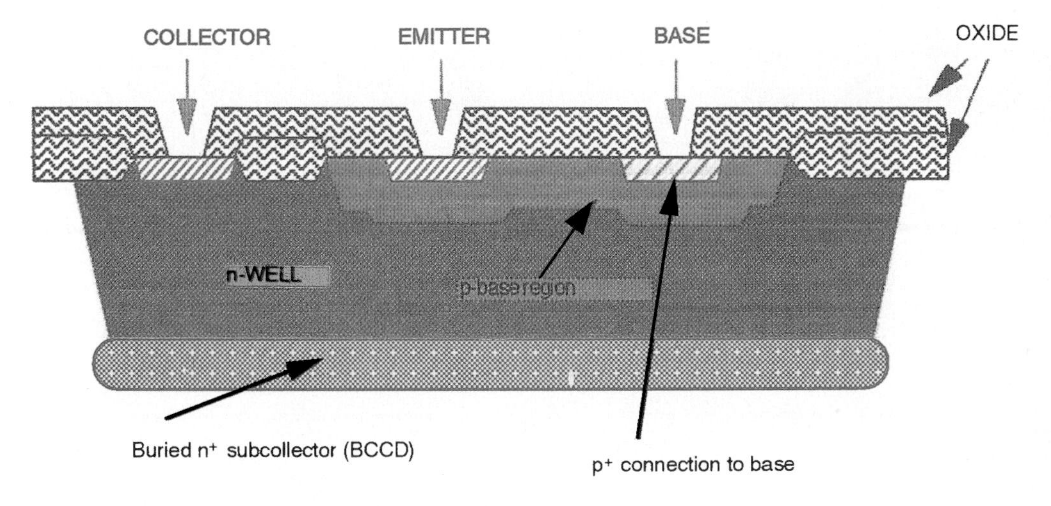

Cross-section through npn transistor (ORBIT 2 μm BiCMOS)

Color plate 6 ORBIT™ 2 μm design rules (f)

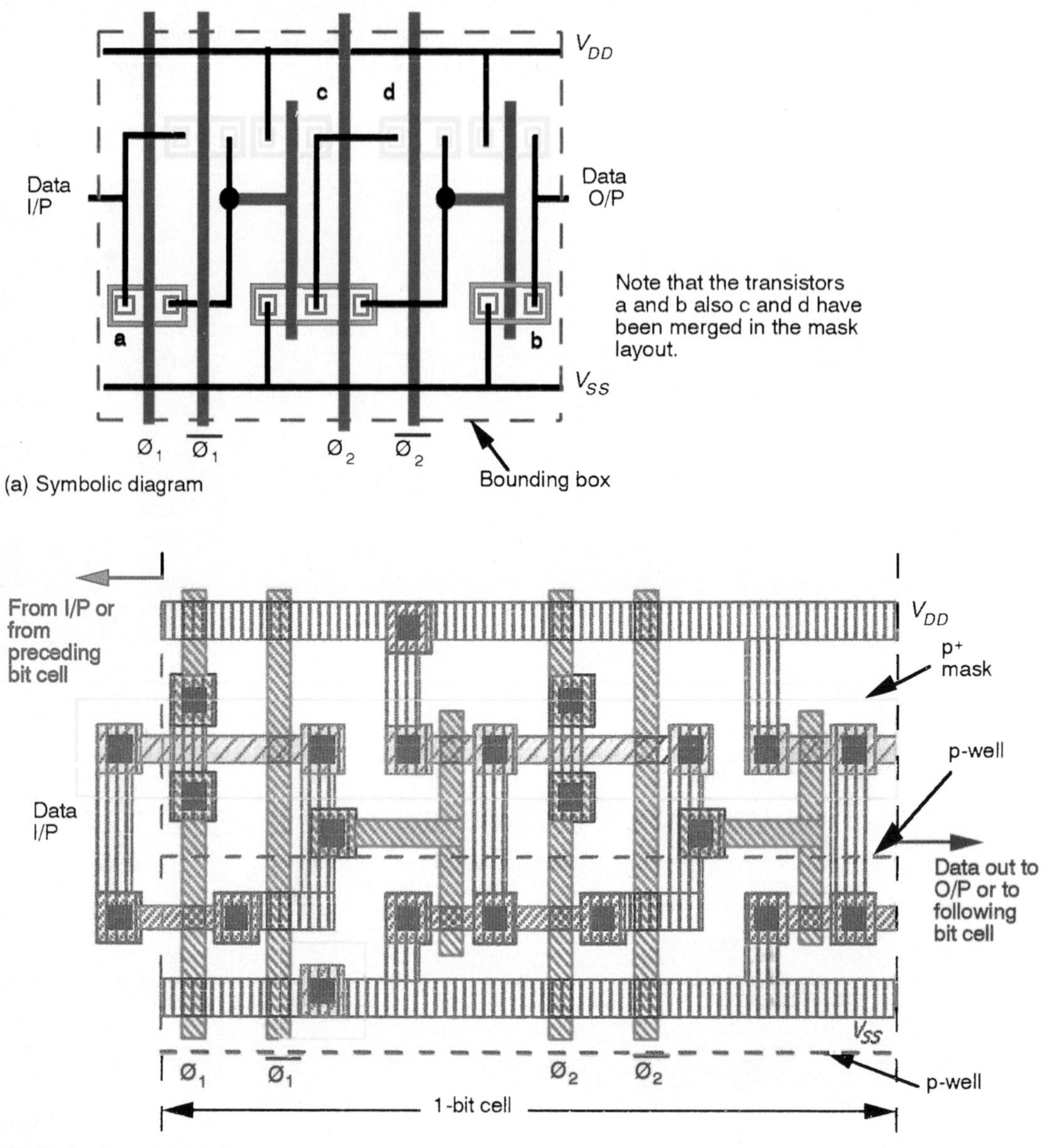

Color plate 7 1-bit CMOS shift register cell

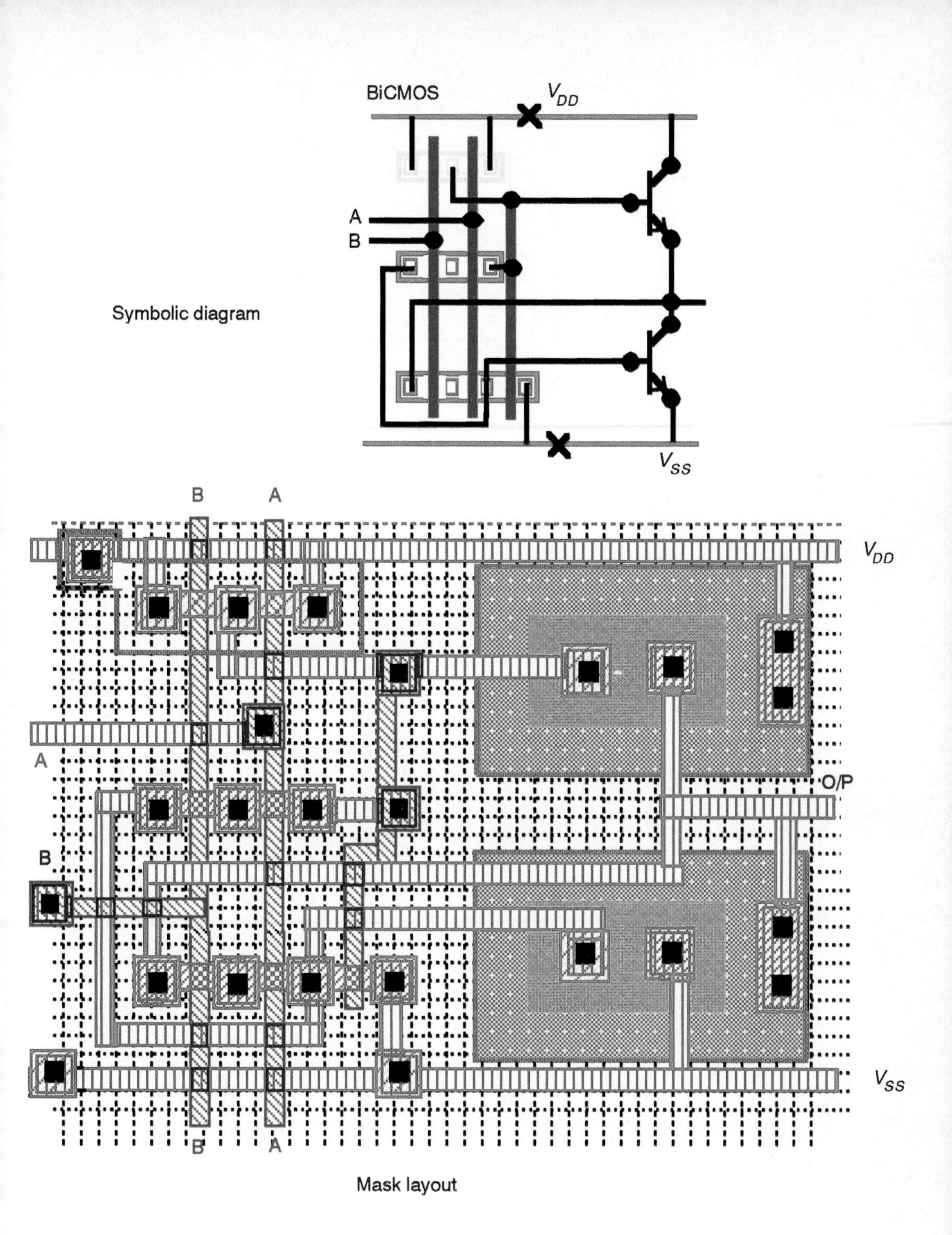

Color plate 8(a) A BiCMOS 2 input *nand* gate

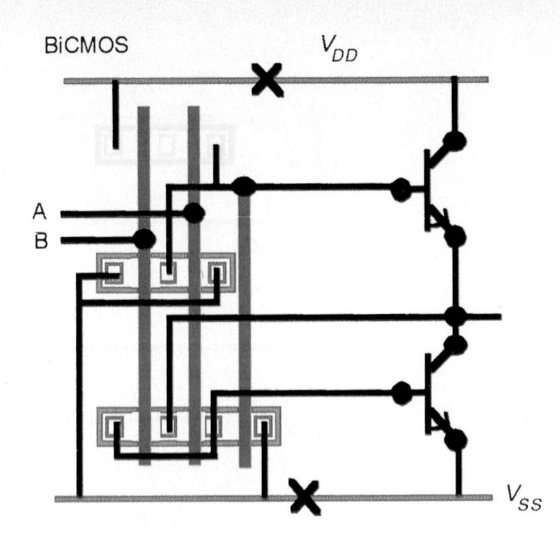

Symbolic diagram

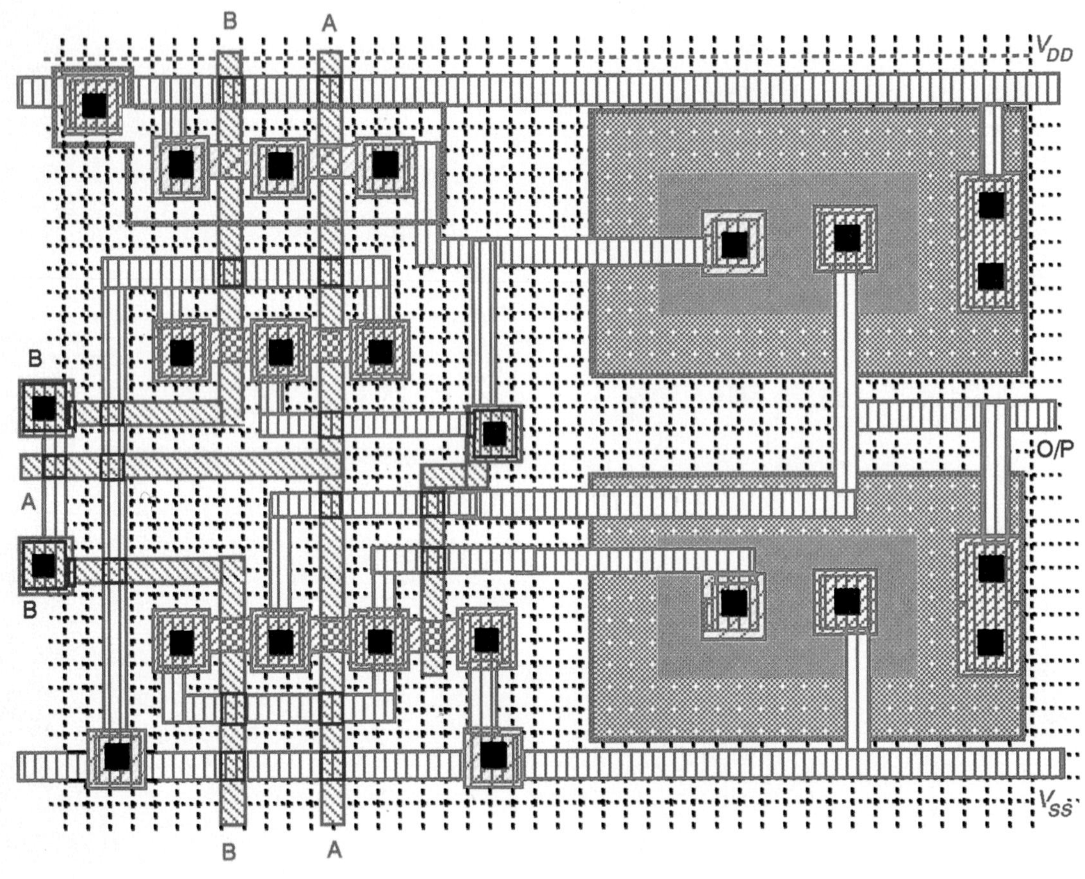

Mask layout

Color plate 8(b) A BiCMOS 2 input *nor* gate

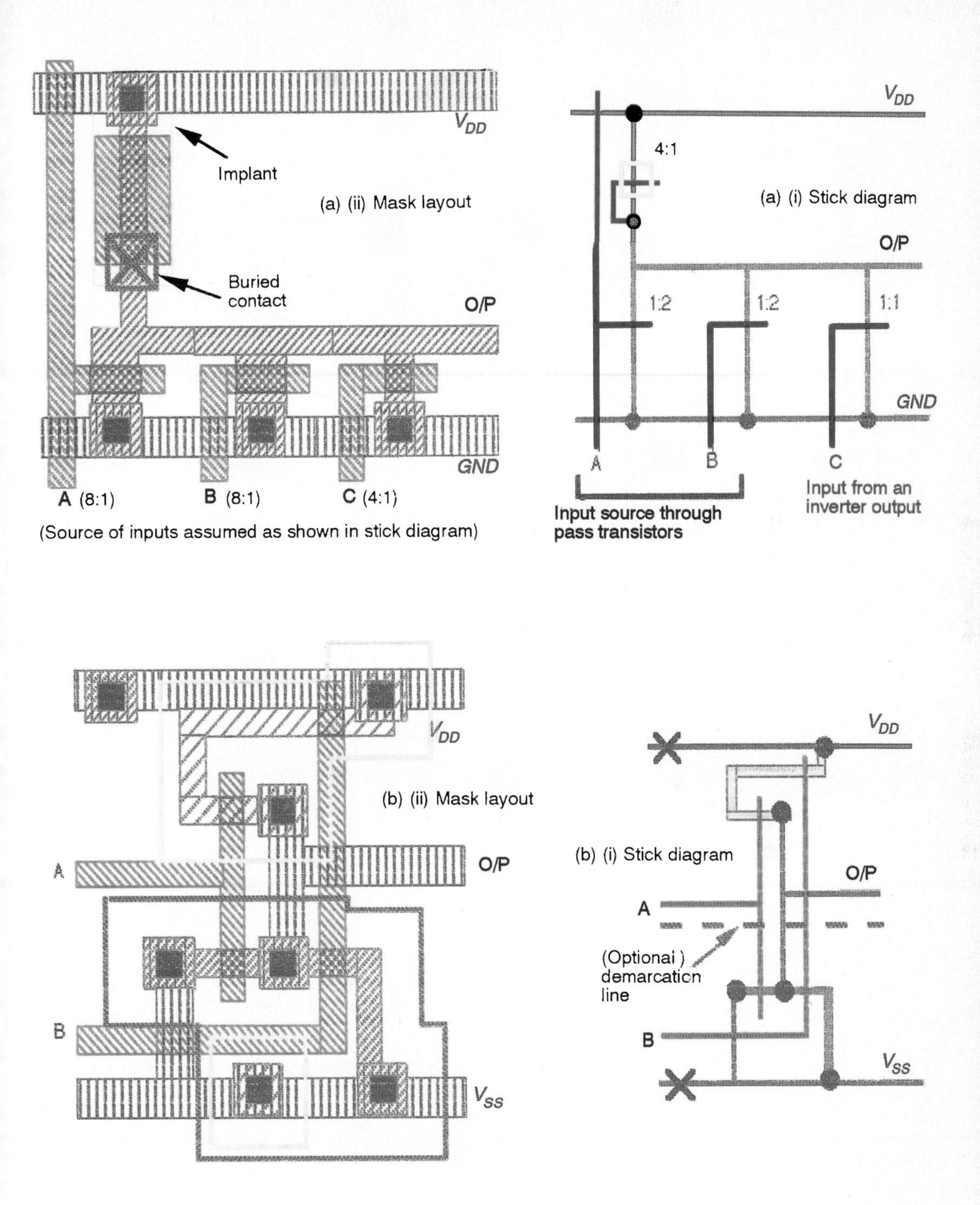

Color plate 9 (a) Three input nMOS *nor* gate; (b) two input CMOS (p-well) *nor* gate

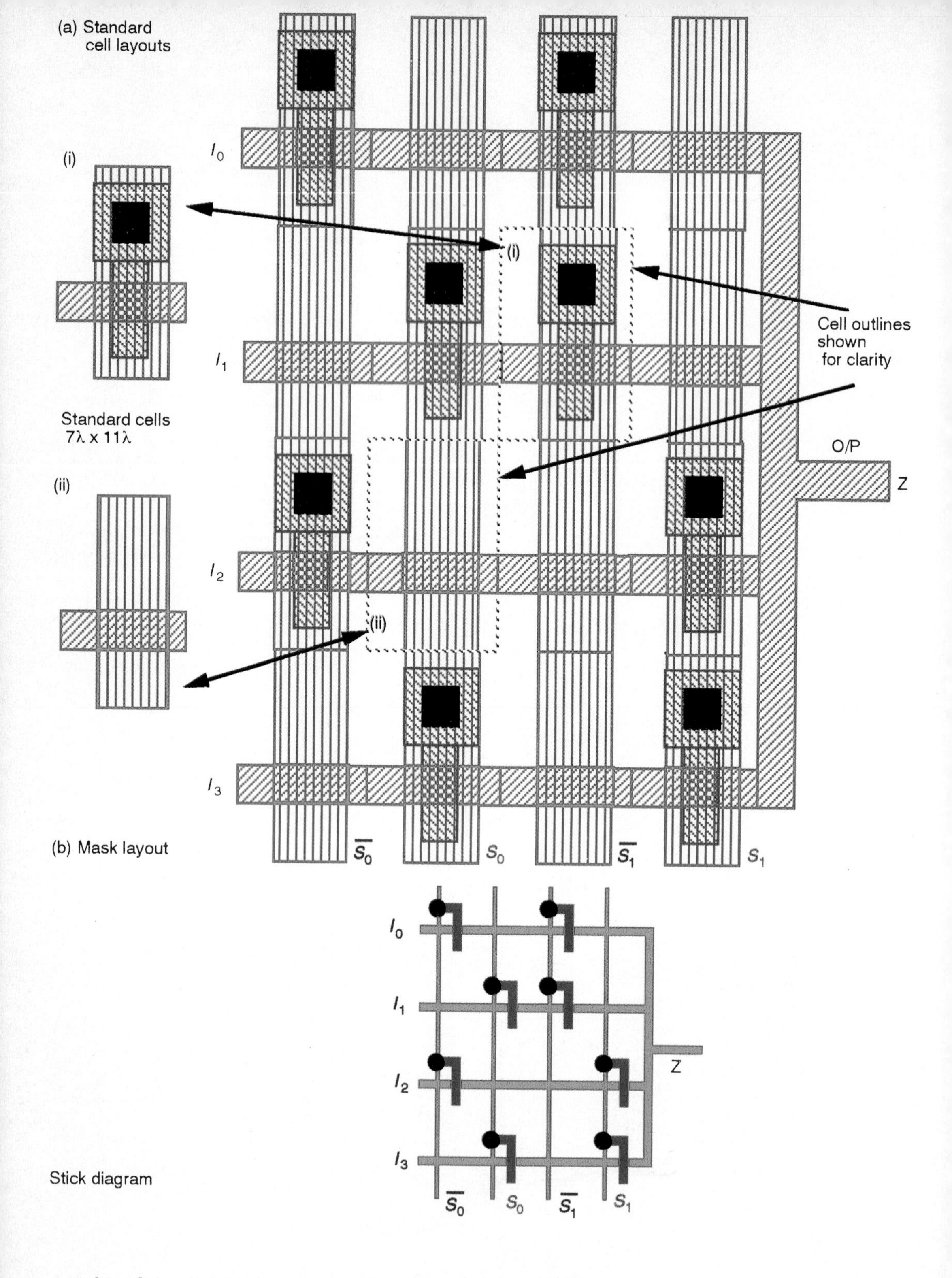

Color plate 10 n-type pass transistor based 4-way MUX

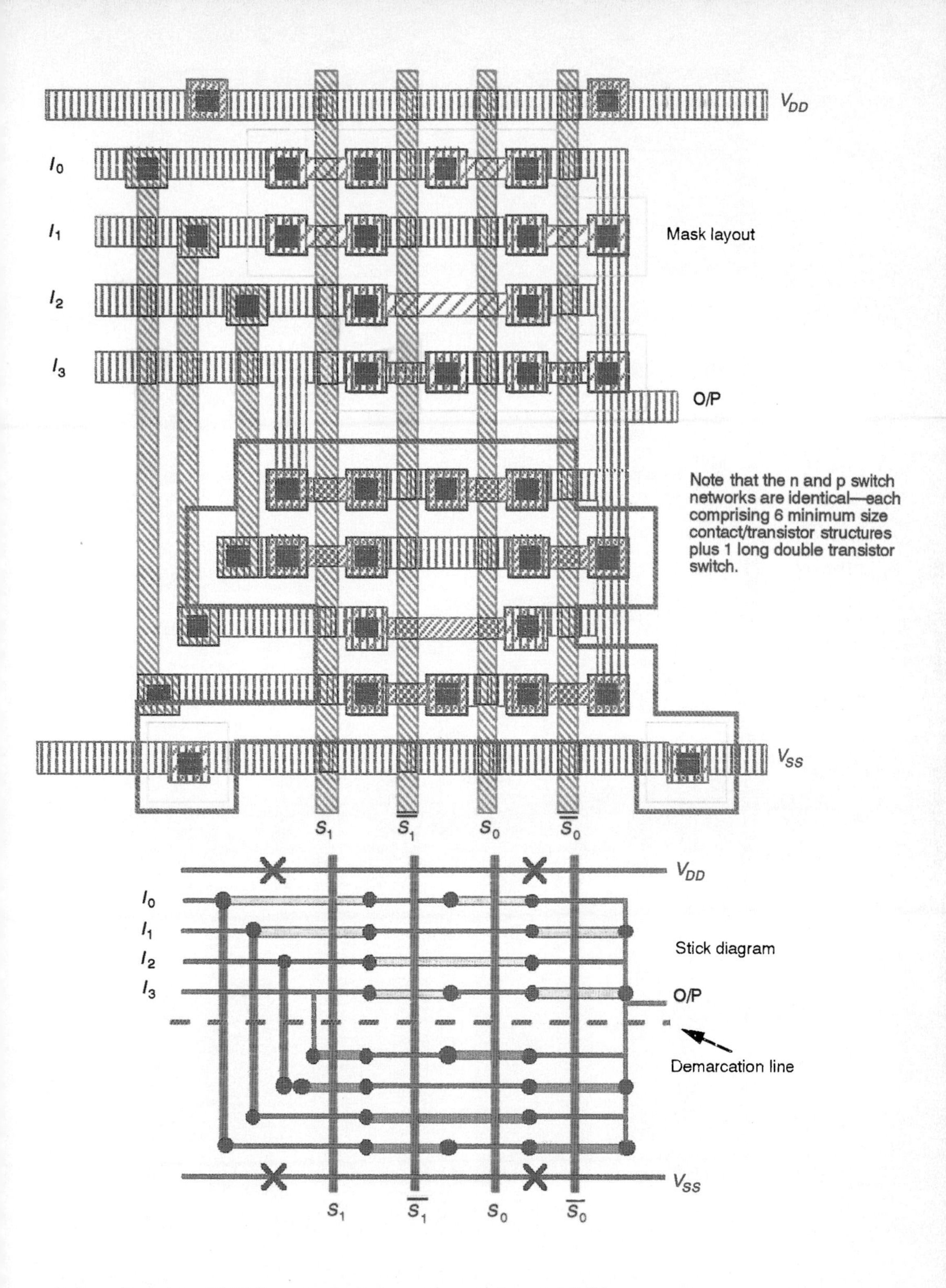

Color plate 11 CMOS transmission gate based 4-way MUX

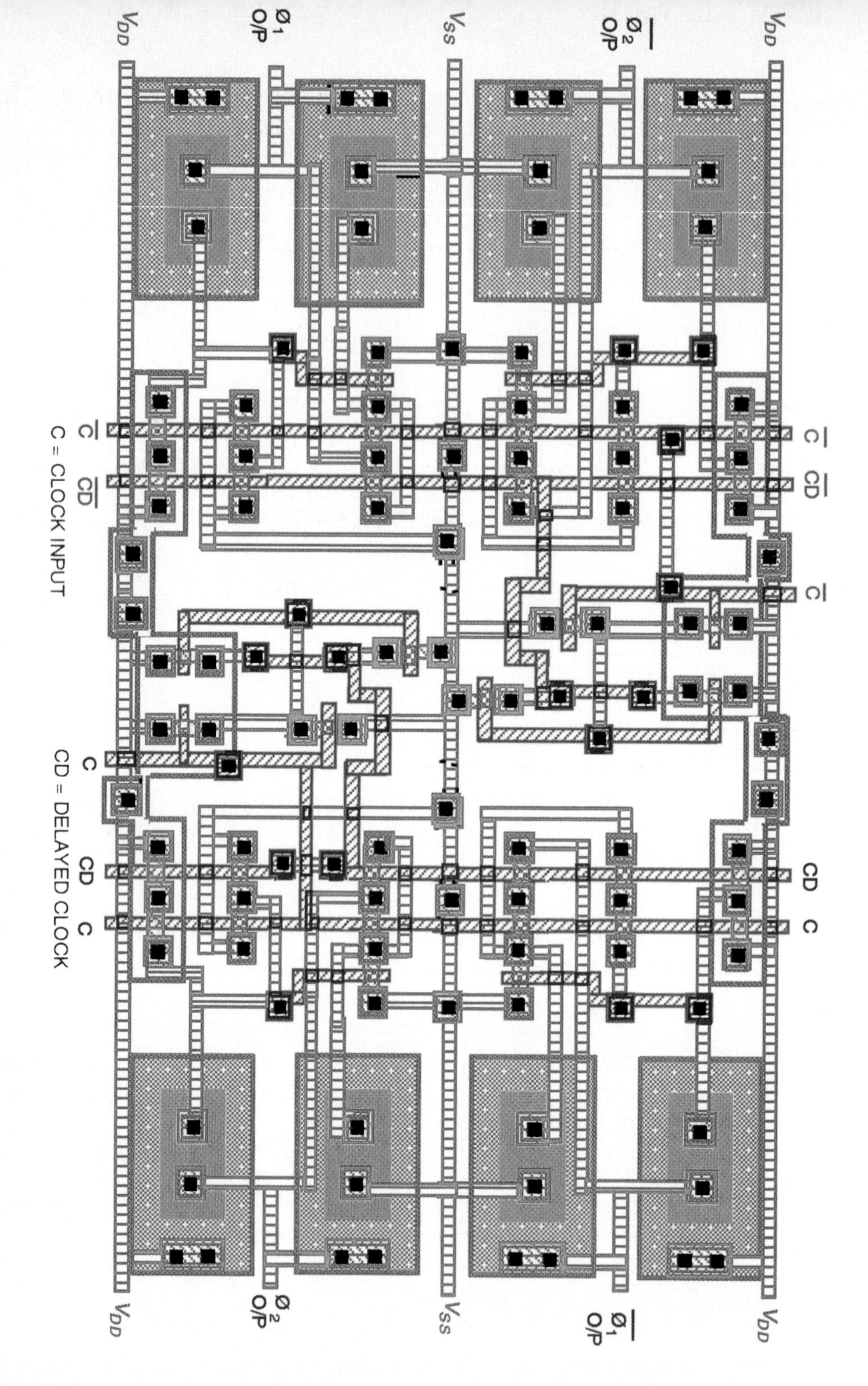

Color plate 12 Mask layout for two-phase (and complements) clock generator (see Figure 5–34)

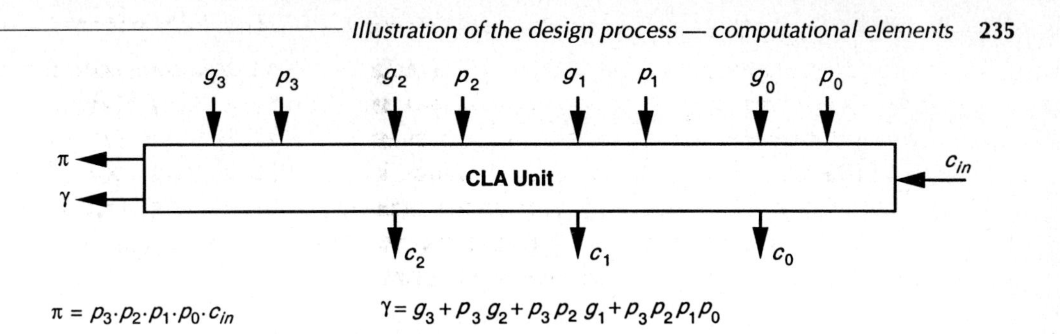

Figure 8–23 4-bit block CLA unit

Following this particular approach, we may now write carry look-ahead expressions in terms of the generate g_k and propagate p_k signals defined earlier. The general form for the carry signal c_k thus becomes

$$c_k = g_k + p_k \cdot g_{k-1} + p_k \cdot p_{k-1} g_{k-2} + \cdots\cdots + p_k \cdots\cdots p_1 \cdot g_0 + p_k \cdots\cdots p_0 \cdot c_{in}$$

Considering a CLA–based adder divided into blocks of 4-bits, as in Figure 8–23, we may write the expressions for the carry circuits in one block as follows:

$$c_0 = g_0 + p_0 \cdot c_{in}$$
$$c_1 = g_1 + p_1 \cdot g_0 + p_1 \cdot p_0 \cdot c_{in}$$
$$c_2 = g_2 + p_2 \cdot g_1 + p_2 \cdot p_1 \cdot g_0 + p_2 \cdot p_1 \cdot p_0 \cdot c_{in}$$
$$c_3 = g_3 + p_3 \cdot g_2 + p_3 \cdot p_2 \cdot g_1 + p_3 \cdot p_2 \cdot p_1 \cdot g_0 + p_3 \cdot p_2 \cdot p_1 \cdot p_0 \cdot c_{in}$$

In order to avoid a sequential propagation of carry signals between the blocks, we may generate additional signals π and γ such that

$$\pi = p_3 \cdot p_2 \cdot p_1 \cdot p_0 \cdot c_{in} \text{ and, } \gamma = g_3 + p_3 \cdot g_2 + p_3 \cdot p_2 \cdot g_1 + p_3 \cdot p_2 \cdot p_1 \cdot g_0$$

An important property of these signals is that c_3, the *carry out* of the block, is

$$c_3 = \gamma + \pi$$

This concept allows CLA techniques to be applied to the carry generation between blocks and for overall carry out as shown in Figure 8–24, which is the overall arrangement of a 16-bit CLA adder.

Further algebraic manipulation allows the expressions for carries within a four-bit block to be written

$$c_0 = g_0 + p_0 \cdot c_{in}$$
$$c_1 = g_1 + p_1 \cdot (g_0 + p_0 c_{in})$$
$$c_2 = g_2 + p_2 \cdot (g_1 + p_1 \cdot g_0 + p_1 \cdot p_0 \cdot c_{in})$$
$$c_3 = g_3 + p_3 \cdot (g_2 + p_2 \cdot g_1 + p_2 \cdot p_1 \cdot g_0 + p_2 \cdot p_1 \cdot p_0 \cdot c_{in})$$

When implementing these circuits in silicon, each carry may be formed by one simple and very regular arrangement as indicated by Figure 8–25, which

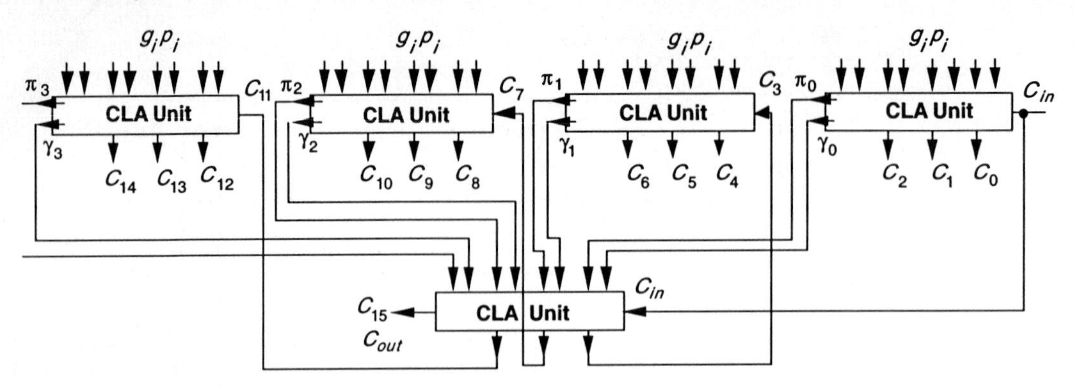

Figure 8–24 A 16-bit, 4 × 4 block CLA adder

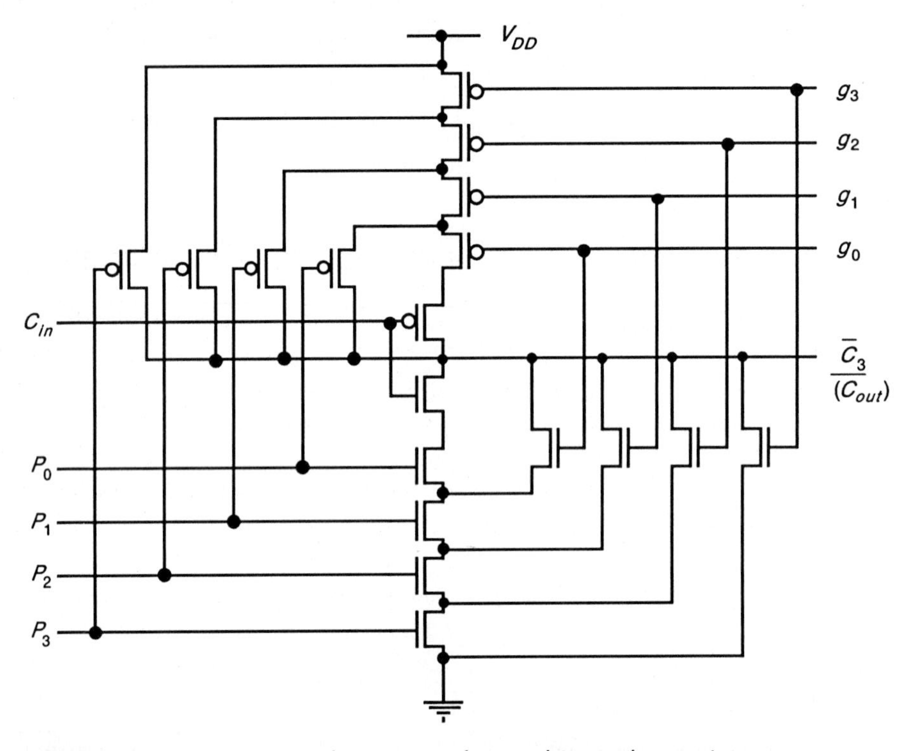

Figure 8–25 Generation of carry out (from 4-bits and carry in)

shows the formation of c_3. For each 4-bit CLA block, four such cells must be implemented, one for each carry c_0 to c_3, and an additional similar circuit is required to form γ.

In order to reduce this complexity, it is possible to use a dynamic logic technique known as 'Multiple Output Domino Logic'. Figure 8–26 illustrates the approach and is, in fact, a four-cell Manchester carry-chain.

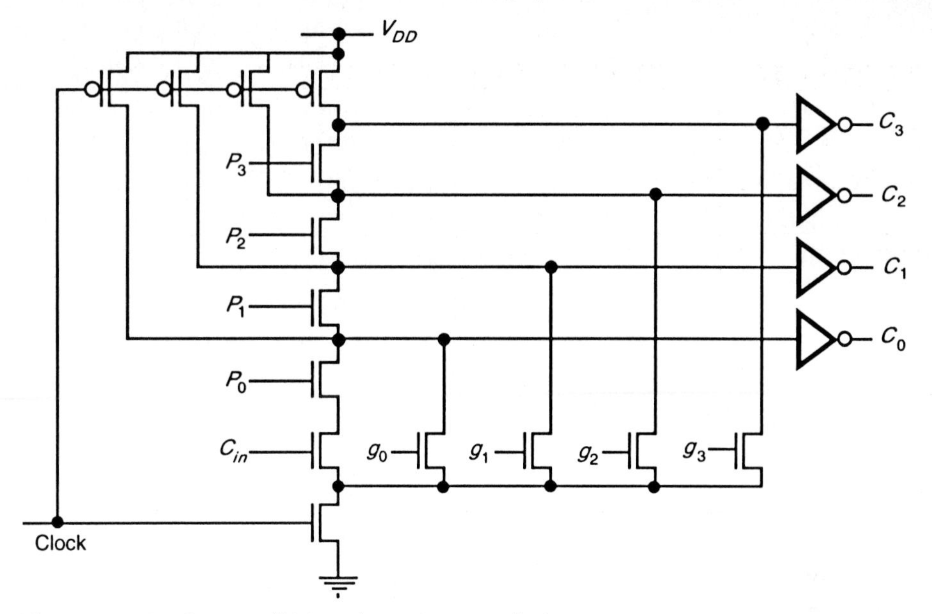

Figure 8–26 Four-cell Manchester carry-chain

8.4.3 A comparison of adder enhancement techniques

This section compares the three enhancement techniques we have discussed from the point of view of area occupied combined with performance. For the purpose of our study, we will compare three 32-bit adders — one carry select, one carry skip and one carry look-ahead. For convenience the carry select and carry skip adders will be assumed to be subdivided into equal size blocks. This must be so as a graduated sizing of blocks relies on an accurate knowledge of the gate delays — information which we do not have for this comparison. The adder cell to be used is required in two versions, one as in Figure 8–14 and a second version — with inverted inputs and carry output — as in Figure 8–27. In both cases, the

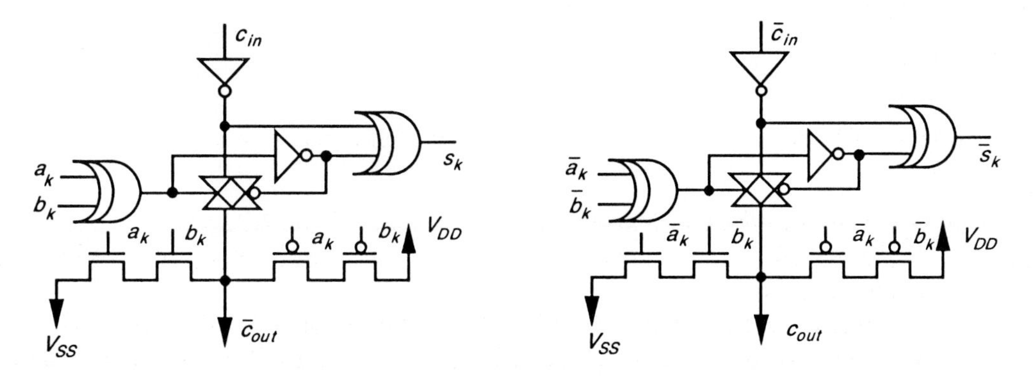

Figure 8–27 Adder cells with alternative input/output arrangement

delay between carry in and carry out is denoted by k_1 (the delay through one adder cell).

8.4.3.1 A 32-bit carry select adder assessment

The multiplexers to be used invert the signal and are based on a simple cell comprising one inverter and one transmission gate as shown for the 2-way multiplexer of Figure 8–28.

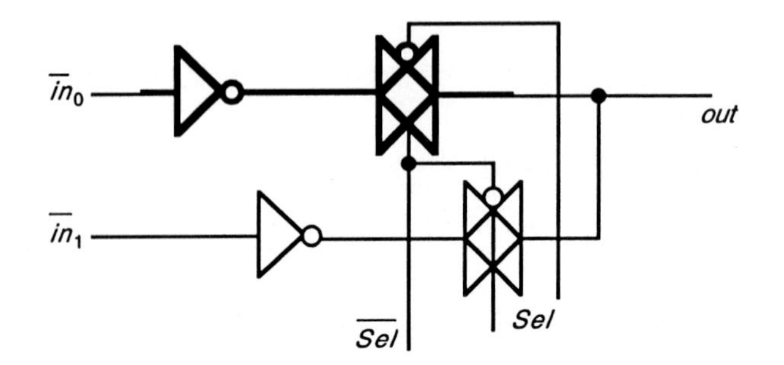

Figure 8–28 2-way multiplexer showing 'multiplexer cell' (bold lines)

Comparing this with the proposed adder cell, we may see that the multiplexer delay k_2 is the same as that for the adder cell so that $k_1 = k_2$ and, in consequence, the optimum block size evaluates as six. This does not divide exactly into 32, but we may choose to use four blocks of five cells and the remaining two blocks will then have six cells each as shown in Figure 8–29. The adder completion time is thus:

$$T = 5k_1 + 5k_2 = 10k$$

where $k = k_1 = k_2$.

The area of this 32-bit carry select adder is roughly twice that of a 32-bit ripple carry adder.

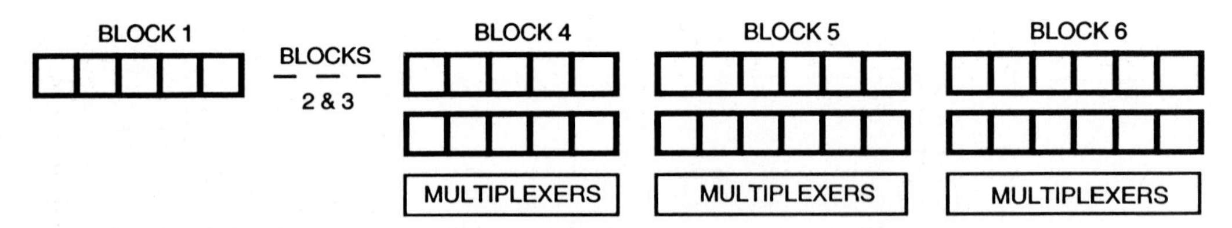

Figure 8–29 Arrangement of a 32-bit carry-select adder

8.4.3.2 A 32-bit carry skip adder assessment

Once again, the cell delay k_1 and the multiplexer (as in Figure 8–28) delay k_2 may be assumed to be equal. In order to simplify the propagation time assessment, we will neglect the time taken to compute all the generate and propagate, as well as the block propagation signals, since they are all computed simultaneously and may be represented as an overhead = $k \approx k_1$.

Care must be taken to allow for the inversion of the carry signal, both in the adder cell and in the multiplexer. For this reason, the block size must be an even number of bits. Again, since the ratio between the cell delay and the multiplexer delay is assumed to be 1:1, we may write

$$k_1 = k_2 = k$$

also, since the ratio $k_1 = k_2$, the optimum block size is four cells so that there will be eight blocks of equal size.

The adder completion time is thus:

$$T = 4k_1 + 4k_2 + 6k_2 + k = 15\,k$$

where $k = k_1 = k_2$.

The area of this 32-bit carry skip adder is roughly one and a half that of a 32-bit ripple carry adder.

8.4.3.3 *A 32-bit carry look-ahead adder assessment*

Figure 8–30 represents the structure of a 32-bit carry look-ahead adder. For reasons of simplicity in presentation, each heavy interconnect line represents the interconnection of two signals, $(g_k.p_k)$ and $(\gamma_k.\pi_k)$. The fine interconnect lines are the carry signals.

Let the delay time of a CLA unit be k_3, then the completion time of the adder may be assessed as follows:

At time k_3 : $(\gamma_k.\pi_k)$ for CLA 0–7 are set.
At time $2k_3$: $(\gamma_k.\pi_k)$ for CLA 8 and 9 are set; c_4, c_8, and c_{12} are set by CLA 8.
At time $3k_3$: c_{16} is set by CLA 10; using c_4, c_8, and $c_{12,}$ CLA 1, CLA 2 and CLA 3 set their carry out.

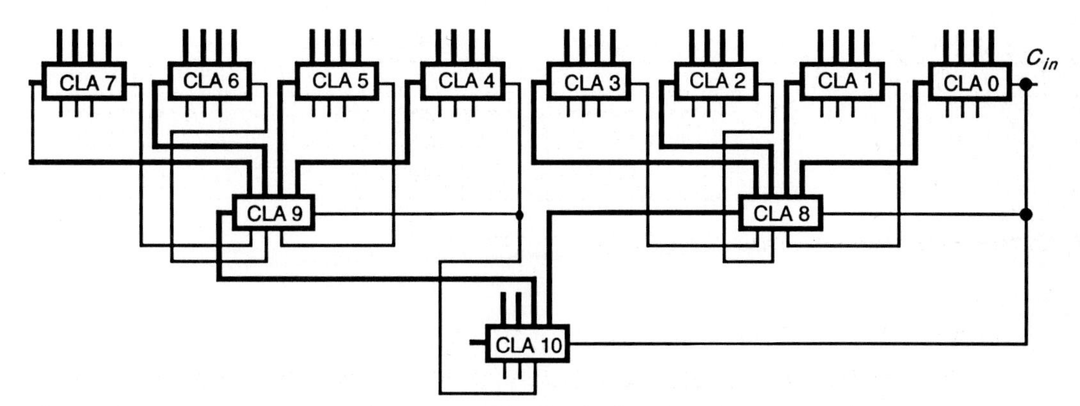

Figure 8–30 Arrangement of a 32-bit carry look-ahead adder

At time $4k_3$: $c_{20,}$ $c_{24,}$ and $c_{28,}$ are set by CLA 9.
At time $5k_3$: Using $c_{20,}$ $c_{24,}$ and $c_{28,}$ CLA 5, CLA 6 and CLA 7 set their carry out.

Therefore, overall time $T = 5k_3$.

The exact value of k_3 depends on the actual CLA adder element arrangement and on the layout used, but, allowing for three levels of logic, it could be conservatively estimated as $1.5k_1$ to $2.0\ k_1$, where k_1 is the delay of the simple adder cell used before in the carry select and skip adders. If this is a reasonable assumption then, in comparison with the other evaluations, overall time T is given by

$$T = 7.5k \text{ to } 10k$$

However, noting the unused inputs of CLA 10 (Figure 8–30), it may be seen that a 64-bit CLA adder could be accommodated within the same overall time delay. Since the CLA cells are considerably more complex than the adder cells used in the carry select and carry skip adders, there will be a penalty in the area occupied. This is difficult to evaluate without detailed design work, but the area occupied will be several times greater than for a 32-bit ripple carry adder.

This concludes the consideration of adder circuitry. In the design of ALUs and digital processors generally, the adder is the most important circuit and is able to directly accommodate additions, subtractions and comparisons, together with a range of logical operations. Another common arithmetic requirement is for multiplication and it will be seen that the adder has an important role to play in the architecture of many multipliers.

8.5 Multipliers

A study of computer arithmetic processes will reveal that the most common requirements are for addition and subtraction, but that there is also a significant need for a multiplication capability. Thus, a brief overview of some common approaches to this problem is given in this section. Although division is obviously useful, it is a much less common requirement and will not be dealt with in this text.

8.5.1 The serial-parallel multiplier

This multiplier is the simplest one, the multiplication being considered as a succession of additions.

If $\quad A = (a_n\ a_{n-1}\ a_{n-2} \cdots\cdots\cdots a_0)\quad$ and

$\quad B = (b_n\ b_{n-1}\ b_{n-2} \cdots\cdots\cdots b_0)$

then the product $A.B$ may be expressed as

$$A.B = (A.2^n . b_n + A.2^{n-1}. b_{n-1} + A . 2^{n-2}. b_{n-2} \ldots\ldots\ldots\ldots A.2^0.b_0)$$

A possible form of this adder for multiplying four-bit quantities, based on this expression, is set out in Figure 8–31. Note that *D* indicates a *D* flip-flop and *F.A.* indicates a full adder — or adder bit slice. Number *A* is entered in the right-most 4-bits of the top row of *D* flip-flops which are connected to three further *D* flip-flops to form a 7-bit shift register to allow the multiplication of number *A* by 2^1, 2^2... 2^n, thus forming the *partial product* at each stage of the process.

In some cases, it may be easier to right shift the contents of the *Accumulator* — (bottom row of *D* flip-flops) rather than left shifting *A*. This approach can be used to eliminate the least significant bits of the product if so desired.

A further reduction in hardware can result from noting that the three most significant bits of the partial product are set to zero initially, and are used only one by one as the shifting of *A* proceeds. These three bits can therefore be used to hold three bits of number *B* initially, thus saving three *D* flip-flops.

The structure under discussion here is suited only to positive or unsigned operands. If the operands are negative and twos complement encoded, then:

1. The most significant bit of *B* will have a negative weight and so a subtraction must be performed as the last step.
2. The most significant bit of *A* must be replicated since operand *A* must be expanded to *2N* bits.

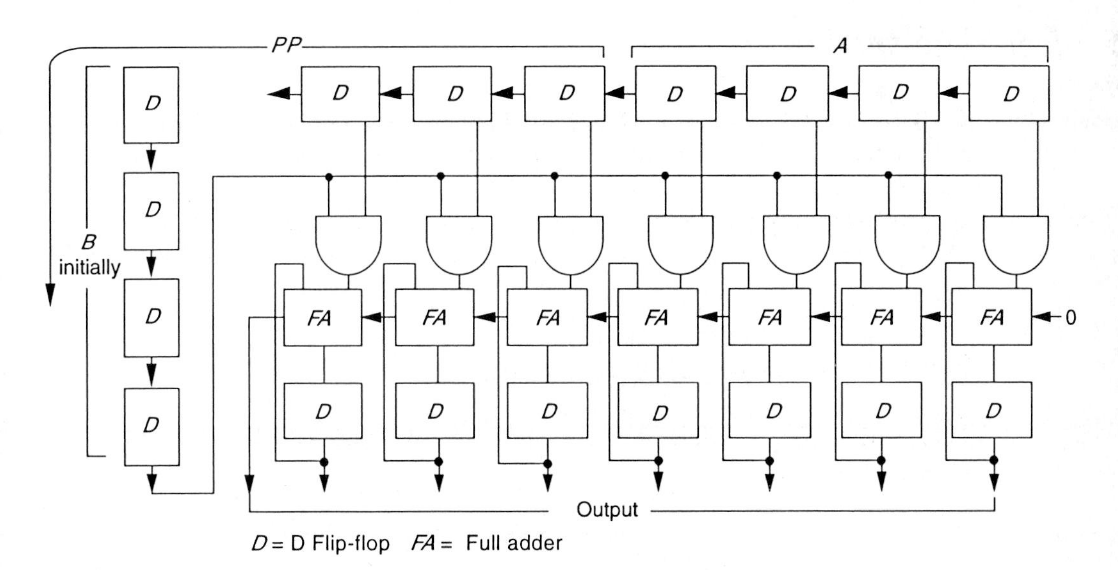

Figure 8–31 Arrangement of a 4-bit serial-parallel multiplier

8.5.2 The Braun array

A relatively simple form of parallel adder is the Braun array (see Figure 8–32). All partial products $A.b_k$ are computed in parallel, then collected through a cascaded array of carry save adders. At the bottom of the array, an adder is used to convert the carry save form to the required form of output. Completion time is fixed by the depth of the array, and by the carry propagation characteristics of the adder. Notice that this multiplier is suited only to positive operands. Negative operands can be handled, for example, by the Baugh-Wooley multiplier which now follows.

8.5.3 Twos complement multiplication using the Baugh-Wooley method

This technique has been developed to design multipliers that are regular in structure and suited for twos complement numbers.

Let us consider two numbers A and B:

$$A = (a_{n-1} \ldots\ldots a_0) = -a_{n-1}.2^{n-1} + \sum_0^{n-2} a_i.2^i$$

$$B = (b_{n-1} \ldots\ldots b_0) = -b_{n-1}.2^{n-1} + \sum_0^{n-2} b_i.2^i$$

The product $A.B$ is given by:

$$A.B = a_{n-1}.b_{n-1}.2^{n-2} + \sum_0^{n-2}\sum_0^{n-2} a_i.b_j.2^{i+j} - a_{n-1}\sum_0^{n-2} b_i.2^{n+i-1} - b_{n-1}\sum_0^{n-2} a_i.2^{n+i-1}$$

If we use this form, it may be seen that subtraction operations are needed as well as addition. However, the negative terms may be rewritten, for example:

$$a_{n-1}\sum_0^{n-2} b_i.2^{n+i-1} = a_{n-1}.\left(-2^{n-2} + 2^{n-1} + \sum_0^{n-2} \overline{b}_i.2^{n+i-1}\right)$$

Using this approach, $A.B$ becomes

$$A.B = a_{n-1}.b_{n-1}.2^{n-2} + \sum_0^{n-2}\sum_0^{n-2} a_i.b_i.2^{i+j} + b_{n-1}\left(-2^{n-2} + 2^{n-1} + \sum_0^{n-2} \overline{a}_i.2^{n+i-1}\right)$$

$$+ a_{n-1}\left(-2^{n-2} + 2^{n-1} + \sum_0^{n-2} \overline{b}_i.2^{n+i-1}\right)$$

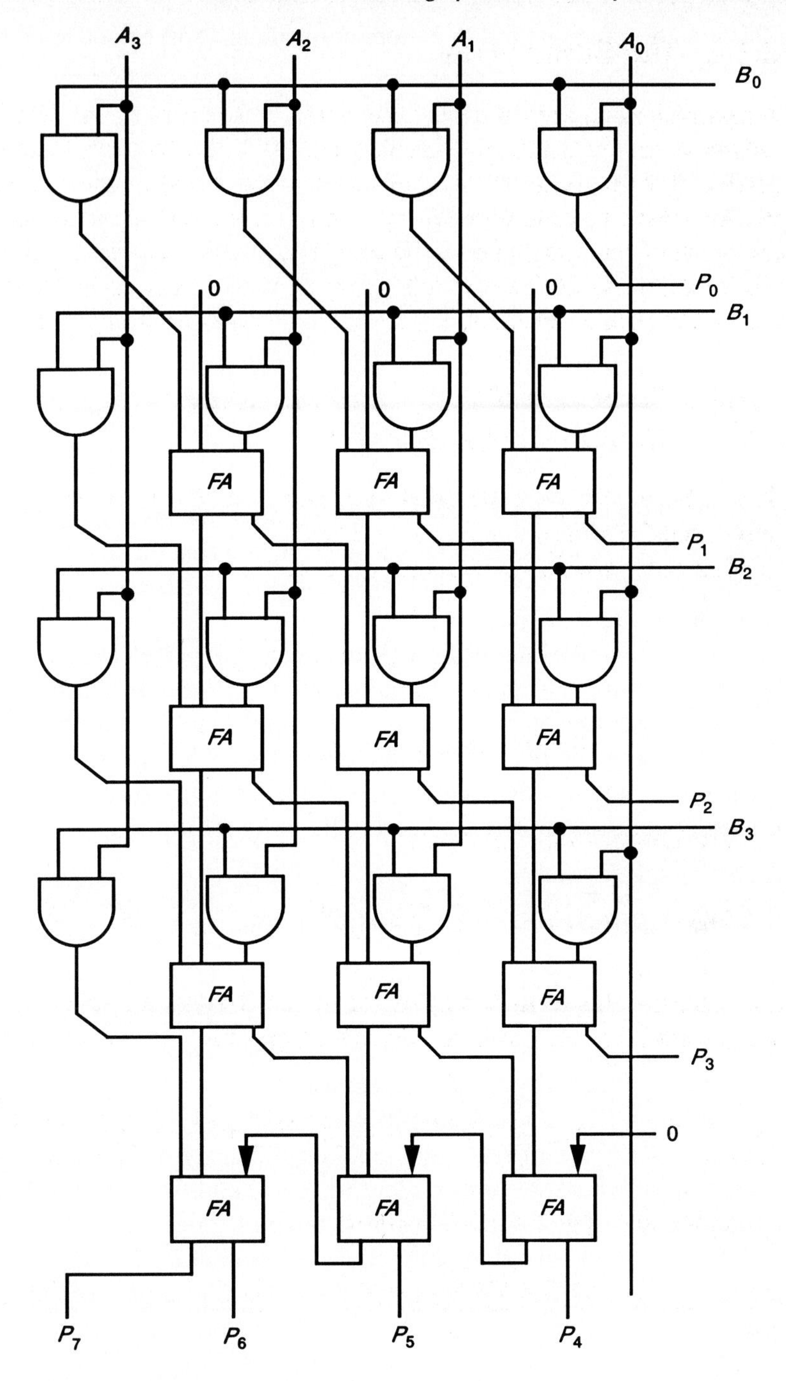

Figure 8–32 A 4-bit Braun multiplier

This equation may be put in a more convenient form by recognizing that

$$-(b^{n-1}+a^{n-1}).2^{2n-2}=-2^{2n-1}+(\overline{a_{n-1}}+\overline{b_{n-1}}).2^{2n-2}$$

Thus, AB is given by

$$A.B=2^{2n-1}+(\overline{a_{n-1}}+\overline{b_{n-1}}+a^{n-1}.b^{n-1}).2^{2n-2}$$
$$+\sum_{0}^{n-2}\sum_{0}^{n-2}a_i.b_j.2^{i+j}+(a_{n-1}+b_{n-1}).2^{n-1}$$
$$+\sum_{0}^{n-2}b_{n-1}.\overline{a_i}.2^{n+1-j}+\sum_{0}^{n-2}a_{n-1}.\overline{b_i}2^{n+i-1}$$

Since A and B are n-bit operands, their product may extend to $2n$-bits. The first, most significant, bit is taken into account by the first term -2^{2n-1} which is fed to the multiplier as a 1 in the most significant cell. The Baugh-Wooley arrangement is set out in Figure 8–33.

In serial-parallel multipliers there are as many idle clock cycles as there are 0s in the multiplicand and the same situation applies in Braun and Baugh-Wooley arrays. For this reason, it may be useful to introduce pipelining concepts between successive lines of the array. The clock speed of the pipeline is limited by the speed of the output adder, but it is possible to introduce further pipelining between the adder cells giving rise to the systolic array multiplier.

8.5.4 A pipelined multiplier array*

Many parallel multipliers are iterative arrays. Some of these are carry-ripple structures with no storage elements, in which a given result must be output before new data words can be input. Such multipliers can be pipelined by introducing latches at appropriate positions in the array.

An example is a parallel multiplier based on *systolic array principles* as in Figure 8–34. It comprises a diamond-shaped array of latched, gated full adder cells, connected only to immediately adjacent cells. This has practical advantages as no broadcasting of data right across the multiplier array occurs.

With multiplicand X, multiplier Y and product P, the *kth* bit of each partial product X_{k-i}, y_i is formed in one of the cells in the *kth* vertical column of the array.

---

* J. V. McCanny and J. G. McWhirter, 'Completely iterative, pipelined multiplier array suitable for VLSI', *IEE Proc,* vol. 129, pt. G, no. 2, 40–46. This structure was designed by P. Evans as part of a VLSI course at the University of Adelaide, South Australia.

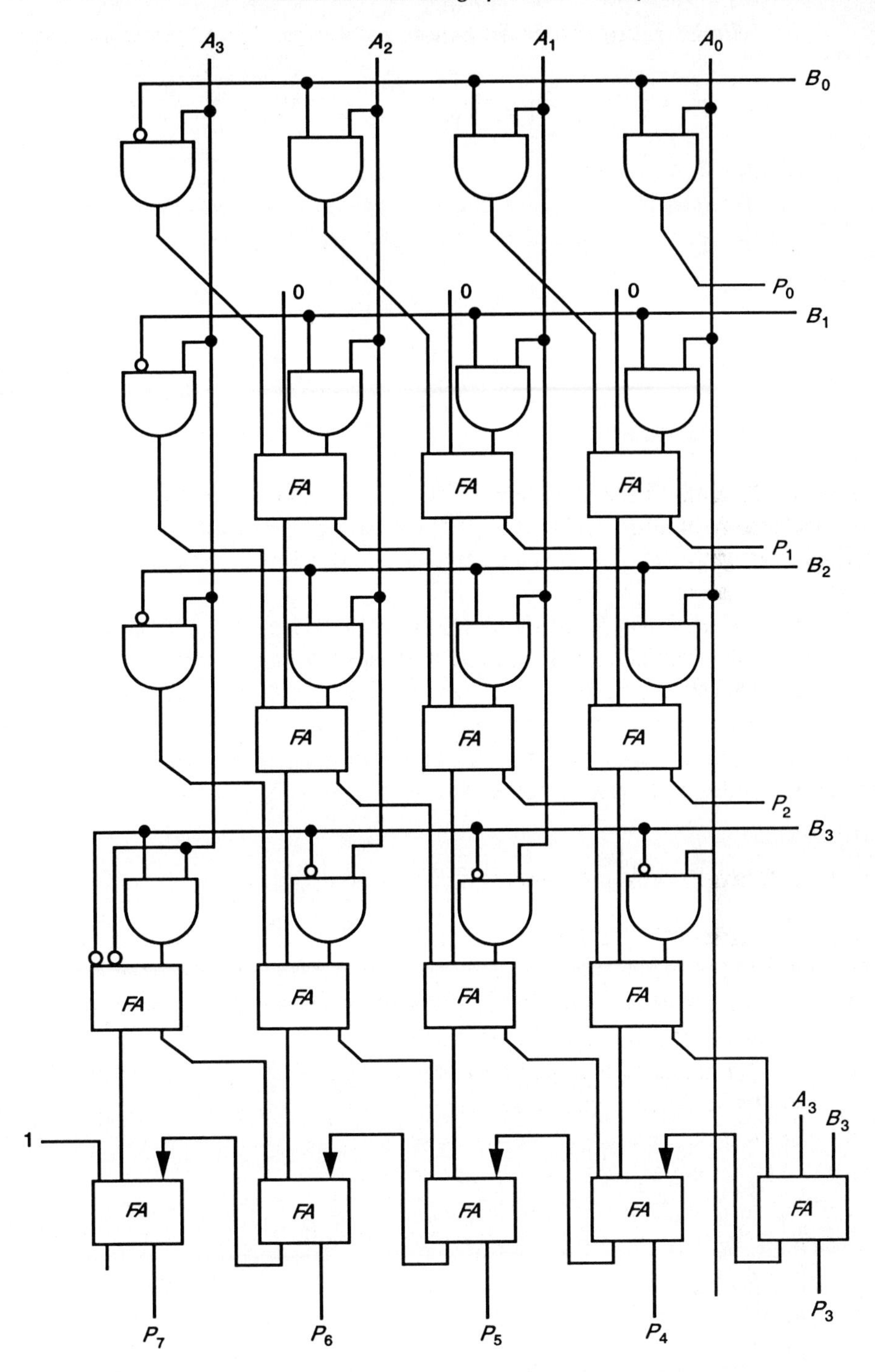

Figure 8–33 A 4-bit Baugh-Wooley multiplier

The *kth* bit of the product

$$P_k = \sum_{i=0}^{k} X_{k-i} \cdot y_i$$

is formed by letting these components accumulate as p_k passes down the column. Carries generated at each stage in the array are passed to the left (next most significant column).

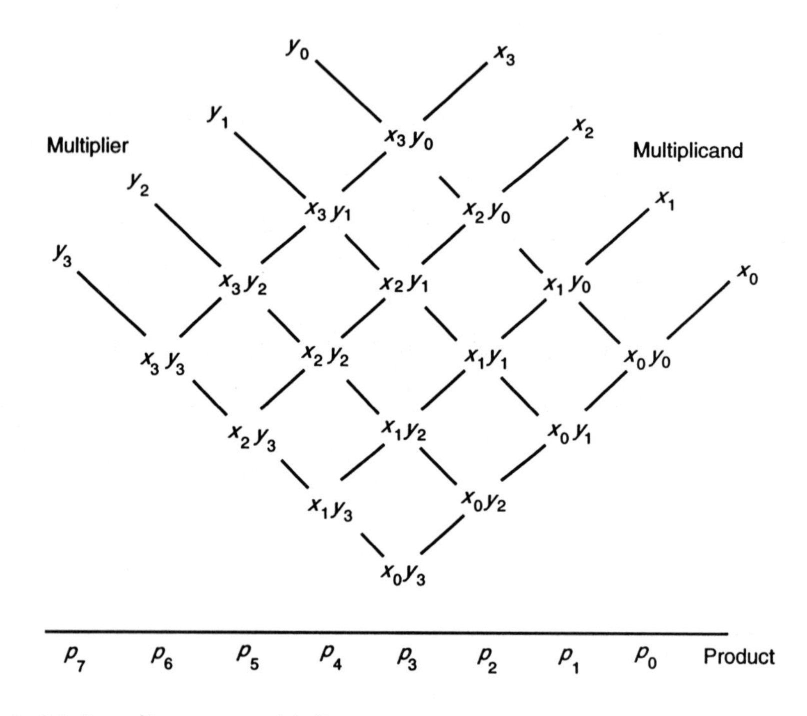

Figure 8–34 Systolic array multiplier

The residual carry bits passing across the lower left-hand boundary of the diamond must be added into the partial product sum to complete the multiplication. This is achieved with half of the above array placed at the lower left-hand boundary, retaining the iterative structure.

This gives the general structure shown in Figure 8–35. For an n-bit × n-bit multiplier, $\frac{1}{2}(3n+1)n$ cells are required. There is a further requirement of $3n^2$ latches to skew and deskew the input and output data. Note that each cell connects to six other cells, provided that it is not on the array boundary. All sum and carry inputs at the array boundary are set to zero.

The structure of the basic cell is shown in Figure 8–36. The gating function for unsigned numbers is $x.y$.

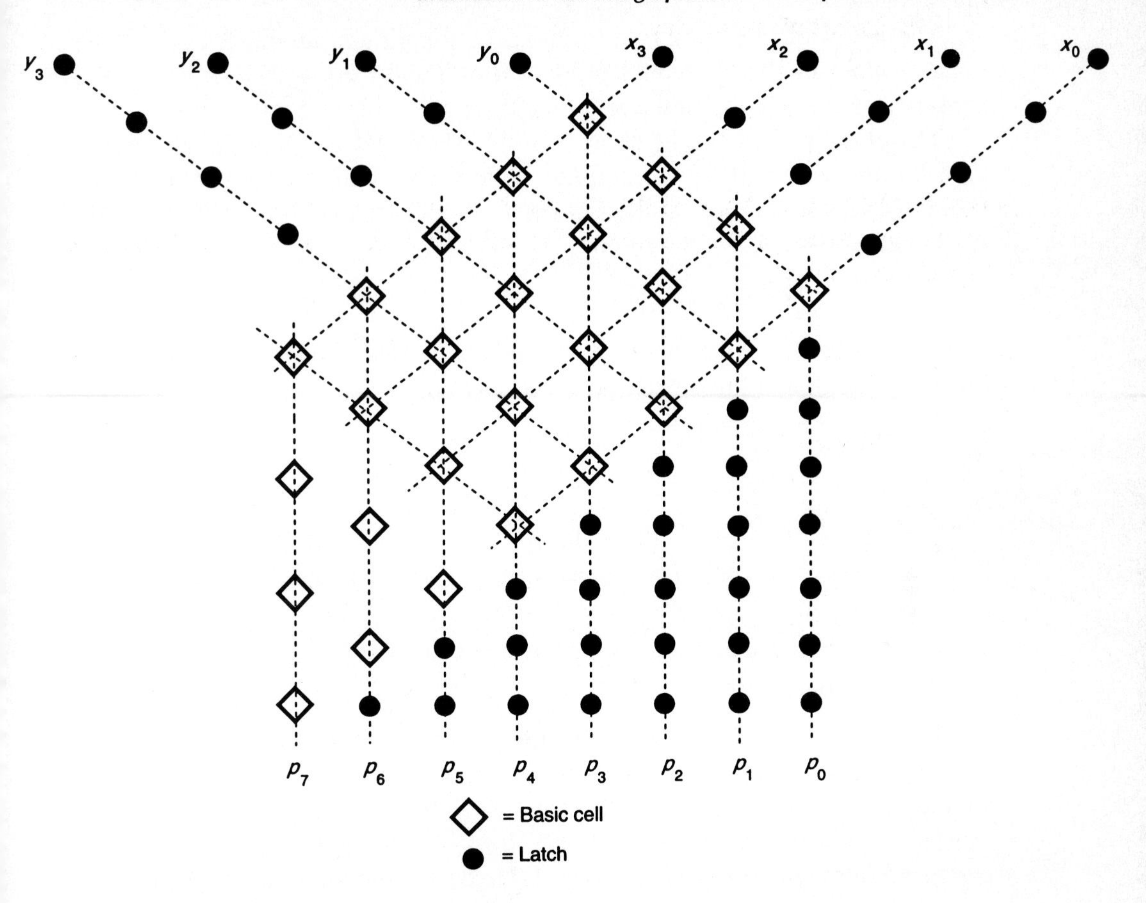

Figure 8–35 Multiplier structure

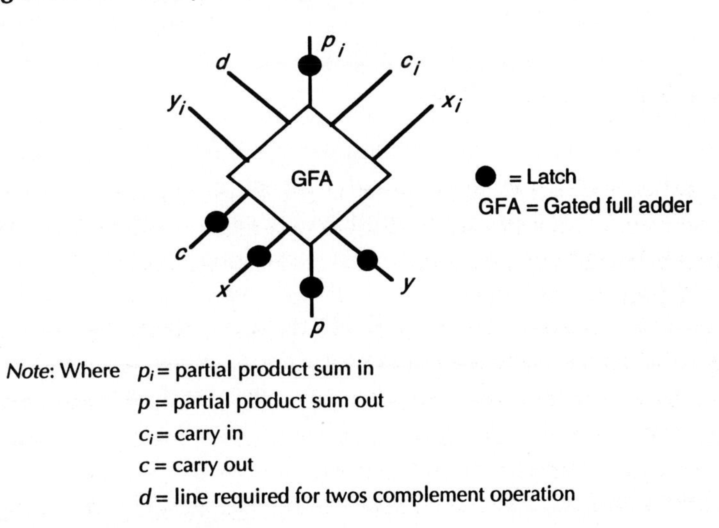

Note: Where p_i = partial product sum in
p = partial product sum out
c_i = carry in
c = carry out
d = line required for twos complement operation

Figure 8–36 Basic cell

The delay of one operation through the pipeline is $3n$ clock cycles (i.e. it takes $3n$ clock cycles to obtain a product after X and Y are input). However, if the pipeline is kept full, a product will be output every clock cycle.

The clock period can be short as it must account for only the propagation time through one cell. The multiplier is thus a very high throughput structure (i.e. low average time per multiplication).

If the product $X.Y$ is rewritten

$$XY = x_{n-1}2^{n-1}.\bar{Y} + x_{n-1}2^{n-1} + \tilde{x}.Y$$

where $\tilde{x}$ is the $(n-1)$ least significant bits of X, then the structure can be used for twos complement numbers, provided that:

1. The gating function is replaced by

   $$(y \oplus d).x$$

   where $d=1$ for all cells on the upper left-hand boundary and $d=0$ elsewhere.
2. The value of x_{i-1} is fed to the carry input c_i as well as to the normal input x_i of the cell in the top row of the the array.
3. Y is sign extended and suitably delayed sign extensions are input to left boundary y_i inputs.

The full adder chosen was a transmission gate adder because of its speed and because it generates the sum and the carry in equal time. The latches chosen were dynamic shift registers as the structure will be continuously clocked.

The timing diagram (Figure 8–37) illustrates the performance of the 8-bit version. After the initial delay of about 1.2 μsec, the output products are available at 50 nsec intervals.

8.5.5 The modified Booth's algorithm

Another approach which avoids having many idle cells in a cellular multiplier as well as reducing the number of cycles compared with the serial-parallel multiplier is the use of the so-called modified Booth's algorithm. In principle, the modified algorithm requires rewriting the multiplicand in such a way that half the bits are 0. Clearly, this is possible only by using a special number system.

This converts a signed standard twos radix number into a number system where the digits are in the set {–1, 0, 1}. In this system any number may be written in several forms, that is, the system has redundancies.

Let us consider a number $B = b_{n-1}\, b_{n-2} \ldots\ldots b_1\, b_0$ written in twos complement form:

$$B = -b_{n-1}.2^{n-1} + \sum_{k=0}^{n-2} b_k.2^k$$

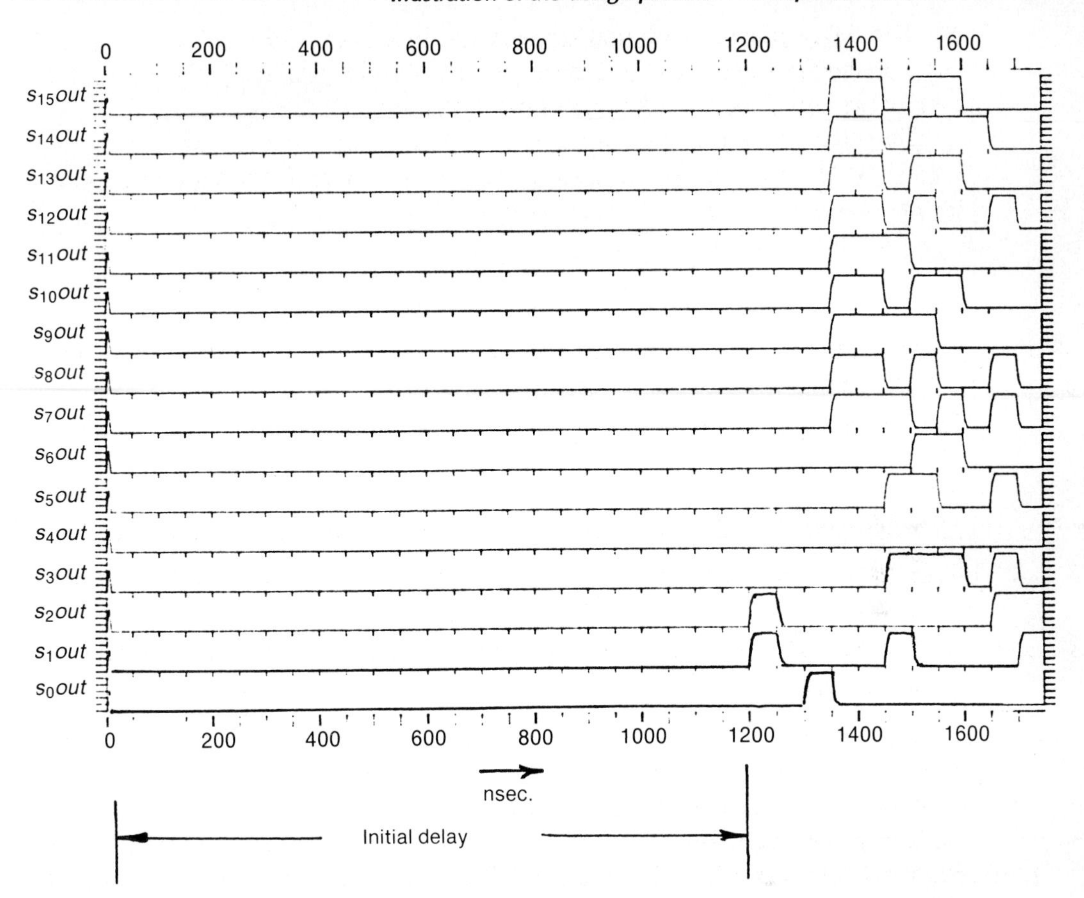

Figure 8–37 Performance of an 8-bit multiplier

which may be rewritten as

$$B = \sum_{k=0}^{n-2-1} (b_{2k-1} + b_{2k} + 2b_{2k+1}) 2^k$$

with $b_1 = 0$.

In this equation, the term in the brackets is in the set {–2, –1, 0, 1, 2}, so it cannot be equal to 3 or –3. In other words, after rewriting B through the modified algorithm, each pair of digits can only take the following forms: [–1, –1], [0, –1], [0, 0], [0, 1], [1, 1], that is (–2, –1, 0, 1, 2). Another consequence of the modified Booth's algorithm is that the sign of the numbers is implicitly taken into account.

8.5.5.1 *Application to multiplication*

Consider two numbers *A* and *B*. Encoding *B* through the modified algorithm converts its form to B' with digits –2, –1, 0, 1, 2. In this form there will be half the number of digits in *B* in B'. The digits of B' are scanned, and at each step, *A* is multiplied by –2, –1, 0, 1, or 2. The different cases are given in Table 8–2. For example, if bit b_{2k} of *B* is 0 and bits b_{2k+1} and b_{2k-1} are 1 and 0 respectively, then we must add $-2A$ to the sum forming the product in the accumulator.

Table 8–2 Modified Booth's multiplication

b_{2k+1}	b_{2k}	b_{2k-1}	*A multiplied by*
0	0	0	0
0	0	1	+1
0	1	0	+1
0	1	1	+2
1	0	0	–2
1	0	1	–1
1	1	0	–1
1	1	1	0

One possible implementation of a circuit to implement the requirements of Table 8–2 is set out in Figure 8–38.

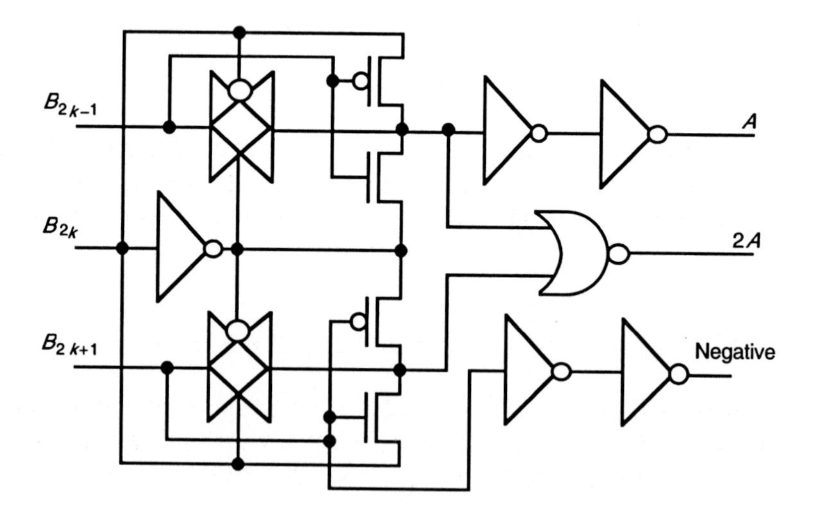

Figure 8–38 Booth encoder

8.5.6 Wallace tree multipliers

Wallace trees were first introduced in 1964 (Wallace, 1964) in order to design multipliers whose completion time grows as the logarithm of the number of bits to be multiplied. The simplest Wallace tree is the full adder cell (three inputs — two outputs). More generally, an n input Wallace tree, as in Figure 8–39, is an n-input operation with $log_2(n)$ outputs, such that the value of the output word is equal to the number of '1'*s* in the input word (consider the full adder in this context). The input bits and the least significant bit of the output word have the same weight, as shown in Figure 8–39.

An important property of Wallace trees is that they may be constructed from adder cells. Furthermore, the number of adder cells needed grows as the logarithm $log_2(n)$ of the number of input bits n. In a Braun or a Baugh-Wooley multiplier with a ripple carry adder, the completion time for multiplication is proportional to $2n$. If the collection of the partial products is made through Wallace trees then the completion time for getting a result, in carry save notation, should be proportional to $log_2(n)$.

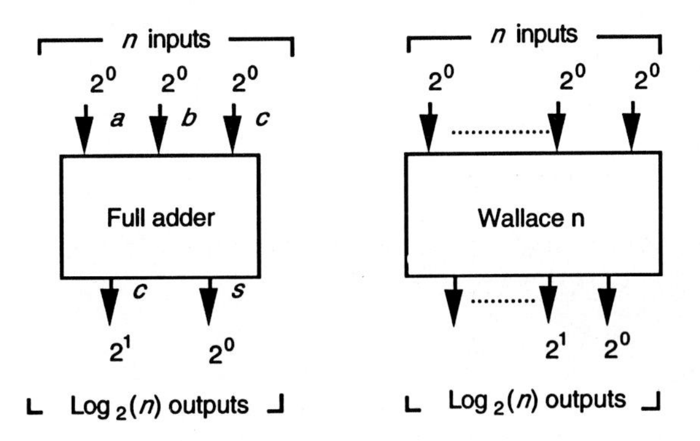

Figure 8–39 Wallace tree elements

Figure 8–40 shows a seven-input adder for each weight and Wallace trees are used until only two bits of each weight remain. These bits are then added using the classical two-input adder. Wallace trees may be applied to multipliers in several ways.

8.5.7 Recursive decomposition of the multiplication

One method, based on recursive decomposition of the multiplication, consists of partitioning the operands. For instance, if A and B are $2p$-bit numbers, then A (also B) may be cut into two parts A_0 and A_1 respectively, so that

$$A = 2^p . A_1 + A_0$$

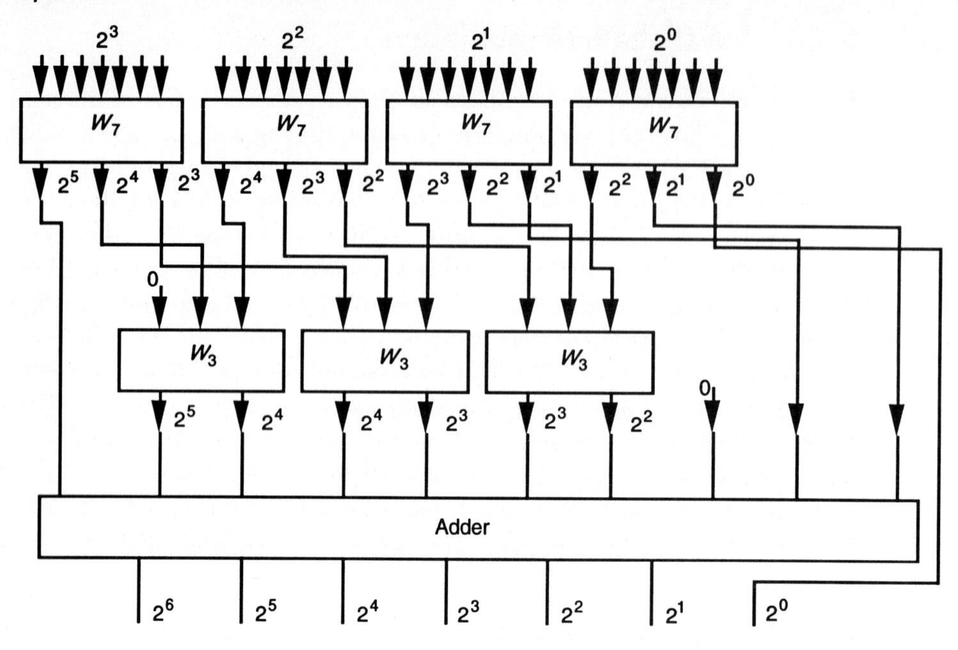

Figure 8–40 Example of the Wallace tree approach

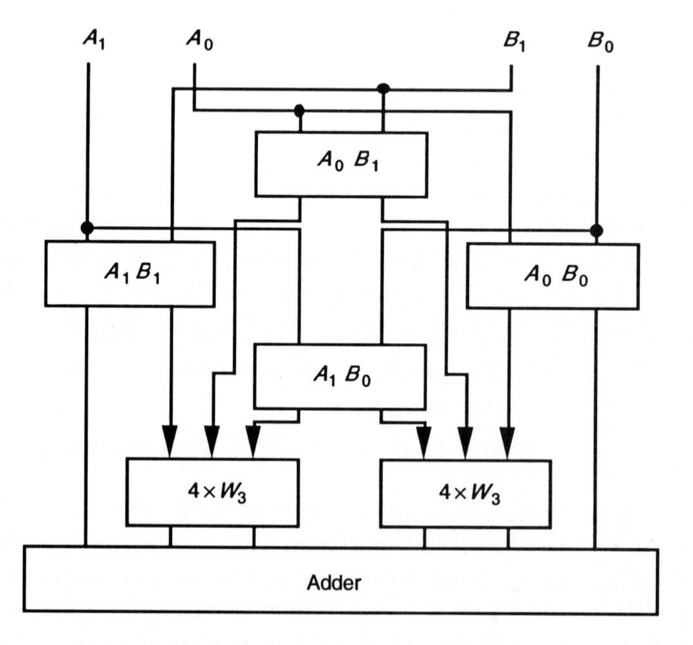

Figure 8–41 8-bit input word multiplier arrangement

$$B = 2^p. B_1 + B_0$$

The product $A.B$ is

$$A.B = 2^{2p}. A_1B_1 + 2^{6p}. (A_1. B_0 + A_0 . B_1) + A_0.B_0$$

Using this method, four p-bit multipliers are used to compute $A_1.B_1$, $A_0.B_1$, $A_1 . B_0$ and $A_0 . B_0$. The results are collected through Wallace trees. The arrangement of a multiplier of this type, with 8-bit input words, is shown in Figure 8–41; the interconnections have been simplified for clarity. A_0, B_0, A_1 and B_1, are in fact 4-bit numbers and the outputs of the multiplier are 8-bit products. In this figure it has been assumed that the multipliers each contain an adder so that each result is not in carry save notation and thus eight adder cells (three-input Wallace trees) are used to collect bits of the same weight. For instance, the multipliers denoted $A_0 . B_1$, $A_1. B_0$ and $A_0 . B_0$ give bits of weight 4, 5, 6, and 7. For each of these weights, three bits (as many as there are multipliers) must be added, and thus an adder cell must be used to reduce the number of bits of the same weight to two.

8.5.8 Dadda's method

Another approach consists in computing all the partial products — like the Braun array — and then collecting all the bits of the same weight through Wallace trees. This is equivalent to partitioning the input operands to work with 1-bit multipliers (i.e. *And* gates). In 1965 L. Dadda developed a technique to build the Wallace layer using the minimum number of adder cells.

Consider k bits of the same weight i coming from k partial products. When adding these k bits by a k-input Wallace tree, bits of weights $i + 1$, $i + 2$, . . . etc. appear which must in turn be added to the bits of weights $i + 1$, $i + 2$, . . . coming from other partial products. Dadda's method consists in handling all bits in the collecting Wallace layer so as to minimize the number of adder cells as well as the critical path between the partial product generation and the final addition. All the developments of this technique may be found in the reference (Dadda, 1965). In conclusion, Wallace tree multipliers should be used only for large operands and where the performance is critical since the arrangement results in poor regularity due to the routing area needed to collect the partial products.

8.6 Observations

This chapter has provided possible designs for the arithmetic subsystem forming part of the complete data path we are designing. Both the subsystems so far designed have comprised only combinational logic with the exception of possible storage requirements at the 'Sum' output of the adder. The third subsystem, to be designed next, will introduce a need for memory or storage and this leads to a review of some possible memory elements and relevant characteristics.

8.7 Tutorial exercises

1. Referring to Figure 8–12, design switches and other logic as necessary to implement the functions performed by the mechanical switches drawn in Figure 8–12. Work out the control lines needed to enable the ALU to perform add, subtract, logical *And,* logical *Or,* logical *Exclusive-Or,* and logical *Equality* operations.
2. Draw a bounding box representation with all inlet and output points shown (as in Figure 8–10) for the *logic circuitry* of an adder, using CMOS multiplexers (Figure 8–4) and CMOS inverters as suggested in Figure 8–9. You may wish to proceed as follows.

   Continue the design of a standard CMOS adder element (as represented in stick diagram form in Figure 8–5) by working out a layout for the complete inverter block and then representing it as a bounding box with inlet and outlet points indicated by layer and position. *Hint:* Design a suitable mask layout for the CMOS inverters and then represent each inverter circuit in bounding box form — with inlet and outlet points — so that only one inverter needs to be drawn in detail in setting out your layout.

   Interconnect the inverter block bounding box with CMOS multiplexer-based adder logic (as in Figure 8–4). Work out an accurate bounding box representation for the complete adder element showing inlet and outlet points, etc., by position and layer.
3. What are the overall dimensions of a 4-bit CMOS adder? Using the bounding box representations draw an accurate floor plan of the whole 4-bit adder (as in Figure 8–11) showing position and layer of inlet and outlet points.
4. Carry out the design of a 4-bit CMOS carry look-ahead adder up to stick diagram form. Then determine what standard cells are needed and design a mask layout for each.

8.8 References

Dadda, L. (1965, March) 'Some schemes for parallel multipliers', *Alta Frequenza*, Vol. 19.

Guyot, A., Hochet, B., and Muller, J. M. (1987, October) 'A way to build efficient carry-skip adders', *IEE Trans. on Computers*, Vol. C–36.

Hotta, T. et al. (1986, October) 'CMOS/Bipolar circuit for 60 MHz digital processing', *IEEE Journal of Solid State Circuits,* 803–13, Vol. 21, No. 5.

Muller, J. M. (1989) *Arithmétique des Ordinateurs,* Editions Masson, Collection Etudes et Recherches en Informatique, Paris.

Wallace, C. S. (1964, February) 'A suggestion for a fast multiplier', *IEEE Trans. on Electronic Computers,* 14–17.

Wang, I. S. and Fisher, A. L. (1989, April) 'Ultrafast compact 32-bit CMOS adders in multiple-output domino logic', *IEEE Journal of Solid State Circuits,* Vol. 24.

9 Memory, registers, and aspects of system timing

Ay, now the plot thickens very much upon us.
George Villiers, 2nd Duke of Buckingham

Objectives

The 4-bit data path design continues with the 4 × 4-bit register array. This raises the subject of memory/storage elements and techniques. Some of the possible dynamic and static memory cells are presented and key properties compared. The concept of an array of memory cells is extended to include RAM arrays and some of the needs for selection and control are explored.

Two of the subsystems of the 4-bit data path (as in Figures 7–3 and 8–1) having already been designed, it is now appropriate to consider the register arrangements in which the 4-bit quantities to be presented to the adder and shifter will be stored. The question of data storage is an important one which has already been mentioned a number of times. It raises the question of the choice of storage elements or memory cells as well as the questions of configuring arrays of such cells and the selection of a given cell or group of cells in an array.

Before looking at register arrangements, we should set out some ground rules for the design of the 4-bit processor. It is essential that such rules should be established early in the piece so that a uniform approach to 'reading, writing and refresh' is adhered to throughout. In practice, such rules would have been set out much earlier than this, but our progress through this text is such that in this case they are most effectively established here and would not have meant much earlier on.

9.1 System timing considerations

1. A two-phase non-overlapping clock signal is assumed to be available, and this clock alone will be used throughout the system.
2. Clock phases are to be identified as ϕ_1 and ϕ_2 where ϕ_1 is assumed to lead ϕ_2.
3. Bits (or data) to be stored are *written* to registers, storage elements, and subsystems on ϕ_1 of the clock; that is, write signals *WR* are *Anded* with ϕ_1.
4. Bits or data written into storage elements may be assumed to have settled before the immediately following ϕ_2 signal, and ϕ_2 signals may be used to r*efresh* stored data where appropriate.
5. In general, delays through data paths, combinational logic, etc. are assumed to be less than the interval between the leading edge of ϕ_1 of the clock and the leading edge of the following ϕ_2 signal.
6. Bits or data may be *read* from storage elements on the next ϕ_1 of the clock; that is, read signals *RD* are *Anded* with ϕ_1. Obviously, *RD* and *WR* are generally mutually exclusive to any one storage element.
7. A general requirement for system stability is that there must be at least one clocked storage element in series with every closed loop signal path.

Strict adherence to a set of rules such as this will greatly simplify the task of the system designer and also help to avoid some of the disasters which will almost certainly occur if a haphazard approach is taken.

9.2 Some commonly used storage/memory elements

Everyone complains of his memory, but no one complains of his judgment.

Duc de la Rochefoucauld

In order to make a comparative assessment of some possible storage elements, we will consider the following factors:

- area requirement;
- estimated dissipation per bit stored;
- volatility.

9.2.1 The dynamic shift register stage

One method of storing a single bit is to use the shift register approach previously introduced in section 6.5.4 (and also Figures 3–14, 3–17, 6–36, 6–37, 6–38, 6–39 and 6–40).

9.2.1.1 Area

This calculation applies to an nMOS design, as in Figure 6–40(a), with buried contacts. Allowing for the sharing of V_{DD} and GND rails between adjacent rows of register cells, each bit stored will require

$$(22\lambda \times 28\lambda) \times 2 \doteqdot 1200\lambda^2$$

For example, for $\lambda = 2.5\ \mu m$

$$\text{Area/bit} = 7500\ \mu m^2$$

To give an idea of what this implies, such area requirements would result in a maximum number of bits stored on a 4 mm × 4 mm chip area ≈ 2.1 kbits.

For a CMOS design, as in Figure 6–40(b), and allowing for the sharing of V_{DD} and V_{SS} rails between adjacent rows of register cells, each bit stored will require

$$(38\lambda \times 28\lambda) \times 2 \doteqdot 2100\lambda^2$$

For example, for $\lambda = 2.5\ \mu m$

$$\text{Area/bit} \approx 13{,}000\ \mu m^2$$

Such area requirements would result in a maximum number of bits stored on a 4 mm × 4 mm chip area ≈ 1.2 kbits.

9.2.1.2 Dissipation

In the case of CMOS designs, the static dissipation is very small and calculation at this stage will not be meaningful since only the switching dissipation will be significant (particularly at high speeds). This dynamic power consumption P_d can be written as

$$P_d = m.\,(C_L.\,V_{DD}^2.\,f)$$

where m is the duty cycle, C_L is effective load capacitance and f is the clock frequency.

In the nMOS case we can readily calculate the static dissipation, noting that in practice the switching dissipation would add to this. Each inverter stage has a ratio of 8:1 and if the layout of Figure 6–40(a) (buried contacts) is used, then, noting that one inverter of the pair must always be 'on',

$$Z_{p.u.} = 4\,R_s$$

and

$$Z_{p.d.} = \tfrac{1}{2}\,R_s$$

Therefore

$$\text{Current} = \frac{V_{DD}}{Z_{p.u.} + Z_{p.d.}} = \frac{5\ \text{V} \times 10^6}{4.5 \times 10^4} = \frac{500}{4.5}\ \mu\text{A} \approx 110\ \mu\text{A}$$

Therefore

$$\frac{\text{Static dissipation}}{\text{Bit stored}} = V_{DD} \times \text{current} = 5\ \text{V} \times 110\ \mu\text{A} = 550\ \mu\text{W}$$

Thus, 2.1 kbits on a single chip would dissipate

$$2.1 \times 10^3 \times 550 \times 10^{-6} = 1.15 \text{ watts}$$

Dissipation can be reduced by using alternative geometry, but this is at the expense of increased area.

9.2.1.3 Volatility

Data is stored by the charge on the gate capacitance of each inverter stage, so that data storage time (without refresh) is limited to 1 msec or less.

9.2.2 A three-transistor dynamic RAM cell

An arrangement which has been used in RAM (random access memory) and other storage arrangements is set out in Figure 9–1.

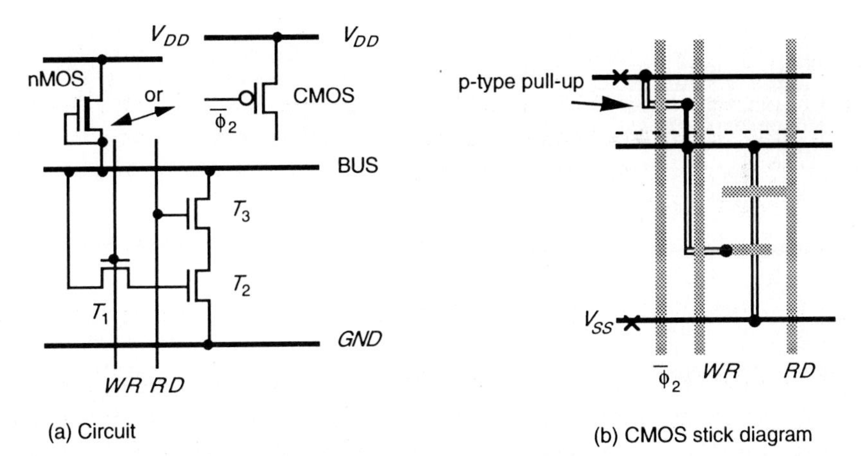

(a) Circuit

(b) CMOS stick diagram

Note: *WR* and *RD* are coincident with ϕ_1.

Figure 9–1 Three-transistor dynamic memory cell

With regard to Figure 9–1(a), the action is as follows:

1. With the *RD* control line in the Lo state, then a bit may be read from the bus through T_1 by taking *WR* to the Hi state so that the logic level on the bus is communicated to the gate capacitance of T_2. Then *WR* is taken Lo again.

2. The bit value is then stored for some time by Cg of T_2 while both RD and WR are Lo.
3. To read the stored bit it is only necessary to make RD Hi and the bus will be pulled down to ground through T_3 and T_2 if a 1 was stored. Otherwise, T_2 will be non-conducting and the bus will remain Hi due to its pull-up arrangements.

   Note that the complement of the stored bit is read onto the bus, but this presents few problems and can be taken care of at some common point in the memory array.

A stick diagram for the cell identified in Figure 9–1(a) is presented as Figure 9–1(b), and possible mask layouts follow in Figure 9–2. Note that this figure gives both nMOS and CMOS designs.

To return to our main theme, it is now appropriate to assess the three-transistor cells in the same manner as the previous one.

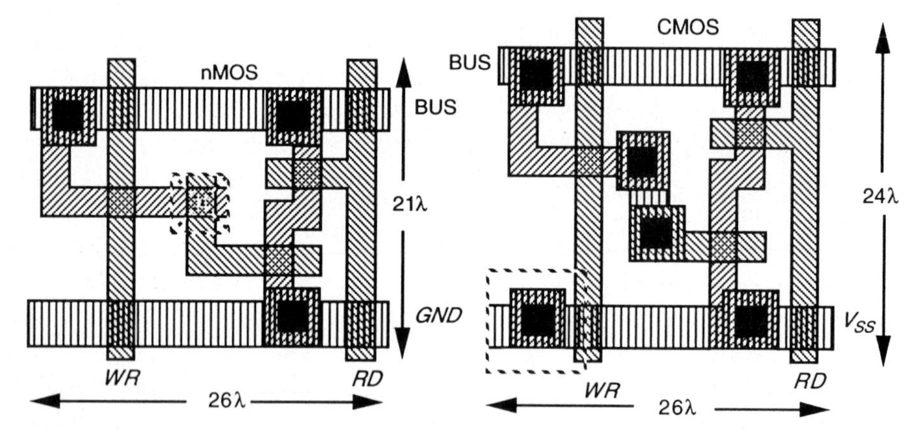

Figure 9–2 Mask layouts* for three-transistor (nMOS and CMOS) memory cell *(pull-ups not shown)

9.2.2.1 *Area*

From the layout it will be seen that an area of more than $500\lambda^2$ is required for each bit stored (less if GND (V_{SS}), and/or bus, and/or control lines are shared with other cells). Thus, for $\lambda = 2.5$ μm

$$\text{Area/bit} \doteqdot 3000\ \mu\text{m}^2$$

Thus, to continue the previous example, the maximum number of bits which could be accommodated on a 4 mm × 4 mm silicon chip is > 4.8 kbits.

9.2.2.2 *Dissipation*

Static dissipation is nil since current flows only when RD is Hi and a 1 is stored. Thus, the actual dissipation associated with each bit stored will depend on the bus pull-up and on the duration of the RD signal and on the switching frequency.

9.2.2.3 Volatility

The cell is 'dynamic' and will hold data only for as long as sufficient charge remains on C_g (of T_2).

9.2.3 A one-transistor dynamic memory cell

The area occupied by each bit stored in each of the previous cases is quite considerable, which clearly limits, say, the number of bits which could be stored on a single chip of reasonable size.

Various approaches have been taken to reduce the area per bit requirements and one such approach is the one-transistor cell as shown in Figure 9–3. The concept of the single transistor cell is quite simple, as may be seen from Figure 9–3(a). It basically consists of a capacitor C_m which can be charged during *'write'* from the *read/write* line, provided that the *row select* line is Hi. The state of the charge C_m can be read subsequently by detecting the state of the charge via the same *read/write* line with the *row select* line Hi again, and a sense amplifier of a suitable nature can be designed to differentiate between a stored 0 and a stored 1.

However, in practice the cell is slightly more complex than first considerations might suggest, since special steps must be taken to ensure that C_m has sufficient capacitance to allow ready detection of the stored content.

The most obvious and readily fabricated C_m in the structure under consideration would be to extend and enlarge the diffusion area comprising the source (S) of the pass transistor in Figure 9–3(b). We would then rely on the junction capacitance between the n-diffusion region and the p-substrate to form $C_{m.}$ However, if we consult Table 4–2 (which gives capacitance values for a typical MOS process), we will see that the capacitance per unit area of diffusion is much less than the capacitance per unit area between gate and channel (i.e. between the channel under the thin gate oxide and the polysilicon gate area).

If we use the diffusion to substrate capacitance alone, a comparatively large area will be required to give any significant value of capacitance; for example, at least $16\lambda^2$ will be needed to give a capacitance equal to $1\square C_g$ (i.e. 0.01 pF in the 5 µm MOS process being considered). A solution is to create a much more significant capacitor by using a polysilicon plate (which is connected to V_{DD}) over the diffusion area. Thus, C_m is formed as a three-plate structure as indicated in Figure 9–3(c). For example, for the area given in Figure 9–3(d), $C_{Diff.\text{-}Poly.} = 10\square C_g$ (= 0.1 pF), while the contribution from the diffusion region to the substrate will be much smaller but will add some 25% to this figure, giving a total C_m of 0.125 pF for the layout considered. Even so, careful design is necessary to achieve consistent readability.

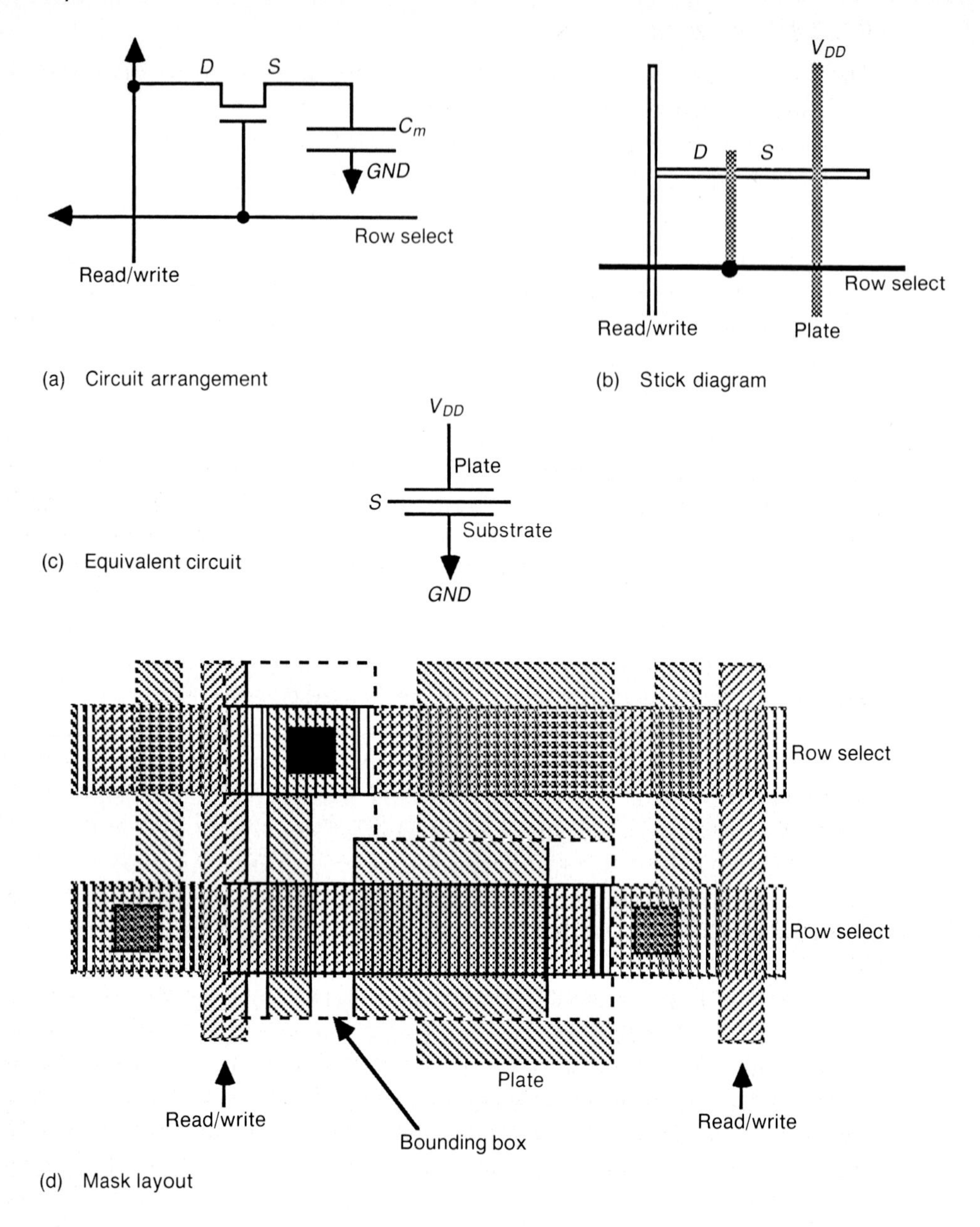

Figure 9–3 One-transistor memory cell

9.2.3.1 Area

The area enclosed to indicate the standard cell in Figure 9–3(d) is $200\lambda^2$. Thus for $\lambda = 2.5\ \mu m$, area/bit stored = $200\lambda^2 = 1250\ \mu m^2$.

Therefore, the number of bits per 4 mm × 4 mm chip area is approximately 12 kbits (allowing some 'overheads' for sensing, etc.).

9.2.3.2 Dissipation

There is no static power associated with the cell itself, but there must be an allowance for switching energy while writing to and reading from the storage elements.

9.2.3.3 Volatility

Quite obviously, leakage current mechanisms will deplete the charge stored in C_m and thus the data will be held for only up to 1 msec or less. Therefore, periodic refresh operations must be provided. It will also be realized that reading the cell is a destructive operation and that the stored bit must be rewritten every time it is read.

9.2.4 A pseudo-static RAM/register cell

So far, all the storage elements considered have been volatile and thus have an implied need to be periodically refreshed. This is not always convenient and it is necessary to consider the design of a static storage cell which will hold data indefinitely. A common way of meeting this need is to store a bit in two inverter stages in series with feedback, say, on ϕ_2 to refresh the data every clock cycle. Circuit arrangements are shown in Figures 9–4(a) and 9–5(a) and it will be seen that a bit may be written to the cell from the bus by energizing the *WR* line. From our system timing consideration of section 9.1, we will assume *WR* to occur in coincidence with ϕ_1 of the clock. Thus, the bit is stored on C_g of inverter 1 and will be reproduced complemented at the output of inverter 1 and true at the output of inverter 2. It will be seen that during every ϕ_2 of the clock the stored bit is refreshed through the gated feedback path from the output of inverter 2 to the input of inverter 1. Thus the bit will be held as long as ϕ_2 of the clock recurs at intervals less than the decay time of the stored bit. To read the state of the cell it is only necessary to energize the *RD* line, which is also assumed coincident with ϕ_1 of the clock, and the bit will be read onto the bus.

Note that:

1. *WR* and *RD* must be mutually exclusive (but are both coincident with ϕ_1).
2. If ϕ_2 is used for refresh, then the cell must not be read during ϕ_2 of the clock unless the feedback path is inhibited during *RD*. If an attempt is made to read

the cell onto the bus during refresh, then charge sharing effects between the bus and input (C_g) capacitances may cause the destruction of the stored bit.

3. Cells must be stackable, both side by side and top to bottom. This must be carefully considered together with the overall strategy to be observed when the layout is drawn.
4. Allow for other bus lines to run through the cell so that register and memory arrays are readily configured.

With these factors in mind, it is possible to draw up stick diagrams as in Figures 9–4(b) and 9–5(b), which show the nMOS and CMOS basic cells.

Mask level layouts follow from this; a possible layout for an nMOS cell

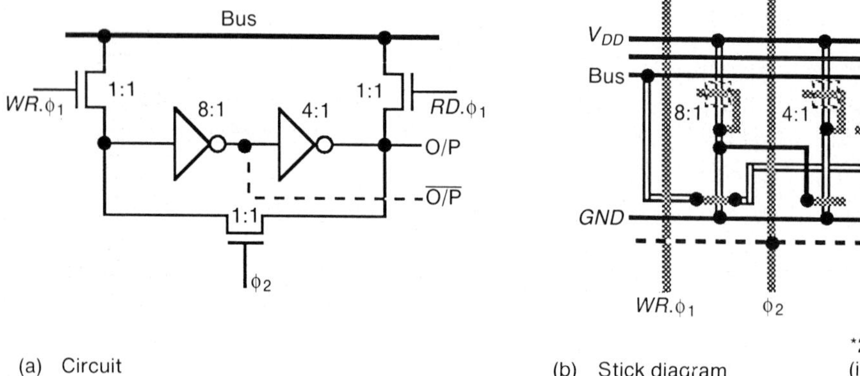

(a) Circuit (b) Stick diagram

Figure 9–4 nMOS pseudo-static memory cell

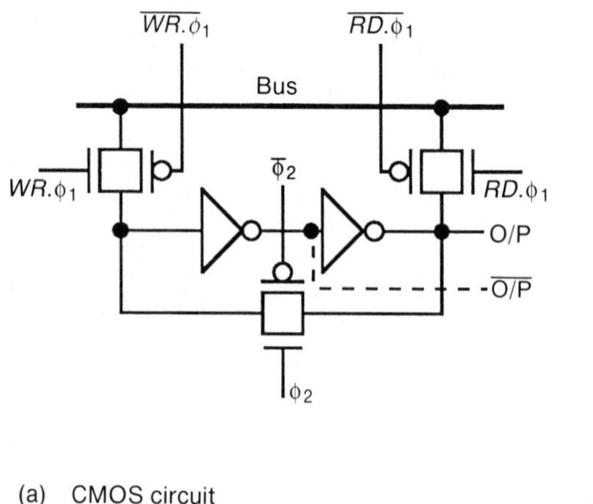

(a) CMOS circuit

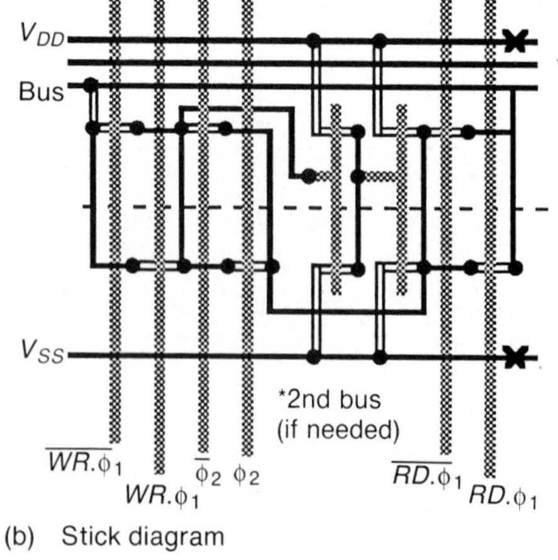

(b) Stick diagram

Figure 9–5 CMOS pseudo-static memory cell

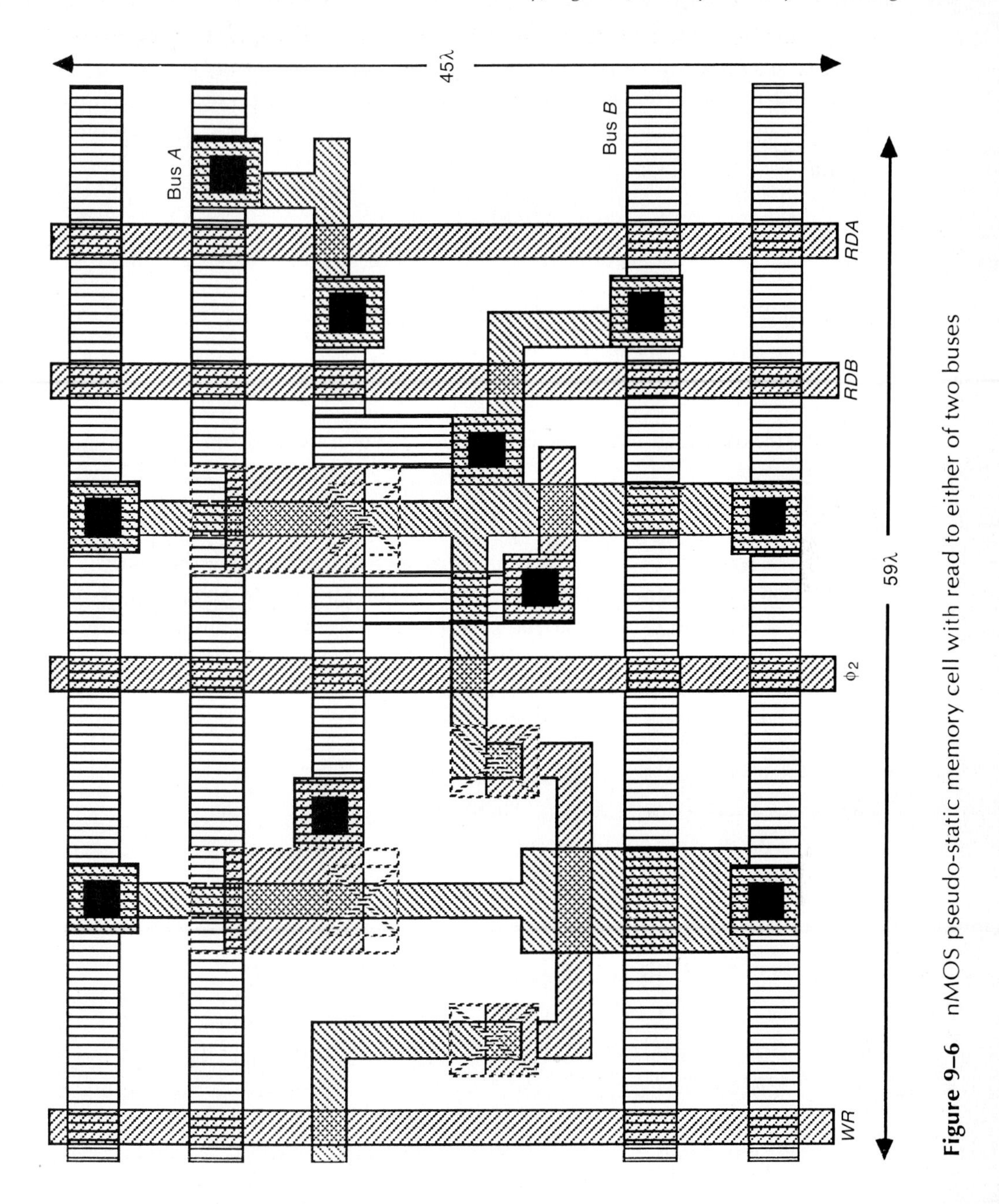

Figure 9–6 nMOS pseudo-static memory cell with read to either of two buses

which can be written to from bus *A* and can be read onto bus *A* or bus *B* is given in Figure 9–6.

The mask layout shown in Figure 9–6 occupies an area of $59\lambda \times 45\lambda = 2655\lambda^2$, but if we are considering a single bus and a more compact layout then the area requirement can be reduced to about $1750\lambda^2$ or less. The CMOS version of this cell is *not* really a practical proposition other than for storing a few bits. Ten transistors are needed per bit stored, which makes the cell too demanding in area to be the basis of larger memories. We will therefore evaluate only the nMOS version of the cell and, to return to the original purpose, we may now set out the relevant parameters for this popular and useful pseudo-static storage cell in the same terms as have been used previously.

9.2.4.1 Area

A typical area for a nMOS single cell with single bus is in the region of $1750\lambda^2$. Therefore, for $\lambda = 2.5\ \mu m$

$$\text{Area/bit} \approx 10\ 000\ \mu m^2$$

Thus, the maximum number of bits of storage per 4 mm × 4 mm chip is approximately 1.4 kbits.

9.2.4.2 Dissipation

The nMOS cell uses two inverters, one with an 8:1 and the other with a 4:1 ratio. Dissipation will depend on the current drawn and thus on the actual geometry of the inverters, but let us assume that inverters are based on minimum feature size gate areas so that the 8:1 stage will present a resistance of 90 kΩ and the 4:1 stage a resistance of 50 kΩ between the supply rails. Now when one stage is off, the other is on so that, say, each spends half-time in the conducting state. Therefore

$$\text{Average current } = 0.5\left(\frac{5\text{ V}}{90\text{ k}\Omega} + \frac{5\text{ V}}{50\text{ k}\Omega}\right) \approx 80\ \mu A$$

Therefore dissipation per bit stored = 80 μA × 5 V = 400 μW. Thus 1.4 kbits on a single chip would dissipate 560 mW.

9.2.4.3 Volatility

The cell is non-volatile provided that ϕ_2 signals are present.

9.2.5 Four-transistor dynamic and six-transistor static CMOS memory cells

Most of the preceding memory cells involved n-type transistors and can therefore be implemented in either nMOS or CMOS designs. The cells about to be described utilize both n-type and p-type transistors and are therefore intended for CMOS systems only (although the dynamic element can be readily adapted to nMOS-only implementation).

Both the dynamic and static elements, set out in Figure 9–7, use a two bus per bit arrangement so that associated with every bit stored there will be a *bit* and

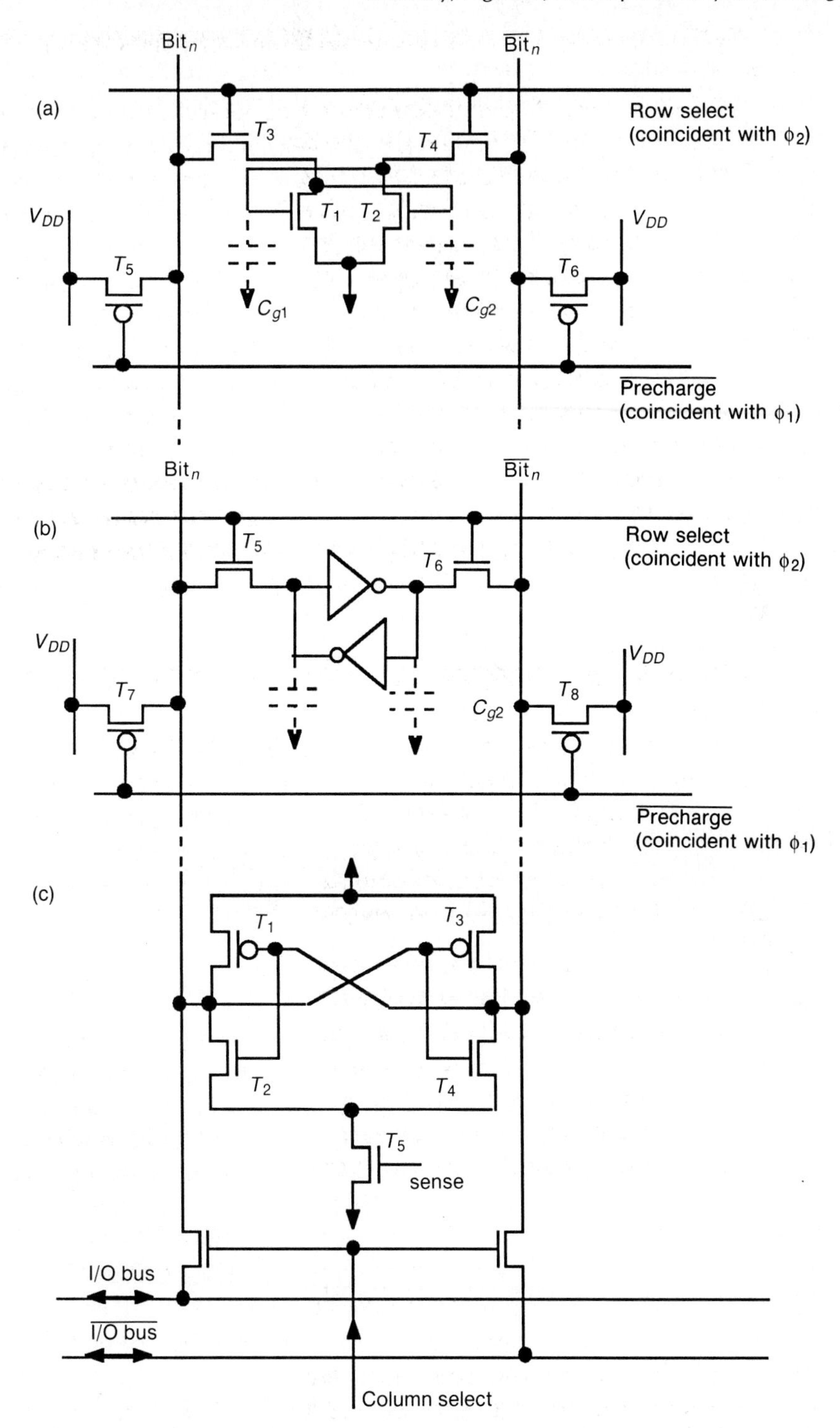

Figure 9–7 Dynamic and static memory cells

a $\overline{bit}$ bus as shown. In both cases the buses are precharged to logic 1 before read or write operations take place.

Figure 9–7(a) gives the arrangement for a four-transistor dynamic cell for storing one bit. Each bit is stored on the gate capacitance of two n-type transistors T_1 and T_2 and a description of the write and read operation follows.

9.2.5.1 *Write operations*

Both *bit* and $\overline{bit}$ buses are precharged to V_{DD} (logic 1) in coincidence with ϕ_1 of an assumed two-phase clock. Precharging is effected via the p-transistors T_5 and T_6 in the diagram. Now (with reference to Figure 9–7(c)), the appropriate *'column select'* line is activated in coincidence with the clock phase ϕ_2 and either the *bit* or $\overline{bit}$ line is discharged by the logic levels present on the I/O *bus* lines, the I/O lines acting in this case as a current sink when carrying a logic 0. The *'row select'* signal is activated at the same time as *'column select'* and the bit line states are 'written in' via T_3 and T_4 and stored by T_1 and T_2 as charges on gate capacitances C_{g2} and C_{g1} respectively. Note that the way in which T_1 and T_2 are interconnected will force them into complementary states while the *row select* line is high. Once the select lines are deactivated, the bit stored will be remembered until the gate capacitances lose enough charge to drop the 'on' gate voltage below the threshold level for T_1 or T_2.

9.2.5.2 *Read operations*

Once again both *bit* and $\overline{bit}$ lines are precharged to V_{DD} via T_5 and T_6 during ϕ_1 so that both lines will be at logic 1. Now if, say, a 1 has been stored, T_2 will be on and T_1 will be off, and thus the $\overline{bit}$ line will be discharged to V_{ss} (logic 0) through T_2 and the stored bit thus reappears on the bit lines.

When such cells are used in RAM arrays, it is necessary to keep the area of each cell to a minimum and transistors will be of minimum size and therefore incapable of sinking large charges quickly. Thus it is important that the charges stored on the bit lines be modest and this may not be the case if they are directly paralleled by the I/O line and other associated capacitances through the column select circuitry. RAM arrays therefore generally employ some form of sense amplifier. A possible arrangement is shown in Figure 9–7(c) in which T_1, T_2, T_3, and T_4 form *a flip-flop circuit.* If we assume the sense line to be inactive, then the state of the bit lines is reflected in the charges present on the gate capacitances of T_1 and T_3 with respect to V_{DD} such that a 1 will turn off and a 0 turn on either transistor. Current flowing from V_{DD} through an on transistor helps to maintain the state of the bit lines and predetermines the state which will be taken up by the sense flip-flop when the sense line is then activated. The geometry of the single sense amplifier per column will be such as to amplify the current sinking capability of the selected memory cell.

Figure 9–7(b) indicates an adaption of the basic dynamic cell, just considered, to form a static memory cell. At the expense of two additional transistors per bit stored, the transistors T_1 and T_2 of Figure 9–7(a) can each be replaced by an inverter as shown in Figure 9–7(b). This arrangement will clearly render the cell static in its data-storing capabilities.

The general arrangement of a RAM utilizing the circuits considered here appears later in this chapter (Figure 9–18).

9.2.6 JK flip-flop circuit

No consideration of memory elements would be complete without the JK flip-flop. The JK flip-flop is a particularly widely used arrangement and is an example of a static memory element. It is also most useful in that other common arrangements such as the D flip-flop and the T flip-flop are readily formed from the JK arrangement. Edge-triggered circuits are conveniently designed with an ASM (algorithmic state machine) approach — see C. Clare, *Designing Logic Systems Using State Machines,* McGraw-Hill, 1983) — and the design equations for a JK flip-flop, as in Figure 9–8, follow from an ASM chart setting out the requirements as in Figure 9–9. It should be noted that the flip-flop is assumed to have an asynchronous clear (Clr) input as well as the clocked J and K inputs, and that J and K are read in during the Hi level of the clock ϕ, and the data thus read is transferred to the output on the falling edge of ϕ.

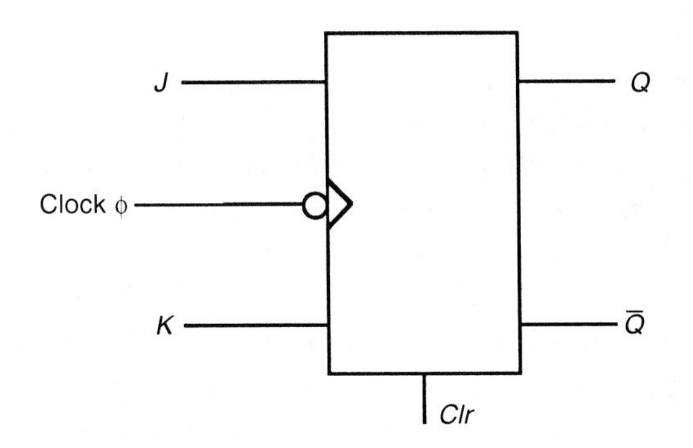

Figure 9–8 JK flip-flop

Design equations are readily derived from the ASM chart of Figure 9–9 and, making the secondary variable assignments (*AB* in the figure), we may express the requirements as follows:

$$A = a.(\overline{Clr}).(\bar{b}+\bar{\phi}+\bar{K})+\bar{b}.(\overline{Clr}).J.\phi$$

$$B = (\overline{Clr}).(a.\bar{\phi}+b.\phi)$$

where output $Q = B$, and a and b are the fed back state of the secondary variables A and B respectively.

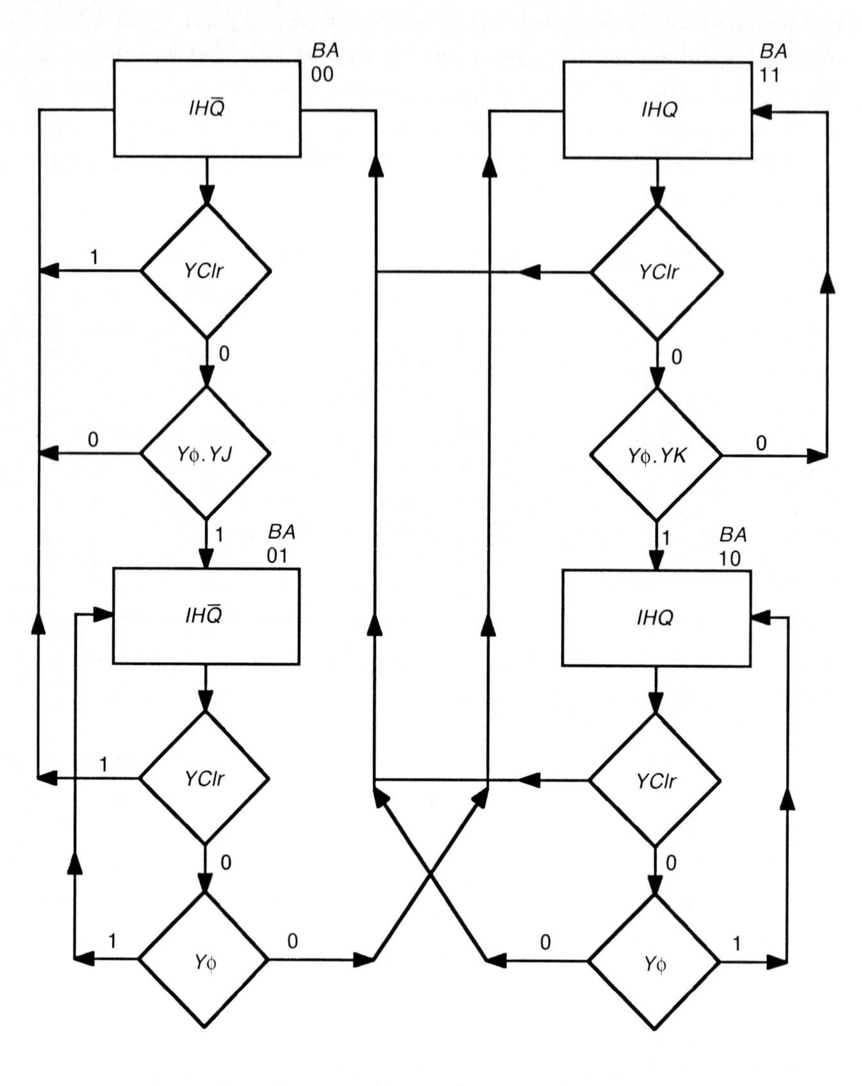

Note: IHQ = Q output 'Hi' (Immediate)
YClr = Yes clear
Yϕ.YJ = Yes clock and J, . . ., etc.

Figure 9–9 ASM chart for JK flip-flop

9.2.6.1 *Logic gate implementations*

We are now faced with a choice of implementations based on *Nand* or *Nor* or switch logic. The expressions for *A* and *B* are readily realized in *Nand* or *Nor* logic, as shown in Figure 9–10, and it will be seen that a master/slave arrangement applies in each case.

However, an initial consideration of each arrangement will reveal that, for nMOS, the *Nand arrangement is impractical,* owing to the relatively large number of gates requiring three or more inputs which will therefore be inherently large in area and slow in performance. The obvious nMOS altemative is a *Nor* gate arrangement which is a practical proposition and can be readily implemented.

For CMOS, both *Nand* and *Nor* gates are suitable although the *Nor* gate is generally slower.

9.2.6.2 *Switch logic and inverter implementation*

In setting out an arrangement of *n* pass transistors to realize the logical requirements, we must bear in mind earlier considerations on the nature of switch logic networks: that is, there should be no more than four pass transistors in series (section 4.9); pass transistors are not to be used to drive the gates of other pass transistors; the logic 0 as well as the logic 1 transmission conditions are to be deliberately satisfied. Thus, we need to implement the expressions for $\overline{A}$ and $\overline{B}$ *as well as* the expressions for A and B given earlier in this section. The resulting arrangement is given at Figure 9–11 and is a realization of the JK flip-flop based on n-pass transistor logic and inverters only.

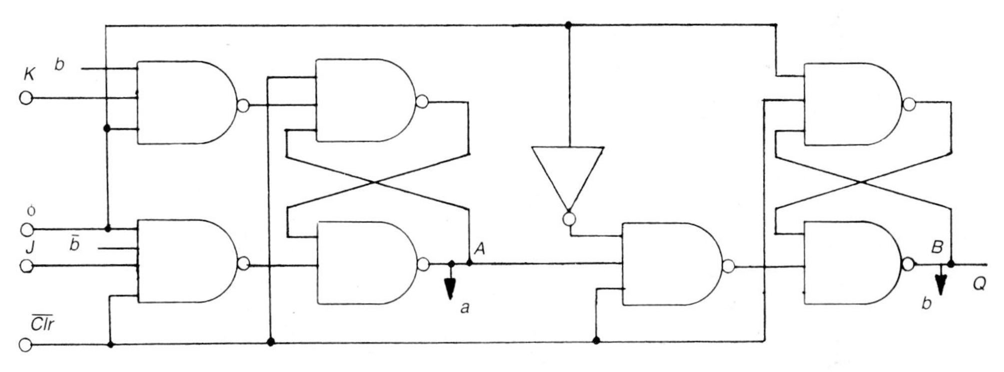

(a) *Nand* gate version

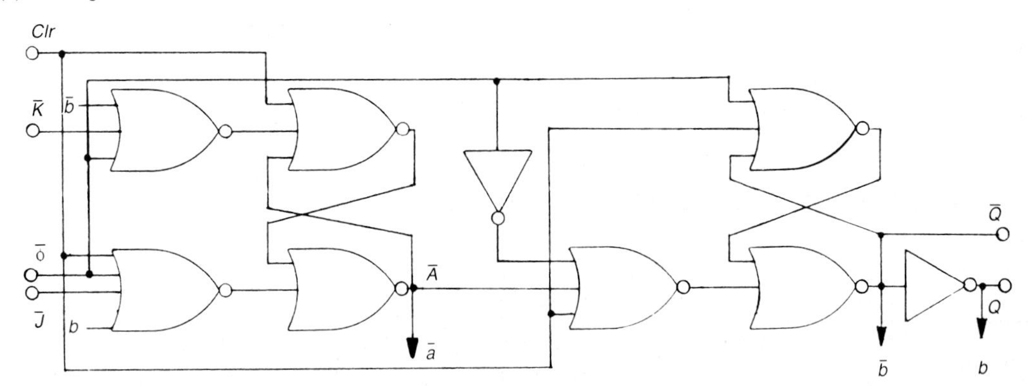

(b) *Nor* gate version

Figure 9–10 Logic arrangement for JK flip-flop

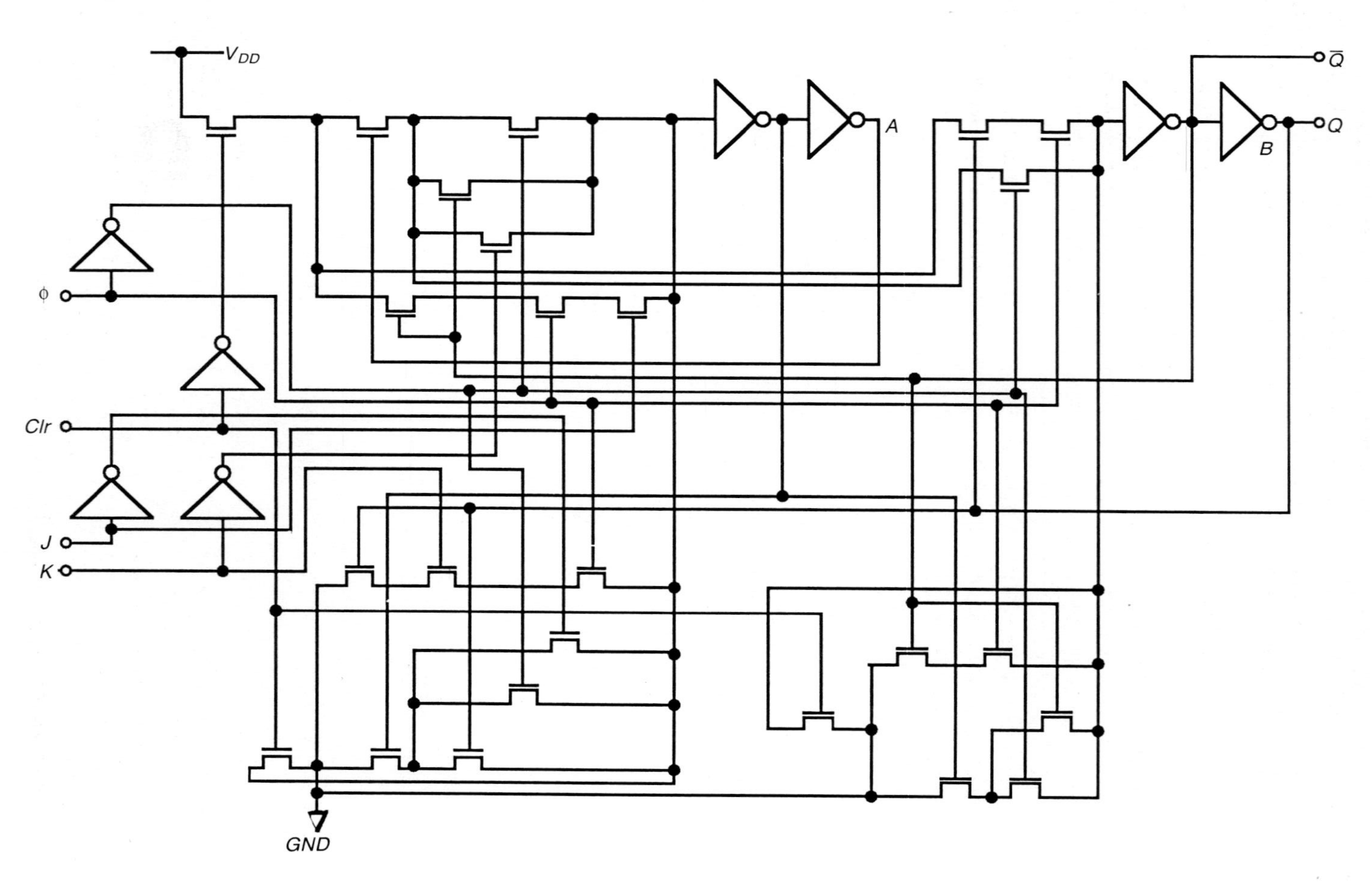

Figure 9–11 Switch logic implementation of JK flip-flop

9.2.7 D flip-flop circuit

A D flip-flop is readily formed from a JK flip-flop by renaming the J input D and then replacing connections to K by $\overline{D}$ (see Figure 9–10). Similarly, a T (Toggle) flip-flop is formed from the JK by making $J = K = E,$ where E is the toggle enabling input.

It should also be noted that the arrangements given may be simplified by the omission of the *Clr* input, or that a *Preset* input can be substituted for or added to the *Clr* input if required. Furthermore, the way in which clock activation takes place may be modified by a reshaping of requirements in the ASM chart of Figure 9–9 and a consequent reformulation of the JK flip-flop design equations given at the beginning of section 9.2.6 of this text.

However, a much simpler version of the D flip-flop is obtained from a pseudo-static approach, as in Figure 9–12 for CMOS. Clearly, an nMOS version is also readily configured.

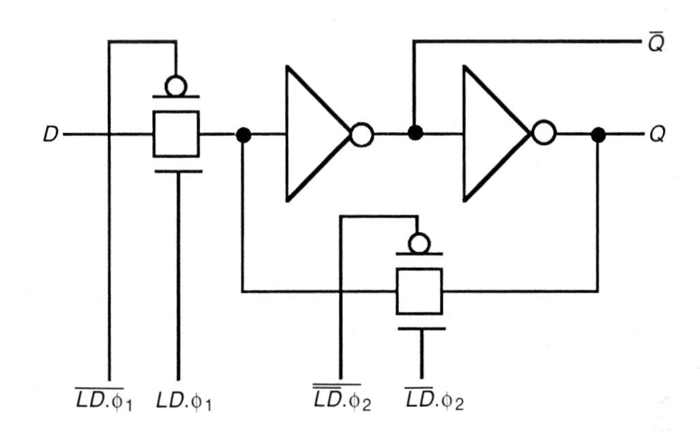

Figure 9–12 A CMOS pseudo-static D flip-flop

9.3 Forming arrays of memory cells

The memory cells discussed in section 9.2 and others will most often be used in arrays of some form or other. Typical arrays are registers and random access memories (RAM) and these arrays will be used as examples in this section. We must not forget, however, that another common application is to use memory elements individually as 'flags' or 'status bits' in system design. In any event, there must be some means of *selecting* a particular cell or group of cells and some means of effecting *read* or *write* operations.

9.3.1 Building up the floor plan for a 4 × 4-bit register array

This will be the third subsystem to be considered for the 4-bit data path, the floor plan of which appeared as Figure 7–5; the first two subsystem minimum bounding box outline dimensions have been given as Figure 7–10 (4 × 4 barrel shifter) and Figure 8–11(b) (4-bit adder). The fourth and final subsystem — the I/O port facilities — will be left for the reader to consider as an exercise in completing a system design (prior to adding inlet and outlet pads through which a chip is bonded to the outside world).

Starting with a bounding box representation of the chosen memory cell — in this case we have presented typical dimensions and connections of a pseudo-static cell with two bus lines, as in Figure 9–13 — we can arrive at a bounding box for a single 4-bit register and hence the floor plan for a 4 × 4-bit register array.

The bounding box representation of the cell is 'stacked' to form a 4-bit register as in Figure 9–14, the overall vertical dimension being about 180λ. Note how the cells stack 'vertically' to form a 4-bit word and note that although a 'vertical' distribution of power has been assumed at the input of the register, power distributes

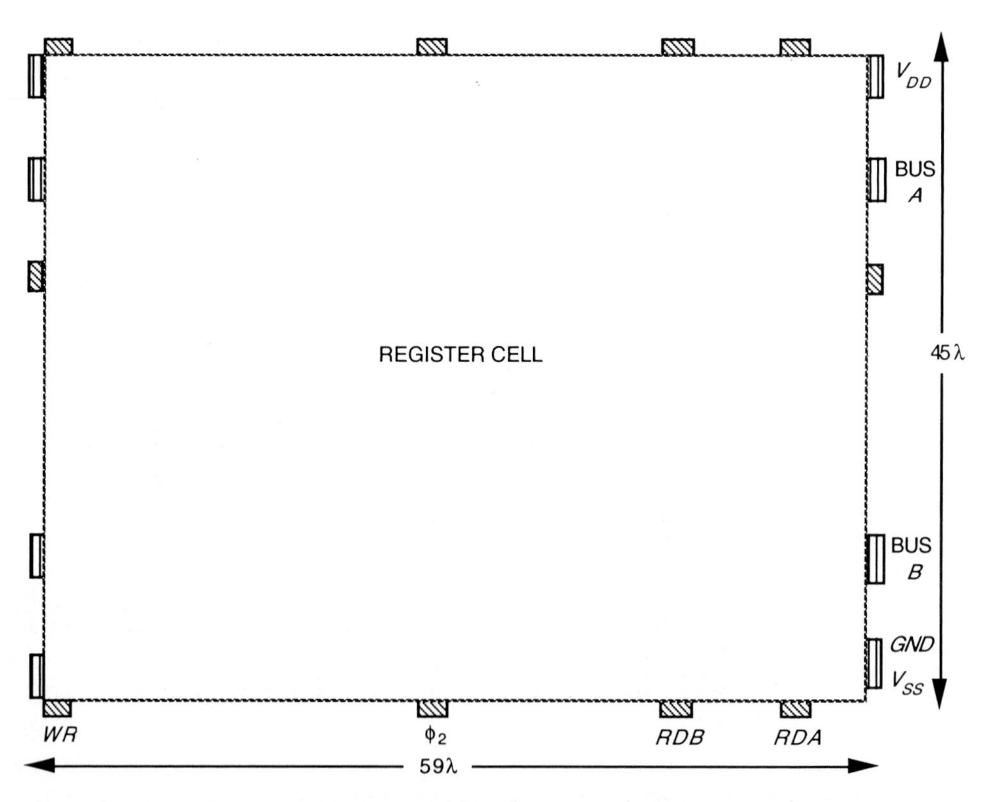

Figure 9–13 Bounding box for register cell

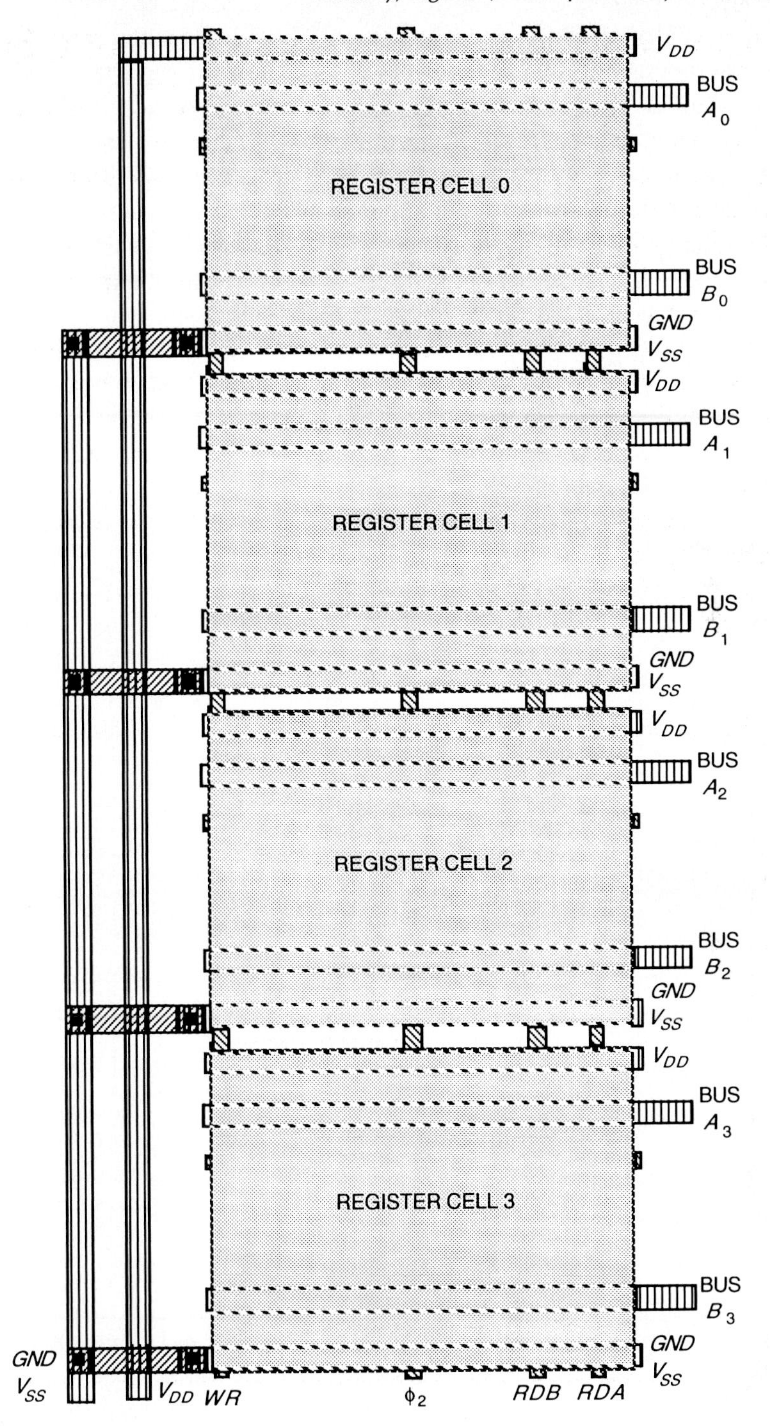

Figure 9–14 4-bit register floor plan

horizontally thereafter. Note that since only a single metal layer has been assumed throughout, short but wide diffusion 'duck unders' have then been used to allow the ground (or V_{SS}) rail to cross the V_{DD} rail. However, it must be stressed that V_{DD} and GND (V_{SS}) connections must *always* be made through metal rails, except where crossovers are unavoidable; such crossovers should then normally be by means of short diffusion 'duck unders' where there is no second metal layer.

The required architecture may then be built up by stacking complete registers, side by side in this case, to form the desired four-register array, the dimensions being around 180λ × 240λ. The floor plan is given in Figure 9–15 and the diagram clearly indicates the direction of data flow and control signal distribution. Note that this floor plan does *not* include the selection and control circuitry.

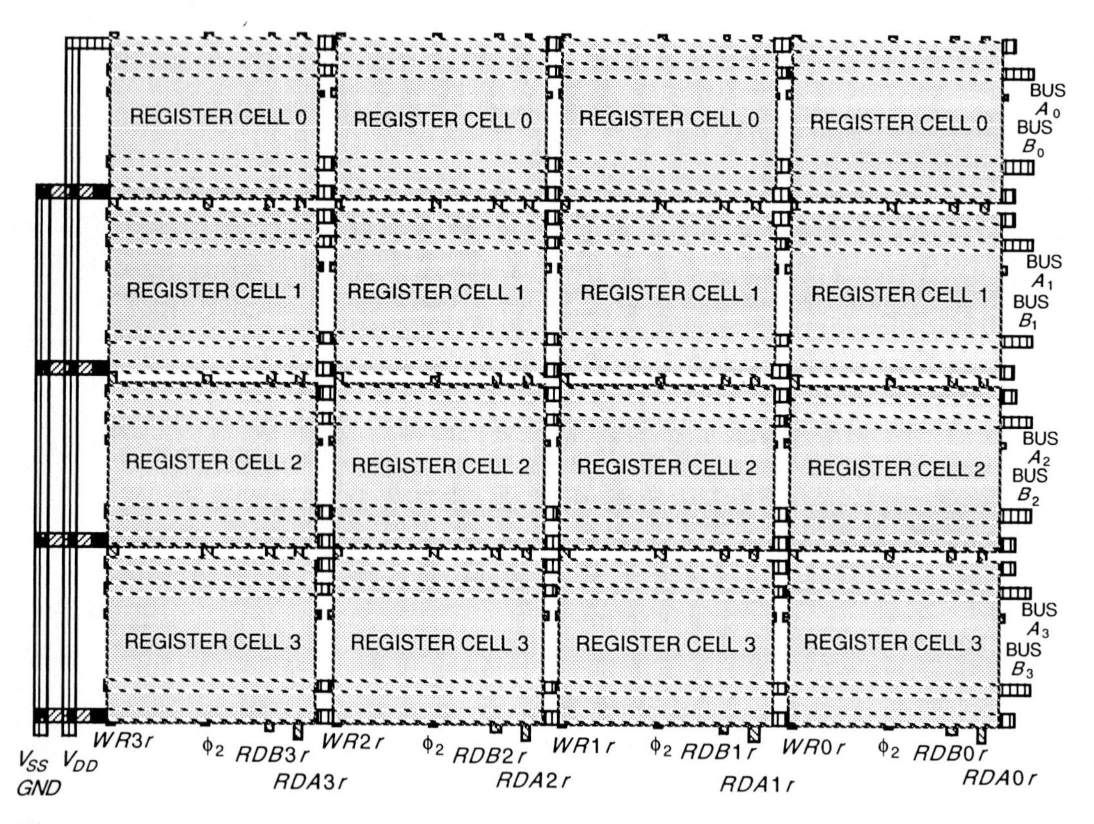

Figure 9–15 4 × 4-bit register floor plan

9.3.2 Selection and control of the 4 × 4-bit register array

Figure 9–15 shows that the register array must be provided with the control signals *WR0r–WR3r, RDB0r–RDB3r,* and *RDA0r–RDA3r,* derived from register select signals so that each register may be selected for read or write. We must also note *that we* need the capability to select two registers simultaneously for connection to the adder and that, in some cases, we may wish to read the contents of a single register to both A and B buses. One approach is to make use of decoder (or demultiplexer) circuits to route the control signals, as suggested in Figure 9–16. (For the reader unfamiliar with demultiplexer circuits, the select lines allow routing of a single input to any one of the output lines, that is, like a multiposition switch, and are the converse of the multiplexer which selects any one of a number of input lines to be routed to a single output.)

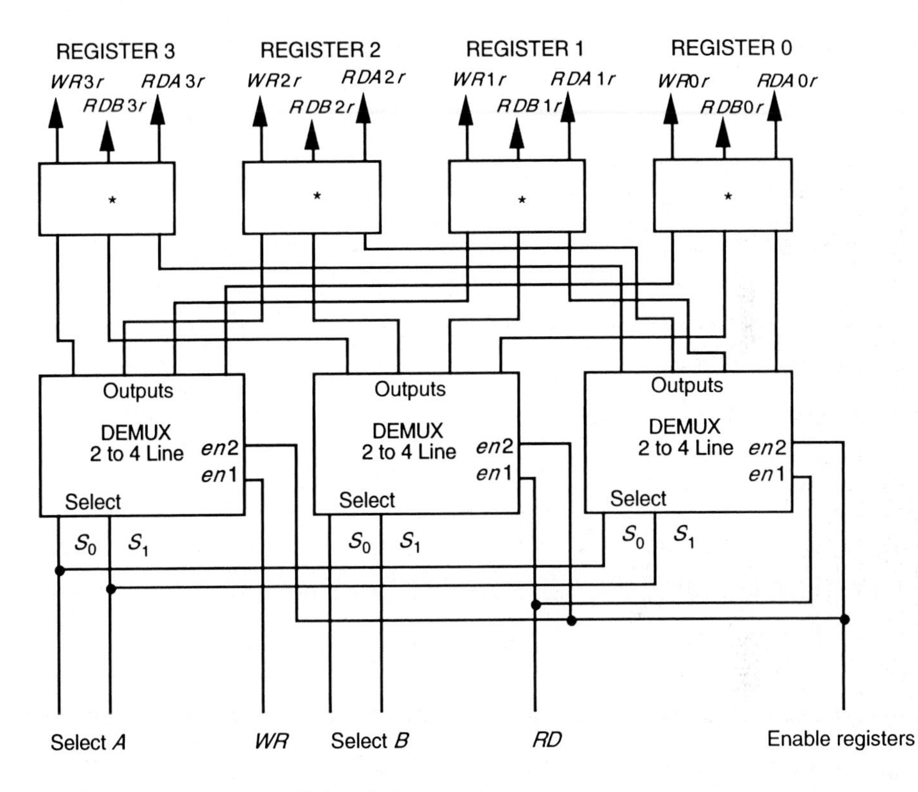

Note: The effective input to each decoder is en1.en2.
* Indicates buffers/drivers as required.

Figure 9–16 Decoder-based selection and control

The whole register array and selection and control circuitry may then be represented in floor plan form, as in Figure 9–17; note that the details of the selection and control circuitry have not been given here.

Note also that this register subsystem is the third of the four main functional blocks of the data path (Figure 7–1), for which we have already designed the shifter (Figures 7–8 to 7–10) and the adder subsystems (Figure 8–11(b)). The completion of the floor planning is discussed in the next chapter.

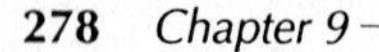

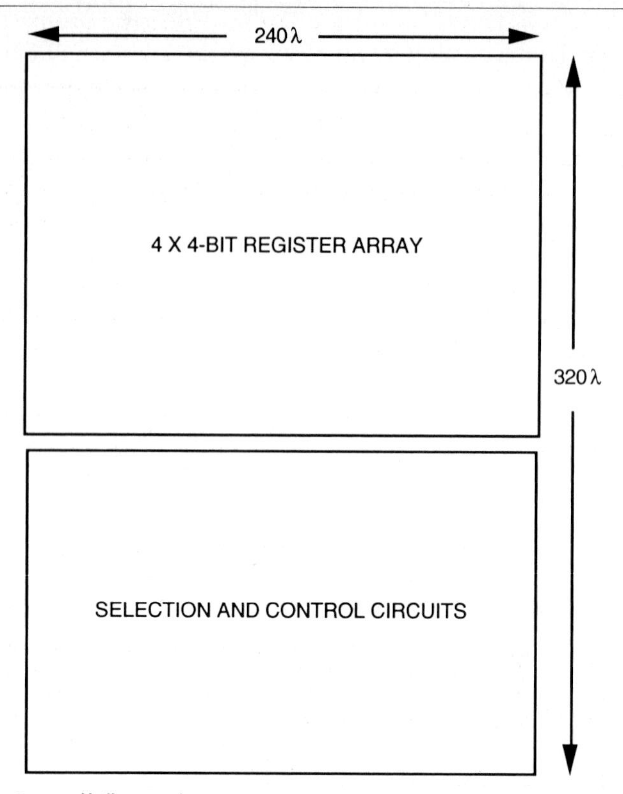

Figure 9–17 Overall floor-plan — register array and select/control circuits

9.3.3 Random access memory (RAM) arrays

Now that we have considered individual memory cells, some of which are quite small in area and low in dissipation, and also the use of memory cells in an array of registers, it is not a large step to consider much larger arrays of which the RAM (random access memory) is the most commonplace.

It is relatively easy to form arrays of memory cells. For example, the CMOS memory cell of Figure 9–7(b) and associated sense circuitry (as in Figure 9–7(c)) will form a RAM array as shown in Figure 9–18. A suitable mask layout for the memory cell used here is suggested in Figure 9–19 to give an idea of overall area for a particular size of array.

Finally, the architecture of a typical RAM array storing 'words' is illustrated by the floor plan and main interconnections for a 16-location × 4-bit word array in Figure 9–20. It will be seen that, in this case, the incoming address lines are decoded into row and column (with *RD* or *WR*) select lines, which are then used to select individual words in the memory. It will be noted that V_{DD} and *GND*

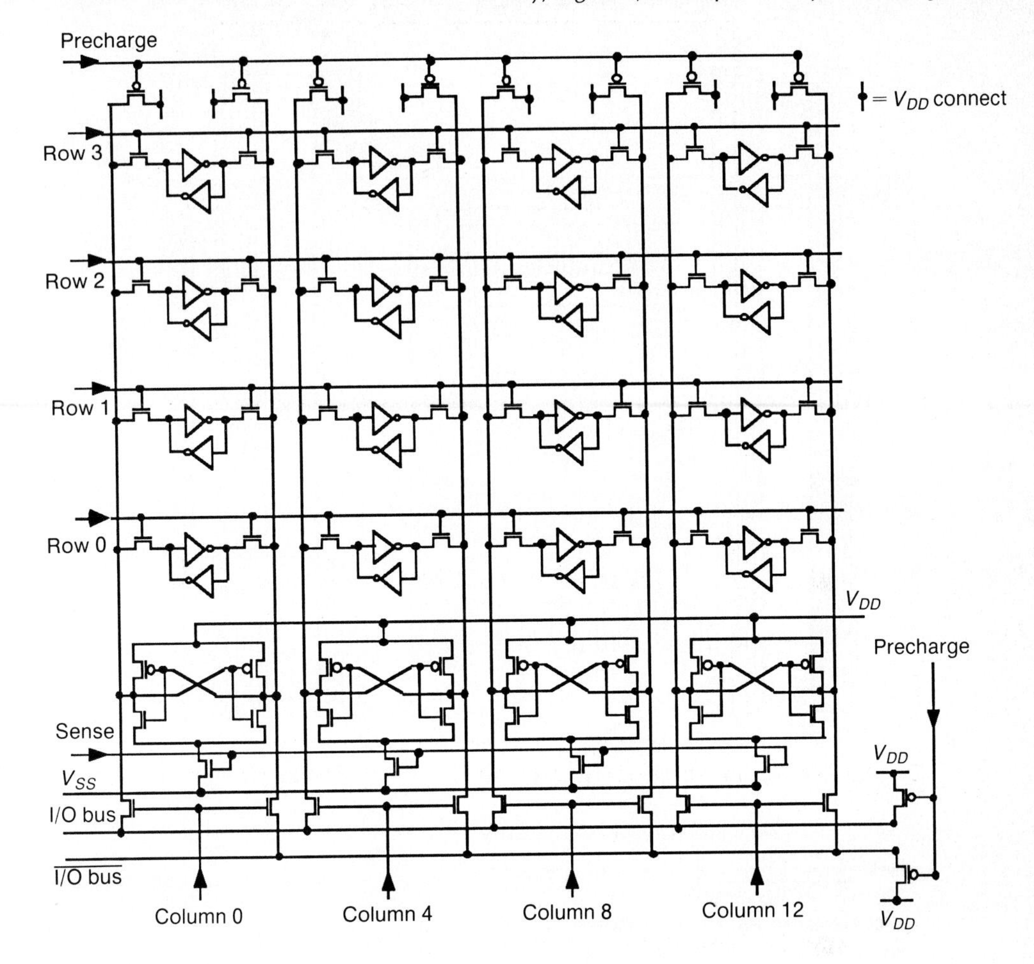

Figure 9–18 16-bit CMOS static memory array

(V_{ss}) rails are not shown, but it is clear that they may be interleaved with the data bus lines. Note that in large arrays in particular, the data bus lines will become relatively long and must therefore be run on the metal layer to avoid excessive capacitance or series resistance. As discussed earlier, the bus capacitance *and* the control line capacitances *must* be allowed for in the design.

To complete this section, Figure 9–21 is a plot of the *metal layer* only for a 16-location × 4-bit memory. The regularity of the memory array, the way in which the buses are run, and the various subsections of the floor plan are clearly evident.

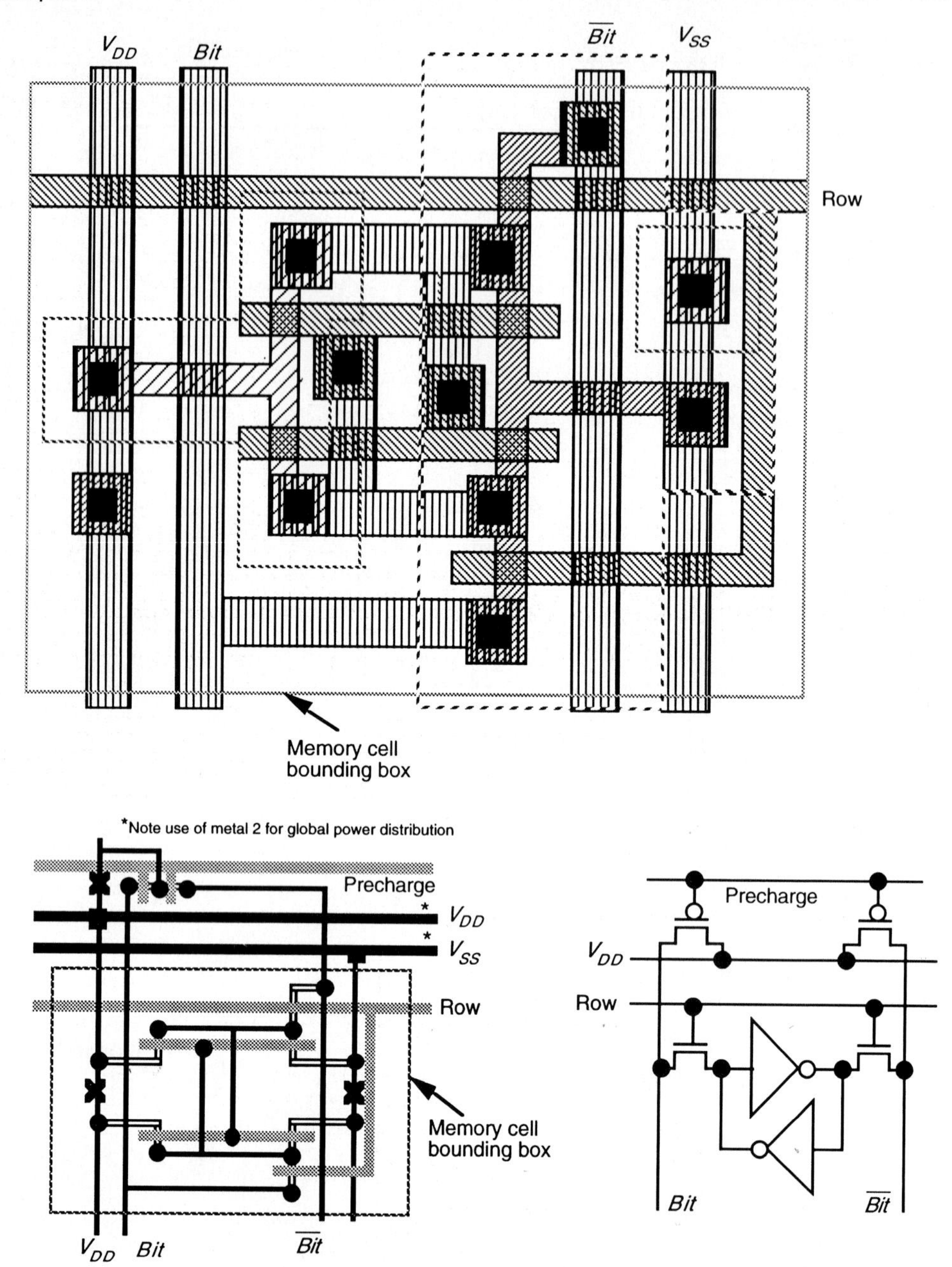

Figure 9–19 CMOS static memory cell-mask and stick layout

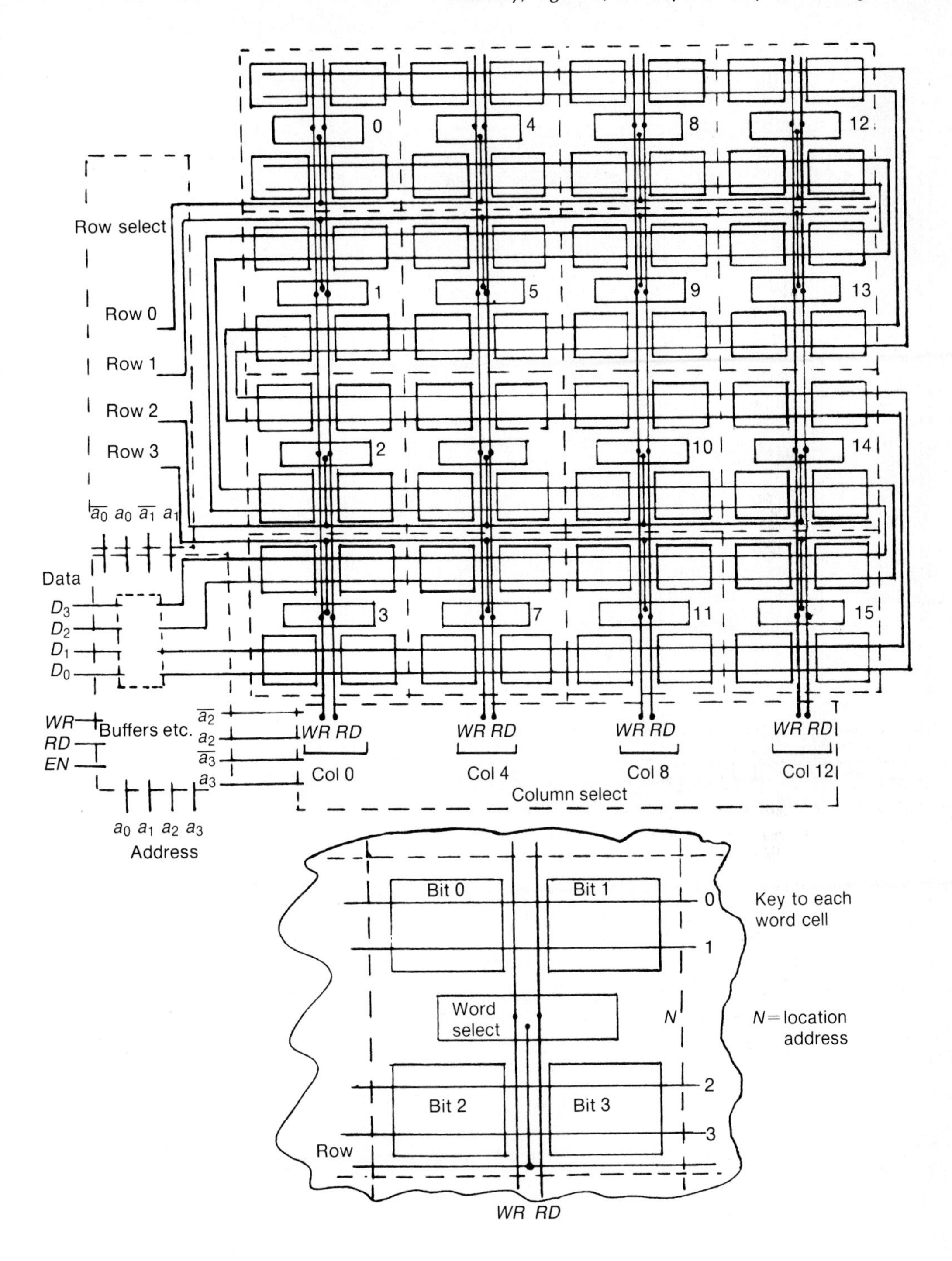

Figure 9–20 Floor plan of 16 × 4-bit RAM

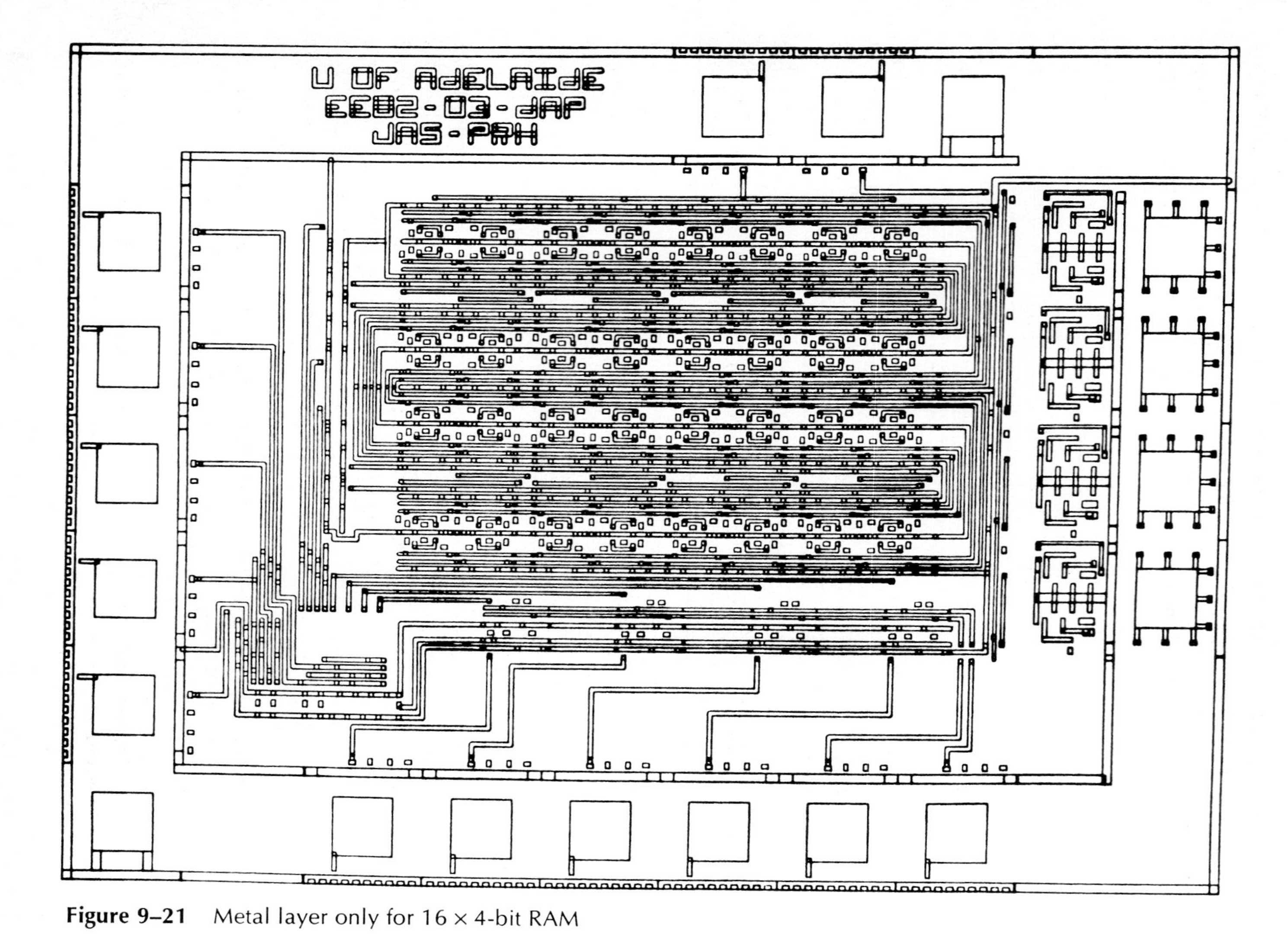

Figure 9–21 Metal layer only for 16 × 4-bit RAM

9.4 Observations

This chapter has completed our introduction to most of the techniques and many of the commonly used circuit arrangements for VLSI design in Silicon.

We have now completed the design of three of the four subsystems comprising the 4-bit data path which has been the 'vehicle' we have used to explore design processes.

We have also begun to see that communications are a very important aspect of design and this will be further emphasized in the next chapter.

9.5 Tutorial exercises

1. Design a two-line to four-line decoder (demultiplexer) circuit to the mask layout level and determine its bounding box. Then work out the arrangement of, and area occupied by, the 4 × 4-bit register select and control circuit of Figure 9–16.
2. Taking a 16-location × 2-bit RAM arrangement as an example, suggest (with sketches only) how such an arrangement could be configured for two-port operation (i.e. data can be read or written to any location from either of two 2-bit data buses).
3. For the 4 × 4-bit register designed in this chapter, and using the select and control circuit suggested in question 1 above, work out all communication paths and interconnections between the three subsystems of the 4-bit data path we have so far designed. Use a bounding box representation for each subsystem and clearly indicate the layers on which interconnections are made. What is a suitable overall area for the processor as so far designed?
4. *nMOS*: Using a 4-bit word arrangement of Figure 9–22, and consulting the RAM arrangement of Figure 9–20, draw a mask level layout for the word select circuit associated with each 4-bit word and thus determine the overall area needed for each word stored. Next, design stick diagram level arrangements for the remaining blocks on the floor plan — that is, the row and column selection circuits and the input buffers and drivers, etc. Then, *estimate* the area needed for the 16 × 4-bit RAM as a whole (without I/O pads).
5. *CMOS*: Starting with the 16-bit RAM array of Figure 9–18, design suitable decoding and control circuitry to allow row and column selection and read and write operation of the array. (You may find it useful to refer to section 9.2.5.)

   Design *one* memory cell as far as the mask layout and determine a suitable area per bit stored.

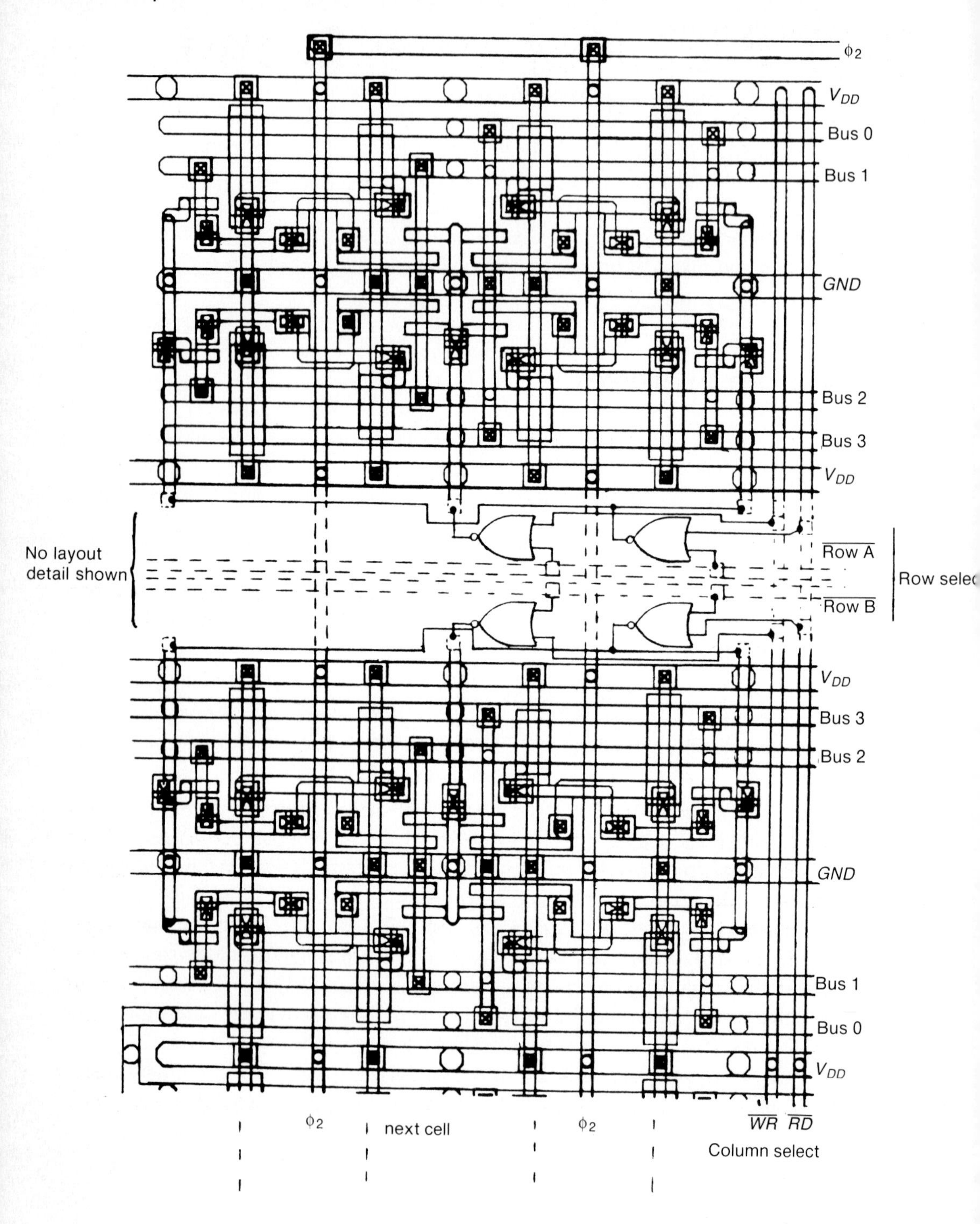

Figure 9–22 Two 4-bit words of nMOS RAM array

Practical aspects and testability

Is it not strange that desire should so many years outlive performance?
Shakespeare: King Henry IV

Objectives

The chapter is intended to round off and summarize much of the preceding text and to discuss some of the practical realities the designer must face. The problems of communication again receive close attention and are illustrated in the context of the 4-bit data path design.

The chapter also includes a section headed 'Ground rules for successful design' and the reader will find that most of the rules, tabulated data and performance parameters are grouped together consecutively in this section. The question of noise margins and other relevant aspects, such as CIF code and CAD tools, are also discussed.

The second half of the chapter is entirely devoted to the very important subject of testability, which must always be a key design requirement for systems of any size.

10.1 Some thoughts on performance

Two important parameters (other than 'does it work at all?') are speed and power dissipation. These factors are generally interrelated; power dissipation and area are also interrelated in MOS technology.

Take, for example, the simple case of an nMOS 8:1 inverter which may be set out with a minimum feature size pull-down transistor (i.e. $2\lambda \times 2\lambda$ pull-down gate area and a minimum width 16λ long $\times$ 2λ wide pull-up channel) giving a total resistance from V_{DD} to GND of 90 kΩ. The maximum power dissipation for this particular design will thus be

$$\frac{(5\ \text{V})^2}{90\ \text{k}\Omega} = 0.278\ \text{mW}$$

An alternative form of 8:1 inverter is to use a pull-down geometry 2λ long and 6λ wide with a 6λ long, 2λ wide pull-up channel giving a V_{DD} to GND resistance of 33.3 kΩ and a consequent maximum power dissipation of

$$\frac{(5\ \text{V})^2}{33.3\ \text{k}\Omega} = 0.744\ \text{mW}$$

that is, about three times the dissipation. However, comparing the total transistor areas for each case we have, in the first case, $2\lambda \times 2\lambda + 16\lambda \times 2\lambda = 36\lambda^2$ area and, in the second case, $2\lambda \times 6\lambda + 6\lambda \times 2\lambda = 24\lambda^2$. In other words, the 3:1 (approximate) reduction in power dissipation is at the expense of a 50% increase in transistor area.

Now consider the aspect of speed (or circuit delays), and take the simple case of one 8:1 inverter driving another similar inverter. The longest delays will occur when the output of the first stage is changing from logic 0 (Lo) to logic 1 (Hi), that is, the Δ transition of the output, and the capacitances associated with the output and the input of the next stage must charge through the pull-up resistance of the first stage as in Figure 10–1. Asymmetry is also present in CMOS devices. It is also obvious that during the complementary ∇ transition the same capacitances must be discharged through the pull-down transistor of the first stage.

For the minimum pull-down feature size nMOS 8:1 inverter, for example

$$R_{p.u} = 8R_s$$
$$R_{p.d} = 1R_s$$
$$C_{IN} = 1\square C_g$$

allow stray and wiring capacitances

$$C_S = 4\square C_g \text{ (say)}$$

Then

$$\Delta \text{ transition delay} = 8R_s \times 5\square C_g = 40\tau$$

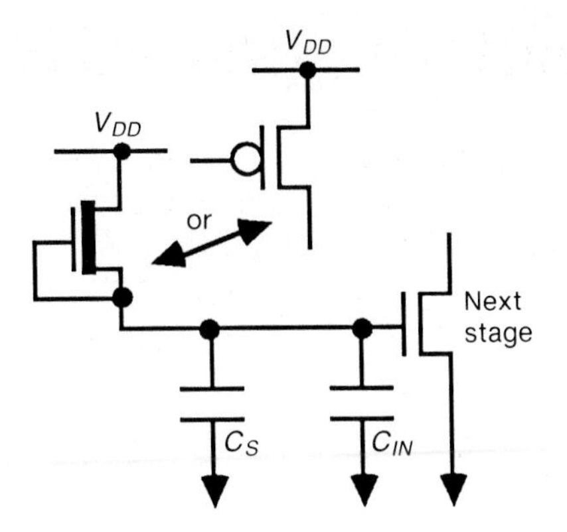

Figure 10–1 Circuit model for inverter driving an inverter on a ΔO/P transition

and

$$\nabla \text{ transition delay} = 1R_s \times 5\square C_g = 5\tau$$

For the alternative 8:1 inverter design discussed earlier, and allowing the same stray and wiring capacitances

$$\Delta \text{ transition delay} = 3R_s \times 7\square C_g = 21\tau$$

and

$$\nabla \text{ transition delay} = \tfrac{1}{3}R_s \times 7\square C_g = 2\tfrac{1}{3}\tau$$

Thus, it may be seen that a speed-up factor of about 2:1 in this case is bought at the expense of a 3:1 increase in power consumption but has the bonus of reducing area by a factor of 2:3. Similar considerations apply to the switching energy of CMOS circuits.

Therefore, as in most engineering situations, there are trade-offs to be made, and it is essential that the would-be designer have a good fundamental understanding of the discipline to be able to make sound decisions.

But remember, in the end there will always be limits imposed by the technology and some specifications will be impossible to meet.

10.1.1 Optimization of nMOS and CMOS inverters*

The approximate calculations presented here should be useful from a qualitative point of view and are intended to give the reader some appreciation of basic CMOS and nMOS circuit optimization problems.

* The authors are indebted to Professor K. S. Trivedi of Duke University for providing this material on inverter optimization.

For a more rigorous treatment of circuit optimization methods, refer to the articles cited at the end of the chapter.

10.1.1.1 The CMOS inverter

The area of a basic CMOS inverter is proportional to the total area occupied by the p- and n-devices.

$$A \propto (W_p L_p + W_n L_n)$$

where

W_p = width of the p-device
L_p = length of the p-device
W_n = width of the n-device
L_n = length of the n-device

Minimum area can be achieved by choosing minimum dimensions for W_p, L_p, W_n and L_n, that is

$$W_p = L_p = W_n = L_n = 2\lambda \text{ (minimum)}$$

Hence

$$\frac{W_p}{W_n} = 1$$

Switching power dissipation, P_{sd}, can be approximated by $C_L V_{DD}^2 f$ where

C_L = load capacitance at the inverter output
V_{DD} = power supply voltage
f = frequency of switching

For fixed V_{DD} and f, minimizing P_{sd} requires minimizing C_L which can be achieved by minimizing the area A since C_L is proportional to the gate areas comprising A.

Asymmetry in rise and fall times, t_r and t_f (transition times between 10% and 90% logic levels), can be equalized by using $\beta_n = \beta_p$. (Notice that t_r and t_f are proportional to the average resistance of the device which is approximately given by $\frac{2}{\beta V_{DD}}$ where $\beta = \beta_n$ or β_p.) This requires that

$$\frac{W_p}{L_p} = \left(\frac{\mu_n}{\mu_p}\right)\frac{W_n}{L_n}$$

to compensate for the lower hole mobility μ_p, compared to electron mobility μ_n.

Assuming $L_p = L_n = 2\lambda$, $\frac{\mu_n}{\mu_p} \doteq 2$, we require $\frac{W_p}{W_n} \doteq 2$. This yields $t_r = t_f$.

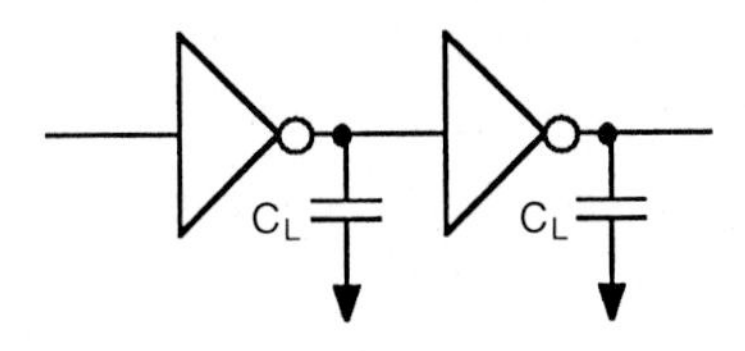

Figure 10–2 Inverter pair

Note that equalizing rise and fall times is not possible in nMOS or pseudo-nMOS inverters because of the ratio requirement.

Asymmetry in noise margins, NM_H and NM_L, can be equalized by choosing $\beta_n = \beta_p$ and hence $\frac{W_p}{W_n} \doteqdot 2$ for $L_p = L_n$. This yields $NM_H = NM_L$. (See Figure 10–4(b).)

Basic inverter pair delay — Consider a basic inverter pair shown in Figure 10–2 where C_L is the capacitive load driven by the two identical inverters, inverter pair delay $D(= t_r + t_f)$ is proportional to $(R_p + R_n)C_L$ where $R_p = 2/(\beta_p V_{DD})$ and $R_n = 2/(\beta_n V_{DD})$ are the average resistances of the p- and n-transistors respectively.

Also

$$C_L = C_E + (W_p L_p + W_n L_n)C_g$$

where

C_E = lumped parasitic capacitance
C_g = gate capacitance per unit area

Hence

$$D = D_0\left[\left(\frac{2}{\beta_p V_{DD}} + \frac{2}{\beta_n V_{DD}}\right)(C_E + (W_p L_p + W_n L_n)C_g)\right]$$

where D_0 is a constant of proportionality. Assuming $\frac{\mu_n}{\mu_p} \doteqdot 2$

$$D = D_0\left[C_E\left(\frac{2L_p}{W_p} + \frac{L_n}{W_n}\right) + C_g\left(2L_p^2 + 2L_p L_n \frac{W_n}{W_p} + L_p L_n \frac{W_p}{W_n} + L_n^2\right)\right]$$

Since D increases with L_n and L_p, for minimum D choose $L_n = L_p = 2\lambda$ (minimum). Minimizing D with respect to W_p yields a solution

$$W_p / W_n = \sqrt{2}\left[1 + \frac{C_E}{C_g L_n W_n}\right]^{1/2}$$

$$W_p / W_n \doteqdot \sqrt{2} \text{ for } C_E << C_g L_n W_n \text{ (normal case)}$$

However, D does not vary significantly with W_p/W_n in the range $1 \leqslant \frac{W_p}{W_n} \leqslant 2$ (see Figure 10–3). Hence simultaneous optimization of various parameters mentioned above seems to be easily achievable in the CMOS inverter, without greatly increasing the delay D.

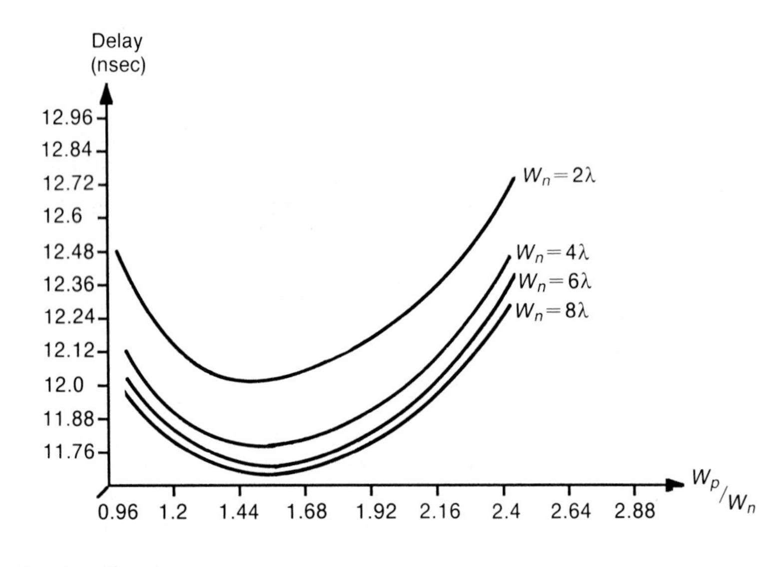

Notes: $L_p = L_n = 2\lambda = 5\ \mu m$
Gate capacitance $C_g = 4 \times 10^{-4}\ pF/\mu m^2$
$C_E = 4 \times 10^{-3}\ pF$

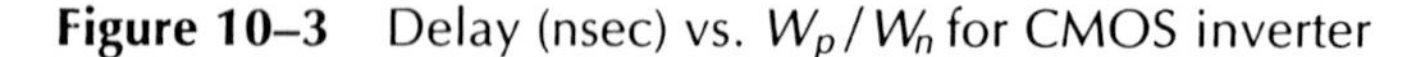

Figure 10–3 Delay (nsec) vs. W_p/W_n for CMOS inverter

10.1.1.2 nMOS inverter

Let $Z_{p.u.}/Z_{p.d.} = \frac{L_{p.u.}W_{p.d.}}{W_{p.u.}L_{p.d.}} = k$ where the subscripts *p.u.* and *p.d.* refer to the pull-up and pull-down transistors respectively. Then area

$$A = A_0(L_{p.d.}\ W_{p.d.} + L_{p.u.}\ W_{p.u.})$$

$$= A_0\left(L_{p.d.}W_{p.d.} + kW_{p.u.}{}^2\,\frac{L_{p.d.}}{W_{p.d.}}\right)$$

where A_0 is a constant of proportionality. For a fixed k, to achieve minimum A, we need $L_{p.d.} = W_{p.u.} = 2\lambda$. Minimizing A with respect to $W_{p.d.}$ yields a solution $W_{p.d\cdot} \sqrt{k} W_{p.u.} = \sqrt{k}\, 2\lambda$. Hence, using $Z_{p.u.}/Z_{p.d.} = k$, we obtain

$$L_{p.u.} = \sqrt{k}.L_{p.d.} = \sqrt{k}.2\lambda$$

This implies $Z_{p.u.} = \sqrt{k}$ and $Z_{p.d.} = 1/\sqrt{k}$. Giving

$$\text{Minimum area} = 8A_0\lambda^2\sqrt{k}$$

Static power dissipation, $P_d = P_0 \frac{V_{DD}{}^2}{(k+1)Z_{p.d.}}$, where P_0 is a constant of proportionality — for fixed k and V_{DD}, P_d is minimized by choosing as large a $Z_{p.d.}$ as possible. However, a large $Z_{p.d.}$ requires a large a $Z_{p.u.}$ ($Z_{p.u.} = kZ_{p.d.}$), and hence the delay D of the inverter pair increases. One has to choose the maximum $Z_{p.d.}$ possible for a given maximum allowed delay D.

If we use $Z_{p.d.} = 1$ with $L_{p.d.} = W_{p.d.} = 2\lambda$, and $Z_{p.u.} = k$ with $L_{p.u.} = 2k\lambda$ and $W_{p.u.} = 2\lambda$, we obtain

$$P_d = \frac{P_0 V_{DD}{}^2}{(k+1)}$$

$$A = 4A_0(k+1)\lambda^2$$

Inverter pair delay — Proceeding in a similar manner to the CMOS case

$$C_L = C_E + C_g\, W_{p.d.}L_{p.d.}$$
$$D = t_r + t_f = D_0(Z_{p.d.} + Z_{p.u.})C_L$$
$$= D_0[Z_{p.d.}C_E(1 + k) + C_g(1 + k)L_{p.d.}{}^2]$$

To minimize D:

1. Choose minimum $L_{p.d.} = 2\lambda$.
2. For maximum $W_{p.d.}$, choose $L_{p.u.} = 2\lambda$, as $W_{p.d.} = 2k\lambda \frac{W_{p.u.}}{L_{p.u.}}$ which yields $W_{p.d.} = kW_{p.u.}$

Choosing large $W_{p.d.}$ to minimize D increases A. Hence for a given area $A(= W_{p.d.}L_{p.d.} + W_{p.u.}L_{p.u.})$ with $L_{p.d.} = L_{p.u.} = 2\lambda$, we must have

$$W_{p.u.} = \frac{A}{2\lambda(k+1)} \qquad W_{p.d.} = \frac{kA}{2\lambda(k+1)}$$

With $W_{p.u.} = 2\lambda$, we have $W_{p.d.} = k2\lambda$. Hence $Z_{p.u.} = 1$ and $Z_{p.d.} = 1/k$ for minimum D.

$$\text{Minimum } D = D_0(1 + k)\,(C_E/k + 4\lambda^2 C_g)$$

Table 10–1 shows the summary of optimization of the three parameters, D, A and P_d. Notice that the solution for minimum power dissipation also gives the lowest power delay product among the three designs.

Table 10–1 Optimum parameters for nMOS inverters

	$L_{p.d.}$	$W_{p.d.}$	$Z_{p.d.}$	$L_{p.u.}$	$W_{p.u.}$	$Z_{p.u.}$
Minimum D	2λ	$2k\lambda$	$1/k$	2λ	2λ	1
Minimum A	2λ	$2\lambda\sqrt{k}$	$1/\sqrt{k}$	$2\lambda\sqrt{k}$	2λ	$\sqrt{k}$
Minimum P_d	2λ	2λ	1	$2\lambda k$	2λ	k

	A/A_0	D/D_0	$P_d/(P_0 V_{DD}^2)$
Minimum D	$4\lambda^2(k+1)$	$(1+k)\,(C_E/k + 4\lambda^2 C_g)$	$\dfrac{k}{k+1}$
Minimum A	$8\lambda^2\sqrt{k}$	$(1+k)\left(\dfrac{C_E}{\sqrt{k}} + 4\lambda^2 C_g\right)$	$\dfrac{\sqrt{k}}{(k+1)}$
Minimum P_d	$4\lambda^2(k+1)$	$(1+k)\,(C_E + 4\lambda^2 C_g)$	$\dfrac{1}{(k+1)}$

10.1.2 Noise margins

Noise margins have been mentioned in the preceding section and it is appropriate now to consider this factor in more detail.

Noise margins are a measure of a logic circuit's tolerance of noise voltages in either of the two logic states; in other words, by how much the input voltage can change without disturbing the present logic output state. In order to examine this, it is convenient to consider a pair of inverters (nMOS or CMOS) and derive the noise margins for signals applied to the input of the second inverter, inverter 2, which is driven from the output of a similar inverter, inverter 1, as in Figure 10–4(a).

Referring now to Figure 10–4(b), we see the transfer characteristics (V_{out} vs. V_{in}) for a pair of CMOS inverters set out in such a way that the output voltage of inverter 1 is applied as the input voltage to inverter 2. By first considering the point at which output 1 starts to enter the transition region (the unity gain point A) and calling this voltage $V_{OH\,min}$ and then considering the input voltage level $V_{IH\,min}$ (point B) at which the transition of the output of inverter 2 commences, we are able to define the high level noise margin of inverter 2 as NM_H where

$$NM_H = V_{OHmin} - V_{IHmin} \text{ (a positive voltage)}$$

Similarly, a consideration of the low logic level conditions gives

$$NM_L = V_{OLmax} - V_{ILmax} \text{ (a negative voltage)}$$

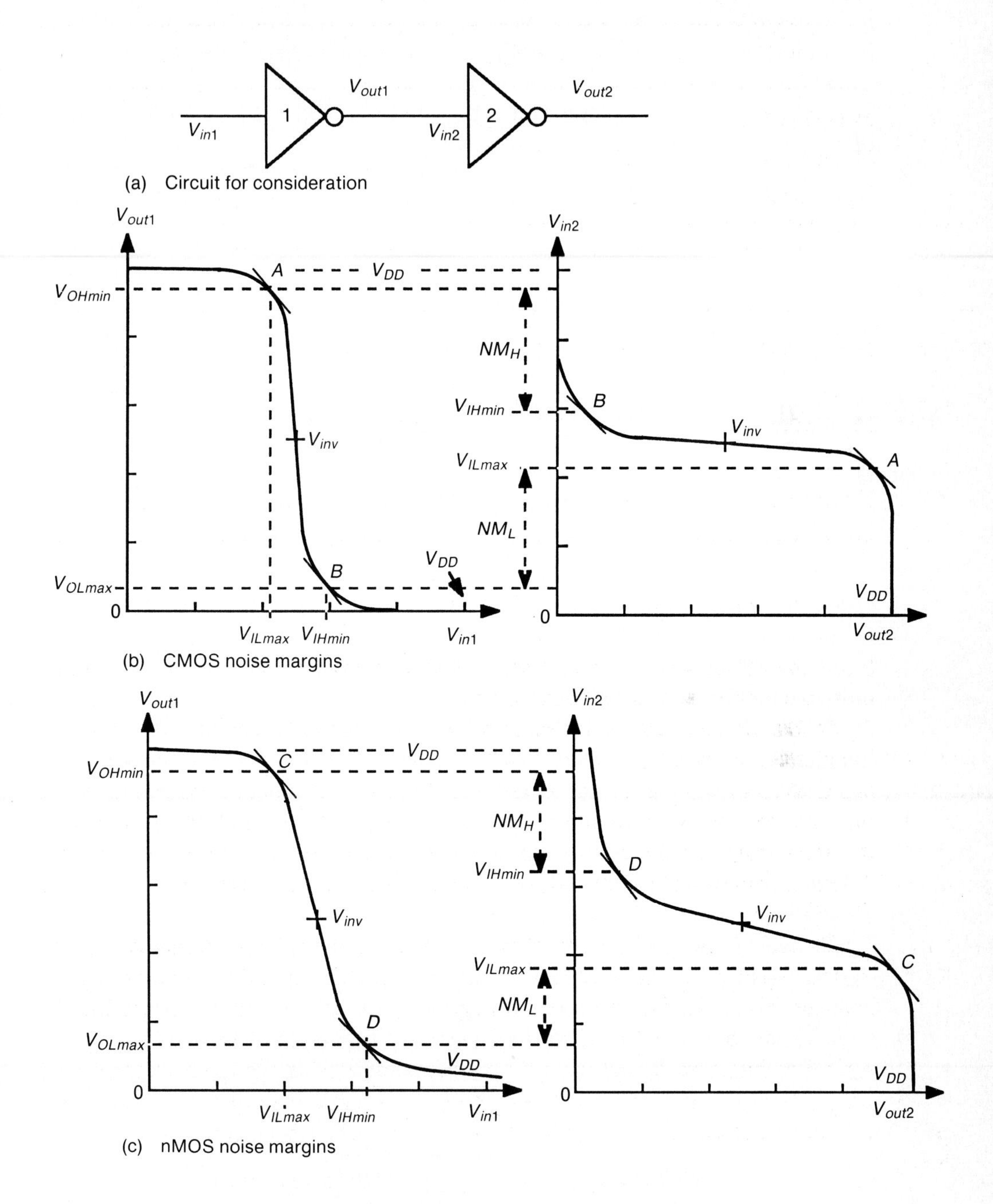

Note: *A* and *B*, *C* and *D* are unity gain points.

Figure 10–4 Inverter noise margins

A similar approach will yield noise margins for the nMOS inverter as shown in Figure 10–4(c). It may be seen that generally the CMOS inverter will have better noise margins than the nMOS inverter, particularly for the low condition.

In both cases, symmetry about V_{inv} is assumed (where V_{inv} is the point at which $V_{out} = V_{in} = V_{DD}/2$). This assumes that $\beta_p = \beta_n$ for CMOS and that the correct ratio of $Z_{p.u.}$ to $Z_{p.d.}$ has been observed for nMOS.

Changes in the β_n / β_p ratio for CMOS or to the $Z_{p.u.}/Z_{p.d.}$ ratio for nMOS will result in a shift in the V_{out} vs. V_{in} characteristics (see Figures 2–7 for nMOS and 2–15 for CMOS) and consequent degradation of one or the other noise margin in each case.

Thus the effect of ratios on noise margins performance must be taken into account in design.

10.2 Further thoughts on floor plans/layout

In considering the layout of the four-bit data path used earlier as a design exercise, we could have waited until we knew the minimum size and disposition of connections to each functional block in order to finalize the floor plan. Indeed, this is a possible approach if communications will allow. Quite accurate floor plans can be set out at an early stage if a library of properly dimensioned and characterized elements/cells is available to the designer.

However, a better approach is to draw up quite specific floor plans at the outset and then design/configure the subsystems to conform to the required floor plan. This approach is more general than the one we have used so far. The same 4-bit processor (Figure 8–1) will be used to illustrate the method and considerations involved.

First (as before) determine an *overall strategy* (perhaps as suggested in Figure 10–5) and then use this to determine the best relative disposition of subsystems in light of data flow and control paths through the system. For the 4-bit data path, a suitable layout is shown in Figure 10–6.

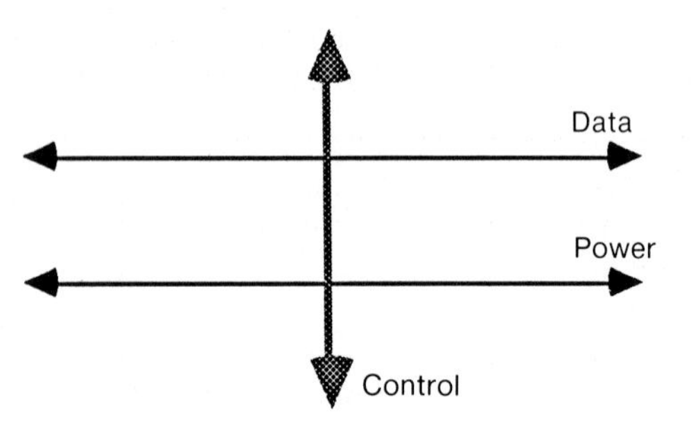

Figure 10–5 A communications strategy

When approached this way, a reasonably well thought-out floor plan can be developed before knowing any real detail of the subsystem/block areas. In the event, features of individual subsystems (Figures 7–8, 7–9 with 7–10, 8–11(b), and 9–15 with 9–17) will, in general, dominate the overall layout and other blocks may then be stretched and/or reconfigured as necessary to conform with the dominant features.

In order to do so it is essential to set out clearly the way in which data will flow on the buses. In this case:

1. Floating bus lines are envisaged.
2. All read and write operations are coincident with ϕ_1.
3. Bus *A connects the* I/O port to the register array and carries one operand (A_k) from the registers to the adder. It will also be used to carry the output of the shifter back to the register array (and I/O port). Bus *A* is therefore *bidirectional.*
4. Bus *B* connects the register array with the other input (B_k) of the adder and may also be used to carry the sum output (S_k) from the adder to the input of the shifter. Bus *B* is *unidirectional.*

Taking the subsystems of the 4-bit data path example (Figures 7–9 with 7–10, 8–11(b), and 9–15 with 9–18), one of the main features is the bus spacing, that is, the spacing between buses A_n and B_n and between A_n and A_{n+1}, etc., and close examination of the interconnection of designs pursued in this text will reveal that the bus spacings of the adder subsystem dominate those of the other subsystems.

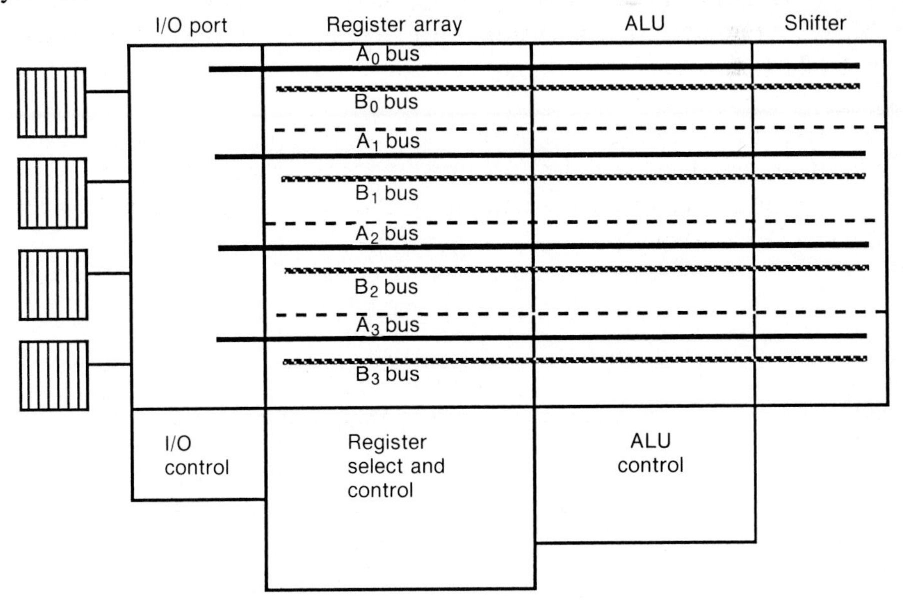

Figure 10–6 Possible floor plan for 4-bit processor

Rearrangements consequent on these considerations affect the barrel shifter (Figures 7–8 to 7–10) in particular. It is necessary to interchange the relative position of the *In* and *Out* bus lines and also make the cell stretchable to match the height of the dominant (adder) block and its bus spacing. Also, to mate with the bus structures of the other blocks, the *In* and *Out* bus lines should be in metal rather than polysilicon and diffusion, as used in our original design of Figures 7–8, 7–9 and 7–10.

The way in which this may be done is indicated in the revised standard cell layout (Figure 10–7); it is necessary to allow for rifts and extensions *and* to cope with optional features which result from the four versions (owing to optional contacts) of the standard cell required, thus ensuring generality.

The concept of the use of a Y RIFT which is extendable from 0λ minimum upward and X EXTN and Y EXTN which are extensions of the cell from 0λ upward make the barrel shifter configurable to match most bus dispositions. Note that rifts and extensions should be placed where they cut a minimum amount of simple geometry; for example, Y RIFT involves the stretching of two wires — one in polysilicon and the other in diffusion. Once such a degree of freedom is available, subsystems may be mated with a smooth flow-through of buses as suggested in Figure 10–8, which, for simplicity, shows the mask layout for an nMOS adder and a shifter which is on the right.

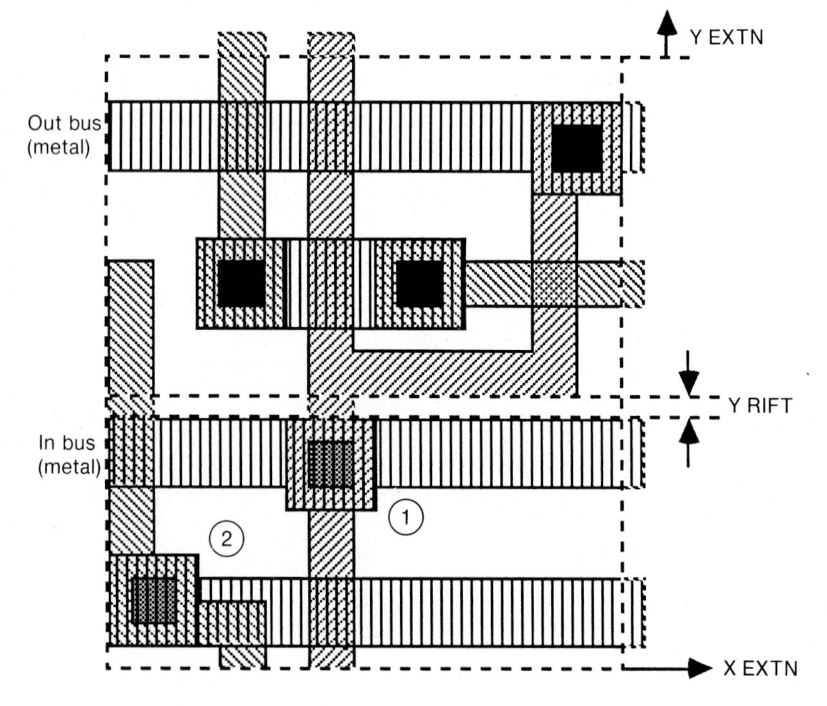

Note: 1 and 2 are optional contacts.

Figure 10–7 Standard cell for barrel shifter

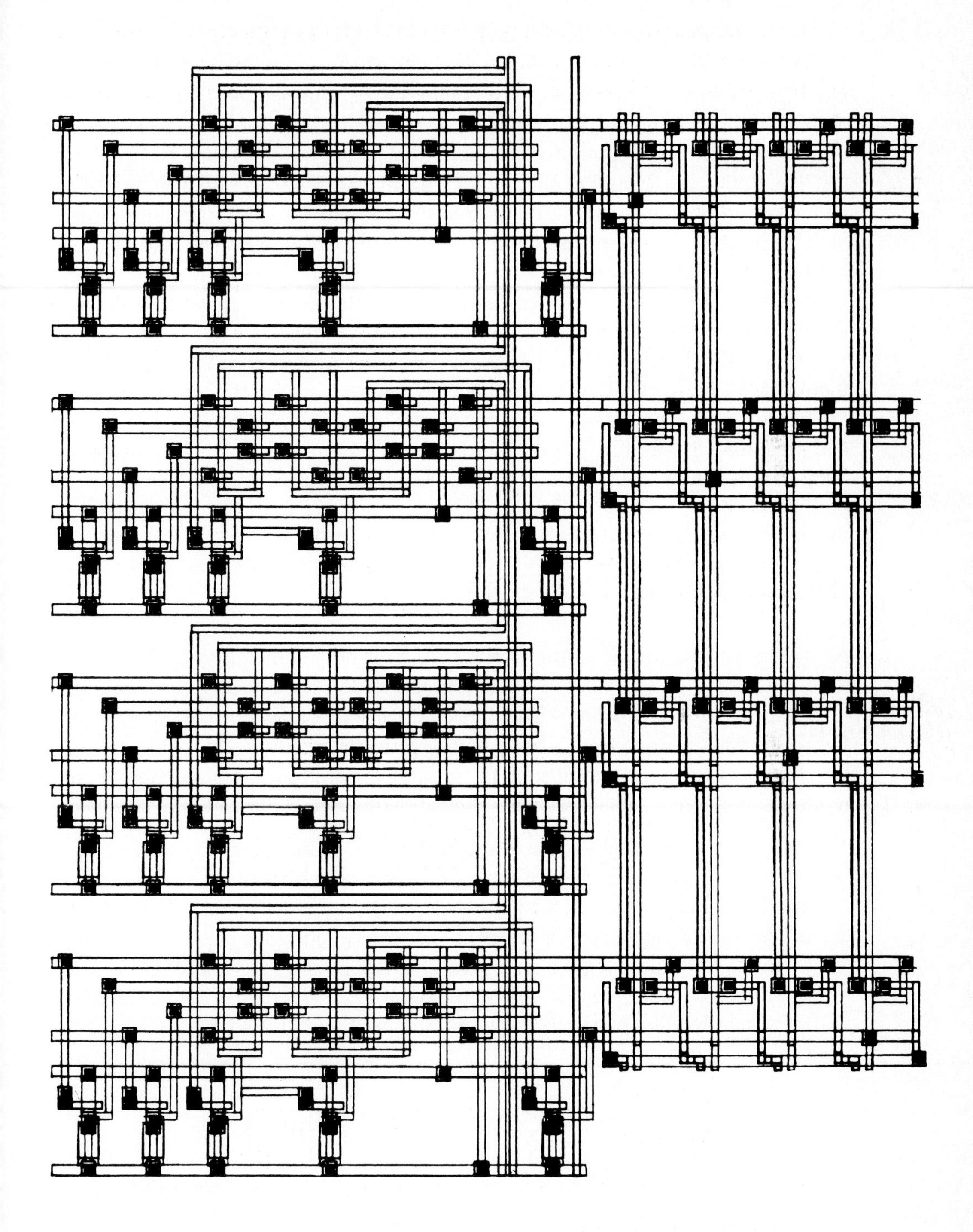

Figure 10–8 A possible interconnection of the adder and shifter subsystems

10.3 Floor plan layout of the 4-bit processor

Having designed the three main subsystems and determined their bounding boxes and interconnection dispositions, we can now envisage a complete system in which they are disposed relative to each other as set out in Figure 10–6.

The dominant feature of the layout (in this case, the interbus spacing of the adder circuit) having already been determined, and the shifter having already been redesigned to allow stretching to match the adder, a consideration of the bounding box and of connections to the register array will reveal a need for some stretching of the basic register cell as well so that an easy interconnection of the subsystem can take place through alignment of the buses in each subunit.

A possible arrangement — one that was fabricated as a student project — is included in Figure 10–9. Although layer encoding is lost in this particular black and white reproduction from a color-pen plotter of the mask layouts, the architecture and placement of the subsystems are quite readily apparent. Connections to and from the outside world are made through input and output pads which allow for bonding.

10.4 Input/output (I/O) pads

As well as allowing the bonding of leads from the chip to the pins on the package, the I/O pads cover a number of other requirements. Consequently, several types of pad are required. It is not within the scope of this text to present designs for a family of pads and, in most cases, pad designs are readily obtainable as basic library cells. However, the purposes served by the circuitry associated with pads require some general observations. The following needs must be met:

1. Protection of circuitry on chip from damage from static electricity and capacitive discharge (ESD) effects: this can be a serious problem, and care must be exercised in handling all MOS (and other integrated) circuits. The problem of 'static zap' may be put in perspective by considering the breakdown voltage of the thin oxide between gate and channel in, say, a 5 μm MOS circuit. Silicon dioxide has a breakdown voltage in the region of 10^9 volts/meter and for a gate oxide thickness of 0.1 μm, the maximum allowable voltage gate/channel is

$$V_{gc\,max} < \frac{10^9 \text{ volts}}{\text{meter}} \times \frac{0.1}{10^6} \text{ meter} = 100 \text{ volts}$$

This may sound generous in light of rail voltages of the order of 5 to 10 volts, but relatively high voltages are readily generated on one's person or on tools and handling equipment. Quite innocent pastimes, such as walking across

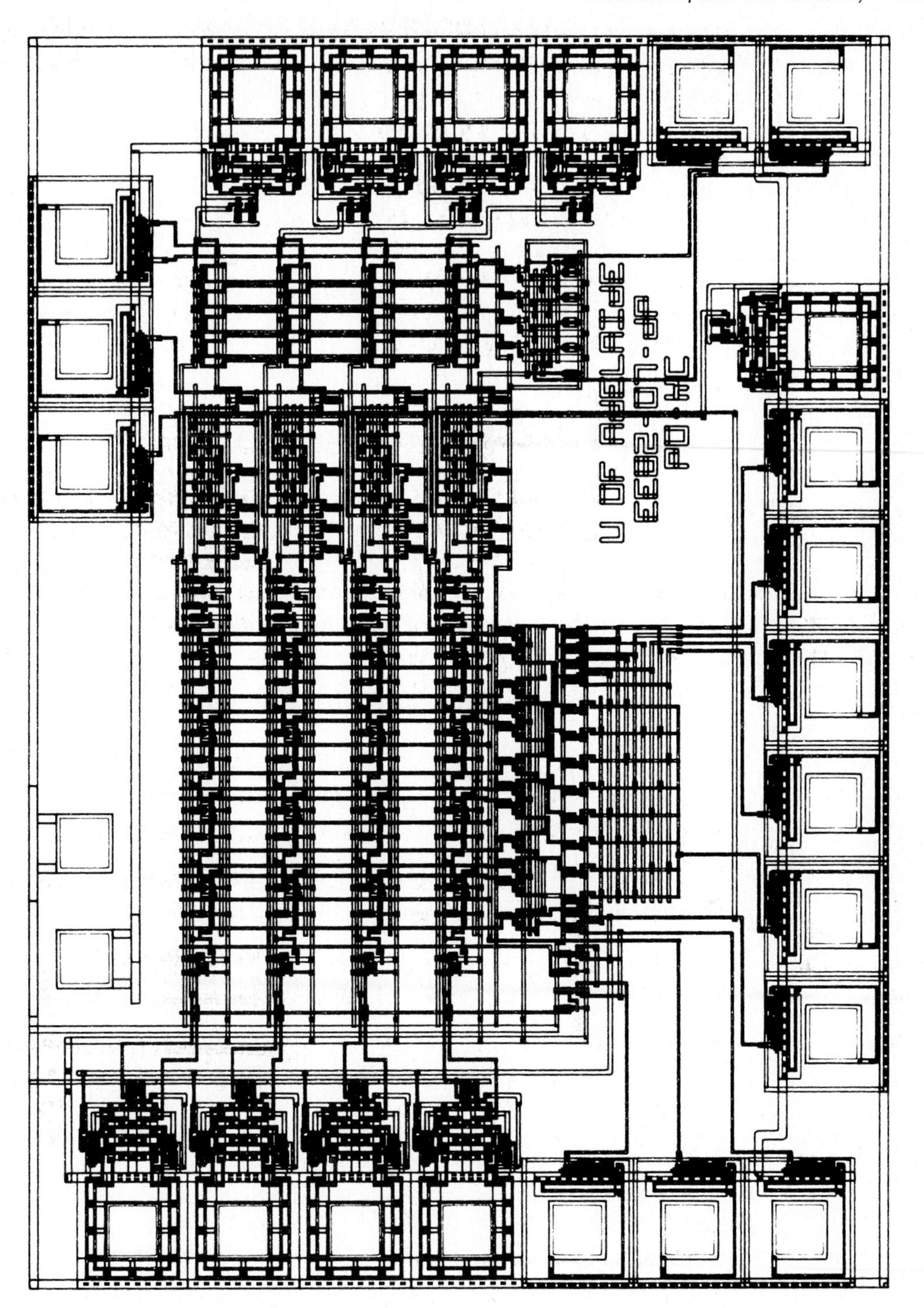

Figure 10–9 Complete layout of 4-bit data path multiproject chip

a vinyl floor or a synthetic carpet, can generate voltages of several hundred volts under conditions of high relative humidity (RH) and more than 10 kV if the RH is low. These voltages are well in excess of 100 volts and, although in some cases immediate failure may not occur, there may be significant degradation of reliability and/or life through 'wounding' of circuits.

2. Provide the necessary buffering between the environments on and off chip. For example, buffers are needed to drive the relatively large capacitances associated with circuits off the chip.
3. Provide for the connection of power supply rails.

A minimum set of pads should include:

1. V_{DD} connection pad;
2. *GND* (V_{SS}) connection pad;
3. input pad;
4. output pad;
5. bidirectional I/O pad (usually tristate logic).

In all cases when input and output (or bidirectional) pad designs from a library are used, the designer *must* be aware of the nature of the circuitry embodied in the pad design, that is:

1. be aware of the ratios/size of inverters/buffers onto which output lines are connected;
2. be aware of how input lines pass through the pad circuit (e.g. are the input signals fed in through pass transistors or do they come from inverter-like stages?).

Unless there are exceptional circumstances pads must always be placed around the *periphery* of the chip area, otherwise bonding diffficulties may be encountered. A sample set of nMOS 5 μm pad designs may be consulted in Hon and Sequin, 1980, and Newkirk and Mathews, 1984. CMOS pad designs are usually available from fabricators.

The designer must allow for the way in which the number of available pads quickly get used up and the very significant area they occupy. Take, for example, a simple processor of the type discussed in this text together with some RAM memory to form a basic microprocessor circuit. A typical arrangement is shown in Figure 10–10. Allowing for eight memory address lines (i.e. 256 locations of RAM), the complete chip as shown will need more than 30 pads which must therefore be accommodated in the layout. Such a number is readily bonded to, say, a 40-pin header, but the designer must also bear in mind that the package to be used will impose an ultimate limitation on the allowable number of pads.

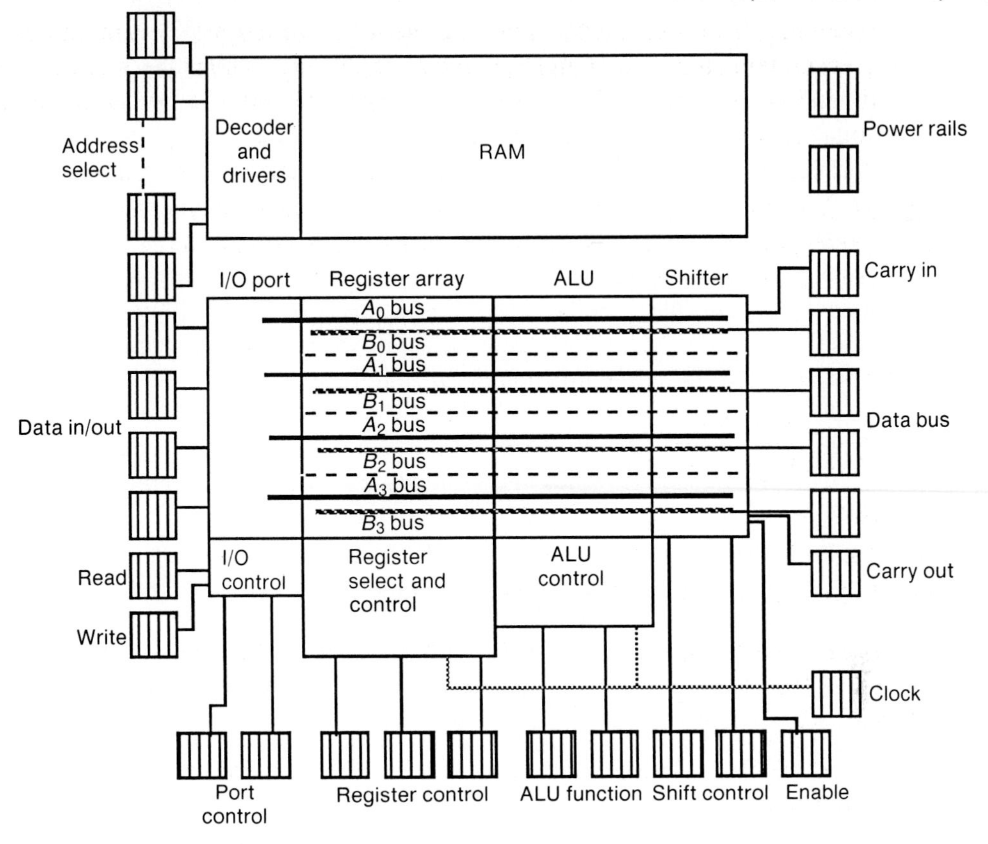

Figure 10–10 4-bit processor — pad utilisation

10.5 'Real estate'

Give me land, lots of land . . .
(words of a popular song of yesteryear)

One of the most common mistakes among beginners is to assume that phenomenal amounts of circuitry occupy very little area on the chip (VLSI = very little silicon indeed?). In order to correct such over-optimism it is necessary to consider only one or two of the practical factors which arise in system design.

For example, consider the area required by the I/O pads for the floor plan of Figure 10–10. The connections shown require 33 pads and typical standard 5 μm pad layouts require an area of 105λ by 100λ to 200λ (depending on the nature of the pad). An average pad then occupies some 105λ by 150λ, say, that is, an area of $15{,}750\lambda^2$. Thus the area required for 33 pads is over $500{,}000\lambda^2$. To put this into perspective, the average area allowance for each student project for a multiproject chip (MPC) design was typically somewhere in the region of $1000\lambda \times 1000\lambda$, that is, $10^6\lambda^2$. Thus, for the floor plan given in Figure 10–10, the pads would occupy one-half of this total area. Certainly, the design given here is somewhat pad-intensive but, as a rule of thumb, the small system designer should allow *one-third* of the chip area for pads.

Having come to terms with this, the budding designer may then consider what to do with the layout of the remaining two-thirds of the chip area (i.e. about $700{,}000\lambda^2$ for an example MPC design). What is the prognosis?

An assessment of what could be fitted into such an area could be approached by considering the basic enhancement mode pass transistor of *minimum size* occupying an area of $4\lambda^2$. If 2λ clearance is allowed all around, then the *on chip area* will be $36\lambda^2$. Dividing this into the available area, one might conclude that almost 20,000 such devices could be fitted into the area under discussion. However, MOS circuitry necessitates the use of inverters or inverter-like circuits. When two transistors are put together and contacts etc. are added, then, typically, a single inverter occupies at least $200\lambda^2$. Viewed from this point, the same area should thus accommodate about 3500 inverters. However, this is also an over-optimistic assessment of the possible circuit density, since one has to consider the significant effect of interconnections even within a leaf-cell. Consider the simple memory cell of Figure 10–11 which we might use to implement the RAM of Figure 10–10. The temptation is to assess area requirements by reasoning thus:

$$\text{two inverters} + \text{three pass transistors} = 2 \times 200\lambda^2 + 3 \times 36\lambda^2 = 508\lambda^2.$$

However, when design rule clearances, buses, power and control wiring are allowed for, this cell can occupy $1500\lambda^2$ or more (i.e. a factor of 3:1 over the 'simple' estimates).

Now, consider the available area on the floor plan and further assume that about half this area (i.e. approximately $350{,}000\lambda^2$) is to be devoted to the RAM. This area will allow no more than 256 bits of storage elements, as in Figure 10–11, and if each RAM location must hold a 4-bit word, then the designer can be no more ambitious than a 64-word RAM. The running of extra bus lines, as in

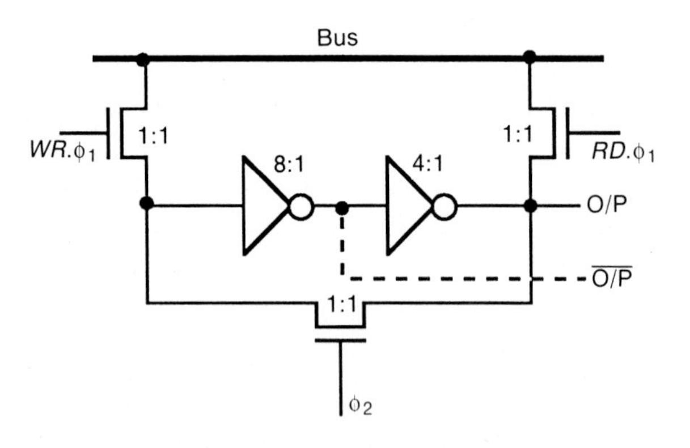

Figure 10–11 Pseudo-static memory cell

the register array, will further substantially increase the area occupied by each memory cell.

10.6 Further thoughts on system delays

10.6.1 Buses

He thought he saw [an operand],
descending from a bus,
he looked again and saw it was
a hippopotamus.

(With apologies to Lewis Carroll)

The use of bus lines is a convenient concept in distributing data and control through a system. However, it is easy to lose sight of what is *really* happening and bus-derived signals tend not to be what were expected.

Bidirectional buses are convenient but conflicts must be avoided since data cannot flow in both directions at once. Clearly, in our data path design, the sum S_k must be stored and then subsequently read onto the bus, since it becomes obvious that two buses cannot carry two input operands and the sum simultaneously. A significant problem which is often underestimated is that of speed restrictions imposed by the capacitive load presented by long bus lines.

The largest capacitance (for a typical bus system) is contributed by C_{BUS} (the bus wiring capacitance), and for small chips with, say, a 1000λ long bus this can be as high as 0.75 pF for a metal layer bus in 5 μm technology. In total, then, the bus and associated circuitry for the system being considered could contribute a capacitive load of about 0.8 pF, which may be *driven* through pull-up (typically 20 to 40 kΩ 'on' resistance) and pull-down (typically 10 kΩ 'on' resistance) transistors and through at least one pass transistor or transmission gate in the series.

Therefore, sufficient time must be allowed to charge the total bus capacitance during, say, ϕ_1 of the clock. In the data path system considered here, the time required for the total bus capacitance to charge to an appropriate level (to, say, $> 90\%$ of V_{DD}) is in the region of 100 nsec. Thus, it may be seen that equal ϕ_1 and ϕ_2 clock periods would result in an upper clock frequency limitation for the processor due to bus loading alone of 5 MHz. This frequency can be increased by using asymmetric ϕ_1 and ϕ_2 periods or by using BiCMOS drivers.

10.6.2 Control paths, selectors, and decoders

A basic operation of a data path is to add together the numbers stored in two registers to produce a sum and a carry at the 'carry out' pad (for cascading, etc.).

In terms of *delays* involved, and in the context of the 5 μm system considered here, the following delay mechanisms are encountered during this process:

1. *Select register* and open pass transistors (or transmission gates) to connect cells to bus. For a particular design, the select logic and associated drivers might have the equivalent circuit as shown in Figure 10–12.

   The overall *delay* of this arrangement may be assessed in terms of τ (where τ is the time constant of $1\square C_g$ charging through a minimum-size n-type pass transistor).

Element(s) contributing	*Delay*
Input pad	30τ (typical)
Three pass transistors ($n^2\tau$) = 9τ	9τ
Driver inverter pair ($\Delta A \rightarrow \nabla B \rightarrow \Delta C$) (Assuming $4\square C_g$ load at C)	34τ
Sum of delays (select register)	$= 73\tau$

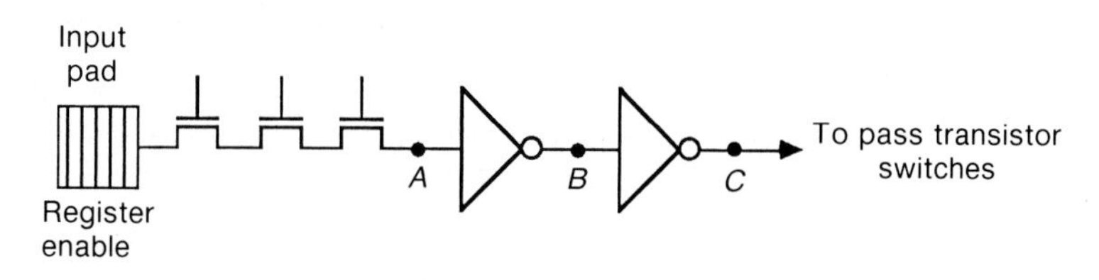

Figure10–12 Register select circuit

2. *Data propagation along bus* — This has already been calculated as 100 nsec.
3. *Carry chain delay* — The longest delay in the particular design of adder used is that of forming the 'carry out' which, in effect, propagates through all bits of the adder and then through the outlet pad as shown in Figure 10–13. Timing simulator results for a 2-bit arrangement is given as Figure10–14. It will be seen that, although the ΔC and ∇C delays are slightly different, an average delay of 65 nsec is a fair assumption for the 2-bit system simulated. We may also deduce the delay per bit ($\doteqdot$ 20 nsec) from the simulation. Overall then, a 5 μm 4-bit ripple-carry adder could be expected to have a delay of about 105 nsec.

Thus, the overall delay = select registers + bus delays + carry chain delays = (73τ) + 100 nsec + 105 nsec.

For $\tau = 0.2$ nsec

$$\textit{Sum of delays} = 14.6 + 100 + 105 \doteqdot 220 \text{ nsec}$$

Thus, ϕ_1 of the clock must have a duration longer than 220 nsec.

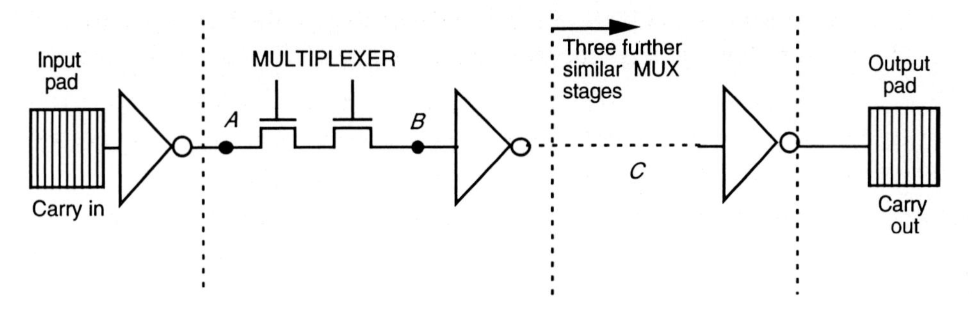

Figure 10–13 Possible carry chain circuit

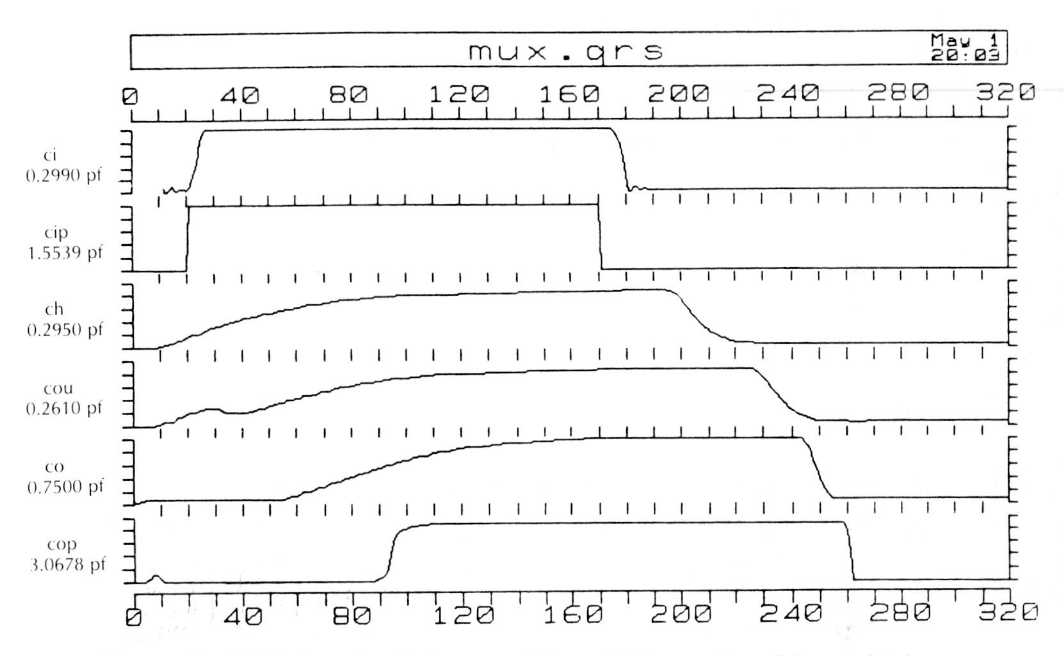

Figure 10–14 Timing simulation result for a 2-bit version of the multiplexer-based adder

10.6.3 Use of an asymmetric two-phase clock

10.6.3.1 Clock period ϕ_2

In many systems, ϕ_2 of the clock is used only to refresh memory/register cells such as that shown in Figure 10–15. From the figure it can be seen that ϕ_2 has to be long enough in duration to allow C_{in} to charge through the pull-up resistance of the second inverter and through the feedback circuit — which may be in the region of 35 kΩ. If time is allowed for C_{in} to charge to within < 10% of its final value, then refresh time $\doteqdot 2.5 \times 10\tau = 25\tau$ which, for the 5 μm system being

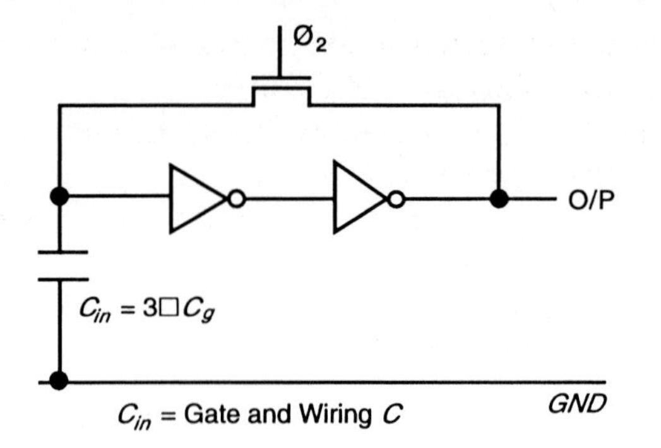

Figure 10–15 Memory cell refresh

evaluated, equates to a minimum 'on' time of 5 nsec for ϕ_2. However, ϕ_2 signals must also propagate through wiring etc., and finite rise- and fall-times must be allowed for so that some extra time should be allowed for the ϕ_2 'on' period. For safety allow, say, 50τ (i.e. 10 nsec) for the ϕ_2 'on' period and also allow 10 nsec underlap between the two phases. Thus

$$\text{total clock period} = 220 + 10 + 10 + 10 = 250 \text{ nsec}$$

Therefore, *in theory,* our simple modeling suggests that the data path chip design should operate on add instructions with a 4-MHz clock.

10.6.4 More nasty realities

Life wasn't meant to be easy.
Malcolm Fraser (former Prime Minister of Australia)

The simple calculations made on the particular processor design seem to indicate that a clock frequency in the region of 4 MHz would be possible. In practice, this may not happen. Why is this so? To answer this it is necessary to consider practical as well as theoretical realities.

From the theoretical aspect, our predictions have been made on very approximate parameter values and on very simple circuit models. We have also mostly ignored the quite significant effects of peripheral capacitance in diffusion regions and fringing field capacitances around conductors on the chip.

Although τ was assumed to be in the range 0.1 to 0.3 nsec for 5 μm technology, the value of τ measured for the fabricated chip may not be within this range.

In fact, the value of τ measured on some 5 μm MPC circuits fabricated and tested for this project was in the region of 0.6 nsec.

The designer, therefore, must be aware of, and allow for, all the significantly nasty realities affecting the performance of the design, and have a good knowledge of the parameters of the processing plant or fabrication line where that design is to be implemented in silicon.

There are two main points of difference between expectations and realization which characterize many of the designs of beginners. They are:

1. The system being designed occupies far more area than was anticipated.
2. The system when manufactured is slower than the designer had estimated.

However, if the first few designs are carefully carried out, are not over-ambitious, and are properly checked for logical and design rule errors, the beginner is usually pleasantly surprised by the fact that the system does in fact function, albeit not quite as fast as intended.

10.7 Ground rules for successful design

This section is intended to provide a convenient focus for design information. From our considerations of system design up to this point a number of ground rules, aspects of philosophy, and some basic data have emerged which help to ease the design process and ensure success. These and one or two other considerations which are important (but have not as yet been formally set out in the text) are presented or referenced here under 19 subheadings.

1. The ratio *rules* (Chapter 2)

   (a) for nMOS inverters and inverter-like stages

   $Z_{p.u.}$:$Z_{p.d.}$ ratio = 4:1 when driven from another inverter

   $Z_{p.u.}$:$Z_{p.d.}$ ratio = 8:1 when driven through one or more pass transistor(s).

   where

   $$Z = L / W \text{ for the channel in question}$$

   (b) for CMOS, a 1:1 ratio is normally used to minimize area, but for pseudo-nMOS inverters etc., a ratio $Z_{p.u.}$:$Z_{p.d.}$ = 3:1 is required.

2. *Design rules* (Chapter 3). Never bend the rules.
3. *Typical parameters for* 5 μm ($\lambda = 2.5$ μm), 2 μm *and* 1.2 μm *feature size MOS* (Chapter 4) including guidelines for signal interconnections.

Table 10–2 (Table 4–1) Typical sheet resistances R_s of MOS layers for 5 μm, and Orbit 2 μm and 1.2 μm technologies

Layer	*R_s ohm per square*	
	5 μm	*Orbit*
Metal	0.03	0.04
Diffusion (n-type or n-active)	10–50*	20–45*
Silicide	2–4	–
Polysilicon	15–100	15–30
n-transistor channel	10^4 †	2×10^4 †
p-transistor channel	2.5×10^4 †	4.5×10^4 †

Note: In some processes a silicide layer is used in place of polysilicon.
*Times 2.5 for p-type.
† These values are approximations only. Resistances may be calculated from a knowledge of V_{ds} and the expressions for I_{ds} given earlier.

Table 10–3 (Table 4–2) Typical area capacitance values

Capacitance	*Value in pF × 10^{-4}/μm² (relative values in brackets)*		
	5 μm	*2 μm*	*1.2 μm*
Gate to channel	4 (1.0)	8 (1.0)	16 (1.0)
Diffusion (active)	1 (0.25)	1.75 (0.22)	3.75 (0.23)
Polysilicon* to substrate	0.4 (0.1)	0.6 (0.075)	0.6 (0.038)
Metal 1 to substrate	0.3 (0.075)	0.33 (0.04)	0.33 (0.02)
Metal 2 to substrate	0.2 (0.05)	0.17 (0.02)	0.17 (0.01)
Metal 2 to metal 1	0.4 (0.1)	0.5 (0.06)	0.5 (0.03)
Metal 2 to polysilicon	0.3 (0.075)	0.3 (0.038)	0.3 (0.018)

Note: Relative value = specified value/gate to channel value for that technology.
*Poly 1 and Poly 2 are similar (also silicides where used).

Table 10–4 (Table 4–3) Typical values for diffusion capacitances

Diffusion capacitance	*Typical values*		
	5 μm	*2 μm*	*1.2 μm*
Area C (C_{area}) (as in Table 42)	1.0×10^{-4} pF/μm²	1.75×10^{-4} pF/μm²	3.75×10^{-4} pF/μm²
Periphery (C_{periph})	8.0×10^{-4} pF/μm	negligible*	negligible*

* Assuming implanted regions of negligible depth.

In order to calculate the total diffusion capacitance we must add the contributions of area and peripheral components.

$$C_{total} = C_{area} + C_{periph.}$$

Standard unit of capacitance $\square C_g$

$1\square C_g$ is defined as the gate-to-channel capacitance of a MOS transistor having $W = L$ = feature size, that is, a 'standard' or 'feature size' square (the concept of $\square C_g$, originated by VTI (USA), has been adapted here).

$\square C_g$ may be evaluated for any MOS process. For example, for 5 μm MOS circuits:

$$\text{standard value } \square C_g = .01 \text{ pF}$$

or, for 2 μm MOS circuits (Orbit) :

$$\text{standard value } \square C_g = .0032 \text{ pF}$$

and, for 1.2 μm MOS circuits(Orbit):

$$\text{standard value } \square C_g = .0023 \text{ pF}$$

The delay unit τ

We have developed the concept of sheet resistance R_s and standard gate capacitance unit $\square C_g$. If we consider the case of one standard (feature size square) gate area capacitance being charged through one feature size square of n channel resistance (i.e. through R_s for an nMOS pass transistor channel), we have:

$$\text{time constant } \tau = 1R_s \text{ (n channel)} \times 1\square C_g \text{ seconds}$$

This can be evaluated for any technology and for 5 μm technology

$$\text{theoretical } \tau = 0.1 \text{ nsec.}$$

and for 2 μm (Orbit) technology

$$\text{theoretical } \tau = 0.064 \text{ nsec.}$$

and for 1.2 μm (Orbit) technology

$$\text{theoretical } \tau = 0.046 \text{ nsec.}$$

However, in practice, circuit wiring and parasitic capacitances must be allowed for so that the figure taken for τ is often increased by a factor of two or three.

Taking account of resistances and total capacitances we may set out practical guidelines on signal path lengths as in the following table (10–5), noting that the figures given are conservative but safe.

4. *Inverter pair delay*

   In general terms, the delay through a pair of similar nMOS inverters is

$$T_d = (1 + Z_{p.u.}/Z_{p.d.})\tau$$

and for a minimum size CMOS complementary inverter pair

$$T_d = 7\tau$$

Table 10–5 (as for Table 4–4) Electrical rules

Layer	*Maximum length of communication wire*		
	lambda-based (5 μm)	*μm-based (2 μm)*	*μm-based (1.2 μm)*
Metal	chip wide	chip wide	chip wide
Silicicide	2,000λ	n.a.	n.a.
Polysilicon	200λ	400 μm	250 μm
Diffusion (active)	20λ*	100 μm	60 μm

* Taking account of peripheral and area capacitances. n.a. not applicable.

5. *Cascaded inverters for driving capacitive load (C_L)*
The approach is to use N cascaded inverters, each one of which is larger than the preceding stage by a width factor f.
It has been shown that the number 'N' of stages required is given by

$$N = \frac{ln(y)}{ln(f)}$$

where

$$y = \frac{C_L}{\square C_g}$$

It can also be shown that total delay is minimized if f assumes the value e (base of natural logarithms); that is, each stage should be approximately 2.7* times wider than its predecessor. This applies to CMOS as well as nMOS inverters. See Chapter 4 for more details.

* *Note*: Usually $f = 3$ will do since the curve is quite flat near the minimum.

6. *Propagation delay through cascaded pass transistors or transmission gates* (Chapter 4)

$$T_d = n^2 rc\ (\tau)$$

where

n = number in series

r = relative series resistance per transistor or per transmission gate in terms of R_s

c = relative capacitance gate to channel per transistor or per transmission gate in terms of $\square C_g$.

Normally, no more than four pass transistors or transmission gates should be connected in series without buffering.

7. *Factors influencing choice of layer for wiring* (Chapter 4)

Table 10–6 (Table 4–5) Choice of layers

	Relative		
Layer	*R*	*C*	*Comments*
Metal	Low	Low	Good current capability without large voltage drop ... use for power distribution and global signals.
Silicide*	Low	Moderate	Modest RC product. Reasonably long wires are possible. Silicide is used in place of polysilicon in some nMOS processes.
Polysilicon	High	Moderate	RC product is moderate; high IR drop.
Diffusion	Moderate	High	Moderate IR drop but high C. Hence hard to drive.

Note: V_{DD} and V_{SS} (or *GND*) rails must always be run in metal, except for very short 'duck unders' where crossovers are unavoidable.
* Not often available — depending on process line.

8. *Subsystem/leaf-cell design guidelines* (Chapter 6)
   (a) Define the requirements properly and carefully.
   (b) Consider communication paths most carefully in order to develop sensible placing of subsystems and leaf-cells.
   (c) Draw a floor plan (alternating with (b) as necessary).
   (d) Aim for regular structures so that design is largely a matter of replication.
   (e) Draw stick diagrams for basic cells, leaf-cells, and/or subsystems or enter the design in symbolic form.
   (f) Convert to a mask level layout.
   (g) Carefully and thoroughly check each mask layout for design rule errors and simulate circuit or logical operation. Correct as necessary, *rechecking* as corrections are made.

9. *Restrictions associated with MOS pass transistors and transmission gates* (Chapter 6)
   (a) No more than four in series without buffering (see Point 6).
   (b) No pass transistor gate must be driven from the output of one or more pass transistors, since logic 1 levels are degraded by threshold voltage V_{tp} (where V_{tp} can be as high as 0.3 V_{DD}).

(c) When designing switch logic networks of pass transistors or transmission gates, care must be taken to deliberately implement *both* the logic 1 and logic 0 output conditions.

Note: An *if, then, else* approach to specifying requirements will help to make sure that this is done.

10. *Storage of logic levels on the gate capacitance of transistors*

(a) Gate/channel capacitance is suitable for storing a bit, but care must be taken to allow for the finite decay time (about 0.25 msec at room temperature).

(b) It is quite allowable to construct pass transistors, etc. *under* metal layers to save space. This is often convenient and is used, for example, in some multiplexer layouts, but care must be taken with *overlying* metal wires where gate/channel capacitance is used for bit storage.

Consider Figure 10–16(a). Three such instances are illustrated here, all of which lie under metal wires. Two of these cases, T_1 and T_3, will operate satisfactorily, since for T_1 the metal wire is actually connected to the gate and for T_3 the metal wire is at a fixed, unvarying potential (that is, V_{DD} in this case). However, T_2 gate region lies under a metal bus which has no connection with the gate of T_2. If a bit is stored on T_2 gate by momentarily connecting *Control A* to the required level, then the bit will be stored but can be disturbed or destroyed by variation of the voltage on the overlying bus, as Figure 10–16(b) reveals.

(c) Restrictions also apply to logic level storage on the input capacitance of a *Nand* gate except for the input *nearest* the *GND* or V_{SS} rail. Conditions are indicated in Figure 10–17.

11. *Enhanced clocking*

One of the basic limitations on the use of simple MOS pass transistors (see Point 9 above) is the degradation of logic 1 levels by V_{tp} and the consequent inability of one pass transistor to drive the gate of a second (or more) pass transistor. This is particularly bothersome in clocking networks and a solution to this problem is to run all clock lines at a voltage level above V_{DD} as shown in Figure 10–18.

Note that the signal propagated through T_1 is V_{DD}, while that propagated through T_2 is $V_{DD} - V_{tp}$.

12. *The maximum allowable current density* in aluminum wires is 1 mA/μm^2. Otherwise, metal migration may occur (Chapter 6). Current density must be particularly carefully considered if the circuit is to be scaled down.

13. *Scaling effects:* see Chapter 5.

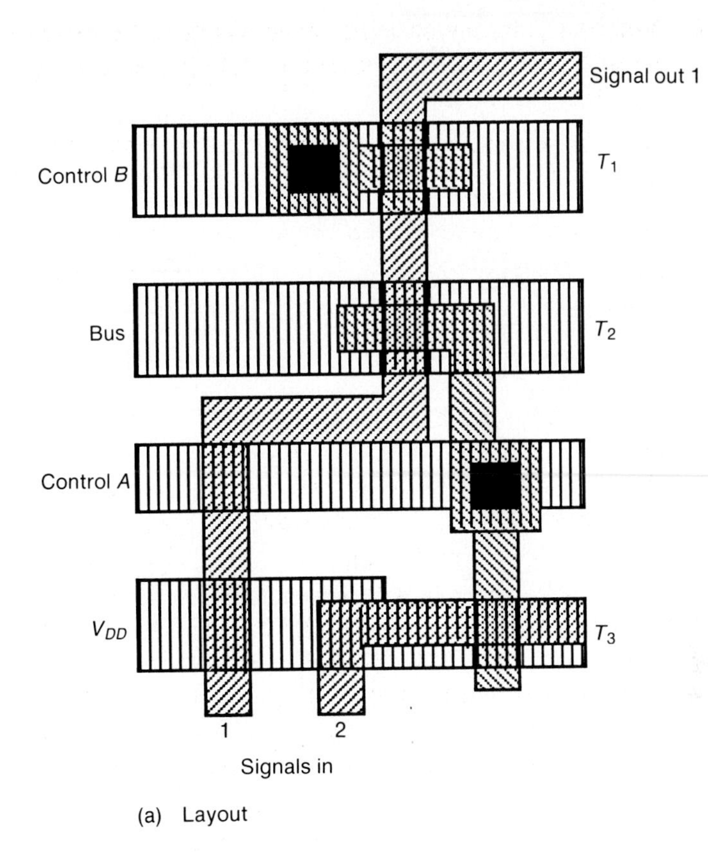

(a) Layout

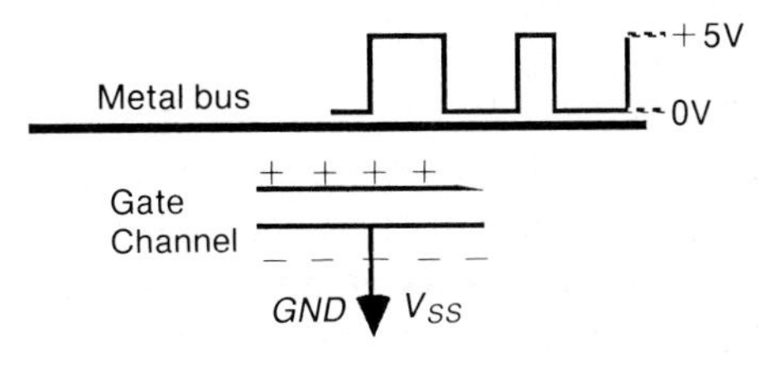

(b) Circuit model

Figure 10–16 Pass transistors under metal wires

14. *System design process* (Chapter 7 — refer also to point 8 in this section)
    (a) Set out a specification together with an architectural block diagram.
    (b) Suitably partition the architecture into subsystems that are, as far as possible, self-contained and give interconnections that are as simple as possible.

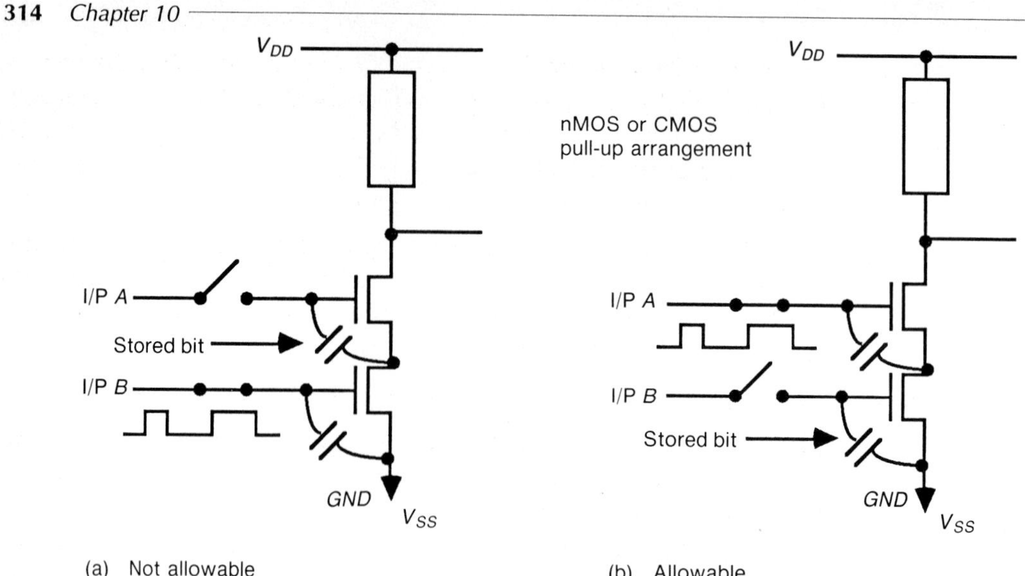

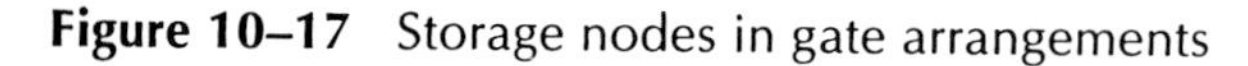
Figure 10–17 Storage nodes in gate arrangements

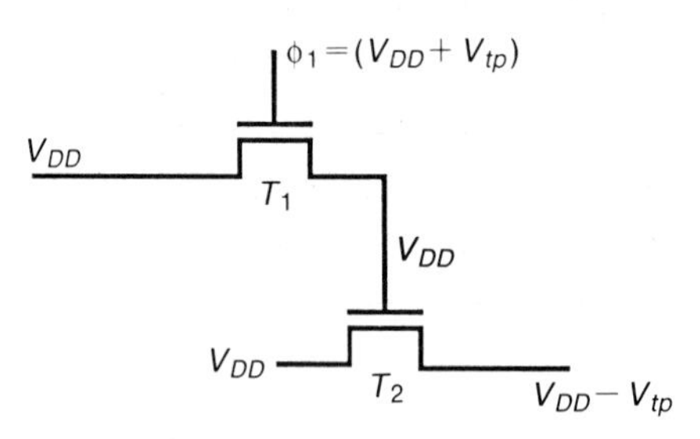

Figure 10–18 Enhanced clocking

(c) Set out a tentative floor plan showing the proposed relative physical disposition of subsystems on the chip.

(d) Determine interconnection strategy.

(e) Revise (b), (c), and (d) interactively as necessary.

(f) Choose layers on which to run buses and main control signals.

(g) Take each subsystem in turn and conceive a *regular* architecture to *conform* to the strategy set out in (d). Set out circuit and/or logic diagrams as appropriate. *Remember* that switch-based logic is such that both logic 1 and logic 0 output conditions must be deliberately satisfied (see Point 9).

(h) Develop stick or symbolic diagrams adopting suitable tactics to meet the overall strategy (d) and choice of layers (f). Determine suitable *leaf-cell(s)* from which the subsystem may be formed.

(i) Produce mask layouts for the leaf-cells *making sure* that cells can be butted together, side by side and/or top to bottom, without design rule violation or waste of space. Carefully check for any design rule errors in each standard cell itself. Determine overall dimensions of each cell and characterize in bounding box form if convenient.

(j) Cascade the replicate leaf-cells as necessary to complete the desired subsystem. This may now be characterized in bounding box form with positions and layers of inlets and outlets. External links, etc. *must* be allowed for. Check for design rule errors.

15. *Further observations on the design process* (based on Chapter 8)

    (a) First and foremost, try to put requirements into words (often an *if, then, else* approach helps to do this) so that the most appropriate architecture or logic can be evolved.

    (b) If a standard leaf-cell(s) can be arrived at, then the actual detailed design work, including simulation, is confined initially to small areas of simple circuitry.

    (c) Aim for generality as well as regularity, that is, leaf-cells, etc. should not be highly specialized unless absolutely necessary.

    (d) Communications dominate any system design.

    (e) A good library of basic leaf-cells and subsystems will speed design and allow accurate floor planning at an early stage.

    (f) A structured and orderly 'top-down' approach to system design is highly beneficial and becomes essential for large systems.

16. *Set out rules of system timing* at an early stage in design. A sample set of such rules is set out in Chapter 9 (section 9.1).

17. *Avoid bus contentions* by setting out bus utilization diagrams or tables, particularly in complex systems and/or where bidirectional buses are used.

18. *Do not take liberties with the design rules* but *do* take account of the ground rules and guidelines.

19. *Remember, IC designers should expect their systems to function first time around** and this will happen if the design concepts are correct and if the rules are obeyed.

    (We do *not* subscribe to the view 'If it works, it's out of date' (Stafford Beer), but we do contend that poorly conceived and badly designed systems may well be out of date before they work!)

* Not necessarily at optimum speed . . . this may take longer and depends on the designer's understanding of the properties of circuits produced in silicon.

10.8 The real world of VLSI design

Knowledge comes, but wisdom lingers.

Alfred Lord Tennyson

The preceding sections of this book have been intended to give the reader an understanding of the way in which system, circuit, and logic requirements may be turned into silicon and a feeling for the nature of silicon circuits. The authors believe that a sound understanding of *cause and effect* is essential if the maximum benefits are to be obtained from VLSI and the fullest range of applications opened up to VLSI realizations. Thus it is without apology that we have dwelt on the fundamental aspects of design in silicon.

From a sound foundation, a VLSI designer can operate with confidence, but must face up to the following requirements when contemplating large system designs in silicon.

1. *CAD.* The VLSI designer will need computer-aided design assistance, not only to assist in the design but also to handle the sheer complexity of the information needed to express the physical aspects of the design in a form suitable for translation into silicon.
2. *Verification tools* are essential to verify that the design is physically and logically correct and will perform correctly at the desired speed.
3. *Testability.* The designer must, from the outset, face up to the requirements of being able to test a system once it is realized in silicon.
4. *Test facilities.* Not only must testability be designed in but complex systems will need sophisticated equipment to actually test for correct operation.

Thus, it is a purpose of this chapter to present an overview of these important topics in order to put them in perspective for the budding VLSI designer. However, it is not our intention to cover comprehensively any of these topics. Although the topics are dealt with separately, it will be readily apparent that they are closely interrelated and can be significantly interdependent.

10.9 Design styles and philosophy

Style, like sheer silk, too often hides eczema.

Albert Camus

When wishing to implement a system design in silicon, various approaches are possible and, of course, a wide range of technologies is available to choose from. The designer must choose an appropriate design style, but at this point it must be stressed that in no case will the choice of style hide the lack of a competent and systematic approach by the designer. However, we may summarize the possibilities into three broad categories:

1. Full custom design of the complete system for implementation in the chosen technology. In this case, the designer designs all the circuitry and all interconnection/communication paths.
2. Semi-custom design using a library of standard leaf-cells together with specially designed circuits and subsystems which are placed appropriately in the floor plan and interconnected to achieve the desired functional performance. In this case, the designer designs a limited amount of circuitry and the majority of interconnections/communications.
3. Gate array (uncommitted logic array) design in which standard logic elements are presented for the designer to interconnect to achieve the desired functional performance. In this case, the design is that of the interconnections and communications only.

Once again the boundaries between these categories may be blurred. For example, full custom design seldom involves the complete design of the entire chip; input/output pad circuits are more or less accepted as standard components and are generally available to the custom designer.

In all cases it is desirable to take a hierarchical approach to the system design in which the principles of iteration or replication (regularity) can be used to reduce the complexity of the design task.

The designer is usually concerned with a number of key design parameters. These will include:

1. performance, in terms of the function to be performed, the required speed of operation and the power dissipation of the system;
2. time taken for the design/development cycle;
3. testability;
4. the size of the die, which is determined by the area occupied by the circuitry and in turn has a marked impact on the likely yield in production and on the cost of bonding and packaging and testing. Large die sizes are generally associated with poor yields and high costs.

Full custom design tends to achieve the best results, but *only* if the designer is fully conversant with the fundamental aspects of design in silicon so that parameters can be optimized. However, full custom design parameter optimization is usually at the expense of parameter 2, the time taken to design.

Semi-custom and gate array designs both have penalties in area and often in speed and this is contributed to by the fact that not all the available logic will be used. This is due to the need for generality in gate array and standard cell geometries. However, it may often be the case that gate arrays will be faster than a prototype full custom design in, say, MPC form and the final custom designs must often be carefully optimized.

Once the approach is chosen, there remains the design philosophy which ranges through the following general possibilities.

1. *Hand-crafted design* in which, for example, the mask layouts are drawn on squared paper with layer encoding and are then digitized to give a machine-readable form of the mask detail. Digitization can be done 'by hand', with entry of coordinates through, say, a keyboard or by more direct digitization of the drawn layout using a digitizer pad and cursor.
2. *Computer-assisted textual entry* of mask detail through a keyboard using some specially developed language employing a text editing program. Such programs may have relatively low-level capabilities, allowing the entry of rectangular boxes, and 'wires', etc. only, or may be at a higher level and allow symbolic entry of circuit elements such as transistors and contact structures.
3. *Computer-assisted graphical entry* of mask geometry through either a monochrome or color graphics terminal, again with the aid of the appropriate entry, display, and editing software.

   In cases 2 and 3 the software usually aids the processes of hierarchical system design in that leaf-cells (or symbols) can be instanced many times, each instance being placed as appropriate in the floor plan. Subsystems thus created may themselves be repeatedly instanced and placed as required to build up the system hierarchy.

   Such tools obviously encourage *regularity* and are generally used with a *generate then verify* design philosophy.
4. *Silicon compiler-based design* in which a high level approach is taken to design, and special languages, analogous to high level programming language compilers, are developed to allow the designer to specify the system requirements in a manner which is convenient and compact. The silicon compiler program then translates this input code into a mask design which will generate a circuit in silicon to meet the specified system requirements. Such programs are the subject of much research and development work at this particular time. Indeed, the work has reached a stage at which silicon compilers have been in use for some time and there are textbooks on the subject (e.g. Ayres, 1983).

10.10 The interface with the fabrication house

Knowledge without practice makes but half an artist.

Proverb

Obviously, real world designs in silicon are intended to be fabricated and there is no doubt that the learning processes associated with VLSI design depend heavily on actually designing systems in silicon, on having them *fabricated* and then on *testing* the fabricated chips. In all cases, then, good *two-way* communications between the fabrication house or silicon broker and the designer must be established.

Communication from the former to the latter usually takes the form of a set of design rules which specify clearances, widths, spacing, overlaps, etc. for the process to be used. The design rules used in this test are examples of such rules. The fabrication house will also supply design parameters relevant to its processes. These include layer resistance values, layer to layer capacitance values, etc., and typical values have been given and used in this text.

In return, the designer must communicate his mask layout designs to the fabricator in a form which is convenient and clearly understandable. Methods of expressing mask geometry are not entirely standardized, but a *de facto* standard appears to be CIF code.

10.10.1 CIF (Caltech. Intermediate Form) code

CIF is a low-level graphics language for specifying the geometry of integrated circuits (Hon and Sequin, *A Guide to LSI Implementation,* Xerox). The purpose of CIF code is to communicate chip geometry in a standard machine-readable form for mask-making. CIF code is reasonably compact and can cope with small and large system geometry. Its format is straightforward and it has the added advantage of being easily read. It has been widely used for the electronic transport of designs between universities and industrial laboratories, using such facilities as ARPANET in the United States and CSIRONET in Australia. Thus, it is appropriate to briefly examine some of the features of CIF so that the reader may appreciate general attributes of the code.

10.10.1.1 Geometric primitives

Various geometric structures such as boxes, polygons, and wires are readily defined. In general, the position, dimensions, and orientation must be specified and, also of course, the layer on which the box exists in the silicon. When examining the attributes of CIF code, the reader should be aware that CIF dimensions and positions are given in X,Y coordinate form but are in absolute dimension units, *not* in lambda form.

A few examples (Figure 10–19) illustrate the features of the representation.

Boxes (B) are specified as

Box Length (L) Width (W) Center (C) Direction (D)

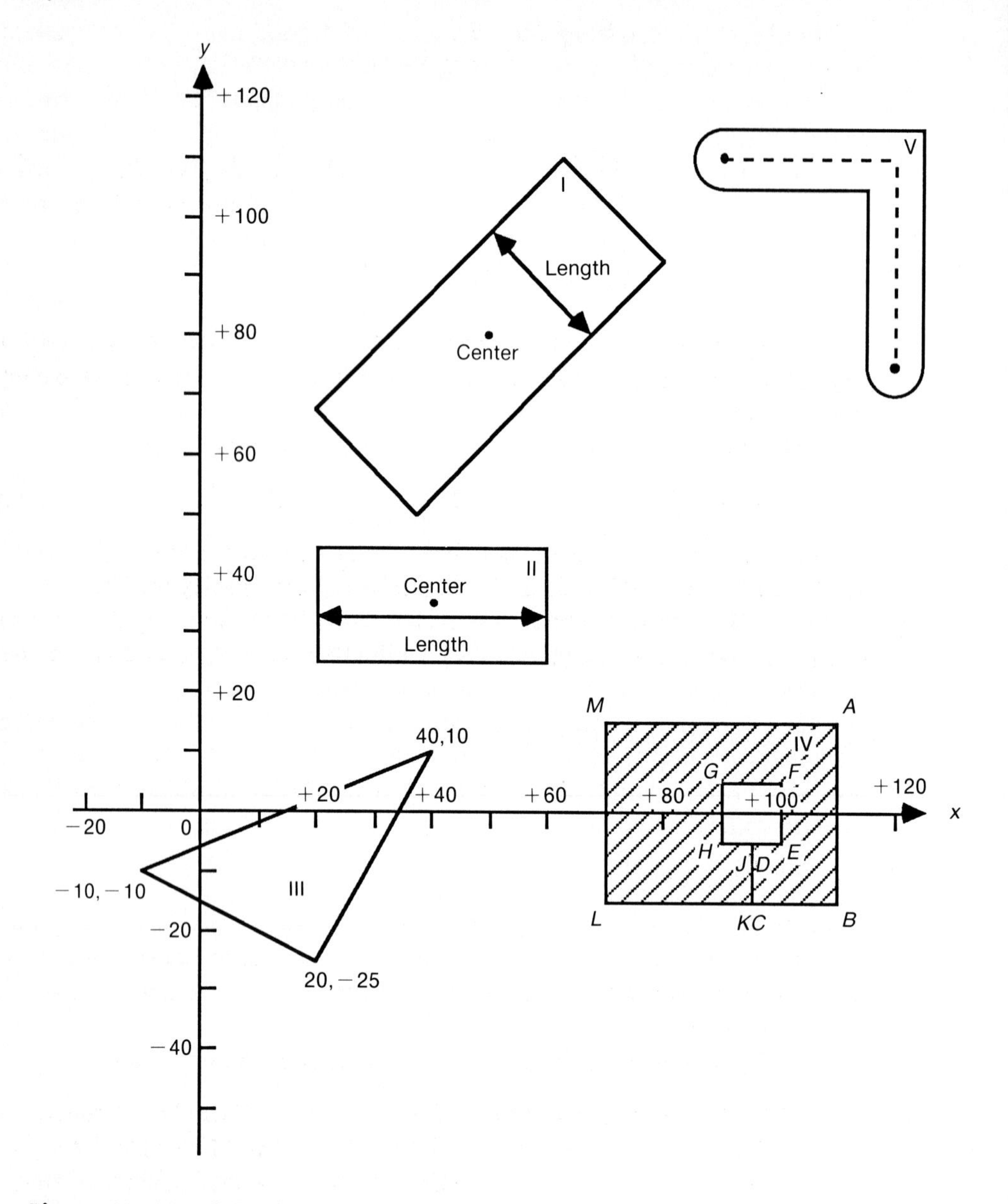

Figure 10–19 CIF Primitives — examples

Note that direction is given as a vector assumed parallel to the length. If not given, then a vector 1,0 (x,y) is assumed (that is, length will be parallel to the x-axis).

Boxes I and II in the diagram would therefore appear in code as

```
B 25        60        50   80        -10   10;                 (box I)
 .           .        \_____/        \______/
 .           .
 .           .
(L)         (W)         (C)             (D)      (L, W, C, D would not
 .           .                                    appear in the actual code.)
 .           .
 .           .        _______
B 40        20        40   35;                                 (box II)
```

Polygons (P) are specified in terms of the vertices in order. An *n-sided* polygon requires *n vertices* and a connection between first and last is assumed to complete the boundary.

Polygon III in Figure 10–19 would therefore appear in code as

```
P - 10 - 10        40   10        20 - 25          (polygon III)
```

In order to represent areas with holes in them, as in polygon (IV), the vertices A, B, C, D, E, F, G, H, J, K, L, M would be used to specify the area.

Wires (W) are specified in terms of their width followed by the center line's coordinates of the wire's path. In Figure 10–19, wire (V) would be specified as follows:

```
W  10  90  110  120  110  120  75
```

Note that each segment of wire ends in a semicircular 'flash' which will overlap any connecting area.

10.10.1.2 Layers

Layer selection and subsequent changes are treated by mode setting prior to or during the entry of geometric primitives. Layer setting must precede the entry of the first piece of geometry and must then precede the geometric inputs on any change of layer.

For the processes in this text the layers are named as follows:

ND (nMOS diffusion/thinox)	CAA	(CMOS diff/thinox)
	CNA	(CMOS nDiff/thinox)
	CPA	(CMOS PDiff/thinox)
NP (nMOS polysilicon)	CPF	(CMOS polysilicon 1)
	CPS	(CMOS polysilicon 2)
NC (nMOS contact cut)	CC	(CMOS contact cut)
NM (nMOS metal 1)	CMF	(CMOS metal 1)
NN (nMOS metal 2)	CMS	(CMOS metal 2)
NI (nMOS implant)		
	CS or CPP	(CMOS p^+ mask)
	CW or CPW	(CMOS p-well)

NV (nMOS Via)	CVA	(CMOS Via)
NB (nMOS buried contact)		
NG (nMOS overglass cuts)	CG or COG	(CMOS overglass cuts)
	CBA	(BiCMOS p-base)
	CCA	(BiCMOS buried collector)

Layer changes are indicated by the letter L followed by the layer name.

CIF also accommodates calls (C) and rotations and translations, etc., but the elementary review given here should convey the essential features. To reinforce this, a simple cell layout is given as Figure 10–20 with the corresponding CIF code given in Table 10–7.

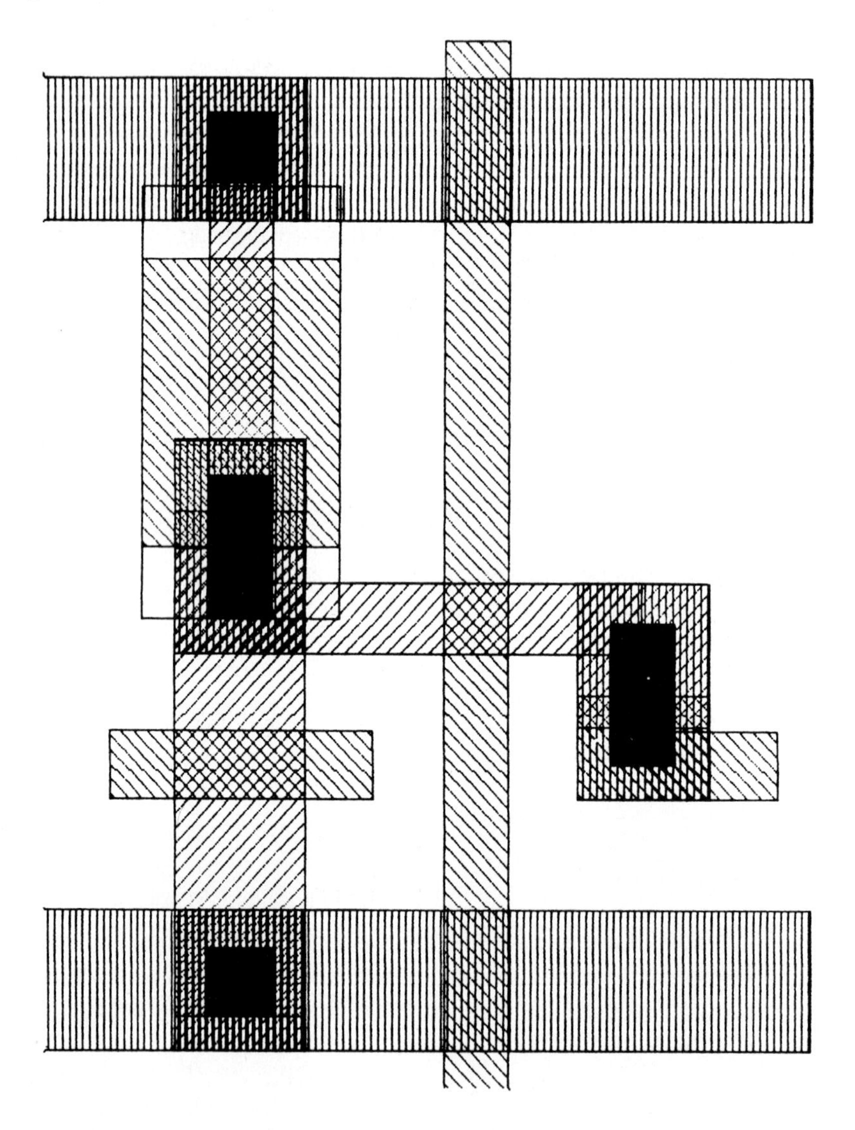

Figure 10–20 Layout of 'SRCELL' (plotted from the CIF code of Table 10–7)

Table 10–7 CIF code for SRCELL

```
25 Lambda = 250;
DS 1001;
9 SRCELL;
42 – 500, – 250 5250,7000;
        L    NM   ;
                        W 1000 0,500 4750,500;
                        W 1000 0,6250 4750,6250;
        L    ND   ;
                        B 1000 1000 1000,500;
        L    NC   ;
                        B 500 500 1000,500;
        L    NM   ;
                        B 1000 1000 1000,500;
        L    ND   ;
                        B 1000 1000 1000,6250;
        L    NC   ;
                        B 500 500 1000,6250;
        L    NM   ;
                        B 1000 1000 1000,6250;
        L    ND   ;
                        B 1000 1000 1000,3250;
        L    NP   ;
                        B 1000 750 1000,3875;
        L    NC   ;
                        B 500 1000 1000,3500;
        L    NM   ;
                        B 1000 1500 1000,3500;
        L    ND   ;
                        W 1000 1000.750 1000,3000;
                        W 500 1000.6000 1000,3500;
        L    NP   ;
                        B 2000 500 1000,2000;
                        B 1500 2000 1000,4500;
        L    NI   ;
                        B 1500 3000 1000,4500;
        L    ND   ;
                        B 1000 1000 4000,2750;
        L    NP   ;
                        B 1000 750 4000,2125;
        L    NC   ;
                        B 500 1000 4000,2500:
        L    NM   ;
                        B 1000 1500 4000,2500;
        L    ND   ;
                        W 500 1250,3000 3750,3000;
        L    NP   ;
                        W 500 4250,2000 4750,2000;
                        W 500 2750,0 2750,6750;
DF;
        C 1001 T 0,0;
End
```

10.11 CAD tools for design and simulation

Efficiency is intelligent laziness.

Arnold Glasgow, *Reader's Digest,* 1974

The design of a chip of reasonable complexity can in time be completed 'by hand' but it is both a hard and inefficient way of doing things. As far as the design of very large systems is concerned, it is *essential* to have computer aids to design so that the design can be completed in a reasonable time and, indeed, so that it can be completed at all. Whatever the size or nature of the design task, there is no doubt that well-conceived tools can make it much easier *and* do it better. Tools are therefore essential to ensure first time (and every time) success in silicon. At the very least, the designer's 'tool box' should include:

1. *physical design layout and editing* capabilities, either through textual or graphical entry of information;
2. *structure generation/system composition* capabilities, which may well be part of the design layout software implementing Point 1;
3. *physical verification.* The tools here should include design rule checking (DRC), circuit extractors, ratio rule and other static checks, and a capability to plot out and/or display for visual checking.
4. *behavioral verification.* Simulation at various levels will be required to check out the design before one embarks on the expense of turning out the design in silicon.

Simulators are available for logic (switch level) simulation and timing simulation. Circuit simulation via such programs as SPICE is also possible, but may be expensive in terms of computing time and therefore impractical for other than small subsystems. Recent advances in simulators have made it possible to use the software as 'a probe' to examine the simulated responses on various parts of the circuit to input stimuli also provided via the simulator. Such a facility, known as a software probe (and analogous to a CRO and associated hardware probe and signal generator), is available in various suites of design programs.

The authors can only stress that the joy of discovering that 'it does what it's supposed to' is only exceeded by the dismay of discovering that 'it doesn't work!' once a chip is fabricated (the designer having failed to carry out proper simulation testing). Some aspects of typical design tools are briefly reviewed next.

10.12 Aspects of design tools

10.12.1 Graphical entry layout

Textual entry of layouts was at one time quite widely used and special textual entry editors are in existence and may well be used for small subsystem layout. However, such tools have been virtually swept aside by a much more convenient and highly interactive method of producing layouts for which monochrome or color graphics terminals are used, and on which the layout is built up and displayed during the design process. Such systems are mostly 'menu driven', in that menus of possible actions at various stages of the design are displayed on the screen beside the display of the current layout detail. Some form of cursor allows selection and/or placement of geometric features, etc., and the cursor may also allow selection of menu items or, alternatively, these may also be selected from a keyboard. Positioning of the cursor may be effected from the keyboard in simple systems and/or cursor position may be controlled from a bitpad digitizer or from a 'mouse', etc.

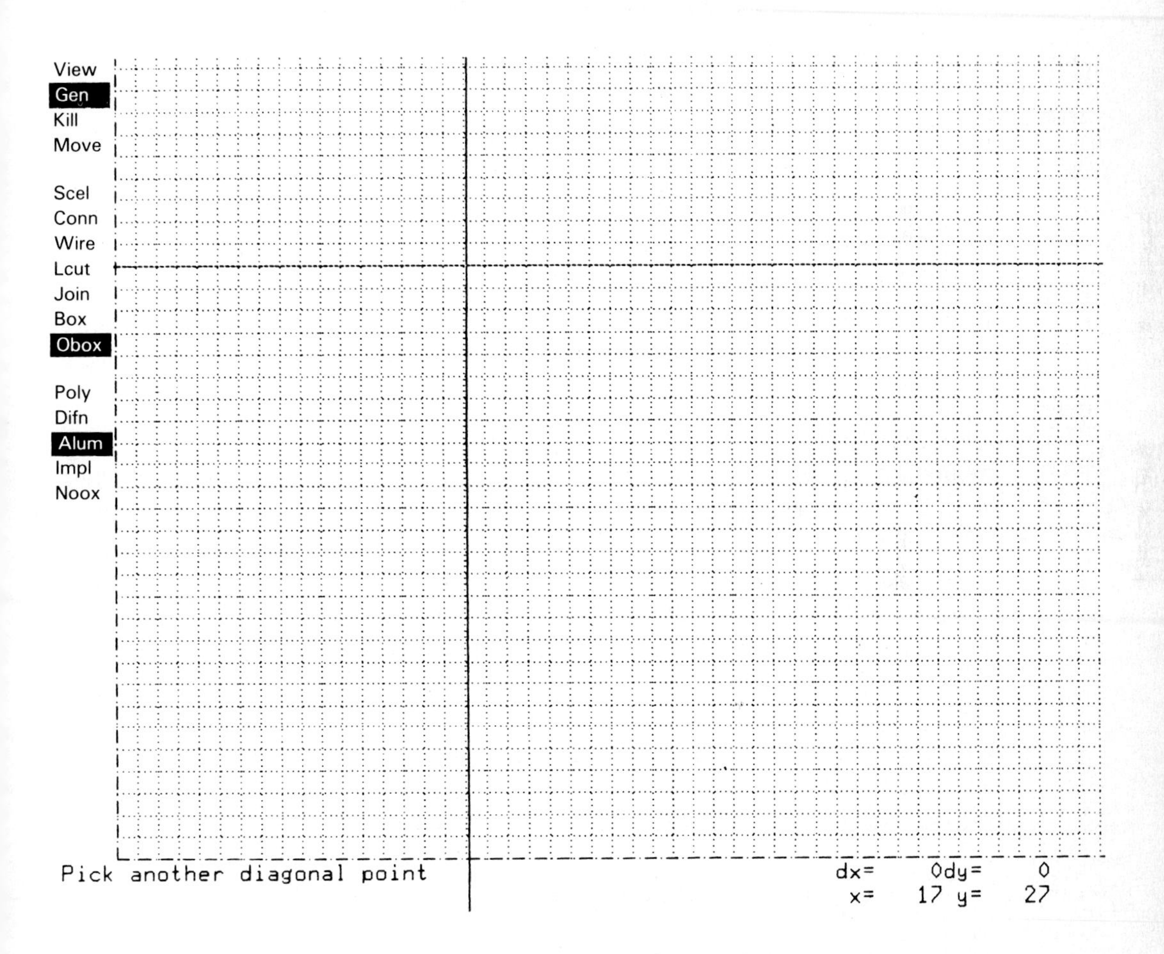

Figure 10–21 Basic PLAN design environment*

* Figure shows λ grid, cross hair cursor, and menu (selected items in inverse video). *x* and *y* values of current or previous cursor position may also be displayed as shown. OBOX is selected to establish an outline (bounding) box. Then a name, (SRCL), is allocated to the enclosed cell.

Two of the earliest available graphical entry layout packages were KIC, developed at the University of California, Berkeley, and PLAN, originally developed at the University of Adelaide. PLAN makes use of low-cost monochrome, as well as color, graphics terminals and is marketed by Integrated Silicon Design Pty Ltd, Adelaide. The use of an early version of PLAN to generate layouts is illustrated in Figures 10–21 to 10–25 and it is hoped that the inclusion of these figures, which show various stages of design, is sufficient to convey an idea of the nature of the layout process using this class of software tools.

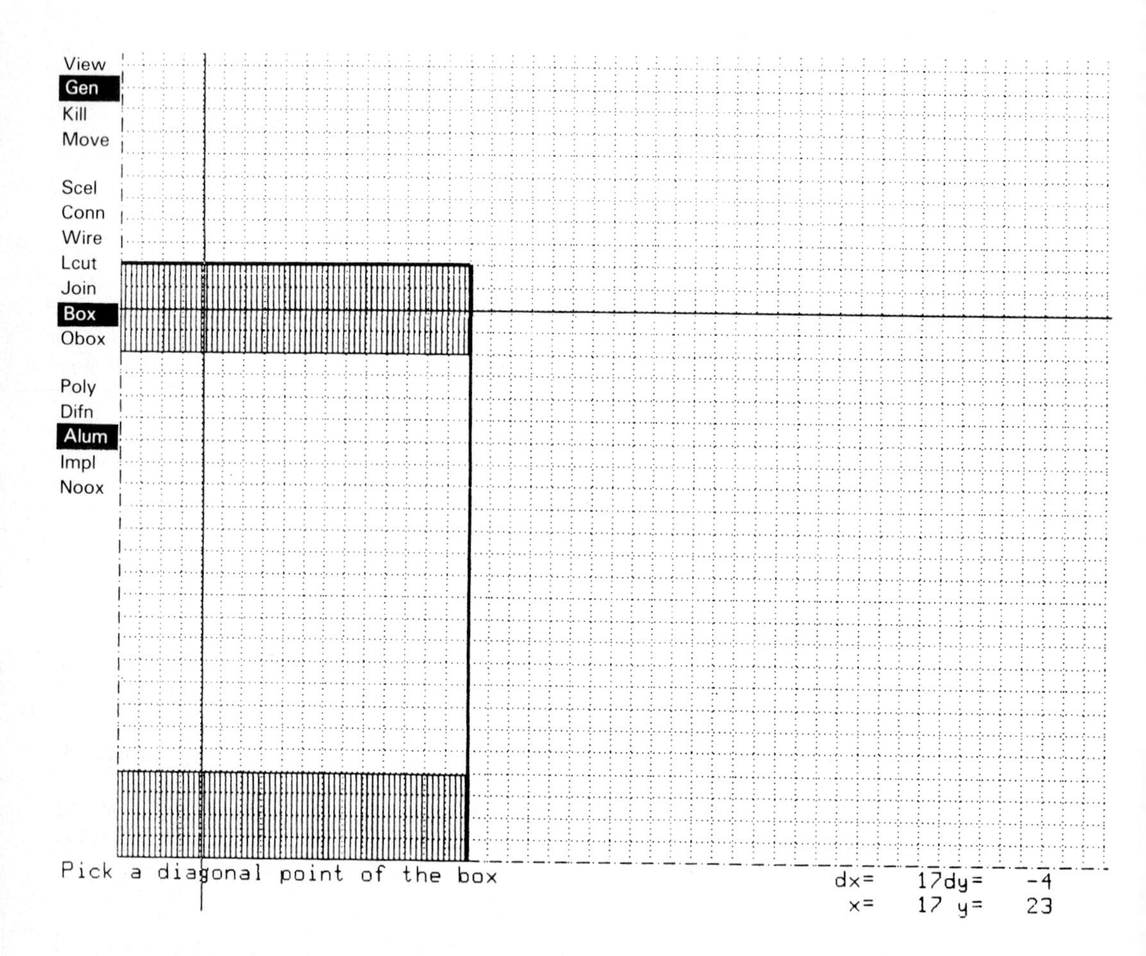

Figure 10–22 Layout of metal geometry using the BOX generate feature*

* V_{DD} and *GND* rails have been drawn by specifying diagonally opposite corners of each box (the Alum or metal layer is selected). *x* and *y* values shown are for the last corner specified and *dx* and *dy* give the relative movement of the cursor between corners of the last box drawn.

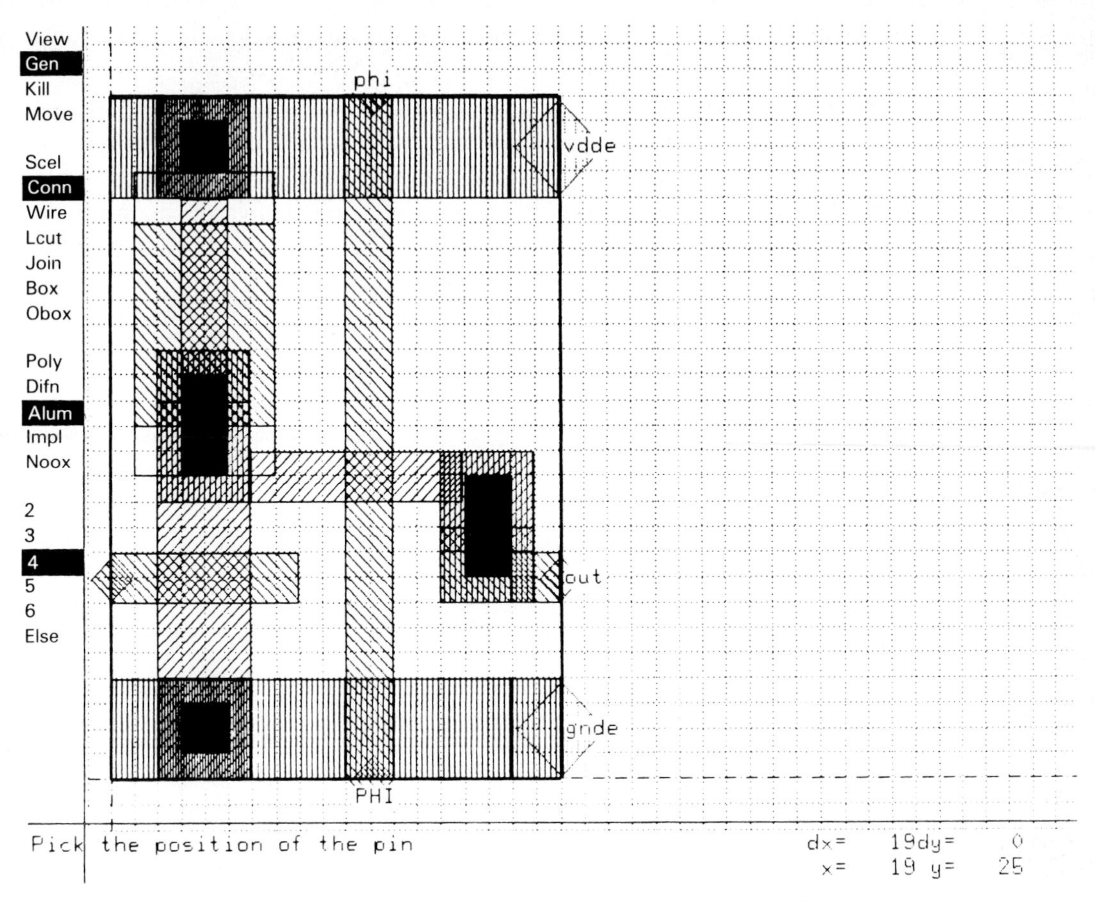

Figure 10–23 Completed layout of shift register cell (SRCL)*

*Identical to that of Figure 10–20. Note the labeled nodes or pins.

10.12.2 Design verification prior to fabrication

Try your skill in gilt first, and then in gold.

Proverb

It is not enough to have good design tools for producing mask and system layout detail. It is essential that such tools be complemented by equally effective verification software capable of handling large systems and with reasonable computing power requirements.

The nature of the tools required will depend on the way in which an integrated circuit design is represented in the computer. Two basic approaches are:

1. Mask level layout languages, such as CIF, which are well suited to physical layout description but not for capturing the design intent.

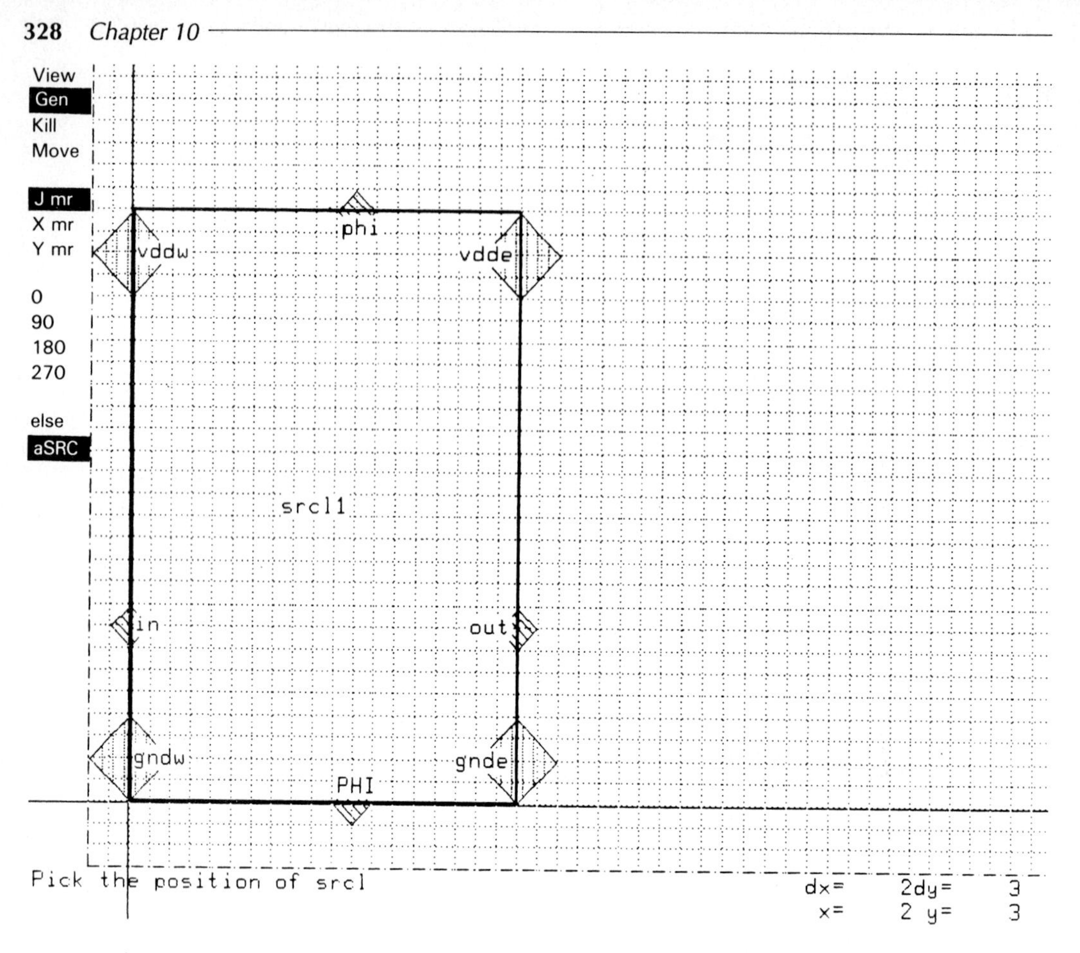

Figure 10–24 Bounding (outline) box representation of SRCL*

*From now on, SRCL may be instanced from the SCEL item on the previous menu and placed as required as shown. Note that the cell is shown now as a bounding box with pins.

2. Circuit description languages where the primitives are circuit elements such as transistors, wires, and nodes. In general, such languages capture the design intent but do not directly describe the physical layout associated with the design.

By and large, therefore, the designer's needs may include the following.

10.12.3 Design rule checkers (DRC)

The cost in time and facilities in mask-making and in fabricating a chip from those masks is such that all possible errors must be eliminated before mask-making proceeds. Once a design has been turned into silicon there is little that can be done if it doesn't work.

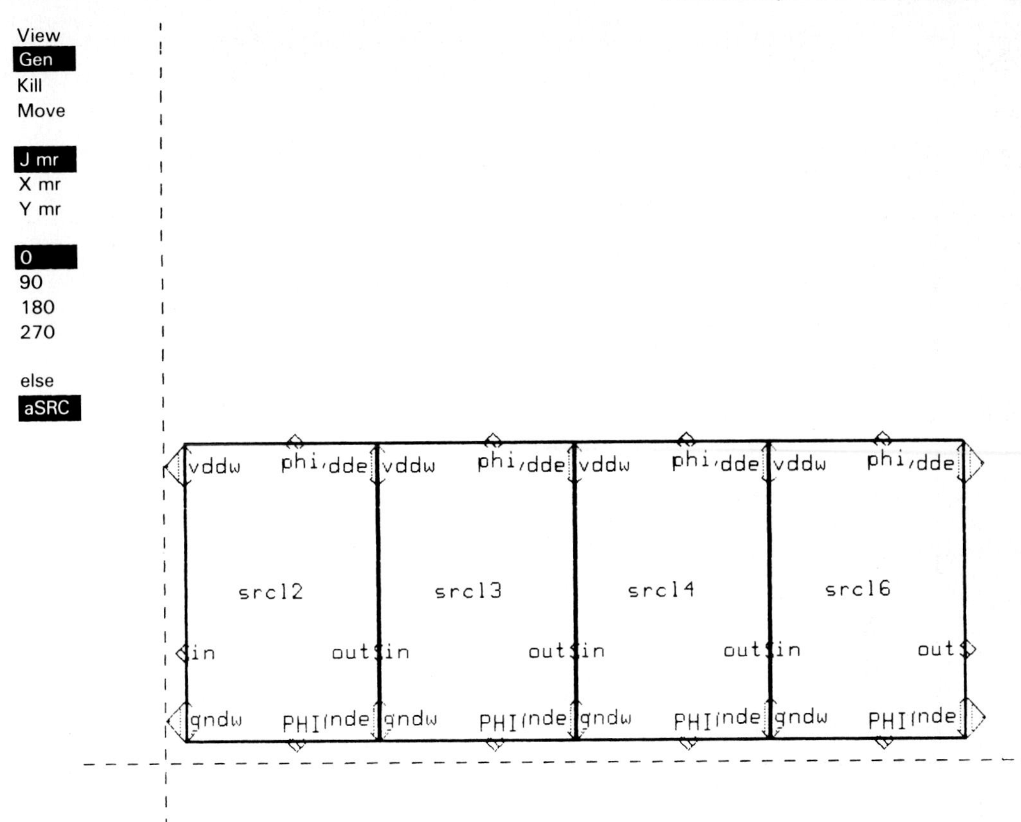

Figure 10–25 Instancing SRCL to form a register*

* Several instances of SRCL may be set out as shown to form a complete register (2-bits only shown) and a bounding box, and a name can be given to the whole structure.

The wise designer will check for errors at all stages of the design, namely:

1. at the pencil and paper stage of the design of leaf-cells;
2. at the leaf-cell level once the layout is complete (e.g. when the CIF code for that leaf-cell has been generated);
3. at the subsystem level to check that butting together and wiring up of leaf-cells is correctly done;
4. once the entire system layout has been completed.

The nature of physical layout verification 'design rule checking (DRC)' software may depend on whether the design rules are absolute or lambda-based, or on whether or not the layout is on a fixed or virtual grid.

A number of DRC programs, based on various algorithms, are available to the designer (e.g. the CHECK program from Integrated Silicon Design Pty Ltd).

10.12.4 Circuit extractors

If design information exists in the form of physical layout data (as in CIF code form), then a circuit extractor program which will interpret the physical layout in circuit terms is required. Although the designer could use the extracted data to check against his or her design intent, it is normally fed directly into a simulator so that the computer may be used to interpret the findings of the extractor. (An example of a circuit extractor program is NET from Integrated Silicon Design Pty Ltd.)

10.12.5 Simulators

In this section we very briefly consider the important topic of simulation prior to the VLSI design being committed to silicon.

From mask layout detail it is possible to extract a circuit description in a form suitable for input to a simulator. Programs that do this are referred to as circuit extractors. The circuit description contains information about circuit components and their interconnections. This information is subsequently transformed by the simulator into a set of equations from which the predictions of behavior are made.

The topology of the circuit determines two sets of equations:

- Kirchoff's Current Law — determining the branch currents; and
- Kirchoff's Voltage Law — determining node voltages.

The electrical behavior is defined by mathematical modeling, the accuracy of which determines two key factors:

- the accuracy of the simulation; and
- the computing power and time needed for the simulation.

We are often interested in relatively simple models to enable the highlighting of key features of performance in the design stage and to be able to observe trends as aspects of a design are changed by means of on-line interactive design.

Various types of simulators are available but generally they fall into the following groups:

- circuit simulators;
- timing simulators;
- logic level (switch level) (functional) simulators;
- system level (functional) simulators.

Circuit simulators are concerned with the electrical behavior of the various parts of the circuit to be implemented in silicon. Simulation programs such as SPICE can do this quite well, but take a lot of computing time to simulate even relatively small sections of a system and are completely impractical for circuits of any real magnitude.

Timing simulators (such as PROBE from Integrated Silicon Design Pty Ltd and QRS (developed at MCNC)) have attempted to improve matters in these respects by concentrating on active nodes and ignoring quiescent nodes in simulation. Work is proceeding in many establishments on improving the nature and performance of simulators; in particular, the way in which devices/circuits are modeled is vital. In all cases, the accuracy of simulation depends on the accuracy of the fabrication house parameters which must be fed into the simulator. In most cases, simulators attempt to predict the electrical performance with an accuracy of 20% or better. Examples of the output form of typical timing simulators have been included at several points in this book.

Timing simulators are becoming increasingly important during the design phase because of their speed and consequent interactive qualities. The structure of these tools ensures that run times are strictly linearly related to the number of devices and nodes being simulated. Speed-up is usually achieved through the use of a simple simulation cycle, a somewhat restricted network model and reasonably simple transistor models.

The simulation cycle is organized around the concept of a timestep. Each node voltage V is updated within each timestep by applying the following relation:

$$V_{new} = V_{old} + \frac{I_{ds}}{C} \Delta t$$

where

I_{ds} = drain to source current
C = node capacitance
Δt = timestep

In order to improve transistor modeling it is possible to include:

- body effect;
- channel length modulation;
- carrier velocity saturation.

The last two effects are particularly important for short channel transistors, that is, channel lengths $\leqslant 3\ \mu m$, and their effects should be taken into account.

Channel length modulation — for voltages exceeding the onset of saturation there is an effective decrease in the channel length of a short channel transistor. For example, the change in channel length ΔL for an n-transistor is approximated by

$$\Delta L = \sqrt{\frac{2\varepsilon_0\varepsilon_{si}}{qN_A}(V_{ds} - V_t)}$$

The resultant drain to source current $I^1{}_{ds}$ is approximated by

$$I^1{}_{ds} = I_{ds}\frac{L}{\Delta L}$$

where I_{ds} is given by the simple expressions developed in Chapter 2.

Velocity saturation — when the drain to source voltage of a short channel transistor exceeds a critical value, the charge carriers reach their maximum scattering limited velocity before pinch off. Thus, less current is available from a short channel transistor than from a long channel transistor with similar width to length ratio and processing.

Logic level simulators can cope with large sections of the layout at one time but, of course, the performance is assessed in terms of logic levels with no or little timing information. However, there may be large sections of a system which can be satisfactorily dealt with and verified this way, provided that leaf-cell elements have been subjected to a more rigorous treatment.

When considering complete systems, logic simulators may be replaced by simulators which operate at the register transfer level.
In all cases, the designer should carefully consider the availability of all such tools when choosing VLSI design software.

10.13 Test and testability

The proof of the pudding is in the eating.

Proverb

Although this topic has been left to last in this chapter, it is by no means least in significance.

Three factors conspire to create considerable difficulties for the test engineer and, indeed, for the designer testing his or her own prototypes:

1. the sheer complexity of VLSI systems;
2. the fact that the entire surface of the chip, other than over the pads, is sealed by an overglass layer and, thus, circuit nodes cannot be probed for monitoring or excitation;
3. with minor exceptions, there is no way that the circuit can be modified during tests to make it work.

It is also essential for faults to be detected as early as possible in the manufacture of a system. A relationship, known as 'the rule of ten', tends to apply as far as test costs are concerned. This rule is concisely put as follows:

> If chip test cost = \$×, then once that chip is soldered into a p.c. board with other components, test cost = \$10×. Further, once that board is integrated into a system/equipment, then the test cost escalates by a further factor of ten to test cost = \$100×. Finally, a factor which is often overlooked is that test costs may escalate by a further factor of ten when the equipment is in service in the field. It is thus essential to test at the chip level as comprehensively as possible.

Thus, chip design/fabrication mistakes can be very costly, both in terms of time and money and, for a complex chip, lack of thought at the design stage may mean that it cannot be properly tested at all. Design for testability (DFT) is an essential part of good design.

The requirements of testing must be considered at the outset and a satisfactory and sufficient measure of *testability* built into the architecture. So important is testability that many designers are prepared to dedicate 30% or more of chip area for this purpose alone.

10.13.1 System partitioning

The problems of testing, particularly at the prototype stage, are greatly eased if the system is sensibly partitioned into subsystems, each of which is as self-contained and independent as possible. To take the example of the four-bit data path chip used earlier in the text, the partitions used — namely, the register array, the adder, and the shifter — are functionally independent to a large extent and have relatively simple interconnections.

At the prototyping stage it is possible to provide special test points (by providing extra pads for probing) at the interface between the subsystems. It is also possible, in a prototype, to provide double pad and fusible link connections in key paths between subsystems. This allows these connections to be open circuited if necessary so that one system can be divorced from another as a last resort in prototype testing.

For production items, also, it helps greatly if subsystems can be checked out individually by providing the appropriate additional inlet/outlet pads for test purposes. The test requirements for exhaustive testing of large digital systems are quite prohibitive if the system is tested as a whole. Take, for example, a finite state machine realized as a mixture of combinational logic and memory elements. Let us assume n possible inputs to the combinational logic and m memory elements, and that m memory elements outputs are fed back as inputs to the combinational logic.

In this case, to fully exercise the system for every possible combination of inputs and internal states would involve the generation of 2^{m+n} test vectors. If, say, $n = 24$ and $m = 20$, the resultant number of test vectors for exhaustive testing is 2^{44} and, even if these are generated at a rate of 10^6 vectors/sec, then testing will take six months at 24 hours per day.

On the other hand, if the system is partitioned for testing, exhaustive testing can be reduced to $2^n + 2^m$ vectors, a much more reasonable proposition (and for the figures given above would result in a test time of less than 20 seconds).

10.13.2 Layout and testability

Although it is impossible to generalize on this topic, common sense and a thoughtful approach to system layout may well considerably ease the problems associated with testing. For example, the inclusion of key point test pads or pads to energize special test modes are possible when the design is evolving.

The designer should also be aware of practical factors which will reduce the likelihood of short and open circuits. In particular, for MOS circuits, it has been shown that short circuits and open circuits in the metal layer and short circuits in the diffusion layer were the dominant faults experienced. Careful observance of design rules and ground rules should help to reduce the incidence of such faults.

10.13.3 Reset/initialization

One simple but very effective aid to testing and testability is to design a reset facility into all digital systems of any complexity. This has the considerable advantage of setting all internal states to known values, and testing may then at least proceed from known conditions.

The simple expedient is quite often overlooked or omitted.

10.13.4 Design for testability*

There are two key concepts underlying all considerations for testability. They are:

1. controllability
2. observability.

Quite simply, these concepts ensure that the designer considers the provision of means of setting or resetting key nodes in the system and of observing the response at key points.

The effects of testability or lack of it are such that it has been predicted that testability will soon become the main design criterion for VLSI circuits. The alternative is to save area by ignoring testability, but the penalties are such that

---

* The authors acknowledge contributions embodied in this section by Dr A. Osseiran of the Swiss Institute of Technology.

even for modest complexity (e.g. 10,000 gates per chip) the test costs could rise by a factor of five to ten, compared with the same system designed for testability. Given that test is already a significant component of LSI chip costs, the effects will be quite dramatic and could well cause the test costs to exceed all other production costs by a significant factor.

Design for testability (observability and controllability) is then reduced to a set of design rules or guidelines which, if obeyed, will facilitate test.

A failure during testing at the chip level may be due to a design defect or a poorly controlled fabrication process.

The inputs of the device under test (DUT) are subjected to a test pattern (or test vector) which supplies a set of binary values, in combination and/or in sequence, to detect faults. The specification of the test vector sequences must involve the designer, while the generation and application of test patterns to a DUT are the problems faced by the test engineer. Test pattern generation is assisted by using automatic test pattern generators (ATPG), but they are complicated to use properly and ATPG costs tend to rise rapidly with circuit size.

Once the application of a test pattern has revealed a fault, the process of diagnosis must be invoked to localize the fault.

10.13.4.1 Test coverage

Detecting all the possible faults in a DUT corresponds to 100% 'test coverage'. In general it is relatively easy to detect the first 80% of faults using various classical test strategies, but when more than an 80% coverage is required, appropriate test strategies must be developed. In any case, it is not generally possible to anticipate 100% of all faults, so that we tend to talk about a set of fault hypotheses which may then be covered 100%.

Faults may be classified using different models, and three such levels of definition are:

- mathematical model
- logical model (stuck-at)
- physical model.

The latter two are most commonly used. The 'stuck-at' model has been widely used and was originaly developed in the testing of p.c. boards but is not in itself sufficient to test actual VLSI CMOS circuits.

A further set of physical fault models is also used:

- Class 0: A single physical defect such as a faulty contact or via, a transistor stuck on or stuck off, an interconnection through any layer open circuit.
- Class 1: Class 0 with a short circuit between metal lines or diffusion lines.
- Class 2: Class 1 with short circuit(s) between two lines on any layer.

10.13.4.2 Nature of failures in CMOS devices

Failures may occur after a CMOS circuit has been fabricated and has successfully passed initial testing. Such failures may be due to poor design, weaknesses introduced during fabrication, ageing, or corrosion (of metallization) mechanisms which, again, may be accelerated owing to poor quality control in fabrication. The MOSFETs used in CMOS circuitry are susceptible to performance defects if there is a change either in the specified threshold voltage or in the transconductance.

Design faults are mostly due to deviations from the design rules specified for the fabrication process and this type of fault is difficult to detect since the manifestation of the fault often occurs later in the life of the device. Such faults mainly take the form of open circuits in conductors or short circuits between conductors. Crosstalk is also a cause of faults which are generally transient and are again due to poor design.

Other failures may be due to oxide breakdown, usually the thin oxide between gate and channel regions. This form of breakdown is often related to the inadequate protection against electrostatic discharge (ESD) and may be traced back to defects in or poor design of input/output pad circuitry.

Another problem is caused by hot carrier injection which causes both threshold voltage shift and transconductance degradation due to charge accumulation in the gate oxide.

10.13.5 Testing combinational logic

The solution to the problem of testing combinational logic is to generate a set of test patterns which will detect all possible fault conditions.

The first approach to testing an N input circuit is to generate all the possible 2^N input signal combinations by means of, say, an N-bit counter (controllability) and observe the output(s) for checking (observability). This is called exhaustive testing and is very effective, but is only practicable where N is relatively small. The reason for this becomes obvious when exhaustive test times are evaluated, taking, say, a relatively high (10 MHz) clock speed to the test pattern generating counter, results for several values of N being as follows:

N inputs	=	*2^N combinations*	=	*$2^N \times 0.1$ µsec.*	=	*total test time*
32 inputs		2^{32}		$2^{32} \times 10^{-7}$sec		$\geq$ 7 minutes
40 inputs		2^{40}		$2^{40} \times 10^{-7}$sec		$\geq$ 30 hours
64 inputs		2^{64}		$2^{64} \times 10^{-7}$sec		$\geq$ 574 *centuries*

10.13.5.1 Sensitized path-based testing

Many of the patterns generated during exhaustive testing may not occur during the application of the circuit. Thus, it is productive to first enumerate the possible faults and then generate a set of appropriate test vectors.

The basic idea is to select a path from the site of the possible fault, through a sequence of gates leading to an output of the logic circuitry under test. The process comprises three steps:

1. *Manifestation*: Gate inputs at the site of an assumed fault, say a 'stuck at' (SA) fault, are specified to generate the opposite value to the assumed SA value (0 for SA1, 1 for SA0).
2. *Propagation*: Inputs of other gates are determined so as to propagate the fault signal along the selected path to the primary output of the circuit. This is done by setting *And/Nand* inputs to '1' and *Or/Nor* inputs to '0'.
3. *Consistency* (or justification): This final step finds the primary input patterns to realize all the necessary values. This is done by tracing backward from the gate inputs to the primary input of the logic.

Examples will help explain the process.

Example 1 : Take an SA1 fault on line 1(L1) in Figure 10–26, then

Manifestation: L1 = 0, then A = 0. In a fault-free situation, output F changes with A if B, C and D are fixed. If L1 is SA1 then F = 0 even if A = 0.

Propagation: For propagation through the *And* gate, line 5(L5) = line 8(L8) = 1 and, since we are propagating the condition L1 = 0, then L10 = 0. So that the propagated fault manifestation can reach the output through the *Nor* gate, then L11 = 0. Output F is thus read and compared with the fault-free value F=1.

Consistency: For the *And* gate, L5 = 1 and thus L2 = input B = 1; also L8 = L7 = 1. So far, we have determined the values of inputs A(= 0) and B(= 1) and also that L7 =1. For this latter contention to be true, we may see that B + C + D = L7 = 1. If L7 is =1 then L9 = 1 and L11 will be 0, which is consistent with the

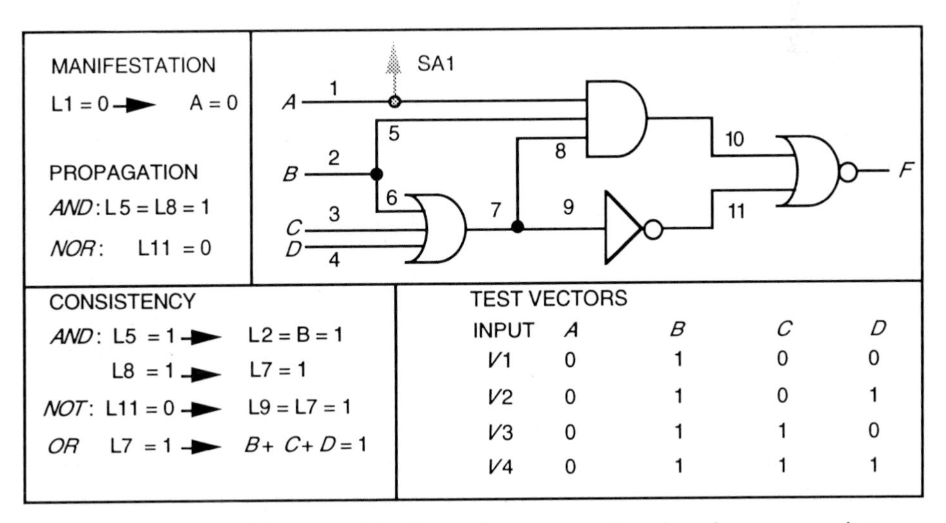

Figure 10–26 Combinational logic testing — sensitized path — example1

Nor gate propagation requirements. A set of test vectors, *V*1 to *V*4, may now be specified as shown in the figure.

These tests will therefore reveal the SA1 fault. However, there are some faults which will be inherently undetectable with the sensitized path approach as illustrated in the following:

Example 2: Take the same circuit as in Figure 10–26, but with a stuck at 1(SA1) fault on line L8 as in Figure 10–27.

Manifestation: L8 = 0.

Propagation: For propagation through the *And* gate, L5 = L1 = 1 and, since we are propagating the condition L8 = 0, then L10 = 0. For the *Nor* gate L11= 0, which means that L9 = L7 = 1.

Consistency: For the *And* gate, L8 = 0 and thus L7 = 0. Clearly this is inconsistent since L7 cannot be set to 1 and 0 at the same time. This conflict cannot be resolved and the fault is undetectable.

10.13.5.2 The D-algorithm [Roth]

Given a circuit comprising combinational logic, the algorithm aims to find an assignment of input values that will allow detection of a particular internal fault by examining the output conditions. In order to do this the algorithm is based on the hypothesis of the existence of two machines — a good machine and a faulty machine. The existence of a fault in the faulty machine will cause a discrepancy between its behavior and that of the good machine for some particular values of inputs. The D-algorithm provides a systematic means of assigning input values for that particular design so that the discrepancy is driven to an output where it may be observed and thus detected. The algorithm is extremely time-intensive

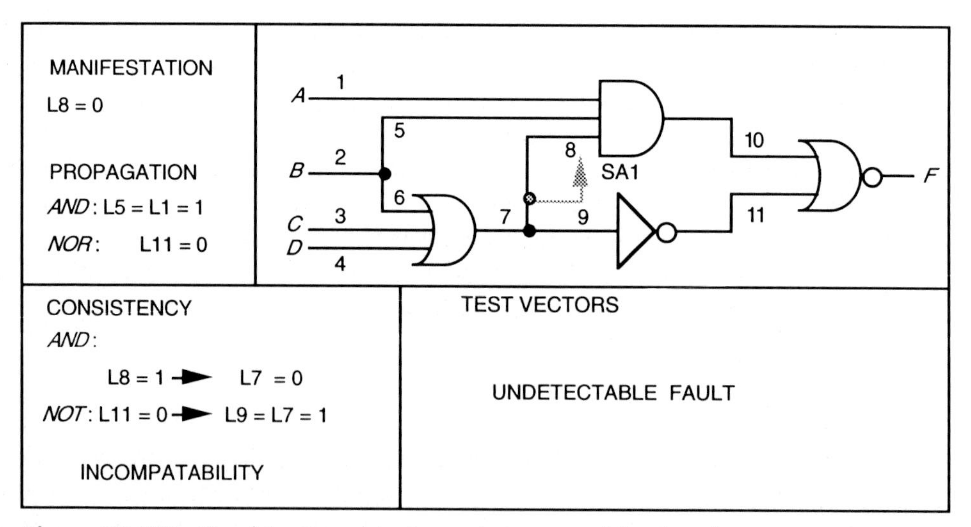

Figure 10–27 Combinational logic testing — sensitized path — example 2

and computing intensive for large circuits and has been the subject of several adaptions, modifications and improvements. LASAR (Logic Automated Stimulus And Response), PODEM (Path Oriented DEcision Making) and FAN (FAN-out oriented test generation) are all improvements on the D-algorithm. Further reading is cited in the reference list which follows this chapter.

10.13.6 Testing sequential logic

Sequential circuits, which may be generally represented as finite state machines, may be modeled as combinational logic with a set of delays and feedback from output to input as shown in Figure 10–28.

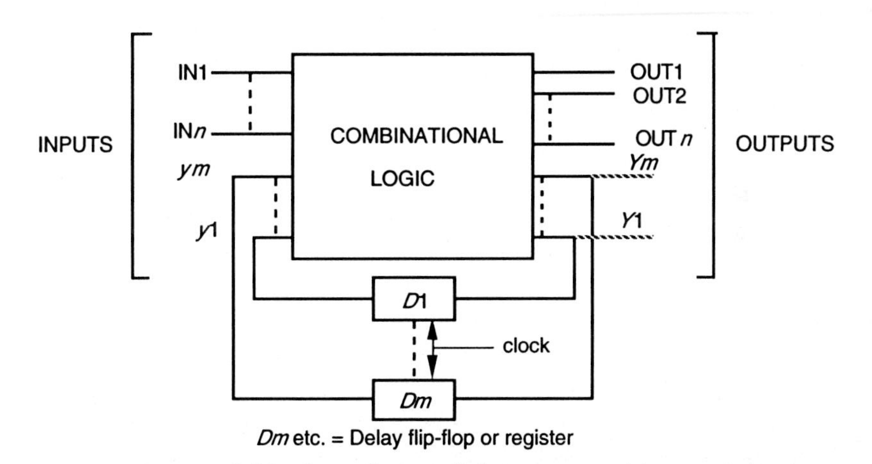

Figure 10–28 Sequential logic testing — finite state machine model

The 'm' feedback variables constitute the state vector and determine the maximum number of finite states which may be assumed by the circuit. In the most general case, the next state and the output are both functions of the present state and the independent inputs. The delay elements are generally assumed to be associated with the feedback path and, for clocked systems, the basic delay elements are flip-flops, although, in asynchronous circuits in particular, the delays may be contributed by circuit propagation delays.

The test generation for a sequential circuit is a very complicated task since the test signals must not only be logically correct but must also occur at the correct time relative to other signals.

10.13.6.1 The effect of memory

All sequential circuits exhibit a memory property since, in deciding what to do next, they take into account (or remember) previous conditions. In testing then, it is not only the test pattern but also the order or sequence in which it is

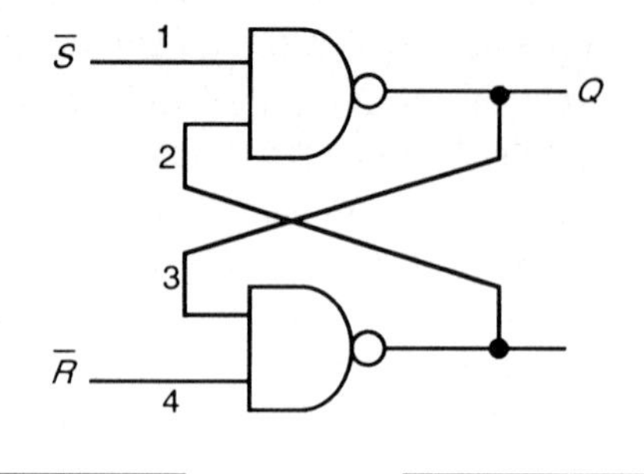

INPUTS		OUTPUT Q				
		GM	Input SA 1			
$\overline{S}$	$\overline{R}$		1	2	3	4
0	0	1	0	1	1	1
0	1	1	0	1	1	1
1	0	0	0	0	0	1
1	1	0	0	0	1	1

INPUTS		OUTPUT Q				
		GM	Input SA1			
$\overline{S}$	$\overline{R}$		1	2	3	4
1	1	?	?	0	1	?
1	0	0	0	0	0	?
0	1	1	0	1	1	1
0	0	1	0	1	1	1

Figure 10–29 Sequential logic testing — effects of memory

applied is significant. Take, for example, a very basic sequential circuit as in Figure 10–29.

The feedback paths are quite obvious but the delay in the feedback path is apparently non-existent, this being a case where the circuit propagation delays contribute the necessary delay elements.

To explain the tabulations under the figure, the input test pattern is applied as shown under the $\overline{S}\ \overline{R}$ heading and working in sequence from top to bottom row. The remaining columns tabulate the state of output Q first of all for a 'good machine' (GM) and then for the 'faulty machine' (FM) for a SA1 fault on each of the four input lines (1, 2, 3, 4).

Note that in the first table with the SA1 fault on line 2 the machine matches the good response and so this particular test sequence will not detect a SA1 fault on line 2.

In the second table, the vectors are applied in exactly the reverse order. In this case '?'s appear owing to the memory property of the circuit, each '?' indicating that Q will retain whatever value it had prior to the application of the test vector for that row. Again, the SA1 fault on line 2 may not be detected if the latches are reset (i.e. $Q = 0$) prior to applying the test sequence.

10.13.6.2 The iterative test generation method

An obvious way of approaching the testing of sequential logic is to 'convert' the logic into combinational logic by cutting the feedback lines, thus creating pseudo inputs and outputs as well as the original primary input and output lines. For an N-state machine, this arrangement is then replicated N times so that an N-state sequential machine is converted into an N-time frame combinational machine.

The main problem of this technique is that a simple fault in the sequential machine is manifest as N multiple faults during test. This is time-consuming for circuits of any complexity. It is also necessary to describe all the initial states of the circuit, which is also time-consuming. For these reasons the iterative test generation (ITG) methods are best suited to logic with few feedback loops as in control logic for example.

10.13.7 Practical design for test (DFT) guidelines

Practical guidelines for testability should aim to facilitate test processes in three main ways:

- facilitate test generation;
- facilitate test application; and
- avoid timing problems.

The discussions in this section address these matters.

10.13.7.1 Improve controllability and observability

Design for test methods must ensure that a design is well enough covered to provide for complete and efficient testing. When a node is difficult to access from primary input or output pads, then a very effective method is to add additional, internal pads to access the desired point. These additional pads may be accessed using a prober.

If the node is a link between blocks of a circuit, as in Figure 10–30, then the most immediately obvious attributes required are to be able to observe the output of block 1 and also to provide for the control of block 2. Some additional circuitry will be required and a possible configuration is set out in the figure. If the $Normal/\overline{Test}$ line is set to 1(*Normal*) then transmission gates T_2 and T_3 are open and T_1 is closed. Normal transmission between the blocks can take place through T_2 but a *control* input to block 2 can also be applied through T_3. When the $Normal/\overline{Test}$ line is set to 0 (*Test*) then transmission gates T_2 and T_3 are closed, there will be no transmission between the blocks, and the output (*observe*) of block 1 can be monitored through T_1 which is now open.

This solution requires three pads and eight transistors in a CMOS environment. This technique must be complemented by other appropriate testing techniques which will depend on the internal structures of blocks 1 and 2.

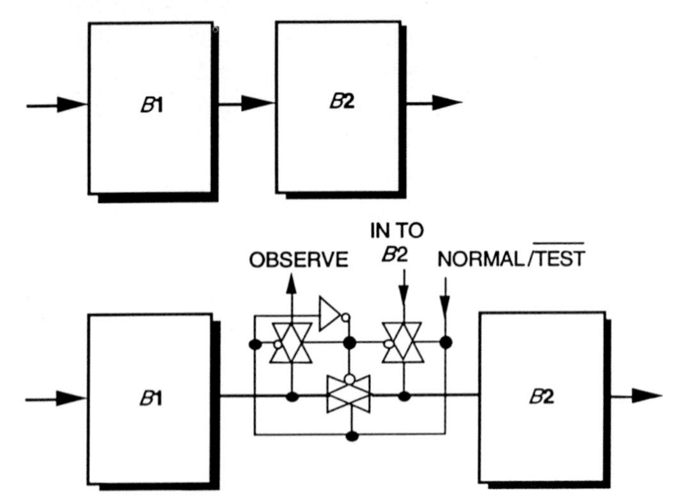

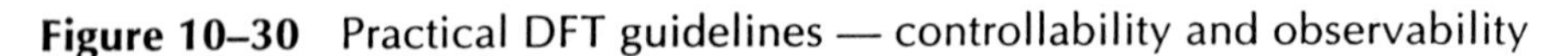

Figure 10–30 Practical DFT guidelines — controllability and observability

10.13.7.2 The use of inter-block multiplexers

Some general attributes are illustrated in Figure 10–31. This arrangement allows the bypassing of blocks. The addition of demultiplexers also improves observability. The major penalties incurred here are the numerous extra devices and the added propagation delays through the multiplexers.

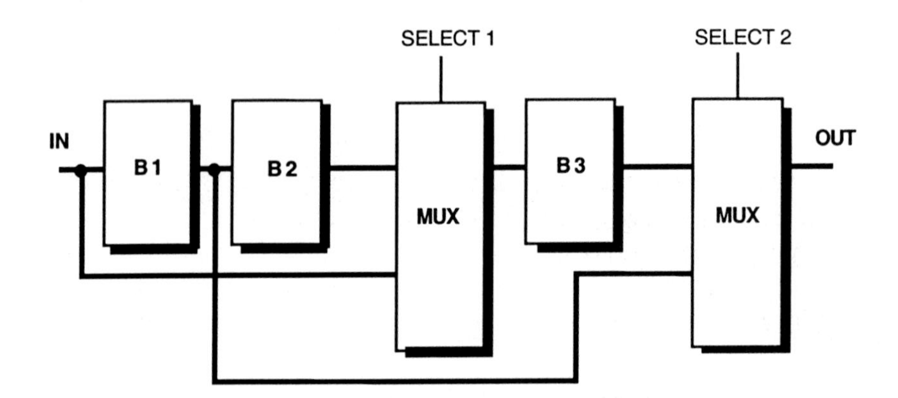

Figure 10–31 Use of multiplexers — increasing internal access

10.13.7.3 The partitioning of large circuits

Partitioning large circuits into smaller subcircuits is an effective way of reducing test generation complexity and test time. It has been shown that test generation effort for an n gate general purpose logic circuit is proportional to somewhere between n^2 and n^3 (Bennetts, 1984). If the circuit is partitioned then the effort is

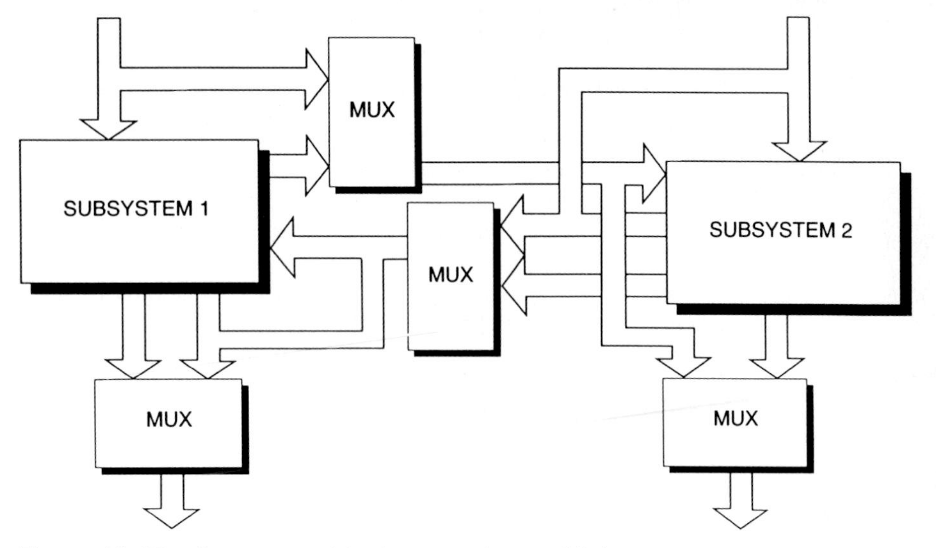

Figure 10–32 System partitioning — using multiplexers

reduced correspondingly. For example, exhaustive testing of the SN7480 adder circuit for SA1 and SA0 faults requires 2^9 (= 512) tests. If the adder is partitioned into four subcircuits, then the number of tests is reduced to 24. Clearly, partitioning should be done on a logical basis into recognizable and sensible subfunctions and can be achieved physically by incorporating clock line isolation and control facilities, reset and power supply lines. Isolation and control are readily achieved through the use of multiplexers as suggested in Figure 10–32.

10.13.7.4 Dividing long counter chains

Counters are sequential and need a large number of input vectors to be fully tested. Partitioning into sub-counters can be very effective in reducing test complexity. For example, the full testing of a 16-bit counter requires the application of 2^{16} = 65,536 clock pulses. Division of the counter into two 8-bit counters, as Figure 10–33, reduces this number to 2×2^8 = 512 clock pulses.

10.13.7.5 Initialization of sequential logic

An important problem in sequential logic testing arises at power-up time where the first state will be quite random if there is no initialization. In this case it is impossible to start a test sequence correctly. The remedy is to design the circuit using elements which have a *preset* and/or *clear* facility (e.g. JK flip-flop elements with *Pr.* and *Clr.* inputs). From a practical viewpoint, this could be very space-consuming and it may often be sufficient to initialize only the first stage. For example, if its first stage only is initialized, a serial in serial out counter will pick up a known state after a few clock pulses.

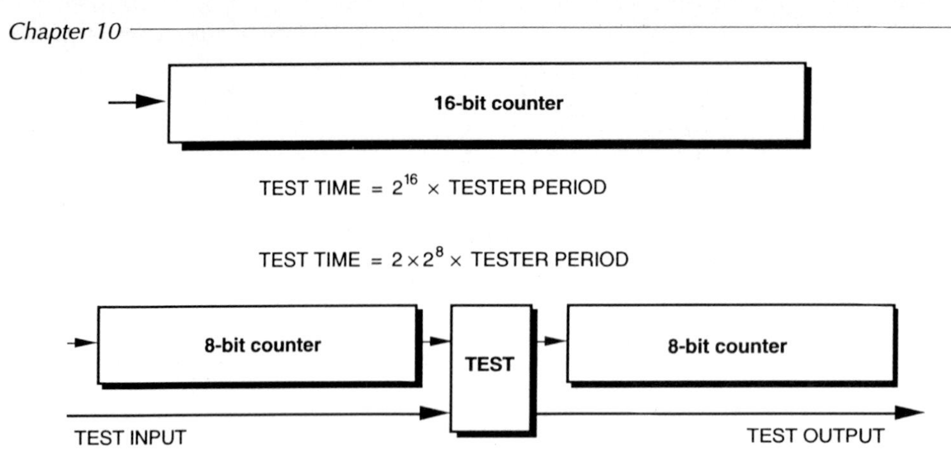

Figure 10–33 Dividing long counter chains

Sometimes it will be necessary for the tester to be able to override the normal initialization state of the logic and the addition of appropriate gating in the *reset/ initialize* control line will achieve this.

10.13.7.6 Asynchronous sequential logic.

Asynchronous logic is driven by self-timing state transitions in response to changes of the primary inputs. This makes it generally impossible to determine when the next state is actually established and in consequence there are large problems to be faced in the timing and also in the memory effects associated with such circuits. Although asynchronous logic is inherently faster than clocked logic it has several serious disadvantages from the test viewpoint as follows:

- testing is difficult;
- sensitivity to tester skew;
- non-deterministic behavior;
- prone to races and other hazards.

The design processes are more difficult than synchronous logic and must be approached with care, taking due account of critical race and other hazard-generating conditions.

10.13.7.7 Avoiding logical redundancy

Logical redundancy may be present by design; for example, in order to mask a static hazard condition, or unintentionally as a design bug. In both cases it is not possible to make a primary output value dependent on the value of the redundant node. Thus, there are certain fault conditions associated with the node which cannot be detected, Take, for instance, the two sets of conditions outlined in Figure 10–34.

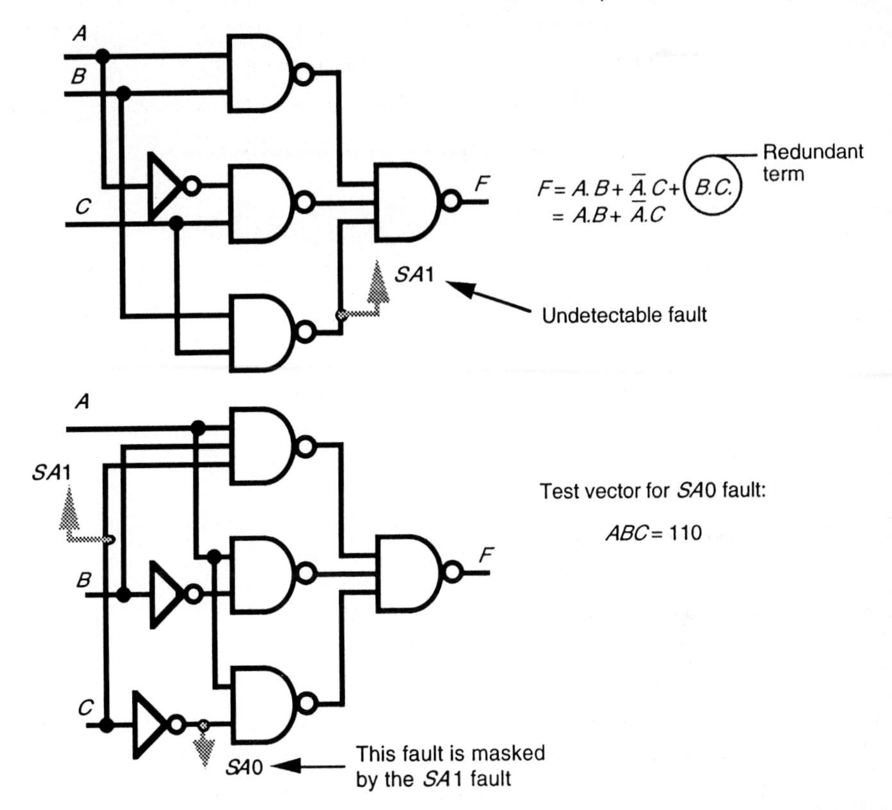

Figure 10–34 Fault masking due to logical redundancy

10.13.7.8 Avoiding delay dependant logic

An example of a delay-dependent circuit is given in Figure 10–35. It will be seen that the presence of a pulse at the *And* gate output depends not on the logical performance of the three inverters but rather on their temporal performance. Automatic test pattern generators (ATPGs) work in the logic domain and view delay-dependent logic as redundant combinational logic. In the case illustrated in Figure 10–35, the ATPG will see the *Anding* of a signal with its complement and will therefore always compute a '0' as the output of the *And* gate — rather than a pulse.

10.13.7.9 Avoiding gating or asynchronous delays in the clock line

When a clock signal is gated with another signal, such as a *load* signal coming from a tester, then any skew (or other hazard) on that signal can cause an erroneous output from the associated logic. This is illustrated in Figure 10–36.

Further, another timing situation to avoid is that illustrated in Figure 10–37

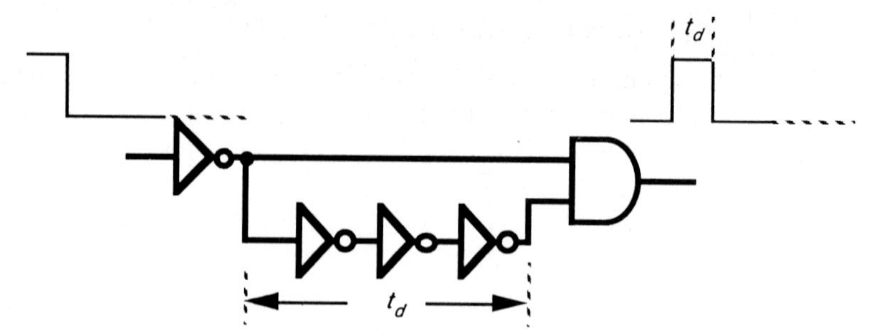

Figure 10–35 Delay-dependent logic

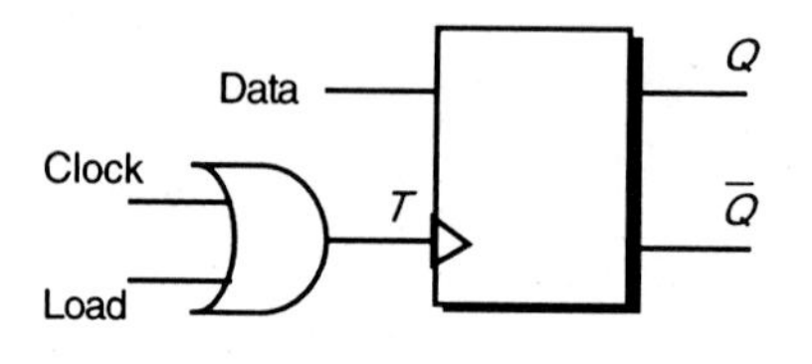

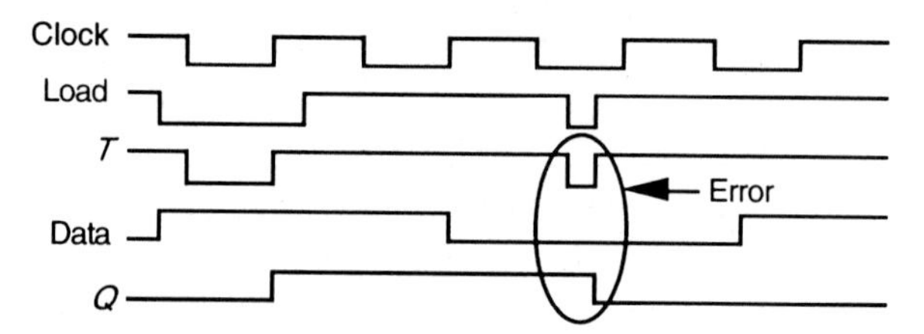

Figure 10–36 Clock line gating hazard

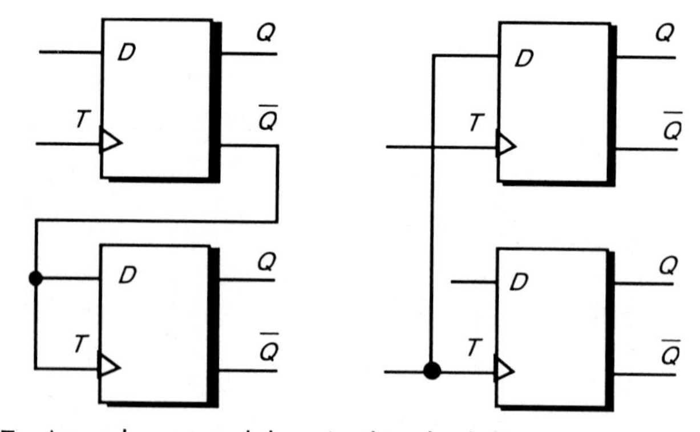

Figure 10–37 Asynchronous delays in the clock line

where the tester could not be synchronized if one or more clock is dependent on asynchronous delays (from the *D*-input to *Q*-output of the flip-flop in the figure), or when a signal is used both as data and as a clock.

10.13.7.10 Avoiding self-resetting logic

The problems here are akin to those in asynchronous logic, since the reset input is independent of the system clock. This can result in an erroneous value being read by the tester. The situation is indicated in Figure 10–38(a).

One solution to this problem is to allow the tester to override by adding an *Or* gate as indicated in Figure 10–38(b). This allows the tester to receive the right response at the right time.

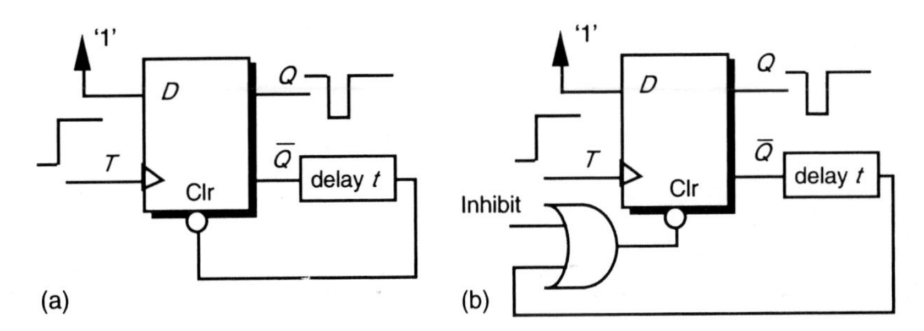

Figure 10–38 Problems associated with self-resetting logic

10.13.7.11 The use of bused structures

This approach is related to the partitioning technique and is very widely used for microprocessor-like circuits as illustrated in Figure 10–39.

Using this arrangement allows the tester access to all the main subsystems and other modules which the buses interconnect. The tester can then effectively disconnect any unit or module from the bus by putting its output into the high impedance state. Test patterns can then be applied to each separately.

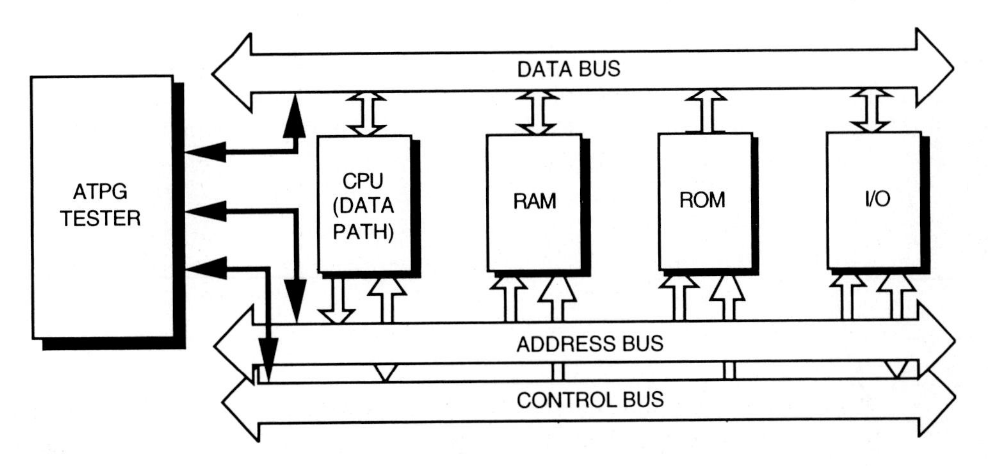

Figure 10–39 Use of bused structures

10.13.7.12 *Separation of analog and digital circuits*

The testing of analog circuits requires a completely different strategy from digital circuits and is therefore incompatible.Furthermore, the fast rise- and fall-times of digital signals can give rise to cross-talk problems in analog signal lines if they are in close proximity. Where it is essential to route digital signals near analog lines, then consideration must be given to balancing and shielding the digital signals.

In the case of analog-digital converters, it is better to bring out the analog signals for observation before conversion. For digital-analog conversion the digital signals may also be brought out for observation prior to the converter as outlined in Figure 10–40.

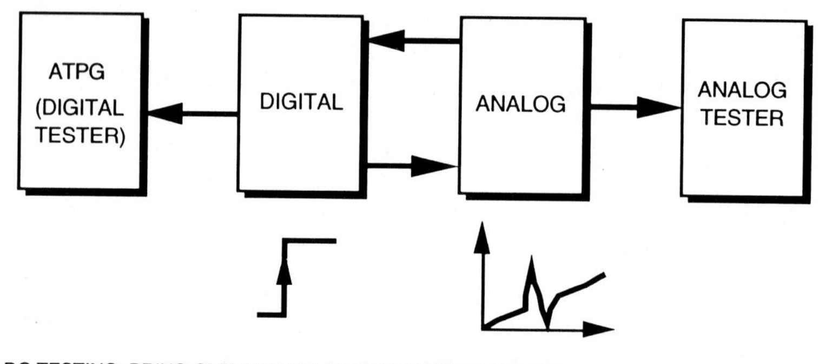

ADC TESTING: BRING OUT ANALOG INPUTS FOR TEST OBSERVE DIGITAL OUTPUT

DAC TESTING: BRING OUT DIGITAL INPUTS FOR TEST OBSERVE ANALOG OUTPUT

Figure 10–40 Separation of analog and digital signals

10.13.7.13 *Bypassing techniques*

Bypassing a subsystem consists of providing the facilities for propagating its inputs directly through to its outputs. The aim is to bypass the sub-system in order to directly access another subsystem to be tested, and, as with partitioning, wide use is made of multiplexers to achieve the bypassing.

Bypassing techniques work well with the following: counters, dividers, RAM, ROM, PLAs, sequential blocks, analog circuits and internal clocks. In the bypassing approach, the subsystems can be tested exhaustively by controlling the multiplexer-based interconnections in the system. To speed up the testing, some subsystems may be tested simultaneously if the propagation paths are associated with other disjoint or separate subsystems.

10.13.7.14 *Some observations on DFT*

The preceding sections have not presented an exhaustive list of DFT techniques but have been intended to present a set of rules which should be respected in design. Some of the guideline goals are to simplify test vector generation and application, and others are intended to avoid timing problems in the design. The references given at the end of this chapter and other related material give variants of the design guidelines for PCBs and for ICs.

10.13.8 Scan design techniques

The testability guidelines so far presented provide ad hoc methods for dealing with random logic designs. The scan design techniques which are now to be discussed are structured approaches to designing sequential circuits so that testability is 'designed in' from the outset.

The major difficulty in sequential circuit testing is in determining the internal state of the circuit. Scan design techniques are directed at improving the controllability and observability of the internal states. The approach aims to reduce the problem of testing a sequential circuit to that of testing combinational logic.

10.13.8.1 The scan path

A sequential circuit comprises combinational logic and storage elements — usually in the feedback path — as illustrated in Figure 10–41. Scan path design techniques configure the logic so that the inputs and outputs of the combinational part can be accessed and the storage elements reconfigured to form a shift register known as the scan path. Thus the internal states of the circuit can be observed and controlled by shifting (scanning) out the contents of the storage elements

The storage elements are usually 'D', JK' or 'RS' flip-flop elements with the classical structure being modified by the addition of a two-way multiplexer on the data input(s). The multiplexer is controlled by an external *mode* signal and allows the scan path reconfiguration to be effected. In Figures 10–41 and 10–42 a basic 'D' flip-flop has been shown with the added input multiplexer. This configuration is commonly known as an 'MD' (multiplexed 'D') flip-flop.
The sequential circuit containing the scan path has two modes of operation — a *normal* and a *test* mode. The configuration associated with each basic mode is set out in Figures 10–42(a) and (b) — *normal* and *test* mode respectively.

A large sequential circuit is generally partitioned into a number of subcircuits each with a combinational section and one associated scan path.. The efficiency of the test pattern generation for the overall combinational circuit is greatly improved by partitioning since its depth is reduced.

Before applying test patterns, the scan path shift register is verified by shifting in all ones then all zeroes.

A general method for testing with the scan path approach is as follows:

1. Set the mode to *test* so that the scan path is configured.
2. Verify the scan path by shifting test data in and out.
3. Set the shift register to a known initial state.

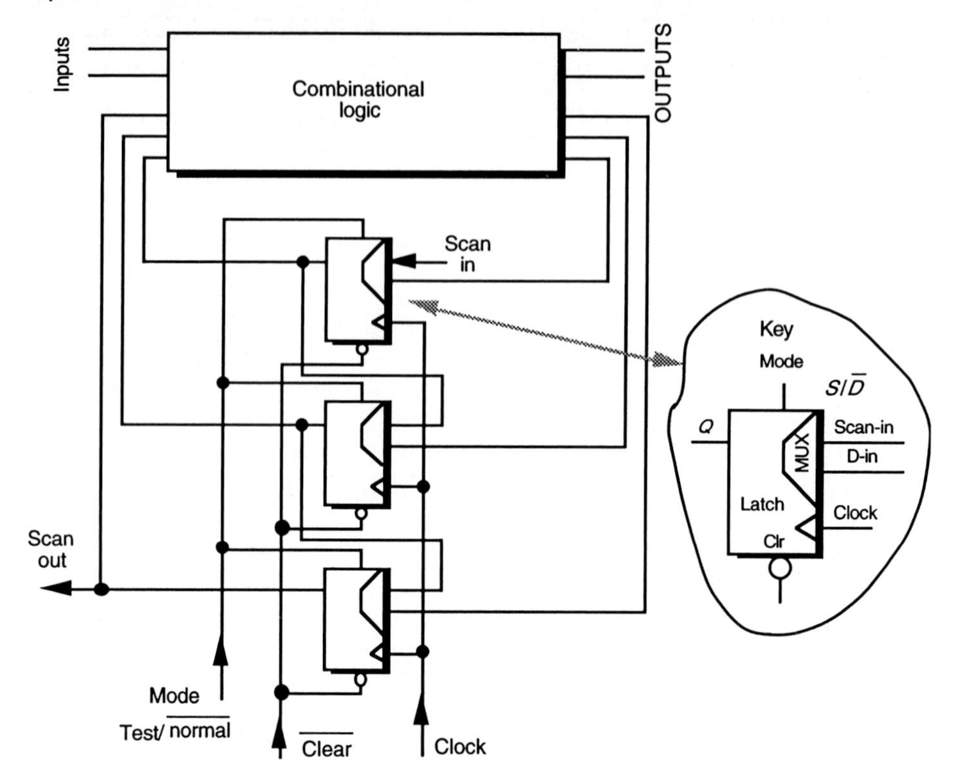

Figure 10–41 Sequential circuit configured for scan path testing

4. Apply a test pattern to the primary inputs of the overall circuit.
5. Set the mode to *normal.* The circuit then settles and the primary outputs are monitored.
6. Activate the circuit with one clock pulse.
7. Return to the *test* mode.
8. Scan out the contents of the scan path registers and simultaneously scan in the next pattern.
9. Repeat from step (4) etc.

10.13.8.2 Level-sensitive scan design (LSSD)

This is a technique, initially developed by IBM, which incorporates two aspects — level sensitivity and a scan path approach (Williams, 1986). The general arrrangement is indicated in Figure 10–43.

The *level-sensitive* aspect means that the sequential network is designed so that when an input change occurs, the response is independent of the component and wiring delays within the network.

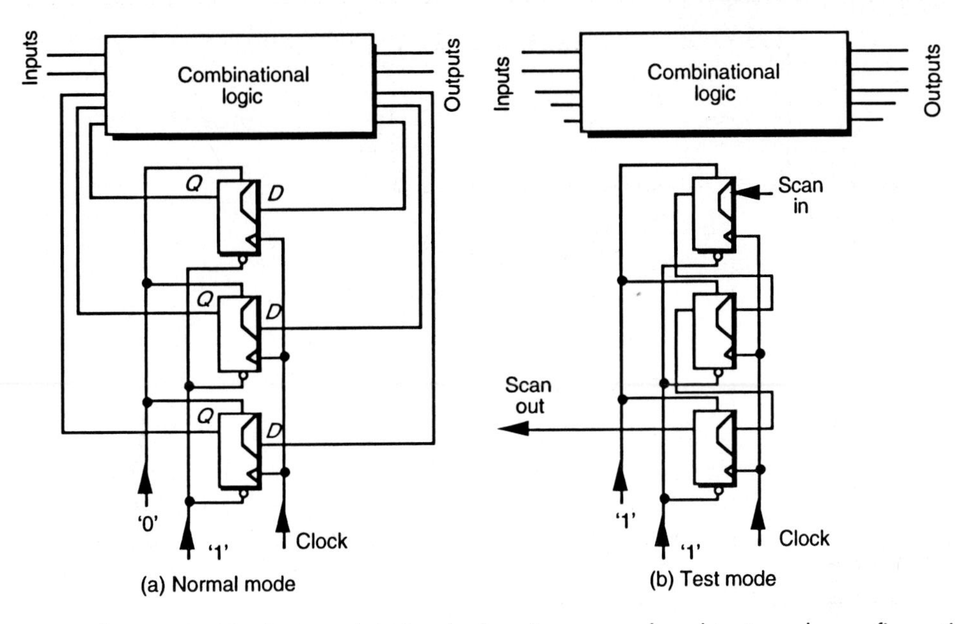

Figure 10–42 Sequential circuit showing normal and test mode configurations

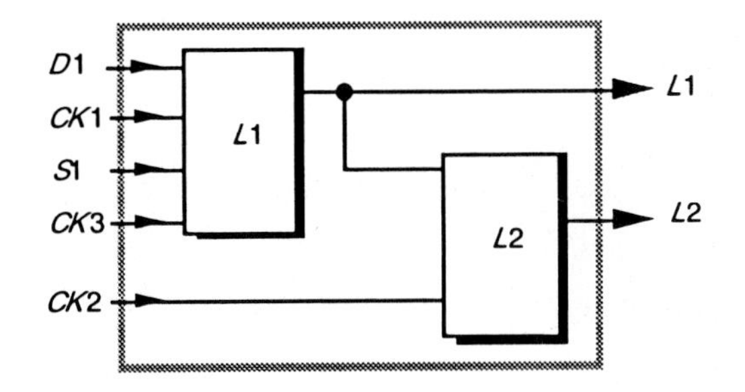

Figure 10–43 Level-sensitive scan design LSSD configuration

The *scan path* aspect is due to the use of shift register latches (SRL) employed as storage elements. In the *test* mode they are connected as a long serial shift register. Each SRL has a specific design similar to a master-slave flip-flop. It is driven by two non-overlapping clocks which can be controlled readily from the primary inputs to the circuit. Input DI is the normal data input to the SRL, clocks CK1 and CK2 control the normal operation of the SRL while clocks CK3 and CK2 control scan path movements through the SRL. The SRL output is derived at L2 in both modes of operation, the mode depending on which clocks are activated. The following advantages are claimed for the LSSD approach:

- The circuit operation is independent of the dynamic characteristics of the logic elements — rise- and fall-times and propagation delays.

- ATP generation is simplified since tests need only be generated for a combinational circuit.
- LSSD methods, when adopted in design, eliminate hazards and races; greatly simplifies test generation and fault simulation.

10.13.8.3 Boundary scan test (BST)

This is a technique involving scan path and self-testing to resolve the problems associated with the testing of boards carrying VLSI circuits and/or surface-mounted devices (SMD). Printed circuit boards (PCBs) are becoming very dense and complex, especially with SMD circuits, so that most test equipment cannot guarantee a good fault coverage.

BST consists of placing a scan path (shift register) cell adjacent to each component pin and to interconnect the cells so as to form a chain around the border of the circuit. The BST circuits contained on one board are then connected together to form a single path. The general idea is illustrated in Figure 10–44.

The boundary scan path is provided with serial input and output pads and appropriate clock pads which make it possible to:

- test the interconnections between the various chips on the board;
- deliver test data to the chips on the board for self-testing;
- test the chips themselves with internal self-test facilities.

BS techniques are grouped by the IEEE standards organization into a 'standard test access port and boundary scan architecture' (namely, *IEEE*, p. 1149.1–1990). The advantages of BST are seen as follows:

- no need for complex testers in PCB testing;
- the test engineer's work is simplified and efficient;
- the time spent on test pattern generation and application is reduced;
- fault coverage is increased.

10.13.8.4 Other scan design techniques

Many other stuctured approaches have evolved over the past few years, for example, *partial scan, scan/set* and *random access scan.*

The partial scan is derived from the scan path technique, but is less area-consuming. The scan path approach needs, on average, a 30% area increase for testing a whole sequential circuit. Using the partial scan approach, only faults not detected by the designer's functional vectors are selected. The test generator decides exactly which flip-flops should be scanned.

In the scan/set method, the storage elements within the circuit are not used to implement a scan path. Instead, a separate register is added whose sole function is to scan test data in and out of the circuit. This allows for the main circuit under

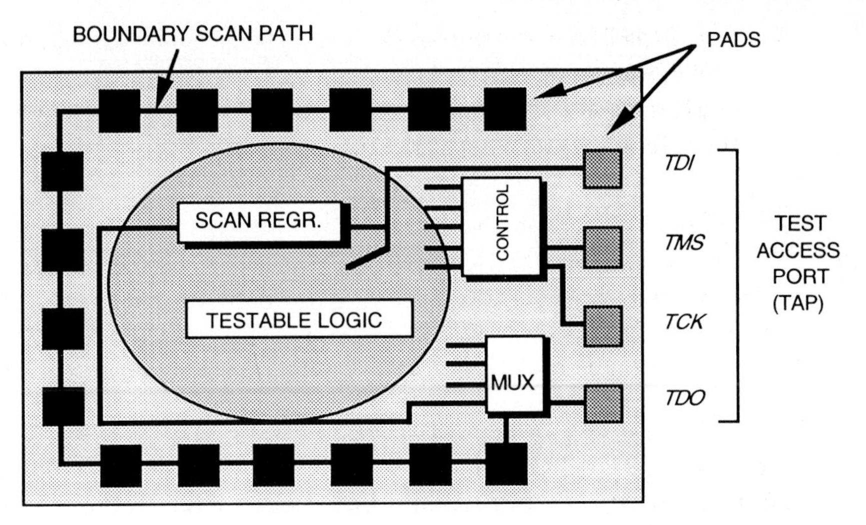

Figure 10–44 Boundary scan test (BST) configuration

test to be of any type — it is not restricted to combinational as before, and the storage elements are not restricted to particular types of latch or flip-flops. The major disadvantage of this technique is the high overhead cost in terms of additional input/output pins.

An overview of the other scan design techniques referred to here is presented in Table 10–8. However, many other scan design approaches exist but are mostly based on one or more of the methods discussed.

Table 10–8 Other scan design techniques

Partial scan	*Scan/set*
Scan: area increase	Separate shift register
Approach targets faults	No interruption to normal operation.
Selected flip-flops	No reduction to combinational test.

Random access scan
No shift register with flip-flops
Matrix of flip-flops addressed, controlled and observed
Disadvantage: The number of I/O pins

10.13.9 Built-in-self-test (BIST)

As the complexity of individual VLSI circuits and as overall system complexity increase, test generation and application becomes an expensive, and not always very effective, means of testing. Further, there are also very difficult problems associated with the high speeds at which many VLSI systems are designed to operate. Such problems require the use of very sophisticated, but not always affordable, test equipments.

Consequently, BIST objectives are:

1. to reduce test pattern generation costs;
2. to reduce the volume of test data;
3. to reduce test time.

BIST techniques aim to effectively integrate an automatic test system into the chip design.

10.13.9.1 Compact test: signature analysis

Data compression techniques are currently used in BIST systems and consist of making comparisons on compacted test responses instead of on the entire test data, which can be huge in some cases. The most important test task — is the circuit fault free? — is hence executed in an efficient manner.

The test compacting scheme currently used most is called *signature analysis*. This was developed by Hewlett Packard in the late 1970s'. Signature analysis performs polynomial division, that is to say division of the data out of the device under test (DUT). This data is represented as a polynomial $P(x)$ which is divided by a characteristic polynomial $C(x)$ to give the signature $R(x)$, so that

$$R(x) = P(x) / C(x)$$

This is summarized in Figure 10.45.

The signature from the DUT is compared with the expected signature to determine if the DUT is fault-free. The differences between the faulty signature and a good signature may also be used to indicate the nature of the fault. Signature analysis has been proved to be a reliable and attractive alternative to full uncompacted testing.

Another technique of data compression — *transition counting* — has been in use for some considerable time. This consists of counting transitions of a specified direction (0 to 1 or 1 to 0) and then comparing this count with the count obtained from the simulation model.

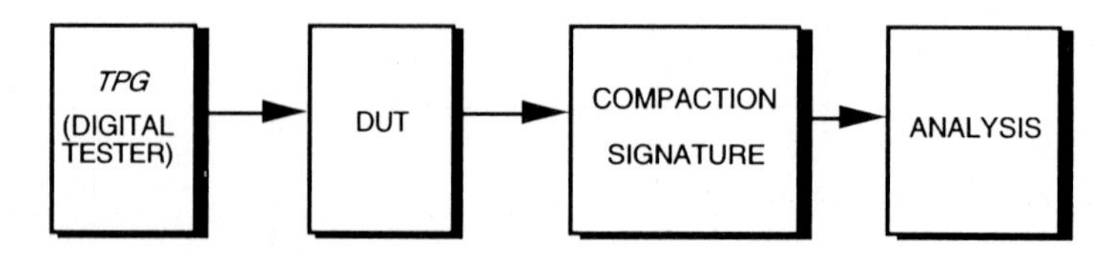

Figure 10–45 Built-in-self-test — signature analysis

10.13.9.2 Linear feedback shift register (LFSR)

The LFSR model is that of a finite state machine comprising storage elements and modulo two adders *(Xor gates)* connected in feedback loops as indicated in Figure 10–46.

LFSR techniques can be applied in a number of ways, including random number generation, polynomial division for signature analysis, and n-bit counting. LFSR can be either series or parallel, the differences being in the operating speed and in the area of silicon occupied — parallel LFSR being faster but larger than serial LFSR.

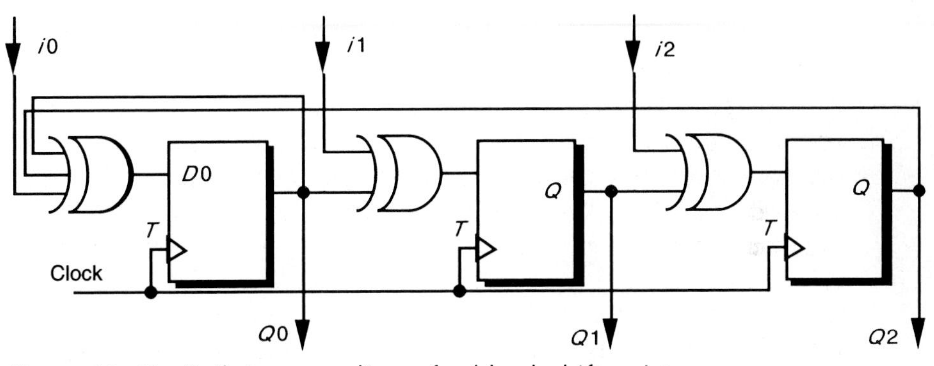

Figure 10–46 Built-in-test — linear feed-back shift register

10.13.9.3 Built-in logic block observer (BILBO)

BILBO is a built-in test generation scheme which uses signature analysis in conjunction with a scan path. It is aimed at integrated modular and bus-oriented systems, such as microprocessor and similar circuits.

The major component of a BILBO is an LFSR with a few gates. In Figure 10–47 the BILBO is controlled by two signals, B1 and B2 which define the modes.

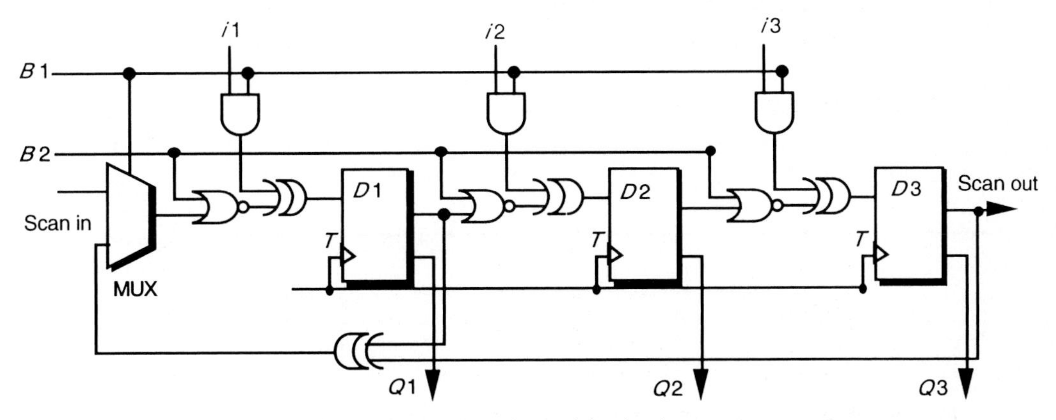

Figure 10–47 Built-in-self-test — Built-in logic block observer (BILBO)

In the *Normal* mode, B1 = B2 = 1 and the storage elements are used independently by the circuit as in Figure 10–48.

In the *Test 1* mode, B1 = B2 = 0 and the storage elements are configured as a scan path, all storage elements being connected as a serial shift register. This is shown in Figure 10–49. Test vectors are then applied to the scan-in input and responses shifted out at the scan path output. The analysis of data is then similar to that for a simple scan-path test.

In the *Test 2* mode, as in Figure 10–50, B1 = 1, B2 = 0 and the circuit is then configured in a LFSR mode and can be used either as a polynomial divider to compact data or as a random test pattern generator.

In the final mode, B1 = 0, B2 = 1 which *resets* the BILBO.

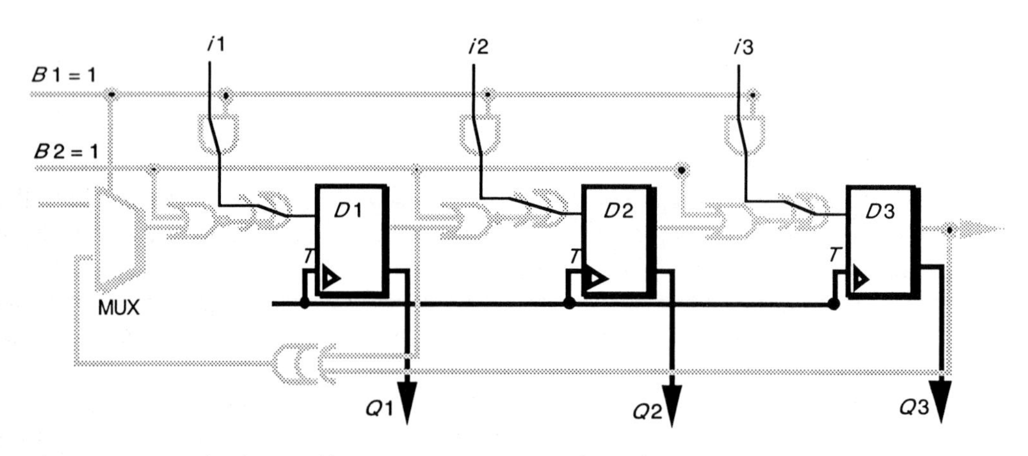

Figure 10–48 built-in-self-test — BILBO normal mode

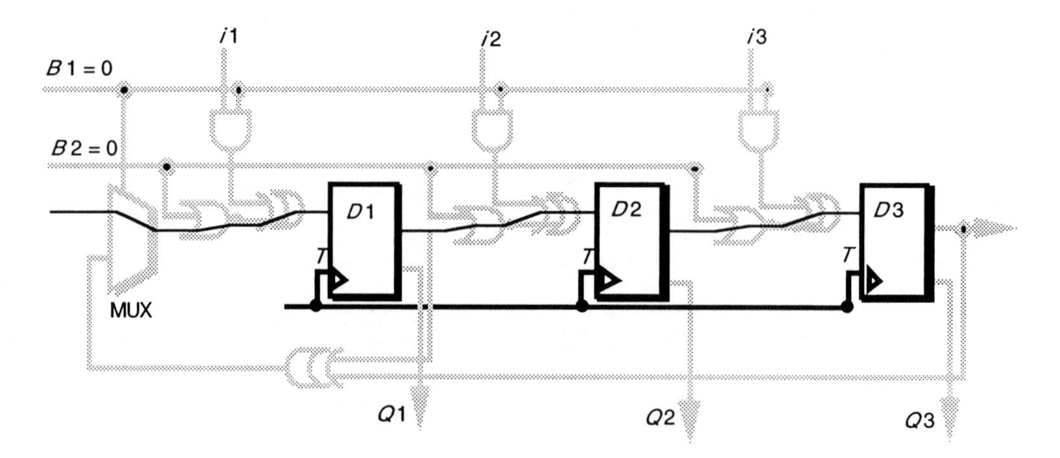

Figure 10–49 Built-in-self-test — BILBO scan path mode

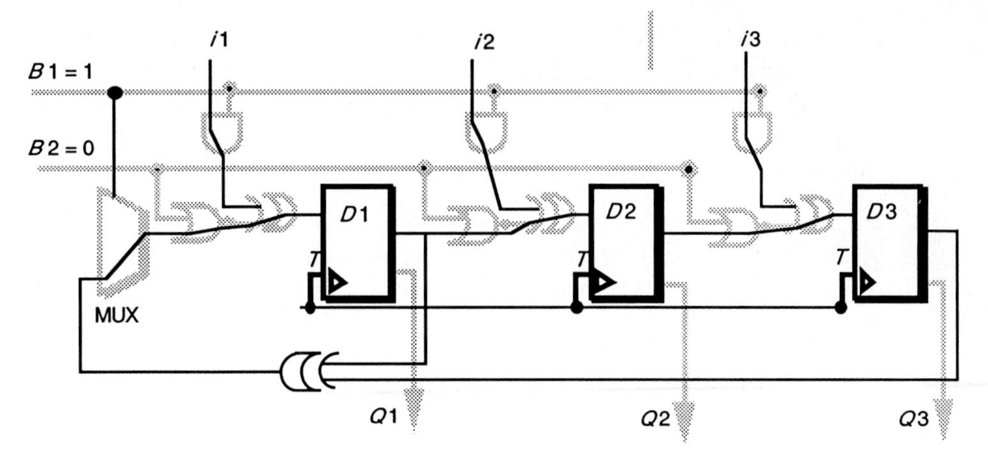

Figure 10–50 Built-in-self-test — BILBO: LFSR mode

10.13.9.4 Self-checking techniques

Data transmission in computer systems commonly makes use of coding to allow for the ready detection of errors. Such error detection techniques have been adapted and extended for built-in test purposes.

Self-checking techniques consist of supplying coded input data to the logic block under test and comparing the output in a checker designed to detect any errors. This is illustrated in Figure 10–51.

The design of logic blocks and checkers should then obey a set of rules in which the logic block is 'strongly fault secure' and the checker 'Strongly Code Disjoint'. A set of hypotheses is used in self-checking design to define the optimal design which allows a test coverage of 100% of these hypotheses.

The code used in data encoding depends on the type of errors that may occur at the logic block output. In general, three types are possible:

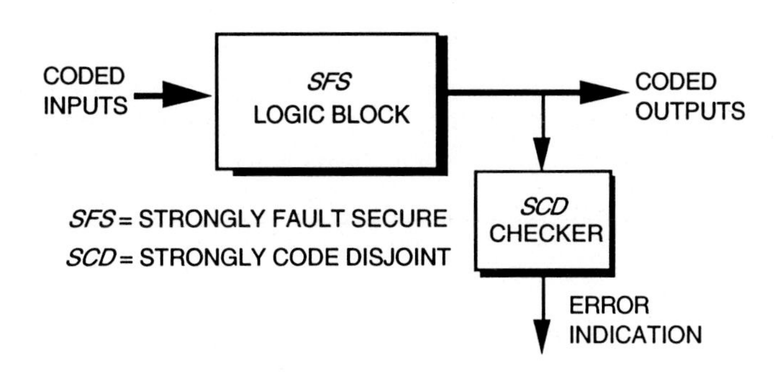

Figure 10–51 Self-checking logic: coding techniques

- Simple error: one bit only affected at a time.
- Unidirectional error: multiple bits at 1 instead of 0 (or 0 instead of 1).
- Multiple error: multiple bits affected in any order.

For each type of error there is a set of appropriate coding techniques. For example, the well-known *parity check* detects simple errors easily by using *Xor* gates. Unidirectional errors may be detected using *Berger code* which consists of additional check bits formed from a binary number which corresponds to the number of 1s in the information bits. Multiple errors are detected by duplication codes which consist of duplicating the information and using its complementary form to give a so-called *double rail* structure.

In this approach, all checkers have a double rail output and are designed to be tested using the normal code inputs. Such checkers can detect errors in their own operation.

Self-checking techniques are applied to circuits in which security is important so that fault tolerance is of major interest. Such techniques will occupy considerably more area in silicon than classical techniques such as functional testing but provide a very high test coverage.

10.13.10 Future trends

Observability and controllability are dramatically improved with new techniques using non-destructive E-beam probes which eliminate any probe capacitive loading. CAD tools are often used, and voltage contrast methods allow an immediate fault diagnosis.

Scanning E-beam microscopes have many test advantages such as the ability to function at very high clock speeds (GHz range) and at very high resolution (e.g. 0.3 μm), and finally with a wide range of CAD tools. The main disadvantage is their very high cost.

Expert CAD tools for IC design are being developed and are based on the hard-earned experience of test engineers, performing test pattern generation in the same manner as an individual test engineer does.

Expert systems generate tests with the objective of detecting all possible faults of interest and using as many as possible of the normal mode circuit operations. Consequently, test programming is acknolwedged as a highly knowledge-intensive task.

Many intelligent systems have been developed and are distinguished by their abilitity to test small and/or structured circuits (Russel and Sayers, 1989).

10.14 References

(a) Chip design

Ayres, R. F. (1983)*VLSI Silicon Compilation and the Art of Automatic Microchip Design*, Prentice Hall, Englewood Cliffs, NJ.

Hon, R. W., and Sequin, C. M. (1980) *A Guide to LSI Implementation*, 2nd edn, Xerox Press, EL Segundo, Calif.

Newkirk, J. A., and Mathews, R. G. (1984) *The VLSI Designer's Library*, Addison-Wesley Publishing Co. Inc., Reading, Mass.

Sung Mo Kang, (1981) 'A design of CMOS polycells for LSI circuits', *IEEE Transactions on Circuits and Systems,* Vol. CAS-28, No. 8, 838–43.

(b) Digital systems optimization and testing

Agrawal, V. D., and Seth, S. C. (1988) *Tutorial-Test Generation for VLSI Chips*, Computer Society Press.

Amersekera, E. A., and Campbell, D. S. (1987) *Failure Mechanisms in Semiconductor Devices*, Wiley.

Antreich, K. J. and Huss, S. A. (1984) 'An interactive optimization technique for the nominal design of integrated circuits', *IEEE Transactions on Circuits and Systems,* Vol. CAS-31, No. 2, 203–12.

Bardell, P. H., McAnney W. H. and Savir, J. (1987) *Built-In Test for VLSI: Pseudorandom Techniques*, Wiley.

Bateson, J. (1985) *In-circuit Testing*, Van Nostrand Reinhold.

Bennetts, R. G. (1982) *Introduction to Digital Board Testing*, Crane Russak.

Bennetts, R. G. (1984) *Design of Testable Logic Circuits*, Addison-Wesley.

Bhattacharya, Debashis and Hayes, John P. (1990) *Hierarchical Modeling for VLSI Circuit Testing*, Kluwer Academic Publishers.

Brayton, R. K. et al. (1981) 'A survey of optimization techniques for integrated circuit design', *Proc. IEEE,* Vol. 69, 1334–62.

Breuer, M. A., (ed.) (1972) *Design Automation of Digital Systems*, Prentice Hall.

Breuer, M. A., and Friedman, A. D. (1976) *Diagnosis and Reliable Design of Digital Systems*, Computer Science Press.

Chang, H. Y., Manning, E. G. and Metze, G. (1970) *Fault Diagnosis of Digital Systems*, Wiley Interscience.

Cortner, J. M. (1987) *Digital Test Engineering*, Wiley.

Davis, B. (1982) *The Economics of Automatic Testing*, McGraw-Hill, London.

Einspruch, N. G. (1985) *VLSI Handbook*, Academic Press.

Fee, W. G. (1978) *Tutorial-LSI Testing*, 2nd edn, Computer Society Press.

Feugate, R. J., and McIntyre, S. M. (1988) *Introduction to VLSI Testing*, Prentice Hall.

Friedman, A. D., and Menon, P. R. (1971) *Fault Detection in Digital Circuits*, Prentice Hall.

Fujiwara, H. (1985) *Logic Testing and Design for Testability*, MIT Press.

Golomb, S. W. (1982) *Shift Register Sequences*, revised edn, Aegean Park Press.

Greason, W. D. (1987) *Electrostatic Damage In Electronics*, Wiley.

Healy, J. T. (1981) *Automatic Testing and Evaluation of Digital Circuits*, Reston Publishing.

Jensen, F., and Petersen, N. E. (1982) *Burn-In*, Wiley, Chichester, UK.

Karpovsky, M. G. (ed.) (1985) *Spectral Techniques and Fault Detection*, Academic Press.

Kohavi, Z. (1978) *Switching and Automata Theory*, McGraw-Hill.

Lala, P. K. *Fault-Tolerant and Fault Testable Hardware Design*, Prentice-Hall, London, UK.

Lightner, M. R. and Director, S. W. (1981) 'Multiple criterion optimization with yield maximization', *IEEE Transactions on Circuits and Systems,* Vol. CAS-28, No. 8, 781–91.

Mahoney, M. (1987) *DSP-Based Testing of Analog and Mixed-Signal Circuits (Tutorial)*, Computer Society Press.

McCluskey, E. J. (1986) *Logic Design Principles with Emphasis on Testable VLSI Circuits*, Prentice Hall.

Miczo, A. (1986) *Digital Logic Testing and Simulation*, Harper & Row.

Miller, D. M. (ed.) (1987) *Developments in Integrated Circuit Testing*, Academic Press.

Needham, W. (1991) *Designer's Guide to Testable ASIC Devices*, Van Nostrand Reinhold (International), London.

Niraj, K. J., Sandip Kundu (1990) *Testing and Reliable Design of CMOS Circuits*, Kluwer Academic Publishers.

Parker, K. P. (1987) *Integrating Design and Test: Using CAE Tools for ATE Programming*, Computer Society Press.

Pradham, D. K. (ed.) (1986) *Fault-Tolerant Computing: Theory and Techniques*, Vols I and II, Prentice Hall.

Pynn, C. (1986) *Strategies for Electronics Test*, McGraw-Hill.

Reghbati, H. K. (1985) *Tutorial: VLSI Testing and Validation Techniques*, IEEE Computer Society Press, North Holland.

Ronse, C. (1984) *Feedback Shift Registers*, Springer-Verlag.

Roth, J. P. (1980) *Computer Logic, Testing, and Verification*, Computer Science Press.

Ruen-wen-liu, (1979) *Testing and Diagnosis of Analog Circuits and Systems*, Van Nostrand Reinhold (International), London.

Russel, G., and Sayers, I. L. (1989) *Advanced Simulation and Test Methodologies for VLSI Design,* Van Nostrand Reinhold (International), London.

Singh, N. (1987) *An Artificial Intelligence Approach to Test Generation*, Kluwer Academic Publishers.

Stevens, A. K. (1986) *Introduction to Component Testing*, Addison-Wesley.
Stover, A. C. (1984) *ATE: Automatic Test Equipment*, McGraw-Hill.
Timoc, C. C. (1984) *Selected Reprints on Logic Design for Testability*, Computer Science Press.
Tsui, F. F. (1986) *LSI-VLSI Testability Design*, McGraw-Hill.
Turino, J. L. (1991) *Design to Test,* 2nd edn, Van Nostrand Reinhold (International), London.
Wilkins, B. R. (1986) *Testing Digital Circuits: An Introduction,* Van Nostrand Reinhold, Berkshire, UK.
Williams, T. W. (1986) *VLSI Testing*, North-Holland, Amsterdam.
Yarmolik, V. N. (1990) *Fault Diagnosis of Digital Circuits*, John Wiley & Sons.

11 Some CMOS design projects

You cannot create experience — you must undergo it.

Albert Camus

Practice makes perfect.

Proverb

And for instructors:

Practice what you preach.

Proverb

Objectives

The 'raison d'être' for this chapter is self-evident. The authors regard project work and the tutorial work which leads up to it as absolutely essential to effective learning. The way in which the five individual projects are approached will, it is hoped, provide further insight into VLSI design processes. The projects have been chosen for their diversity and also to provide designs for useful and practical circuits.

11.1 Introduction to project work

The design exercises tackled earlier in this text were chosen to illustrate the design processes and to introduce the reader to the type of problems suitable for introducing design in silicon and relevant to everyday applications.

Following on from the design processes introduced, it is now instructive to formally tackle CMOS design work on some complete subsystems, and to this end various projects are now tackled in this chapter.

11.2 CMOS project 1 — an incrementer/decrementer

The design to be pursued is that of a 4-bit incrementer/decrementer, but the design is general in that the standard cell envisaged can be cascaded at will to n-bits.

11.2.1 Behavioral description

The truth table for a binary 1-bit incrementer is shown in Figure 11–1, where C_i is the carry bit from the previous stage, Cl is the clock input, C_{i+1} is the carry bit output, and Q_n is the stage output.

The logic expressions for the incrementer are as follows:

$$Q_n = C_i \oplus Q_{n-1} \tag{11.1}$$

$$C_{i+1} = C_i \cdot Q_{n-1} \tag{11.2}$$

The n stages are isolated by the clock signal Cl, and it will be seen that the truth table assumes positive-edge clocking. A reset signal (Res) should also be provided for the incrementer to be able to start from zero at any instant in time.

For the incrementer to function as a decrementer the additional equation that needs to be implemented is as follows:

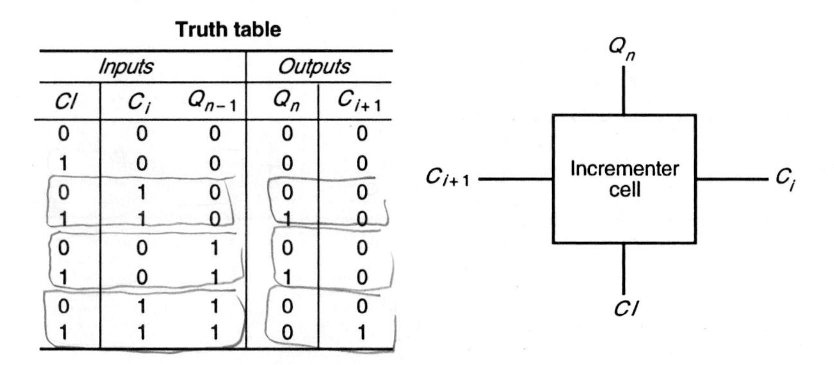

Truth table

Inputs			Outputs	
Cl	C_i	Q_{n-1}	Q_n	C_{i+1}
0	0	0	0	0
1	0	0	0	0
0	1	0	0	0
1	1	0	1	0
0	0	1	0	0
1	0	1	1	0
0	1	1	0	0
1	1	1	0	1

Note: Where Q_{n-1} is state of output prior to clocking.

Figure 11–1 1-bit incrementer cell

$$C_{i+1} = C_i \cdot \overline{Q_{n-1}} \tag{11.3}$$

A particular, but not the only possible, approach to designing this subsystem follows. For those readers who wish to 'fly solo' in tackling this design, the next project follows in section 11.3 on p. 367.

11.2.2 Structural description

11.2.2.1 Logic representation

An incrementer/decrementer cell is realized by direct implementation of expressions (11.1), (11.2), and (11.3) as in Figure 11–2, for example. Note that a reset control line may also be added using the *'clear'* input of the flip-flop to enable the circuit to start from zero at any time, but this is not shown in the figure. The control line which is required to set the circuit operation to that of an incrementer or a decrementer is shown in the figure.

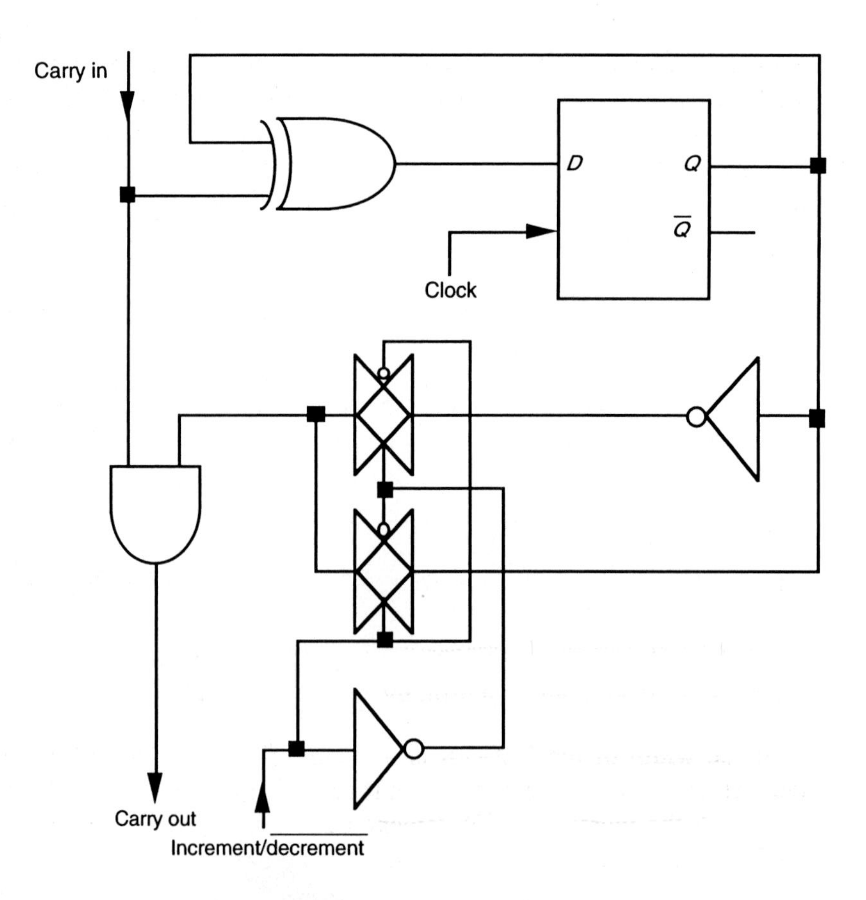

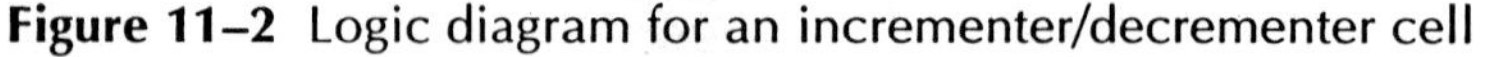

Figure 11–2 Logic diagram for an incrementer/decrementer cell

11.2.2.2 Operation of the circuit

The circuit functions like an adder or a subtractor with one of its three inputs set to zero. The cell uses its current state as one input and the carry in from the previous stage as the other input. The current state and the carry out are modified according to the two inputs on clocking.

11.2.2.3 Critical paths

The critical delay in this circuit is the propagation delay of the carry bit — analogous to the adder situation. Since the circuit is clocked, the minimum allowable clock period is set by the maximum circuit delay; in this case the time that the carry bit needs to propagate from the first to the last stage. This will be reasonably fast as the carry bit passes through only one *And* gate per stage.

11.2.3 Physical description

11.2.3.1 Floor plan of a 4-bit incrementer/decrementer subsystem

The 4-bit incrementer/decrementer is realized by abutting four identical cells. The height of the complete subsystem remains constant while the width grows linearly with n — the number of bits. Therefore the width of each cell should be made as small as possible. The control lines run right across the whole structure and adequate driving capability should be supplied when n is significant. The resulting floor plan is shown in Figure 11–3.

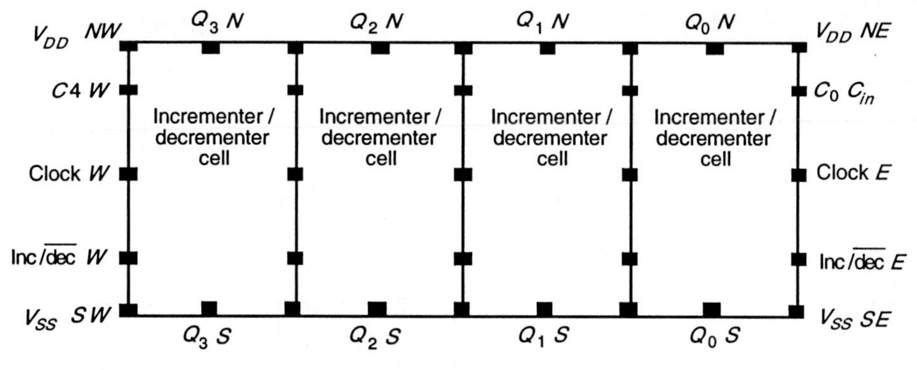

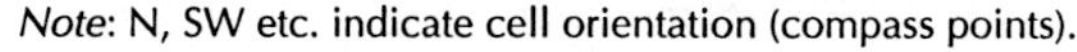

Figure 11–3 Floor plan for a 4-bit incrementer/decrementer

If the width of the leaf-cell is w, then the width of an n-bit incrementer/decrementer is nw. This dimension must be pitch-matched to the rest of the system into which the incrementer/decrementer is to fit (e.g. a VLSI processor, etc.), which may be assumed to be of width W. Therefore

$$w = W/n$$

provided that

$$w_{min} \leqslant W/n$$

where w_{min} is the minimum width of a cell. In the event that $w_{min} > W/n$, then the design must be adjusted to be thinner and taller, otherwise the width W of all mating subsystems may have to be increased.

11.2.3.2 Leaf-cell floor plan

The floor plan of the 4-bit incrementer/decrementer basically determines the floor plan of the leaf-cell which is given in Figure 11–4.

The width w of each cell is set by the total allowable maximum incrementer/decrementer width W which cannot be exceeded if the circuit is to be properly pitch-matched to the rest of the system, e.g. data path, for which it is being designed. The minimum height h of the leaf-cell is set by its complexity once the width w has been fixed. The decision about the output connection and the power rail placements is made at the subsystem level (the subsystem here being the four-bit incrementer/decrementer).

In a complex design, the number of leaf-cells should be kept to the absolute minimum, which implies that the complexity of the leaf cells should be as high as possible. This greatly simplifies the global floor plan, but it must be recognized that the available design tools will determine the maximum size of leaf-cell which can be readily handled. In general, a 50 to 100 transistor leaf-cell can be readily realized with available design tools. Since the incrementer/decrementer leaf-cell is of a medium complexity it should not be further subdivided into sub-leaf-cells, and the design of a mask layout for the circuit of Figure 11–2 may be pursued for an appropriate technology using available design tools.

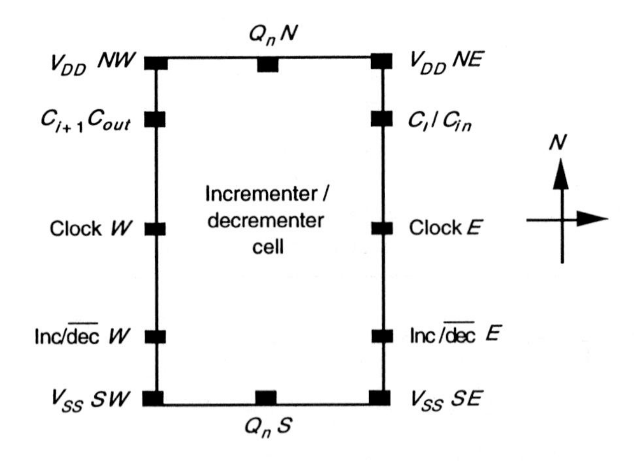

Figure 11–4 Floor plan of incrementer/decrementer leaf-cell

11.2.4 Design verification

The leaf-cell circuit was designed using 5 μm p-well CMOS technology and a mask layout arrived at. The detail present in the CIF code specification for the mask layout was extracted with a circuit extractor (NET) and then a two-bit subsystem simulated with a circuit simulator (PROBE). The simulation results are given in Figure 11–5.

11.3 CMOS project 2 — left/right shift serial/parallel register

This project is concerned with the design of a general purpose shift register cell capable of expansion to form an *n*-bit register.

11.3.1 Behavioral description

Table 11–1 defines the shift register connections that apply to Figures 11–6, 11–7 and 11–8. The logic circuit for a suitable single shift register leaf-cell is shown in Figure 11–7 and in block diagram form in Figure 11–8.

Table 11–1 Shift register control functions

Controls	*Function*	*Conditions required*
dp	parallel data input	latched when *dprl* is asserted
dprl	parallel input data control	$\overline{left} \cdot \overline{right}\ \phi_1$
qp	parallel data output	valid when *qprl* is asserted
qprl	parallel output data control	data valid on ϕ_2 of clock
ds	serial right data input	valid when *right* is asserted
qs	serial right data output	valid when *right* is asserted
right	shift right control	$\overline{dprl} \cdot \overline{left} \cdot \phi_1$
leftin	serial left data input	valid when *left* is asserted
leftout	serial left data output	valid when *left* is asserted
left	shift left control	$\overline{dprl} \cdot \overline{right} \cdot \phi_1$
fb	internal refresh control	$\overline{dprl} \cdot \overline{right} \cdot \overline{left} \cdot \phi_1$
ϕ_2	second clock phase	data latch to output node

11.3.2 Structural description

11.3.2.1 Logic representation

The complete 4-bit shift register is made up of single shift register cells abutted as shown in part in Figure 11–6.

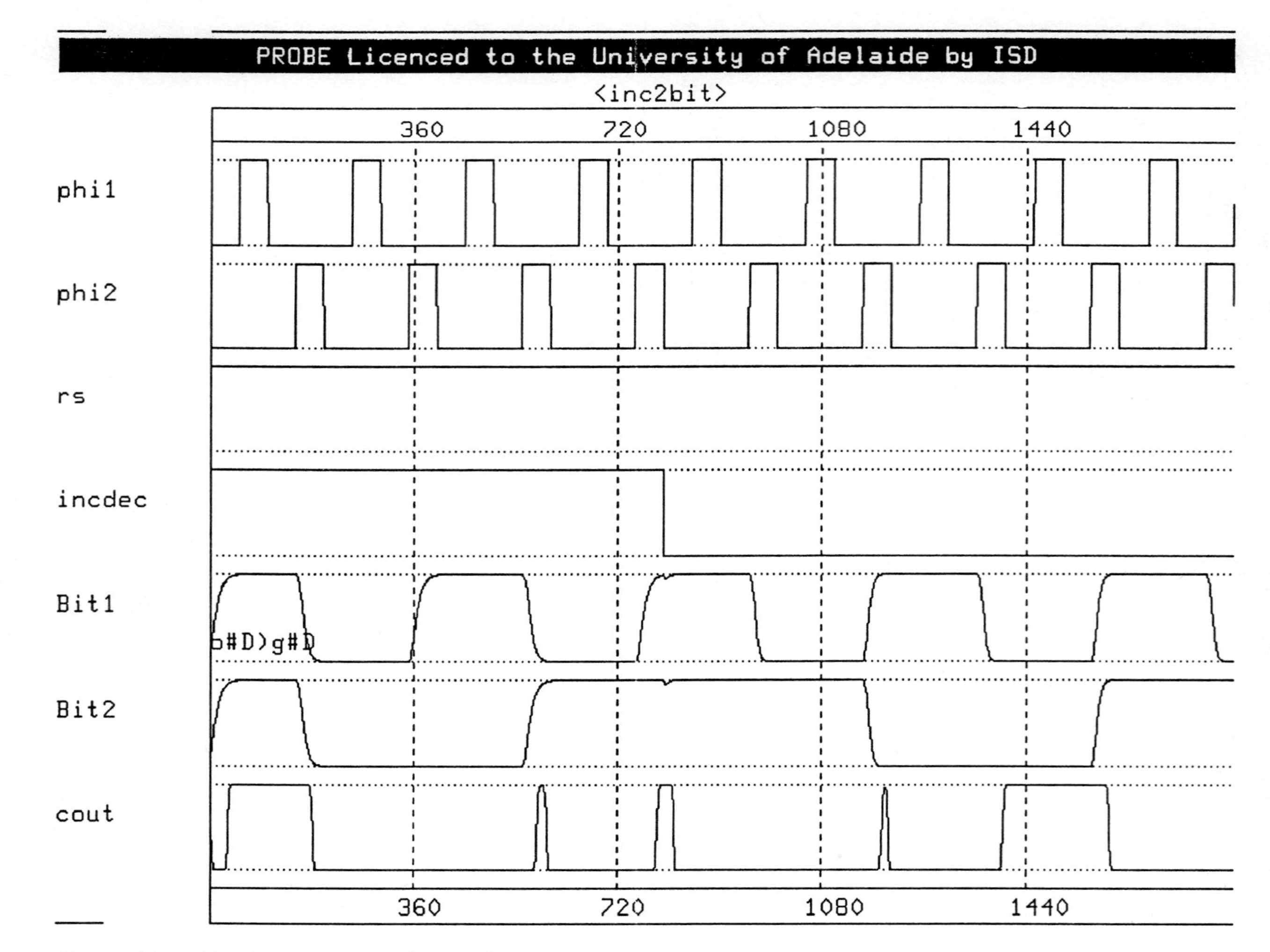

Figure 11–5 Simulation results for a 2-bit system

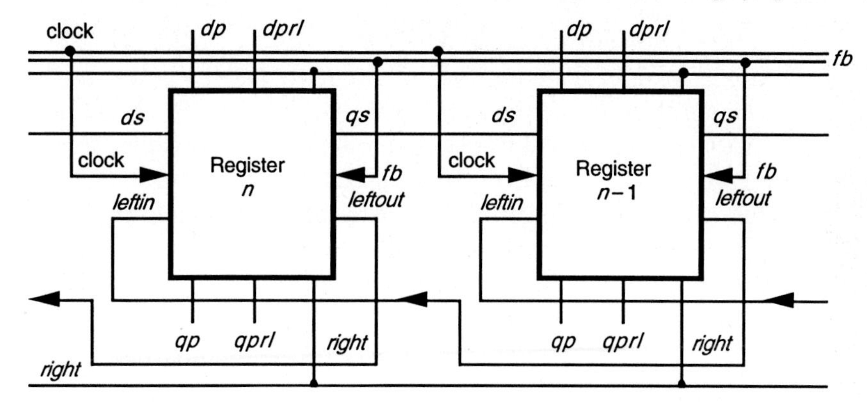

Figure 11–6 Two-bit shift register block diagram

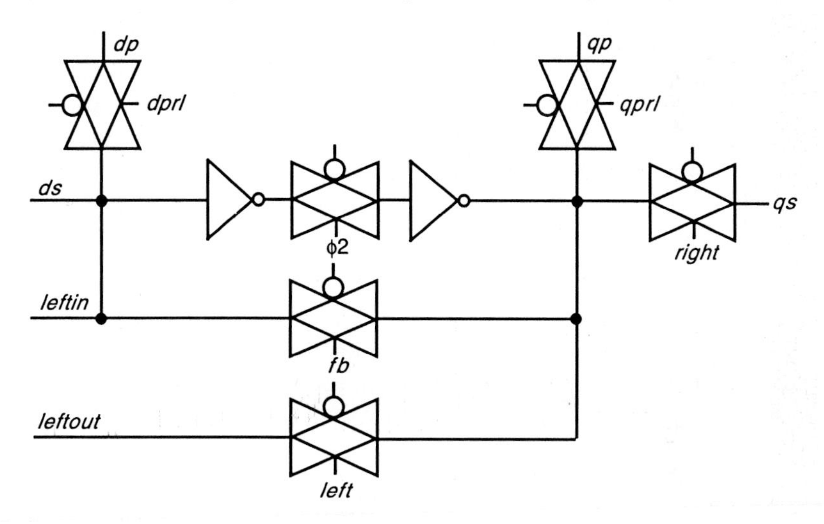

Figure 11–7 Shift register cell logic diagram

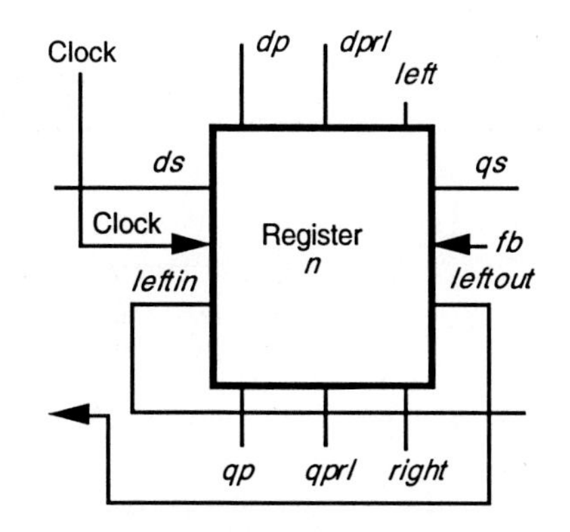

Figure 11–8 Shift register cell block diagram

11.3.2.2 Operation of the circuit

The operation of the complete shift register may be understood by considering the single shift register cell of Figures 11–7 and 11–8. The advantage of this cell is that it may be loaded or read in parallel and the bits may be shifted either left or right within the shift register and an output thus obtained in serial form at either end of the register. The register also uses a two-phase non-overlapping clock of which ϕ_1 allows loading, shifting, and refreshing to occur while ϕ_2 isolates the two inverters so that the cells may be loaded.

The operations of the shift register (Figure 11–7) in detail are as follows:

1. *The refresh loop.* The refresh signal *fb* (or feedback) occurs in coincidence with ϕ_1 and when no other control is asserted (namely *dprl, right,* and *left).* The transmission gate takes the output of the second inverter and uses it to refresh the logic level stored on the input gate capacitances of the first inverter.
2. *In parallel load mode.* The inputs *dp* and control *dprl* are used to load the registers in parallel. Asserting *dprl* when ϕ_1 is at logic level 1 will cause the input of the first inverter to assume the state of *dp.* At this time $\phi_2 = 0$ and the inverters are isolated. Subsequently $\phi_2 = 1$ and the second inverter output assumes the state of *dp* which has been stored dynamically at the first inverter input.
3. *In shift right mode.* The signals associated with the shift right operation are *right, qs,* and *ds.* Asserting *right* when ϕ_1 is at logic level 1 effectively loads the subsequent register with *qs,* while the *qs* output of the register cell to the left of the current one is connected through a transmission gate to the *ds* input of the present cell. Hence the cell is loaded in the same manner as with a parallel load, but with the data input coming from the adjoining cell to the left (that is, a shift right operation).
4. *In shift left mode.* The signals associated with the shift left operation are *left, leftout,* and *leftin.* Asserting *left* when ϕ_1 is a logic level 1 effectively loads the previous register with *qs* via the line *leftout.* The register cell to the right of the current one has its *leftout* connected through a transmission gate to *leftin* of the present cell. Hence the cell is loaded in the same manner as with a parallel load but the data input comes from the adjoining cell to the right (that is, a shift left operation).
5. *For parallel output.* The output data is correctly read at the end of ϕ_2 when there can be no change to the input. This is achieved by asserting *qprl,* in which case *qp* assumes the state of the cell and all outputs are then read in parallel.
6. *Isolation of the inverters by* ϕ_2. The second phase of the clock (ϕ_2) is used to isolate the inverters during a write operation so that the register array does not become 'transparent'. Consider a shift right operation but allow $\phi_2 = 1$.

Here the first inverter output would become $\overline{ds}_{i-1}$ (from the next left cell). The second inverter output would thus become ds_{i-1}. However, since ϕ_2 is logic 1, ds_{i-1} can now be passed on to cell $i+1$, since *right* is asserted and $qs = ds_{i-1}$. Hence the register has become transparent and ds_{i-1} would ripple throughout the entire array. This undesirable effect is eliminated by loading and coupling inverter pairs on separate clock phases.

11.3.2.3 Critical paths

The system is restricted to shifts of 1 bit only in either direction and hence any shifts of more bits will take proportionally more time. In this case, there is a minimum time t_1 for which ϕ_1 must be asserted to allow the data to be stored at the first inverter input gate. After this delay the data is passed to the output on ϕ_2 which must have time duration t_2 for the second inverter input capacitance charge to change its state if required. The total delay (T) is governed by the sum of t_1 and t_2 and the number of shifts n required (i.e. $T = n \cdot (t_1 + t_2)$). To reduce this delay a fast shifting cell is required.

The most critical path at the leaf-cell level is associated with the output of the second inverter which must drive four transmission gate input capacitances. For this reason the second inverter is not usually made minimum size. Note, however, that the second inverter cannot be made too large since the first inverter (which is minimum sized) must drive its input when $\phi_2 = 1$. The final sizing of the transistors may be determined after a series of simulations following circuit extraction from the mask layout.

11.3.3 Physical description

11.3.3.1 System floor plan

The 4-bit shift register may be formed by abutting four identical 1-bit register cells. The most convenient arrangement for an n-bit shift register is to have the parallel data inputs and outputs running perpendicular to the direction of the register array. The control lines are also conveniently run perpendicular to the register array but, on exiting a register cell, may be run along the array with appropriate connections made to adjoining cell control signals. The power rails must be implemented in metal and also run perpendicular to the parallel input/output data. The resulting floor plan is shown in Figure 11–9.

If the width of the leaf-cell is w, then the width of an n-bit register is nw. This dimension must be pitch-matched to the rest of the system (e.g. a VLSI processor, etc.) of assumed width W. Therefore

$$w = W/n$$

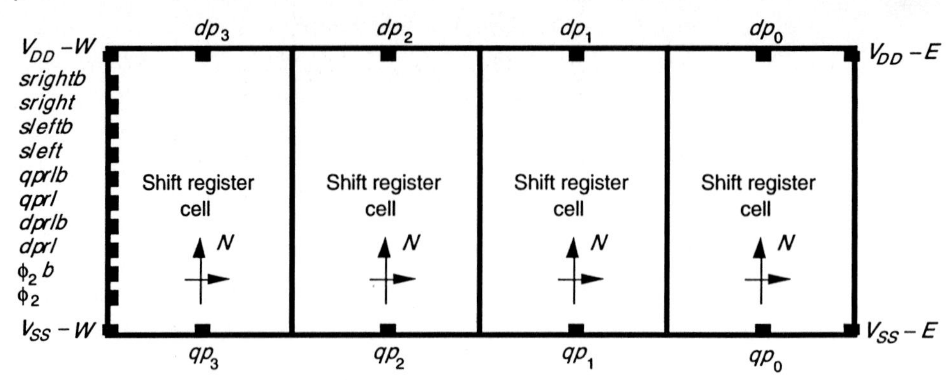

Figure 11–9 Proposed floor plan — 4-bit shift register

11.3.3.2 Leaf-cell floor plan

The floor plan of the 4-bit shifter basically specifies the floor plan of the leaf-cell. The width w is set by the total maximum register width, and this cannot be exceeded if the register is to be properly pitch-matched, for example, to a processor. The minimum height h of the leaf-cell is set by its complexity once the width has been fixed. The decision about the input/output connection and the power rail placements is made at the system's level.

In a complex design, the number of leaf-cells should be kept to the absolute minimum, which implies that the complexity of the leaf-cells should be as high as possible. This greatly simplifies the global floor plan. As stated earlier, a 50 to 100 transistor leaf-cell can usually be readily realized with commonly available design tools. The register leaf-cell described here is of small/medium complexity and thus should not be further subdivided into sub-leaf-cells. The shift register leaf-cell floor plan is shown in Figure 11–10.

11.3.4 Design verification

Simulation results for a 4-bit register realized in 5 μm p-well CMOS technology (using PROBE software) are presented in Figure 11–11.

11.4 CMOS project 3 — a comparator for two *n*-bit numbers

This section describes the design methodology, layout strategy, and simulation results for cascadable comparator cells. A 4-bit comparator was designed using these cells, the general arrangement being as suggested in Figure 11–12.

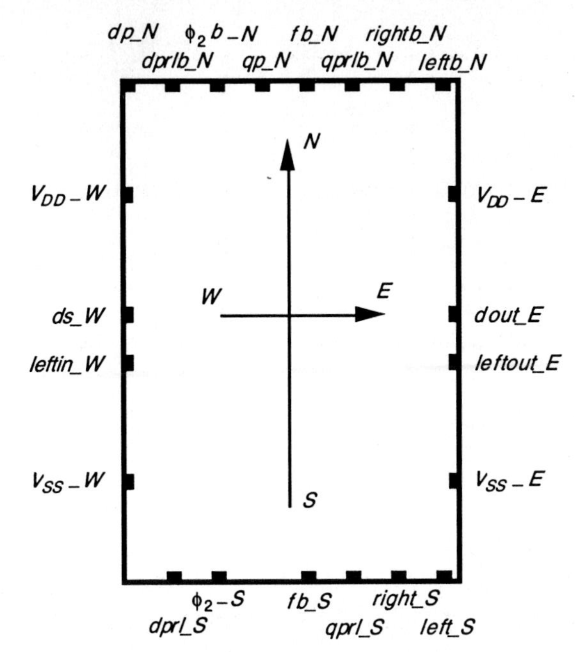

Compass points, used to indicate orientation of cell, may also be appended to signals to indicate position as shown.

Figure 11–10 Shift register cell floor plan

11.4.1 Behavioral description

The truth table and general arrangement for a binary 1-bit comparator bit-slice is shown in Figure 11–13 where A_i and B_i are the two numbers to be compared, C_{i+1} and D_{i+1} are the inputs from outputs of the previous stage, and C_i and D_i are the outputs of the current stage. $C_i = 1$ if $A_i > B_i$; $D_i = 1$ if $A_i < B_i$; and $C_i = D_i = 0$ if $A_i = B_i$.

The logic expressions for the two output signals in terms of the four input signals are as follows.

$$C_i = C_{i+1} + \overline{C}_{i+1} \cdot A_i \cdot \overline{B}_i \cdot \overline{D}_{i+1} \tag{11.4}$$

$$D_i = D_{i+1} + \overline{D}_{i+1} \cdot \overline{A}_i \cdot B_i \cdot \overline{C}_{i+1} \tag{11.5}$$

The two logic expressions may be rearranged into the form

$$\begin{aligned} C_i &= \overline{\overline{C}_{i+1} \cdot \overline{(\overline{C}_{i+1} \cdot \overline{D}_{i+1} \cdot A_i \cdot \overline{B}_i)}} \\ &= \overline{\overline{C}_{i+1} \cdot \overline{\overline{(C_{i+1} + \overline{A}_i + B_i)} \cdot \overline{D}_{i+1}}} \end{aligned} \tag{11.4a}$$

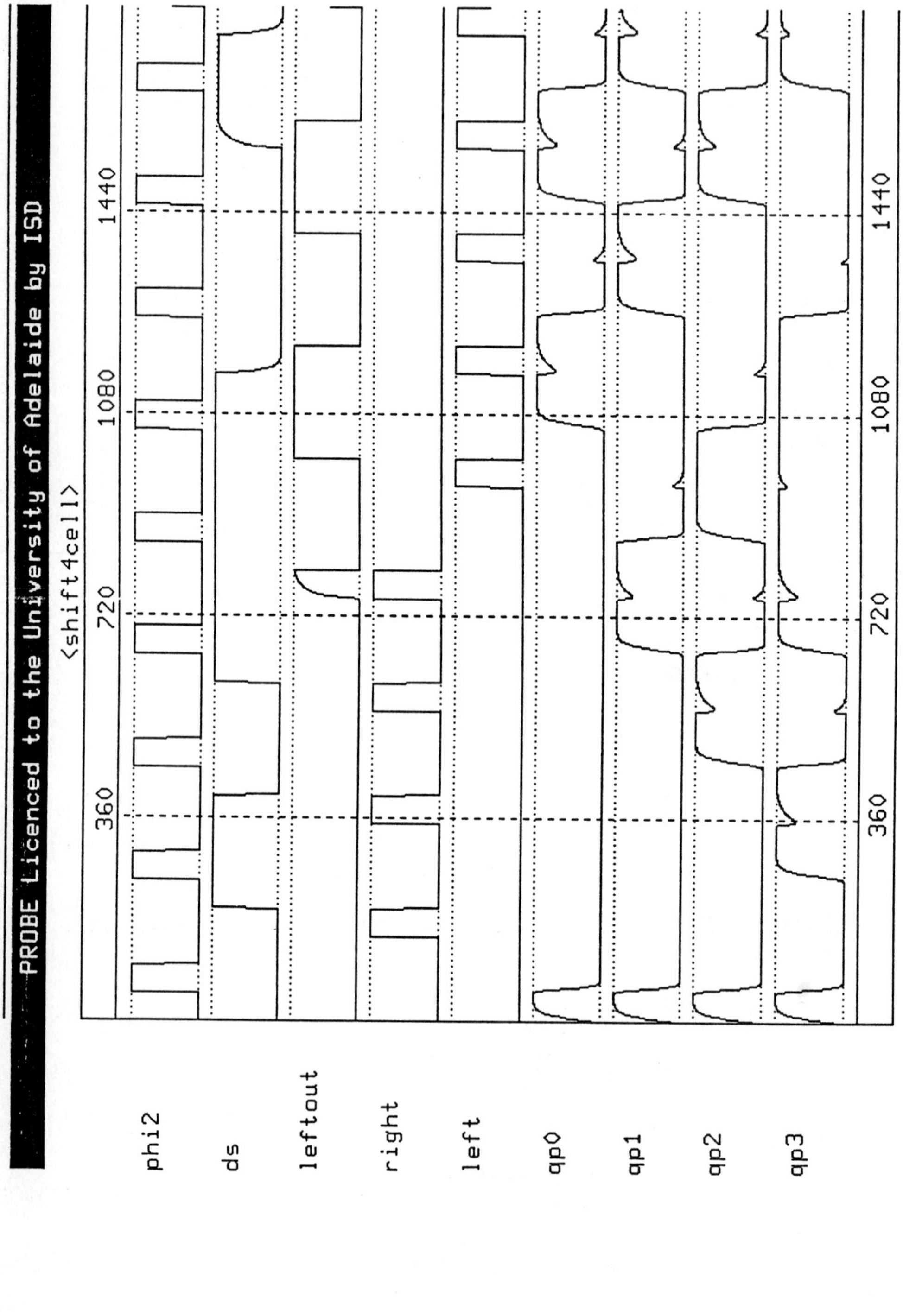

Figure 11–11 Simulation over four shift register cells

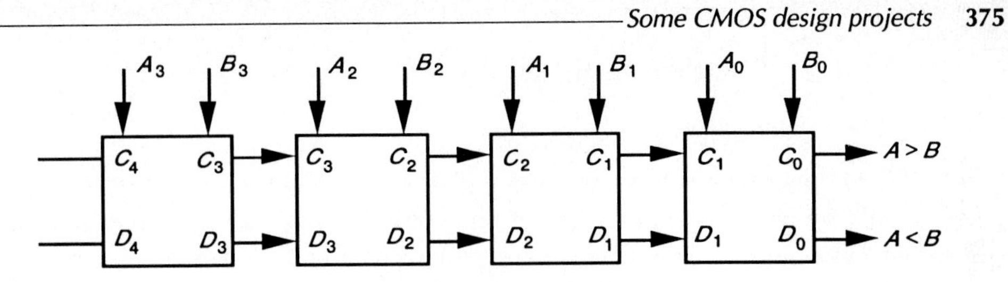

Figure 11–12 4-bit comparator — block diagram

Truth table

Inputs				Outputs	
A_i	B_i	C_{i+1}	D_{i+1}	C_i	D_i
X	X	1	0	1	0
X	X	0	1	0	1
0	0	0	0	0	0
0	1	0	0	0	1
1	0	0	0	1	0
1	1	0	0	0	0

X signifies 'don't care'

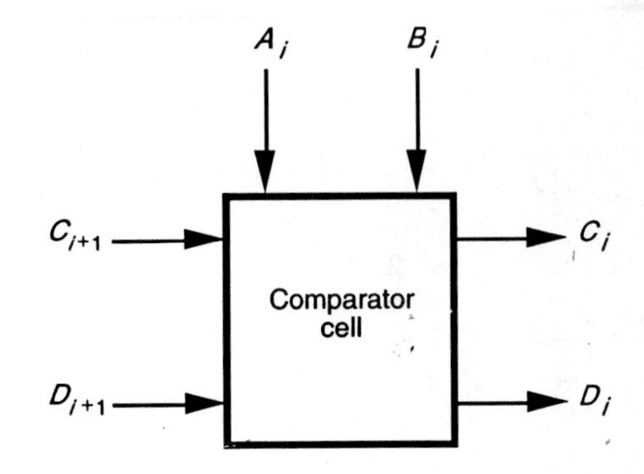

Figure 11–13 Comparator cell behavior

$$D_i = \overline{\overline{D}_{i+1}\cdot\overline{(\overline{D}_{i+1}\cdot\overline{C}_{i+1}\cdot\overline{A}_i\cdot B_i)}}$$
$$= \overline{\overline{D}_{i+1}\cdot\overline{\overline{(D_{i+1}+\overline{B}_i+A_i)}\cdot\overline{C}_{i+1}}} \tag{11.5a}$$

A further simplification may be achieved if alternate logic is used between subsequent cells. Stage i implements

$$\overline{C}_i = \overline{C}_i = \overline{C_{i+1}+\overline{D_{i+1}+\overline{A}_i+B_i}} \tag{11.4b}$$

$$\overline{D}_i = \overline{D_{i+1}+\overline{C_{i+1}+A_i+\overline{B}_i}} \tag{11.5b}$$

and stage $i-1$ implements

$$C_{i-1} = \overline{\overline{C}_i\cdot\overline{\overline{D}_i\cdot A_{i-1}\cdot\overline{B}_{i-1}}} \tag{11.4c}$$

$$D_{i-1} = \overline{\overline{D}_i\cdot\overline{\overline{C}_i\cdot\overline{A}_{i-1}\cdot B_{i-1}}} \tag{11.5c}$$

11.4.2 Structural description

11.4.2.1 Logic representation

The comparator is implemented with complementary cells, that is, the *ith* stage has true inputs and inverted outputs while the $(i+1)th$ stage has inverted inputs and true outputs. The two cells are realized by direct implementation of expressions (11.4c) and (11.5c) (COMPCELLA) and expressions (11.4b) and (11.5b) (COMPCELLB) as shown in Figures 11–14 (a) and (b) respectively.

11.4.2.2 Operation of the circuit

The operation of the complete circuit is as follows:

- The two numbers are compared starting with the most significant bits. The outputs from this comparison are connected to the next most significant bit stage inputs etc. The two output signals C_i and D_i remain at zero as long as the two bits being compared are the same.
- As soon as a difference is detected, the two outputs are set to one of two possible states: if $A_i > B_i$ then $C_i = 1$ and $D_i = 0$; if $A_i < B_i$ then $C_i = 0$ and $D_i = 1$.
- All the remaining pairs of less significant bits then have no further effect on the state of subsequent outputs C_i and D_i.
- If all pairs of bits of the two numbers being compared are equal, then the outputs stay at zero signifying equality.

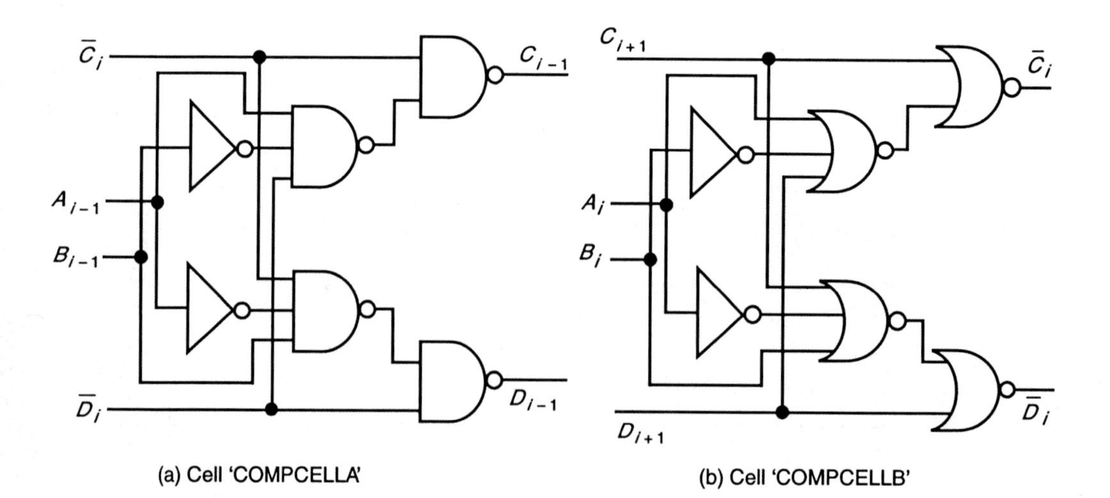

Figure 11–14 Comparator — logic diagram

11.4.2.3 Critical paths

The critical delay in this circuit is the propagation delay of the two outputs through all the stages. The gates passing both outputs should be sized appropriately. The delay is only one gate per stage and should not be the limiting factor on a system's scale. The final sizing of the transistors is usually determined after a series of simulations.

11.4.3 Physical description

11.4.3.1 System floor plan

The 4-bit comparator is realized by abutting cells of each type on an alternate basis. One possibility would be to have both bit inputs on the same side of a cell with the two outputs propagating at right angles to the input data path. Another possible layout would be to have the two bit inputs on opposite sides of a cell. The second approach was adopted here. The height of the comparator remains constant while the width grows linearly with n — the number of bits. Therefore the width of each cell should be made as small as possible.

A possible floor plan is shown in Figure 11–15: the inputs A_i and B_i come in at the top and bottom of each cell respectively, and C_i and D_i propagate horizontally. V_{DD} and V_{SS} rails may also propagate horizontally in global terms but may be distributed at right angles within a cell if convenient.

If the width of the leaf-cell is w, then the width of an n-bit comparator is nw. This dimension must be pitch-matched to the rest of the system (e.g. a VLSI processor, etc.) of width W. Therefore

$$w = W/n$$

11.4.3.2 Leaf-cell floor plan

The floor plan of the 4-bit comparator basically specifies the floor plan of the leaf-cells as shown in Figure 11–16. The width w is set by the total maximum

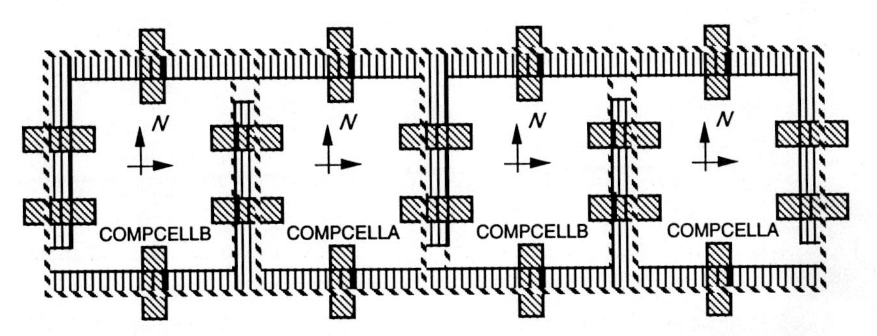

Figure 11–15 Proposed floor plan — 4-bit comparator showing shared power rails

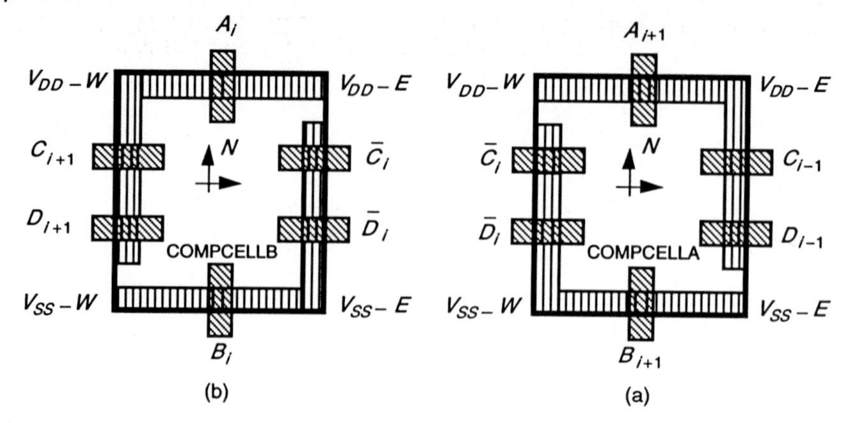

Figure 11–16 Comparator leaf-cells — floor plan

comparator width W. The minimum height h of the leaf-cell is set by its complexity once the width w has been fixed. The decision about the input/output connection and the power rail placements is made at the system's level (the system here being the 4-bit comparator).

In a complex design the number of leaf-cells should be kept to the absolute minimum, which implies that the complexity of the leaf-cells should be as high as possible. This greatly simplifies the global floor plan. A 50 to 100 transistor leaf-cell can usually be readily realized with available design tools. The comparator leaf-cell is of medium complexity and does not require any further subdivision.

11.4.4 Symbolic or stick representation to mask transformation

A mask representation is generally obtained from a symbolic form of cell specification by the process of compaction. A compactor is a tool that takes a symbolic representation of the given cell and produces a mask description of the cell according to some predefined set of process design rules. A mask description of the cell may also be obtained by direct mapping from a stick diagram using a mask level graphics editor.

A few basic rules should be observed when designing a circuit:

1. Start the design by placing an imaginary demarcation line (for p-well CMOS, this is closely related to the top edge of the well, and for n-well CMOS, the bottom edge of the well). This line separates the p-type devices, which are placed above it, from the n-type devices, which are placed below it; that is, the two types of transistors should not be intermixed. This style of design allows easy placements of the well and the p+ or n+ masks (Figure 11–17).
2. Keep the V_{DD} and V_{SS} supply rails well separated. This allows all the devices to be placed close to the required rail and be completely within the V_{DD} to V_{SS} boundaries, greatly simplifying the inter-cell connections.

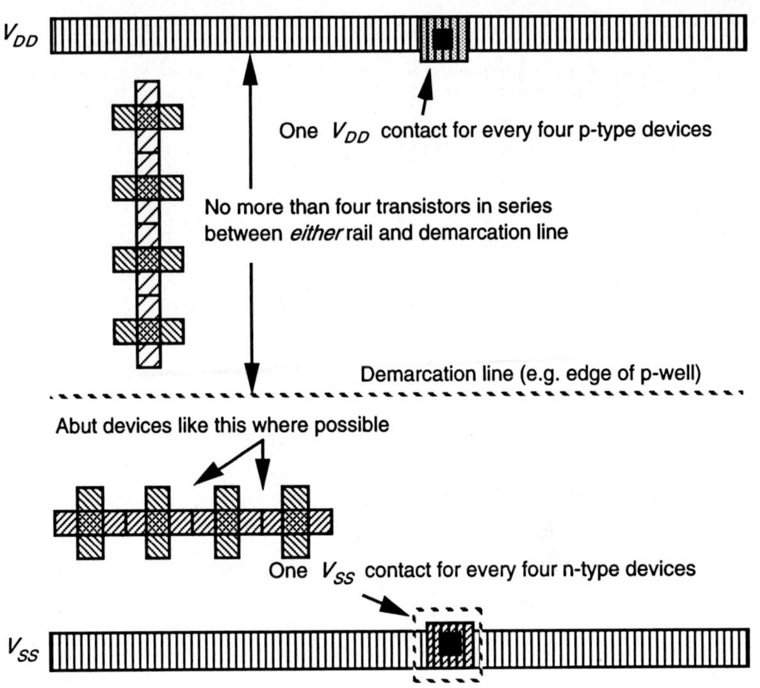

Figure 11–17 Layout design style

3. Abut as many devices as possible to minimize the interconnect resistance and capacitance between them.
4. Do not use more than four levels of devices between a rail and the demarcation line (as shown).
5. Place one V_{SS} contact for every four n-type devices, and one V_{DD} contact for every four p-type devices.

A possible embryo mask layout for COMPCELLA (*Nand* gate-based) is given in Figure 11–18. Note that input A_i defines the top of the cell and B_i the bottom. This layout is readily adapted to form cell COMPCELLB by exchanging *Nand* for *Nor* gates. Dimensions and separations, etc. will be fixed by the chosen technology to give the final working mask layout.

11.4.5 Design verification

Before the actual layout can be submitted for fabrication, the whole design must be carefully checked and verified. A simulator (e.g. PROBE) is used during the design process to verify and improve the timing behavior of each leaf-cell. When

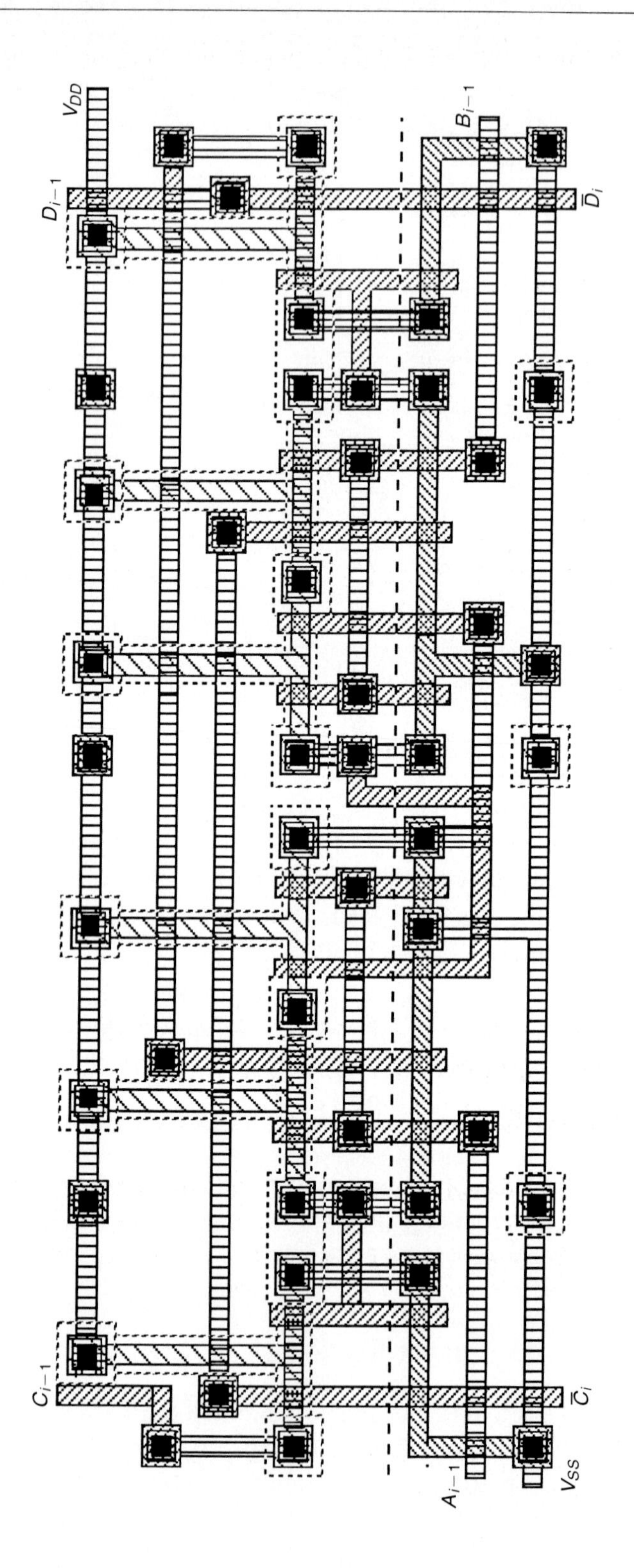

Note: Cell orientation in relation to the floor plan (Figure 11–16).

Figure 11–18 COMPCELLA mask layout

the layout is completed, it must be passed through a design rule checker to verify its compliance with the design rules of the fabrication process to be used and an electrical rules checker to test for the number of V_{DD} and V_{SS} contacts etc. The necessary steps are:

- Pass the design through a design rule checker and correct any design rule errors.
- Check the circuit for any electrical rule errors using an electrical rules checker. Correct as necessary.
- Extract the circuit from the mask layout. This produces a file which is a circuit description containing connectivity information obtained from the CIF description.
- Simulate the cell and make any necessary changes to the transistor dimensions etc. to improve the performance. Remember to apply correct capacitance loadings to the output terminals when simulating. Simulation results obtained in this case for 5 μm technology are presented in Figures 11–19 and 11–20. Correct as necessary.
- Recheck for design rule errors.
- Carry out the final simulation at cell level.
- Assemble the complete 4-bit comparator using a suitable editor. Simulate and design rule check the complete subsystem.
- Place the input and output pads around the circuit as suggested in the floor plan in Figure 11–21. Note that the overall system, including the pads, should now have its operation checked by simulation.

11.5 CMOS/BiCMOS project 4* — a two-phase non-overlapping clock generator with buffered output on both phases

This project differs from the previous three since it is not concerned with the design of a cascadable architectural subsystem bit-slice. Rather, it is a control signal (two-phase clock) generator and bus driver which must perform well when driving relatively large capacitive loads.

* The work reported here was carried out by Brenton Cooper as part of a final year project at the University of Adelaide.

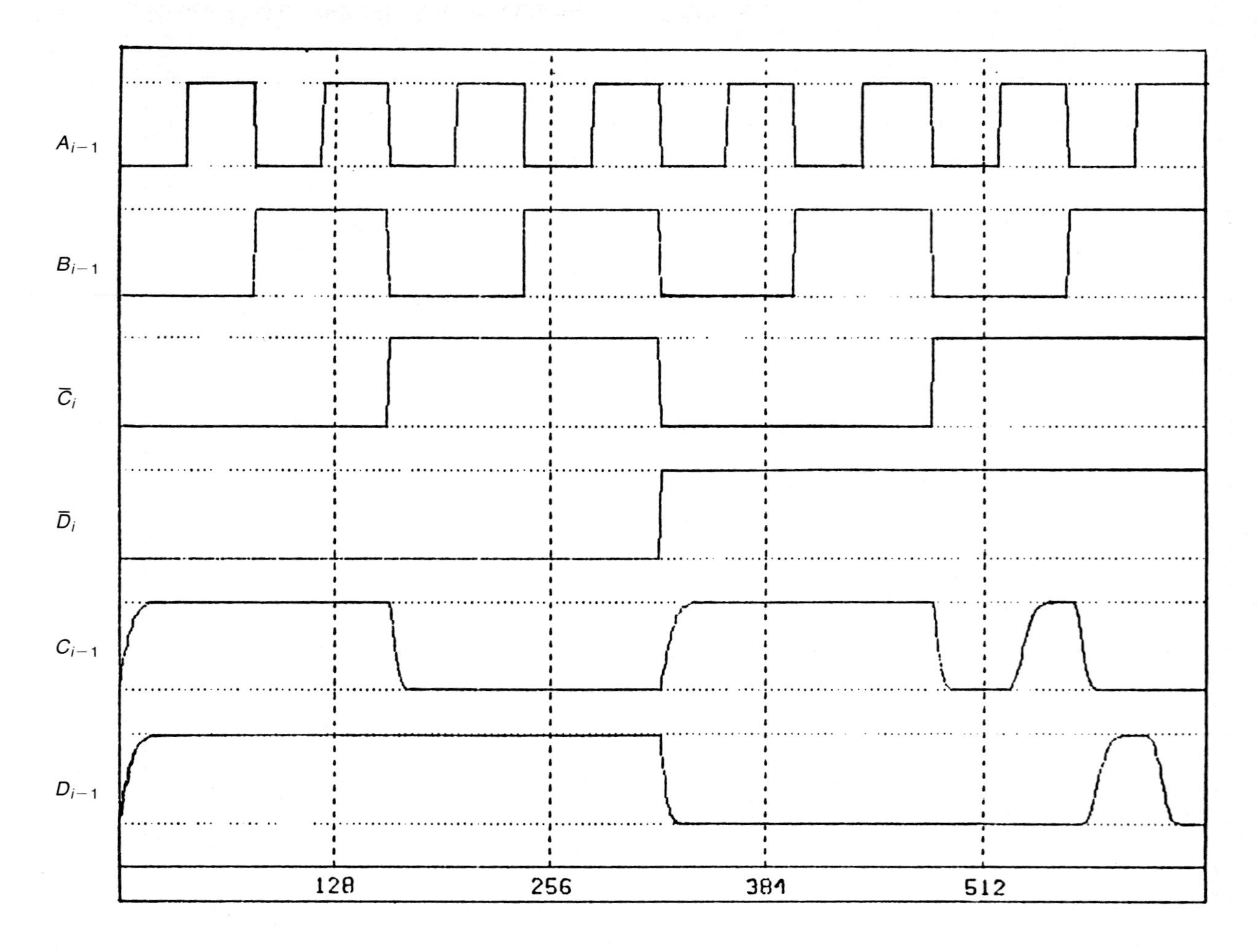

Figure 11–19 PROBE simulation results — COMPCELLA

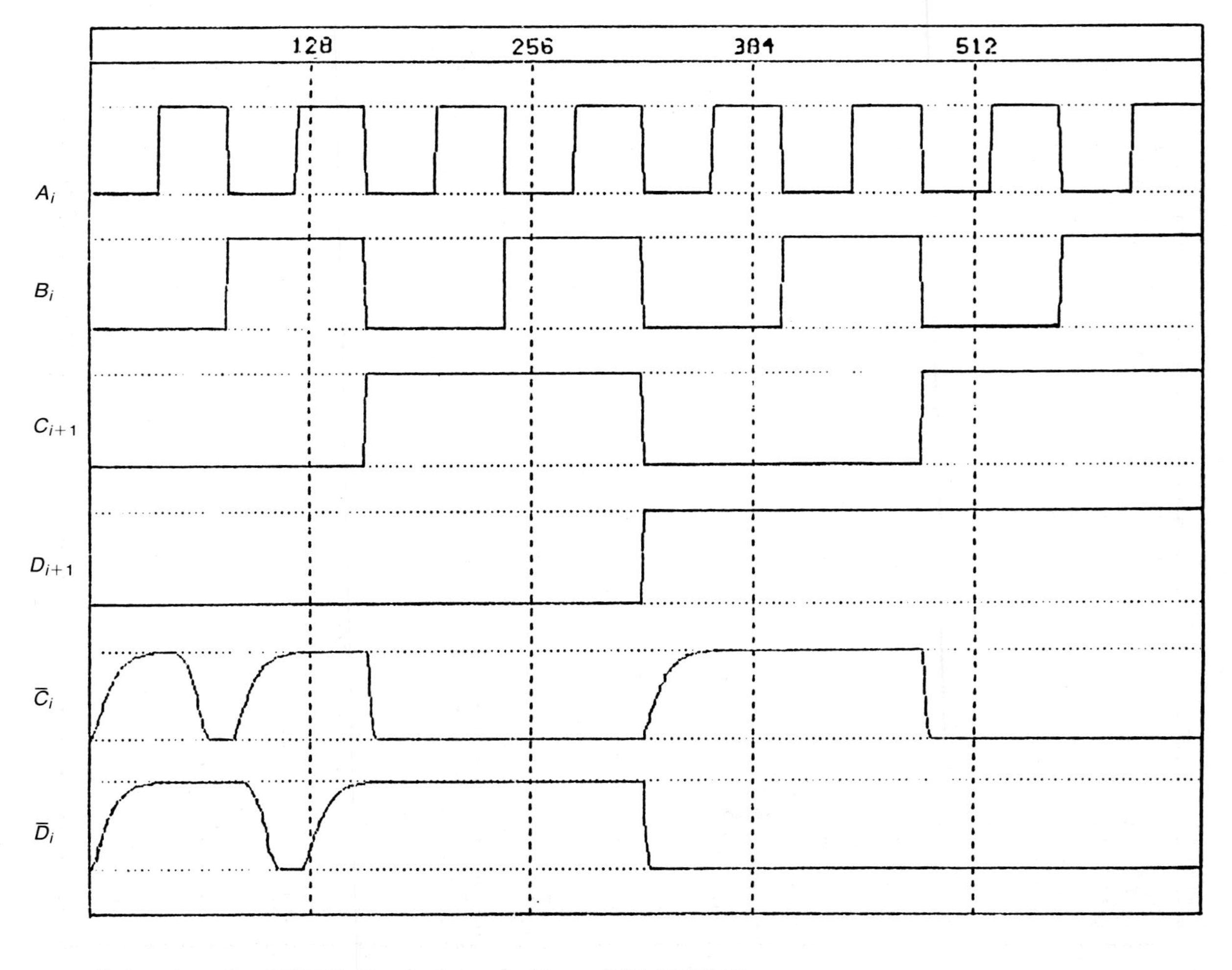

Figure 11–20 PROBE simulation results — COMPCELLB

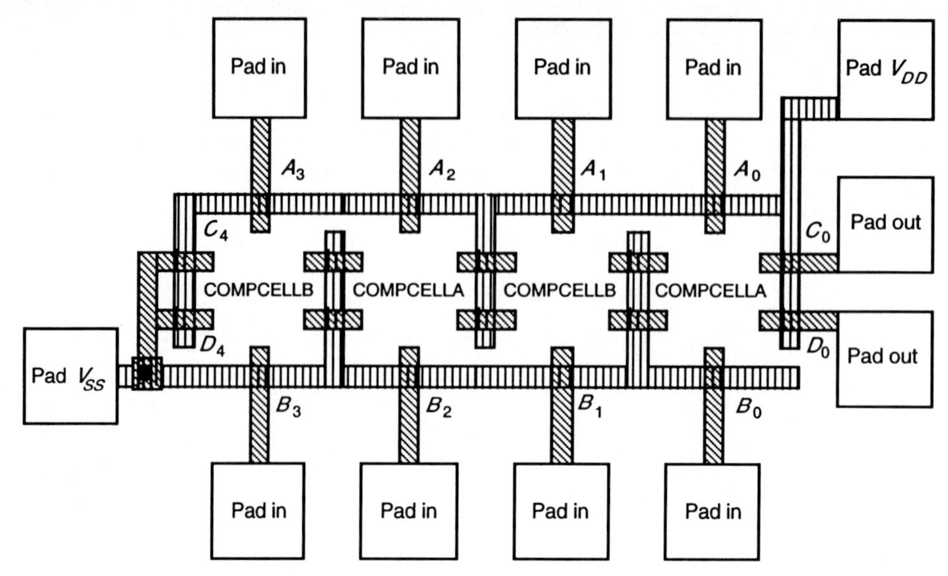

Figure 11–21 4-bit comparator with illustrative placements

11.5.1 Behavioral description

The circuit is required to accept a single-phase input clock signal of 10 MHz maximum frequency and, from this, generate two-phase non-overlapping clock signals at the input clock frequency. The two-phase clock signals are to be good, clean 'square' waves, and each phase should be capable of driving a load capacitance of 0.33 pF without undue waveform degradation. The approach taken is one of circuit development through the design of mask layout, simulation of performance, improvement where necessary, modified mask layout, simulation, and so on.

11.5.2 Structural description

The structure of a suitable basic circuit arrangement, previously introduced in Chapter 6 (Figure 6–33), is quite simple, comprising two gates and two inverters repeated here in Figure 11–22. Output buffers will be added to the two-phase outputs to cater for the specified capacitive load.

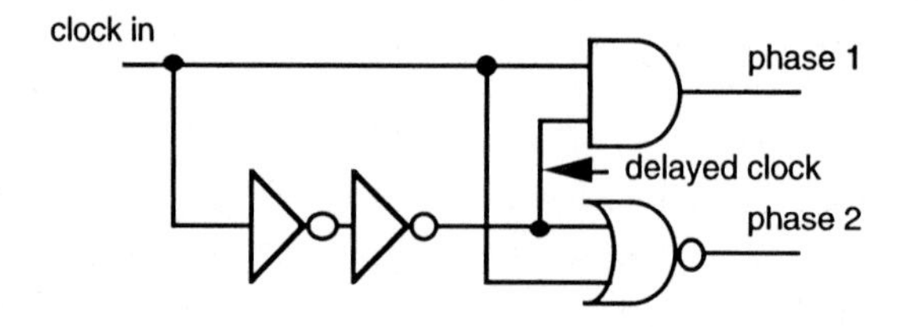

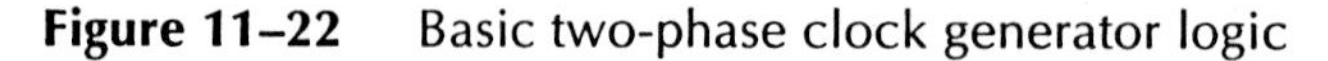
Figure 11–22 Basic two-phase clock generator logic

11.5.3 Design process

11.5.3.1 Version 1

In order to achieve the waveform requirements, it was decided to first complete a mask layout for the basic arrangement without output buffers and simulate this before proceeding further. The circuit realized is shown in Figure 11–23 and is a straightforward translation of the logic of Figure 11–22.

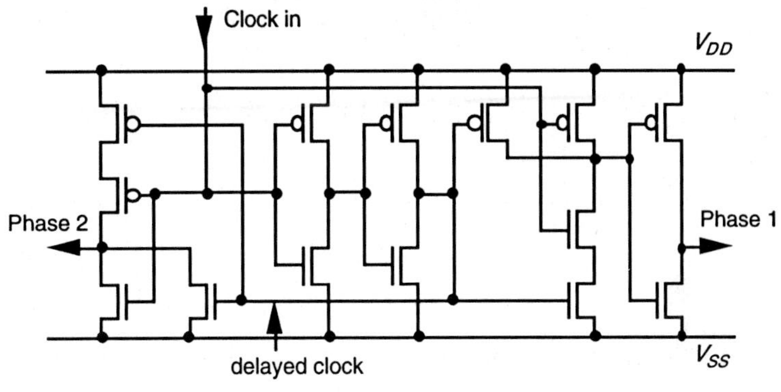

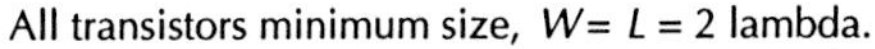

Figure 11–23 Basic two-phase clock generator circuit

The design rules used will be lambda-based with a value of $\lambda = 2\ \mu m$ for fabrication in single poly., single metal, p-well CMOS technology. All transistors are of minimum size, that is, $W = L = 2\lambda = 4\ \mu m$. The initial mask layout is given in Figure 11–24 (B&W copy of color pen plotter output) and corresponding 'H-Spice' simulation results at 10 MHz on zero external load in Figure 11–25.

The following observations are relevant:

1. The amount of 'underlap' between the phase 1 and phase 2 waveforms is barely adequate.
2. Phase 1 output rises faster than phase 2, and phase 2 peak voltage does not quite reach + 5 V owing to the time required for the *Nor* gate output to rise.
3. The square waves produced at each output are not particularly good and there is a noticeable 'glitch' on the phase 2 output.

11.5.3.2 Version 2

Clearly, all the above performance features could be improved by:

1. increasing the delay presently introduced by the two inverters in series;
2. reducing the output resistance of the final delay generating inverter so that the gate capacitance of the *Nor* gate will be charged faster, and also reducing the output resistance by widening the channels of the two p-type pull-up transistors of the *Nor* gate; and

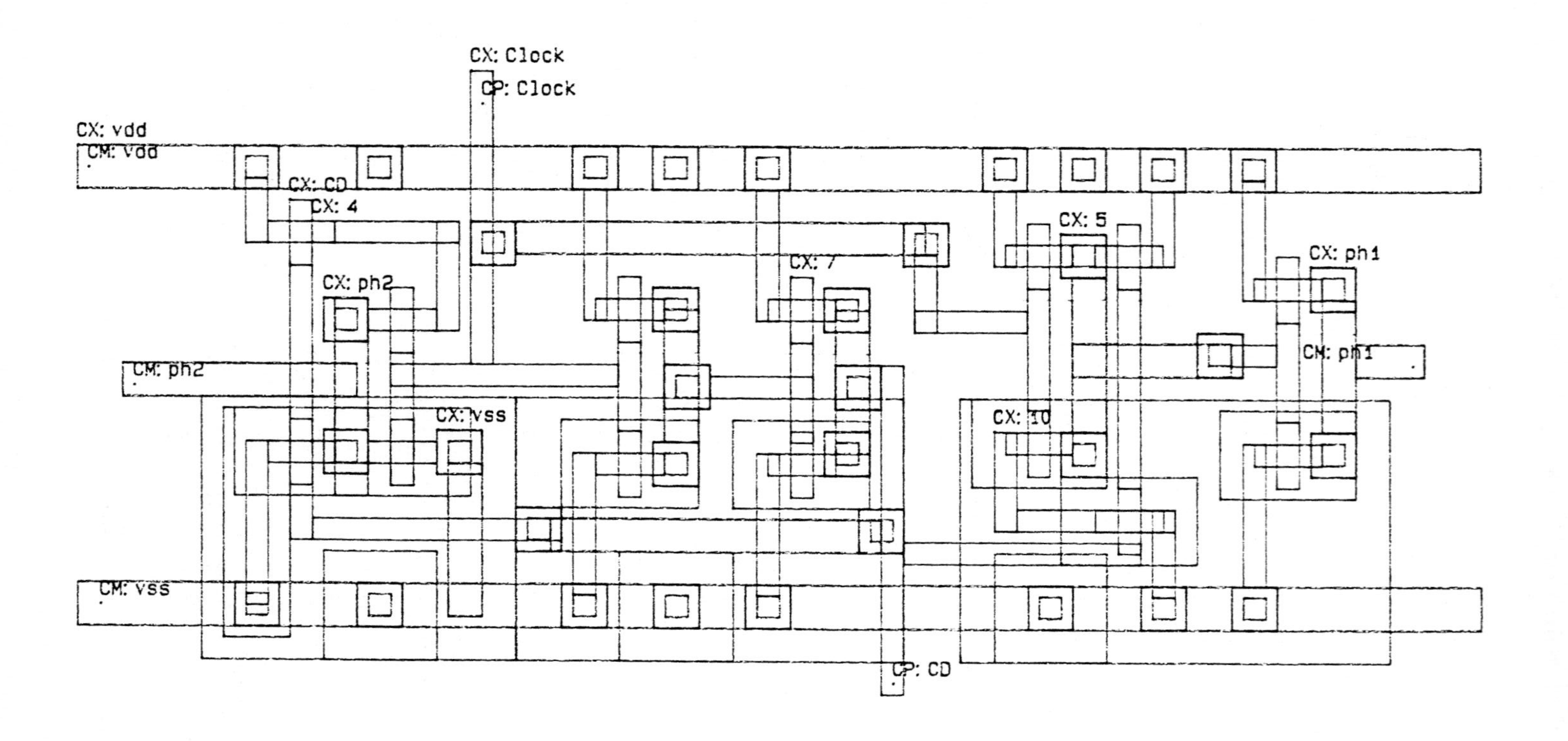

Figure 11–24 Mask layout (version 1) for two-phase clock generator circuit

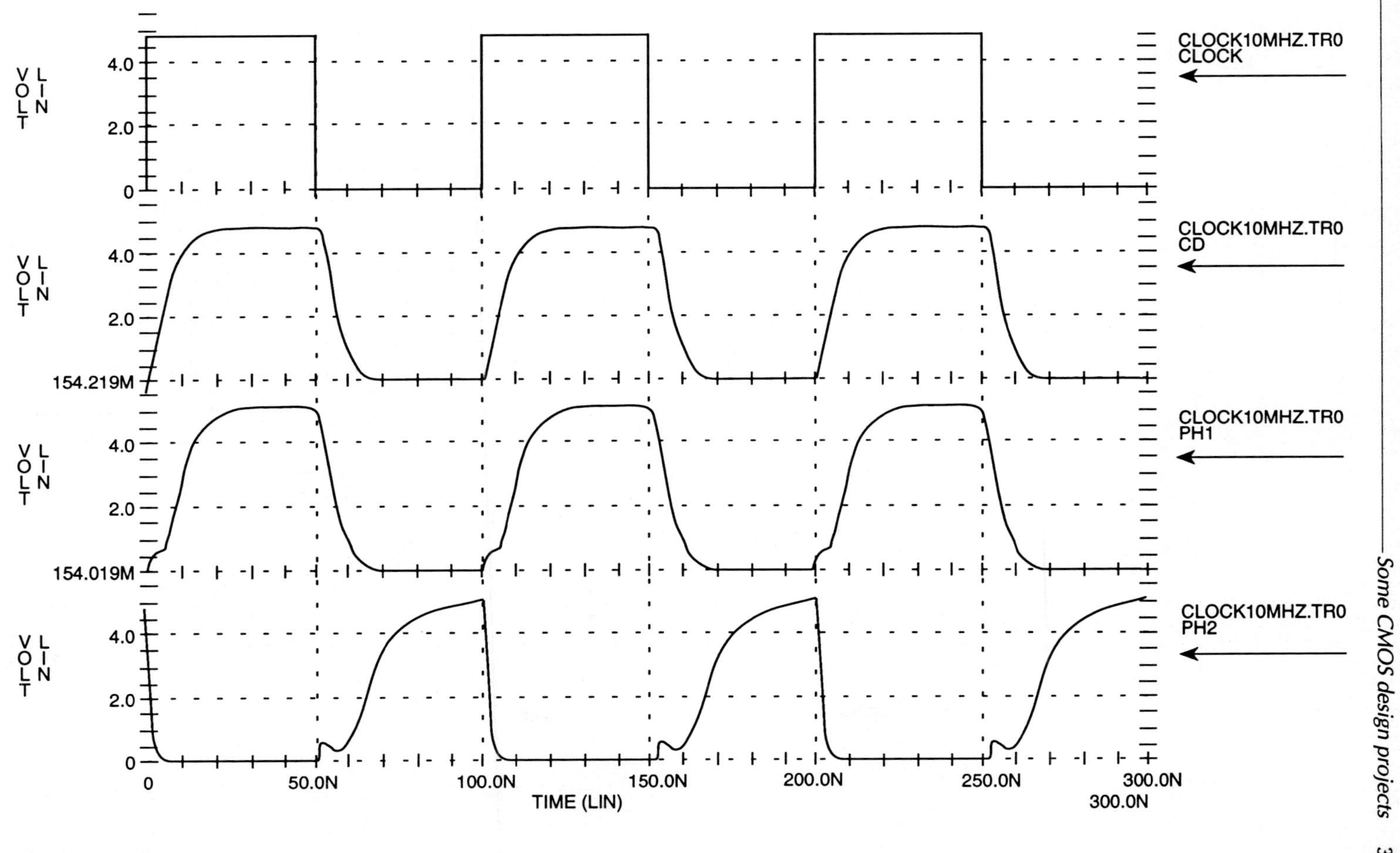

Figure 11–25 Simulation results for version 1

3. routing the delayed clock waveform to the p-type transistor closest to the output node of the *Nor* gate.

All the above points have been taken into account in version 2, noting that (1) the delayed clock waveform is now generated by four inverters in series and (2) that the driving capability has been improved by progressively decreasing the L:W ratio for the transistor channels in each inverter. Improvement (3) has been taken account of by rearranged connections to the *Nor* gate pull-up gates.

The circuit implementation now appears as Figure 11–26, the revised mask layout as Figure 11–27 and the corresponding simulation results as Figure 11–28.

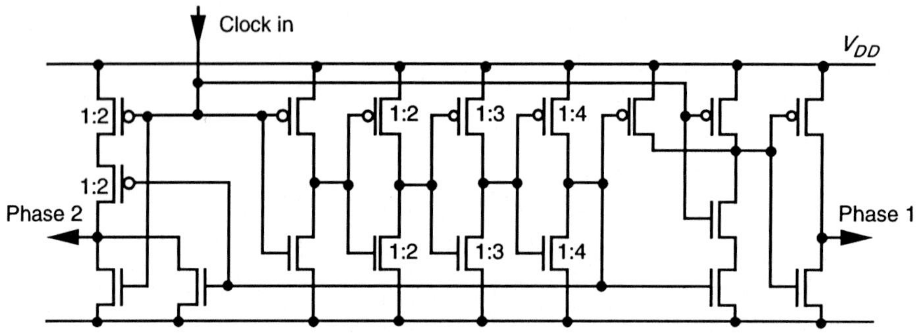

Figure 11–26 Circuit (version 2) for two-phase clock generator circuit

11.5.3.3 Version 3

Waveforms and delay are now predicted to be within acceptable limits and it now remains to add the output buffer at each output. An acceptable approach is to cascade inverters of increasing channel width as set out in Chapter 4, section 4.8.

In this case, the ratio

$$y = \frac{C_L}{\square C_g} = \frac{0.33 \text{ pF}}{0.01 \text{ pF}} = 33$$

(An approximate value of .01 has been assumed for $\square C_g$.)

The number N of cascaded stages is given by $N = \ln(y)$; thus in this case.

$$N = 3.5 \text{ (say, } N = 3).$$

Thus, we need three inverters in series, each one being ≑ 2.7 times (say 2.5) its predecessor's width. Noting the existing output inverter stage for phase 1, we need two additional buffer inverter stages to provide the phase 1 output. Three

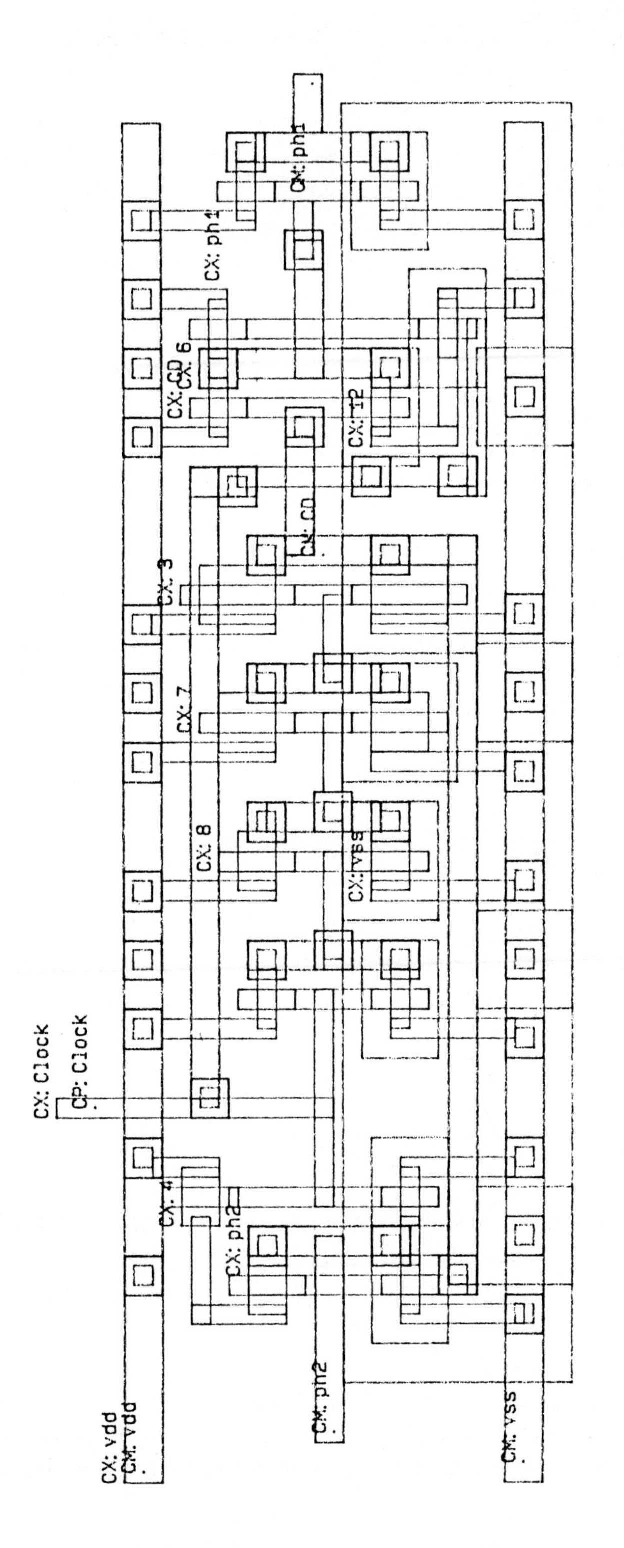

Figure 11–27 Mask layout (version 2) for two-phase clock generator circuit

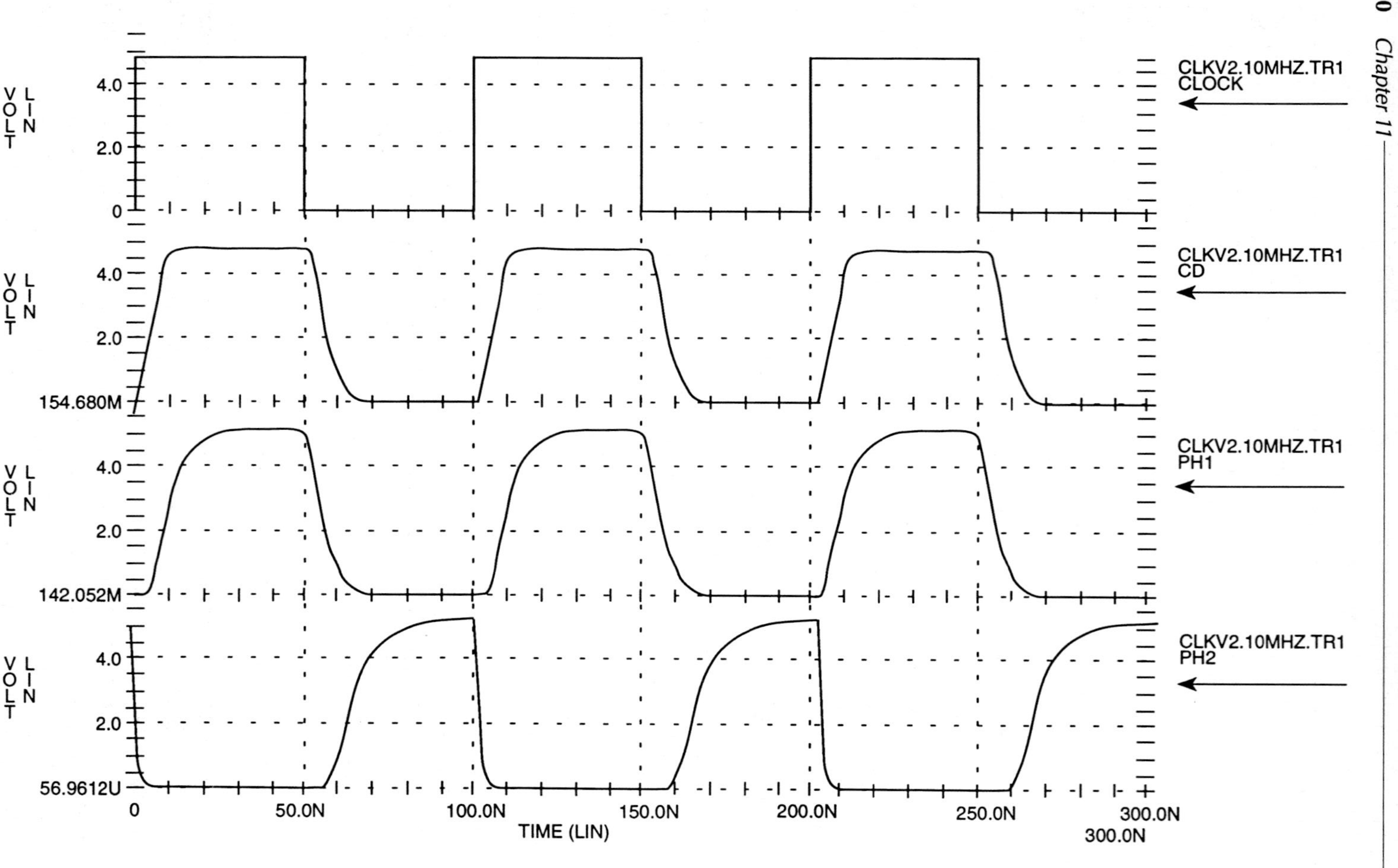

Figure 11–28 Simulation results for two-phase clock generator (version 2)

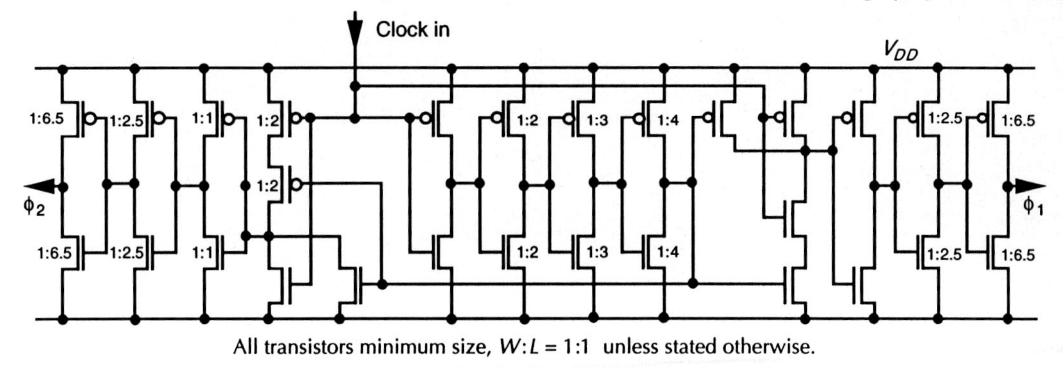

Figure 11–29 Circuit (version 3) for two-phase clock generator circuit

will be needed for phase 2 but a fourth is added to maintain the original phase relationship. The circuit is shown in Figure 11–29 and the modified mask layout is shown in Figure 11–30.

11.5.4 Final test (simulation) results

The final version (version 3) of the mask layout was first simulated on no external load and the waveforms generated at 10 MHz clock frequency. Near ideal non-overlapping square waves were observed as shown in Figure 11–31.

In order to assess the effect of increasing this maximum operating frequency, the input clock rate was doubled and results for a 20 MHz input clock were observed as in Figure 11–32. It will be seen that the performance of the circuit is still very good.

Finally, the outputs were loaded with load capacitance C_L which was increased in value until the slope of the clock edges began to erode the underlap between the phases. Figure 11–33 shows that acceptable waveforms are still generated even if the originally specified C_L value is exceeded by a factor of six times, indicating a very conservative design.

11.5.5 Further thoughts

The mask layouts presented here are those from which the simulation results were obtained. They should, however, be used with care since the yellow lines defining p+ areas do not copy in monochrome (B&W) form and are thus not apparent in the layouts reproduced here. Also, actual details of the appropriate technology design rules should be applied to a layout as necessary.

If larger capacitive loads are required to be driven, such as an output pad with associated off chip wiring etc., then two possibilities are:

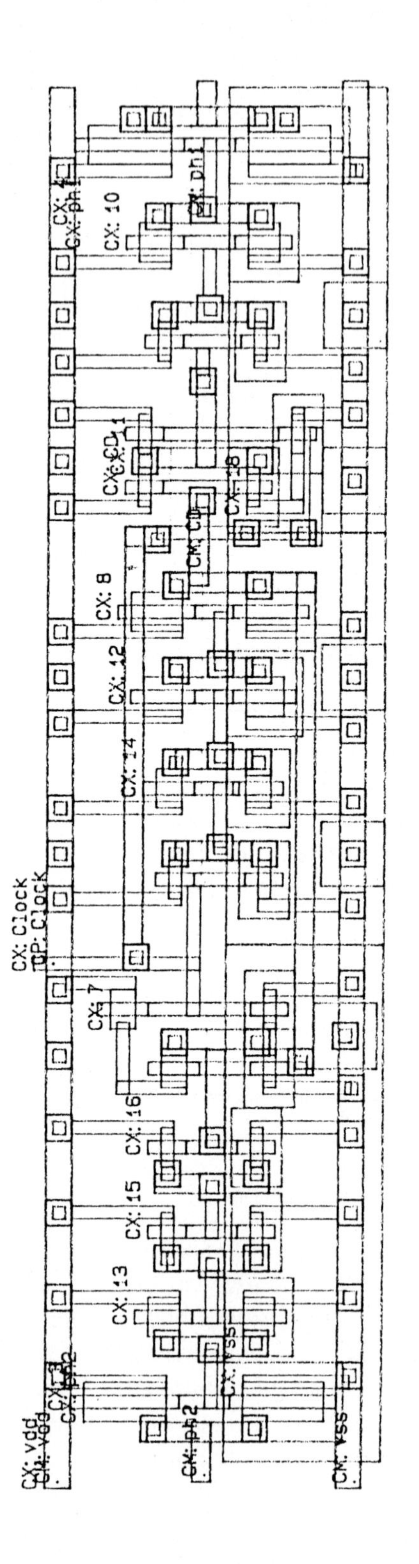

Figure 11–30 Mask layout (version 3) for two-phase clock generator circuit

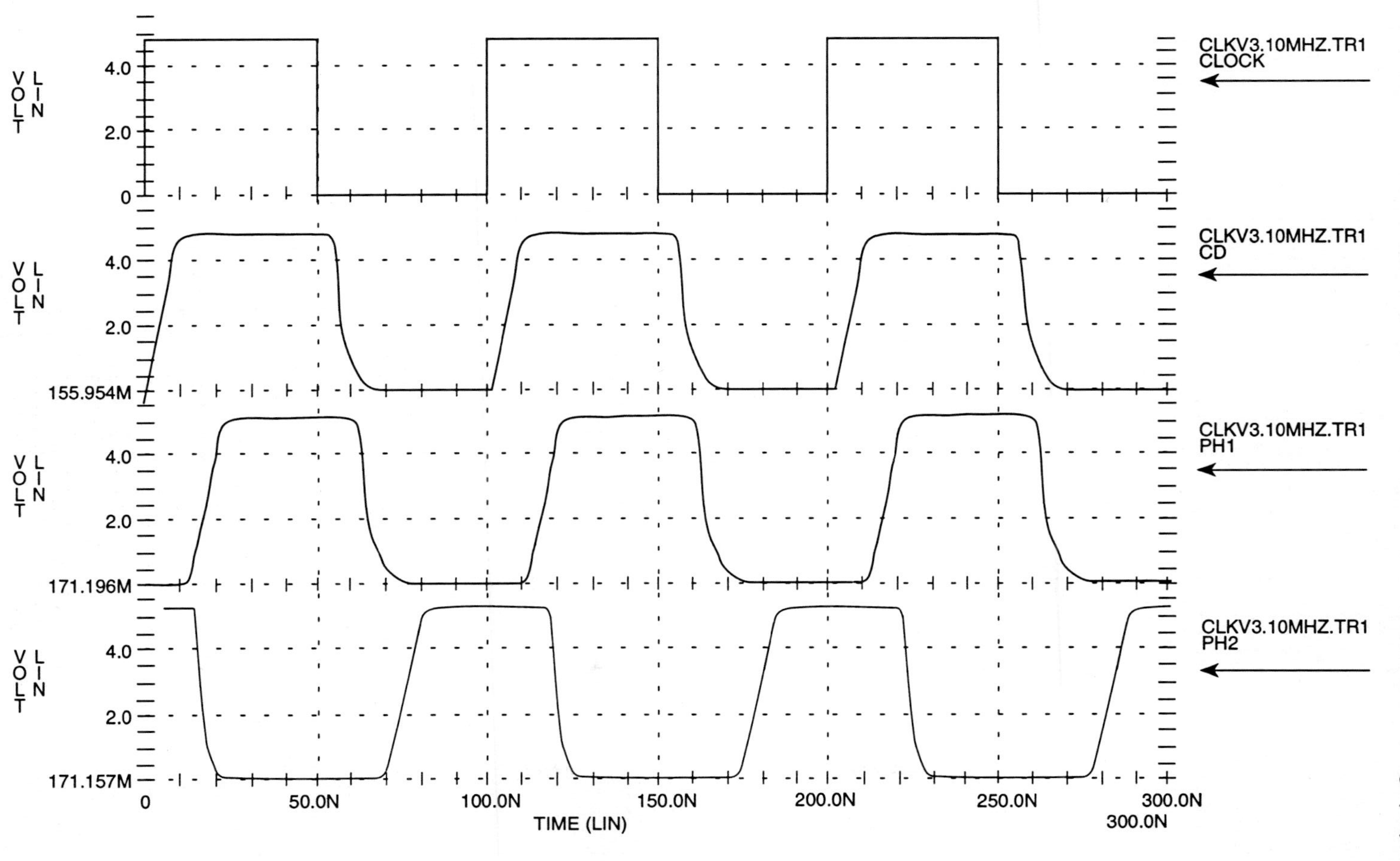

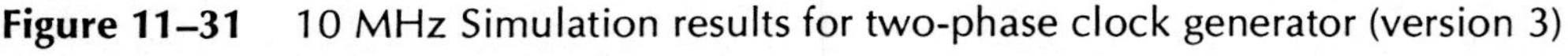

Figure 11–31 10 MHz Simulation results for two-phase clock generator (version 3)

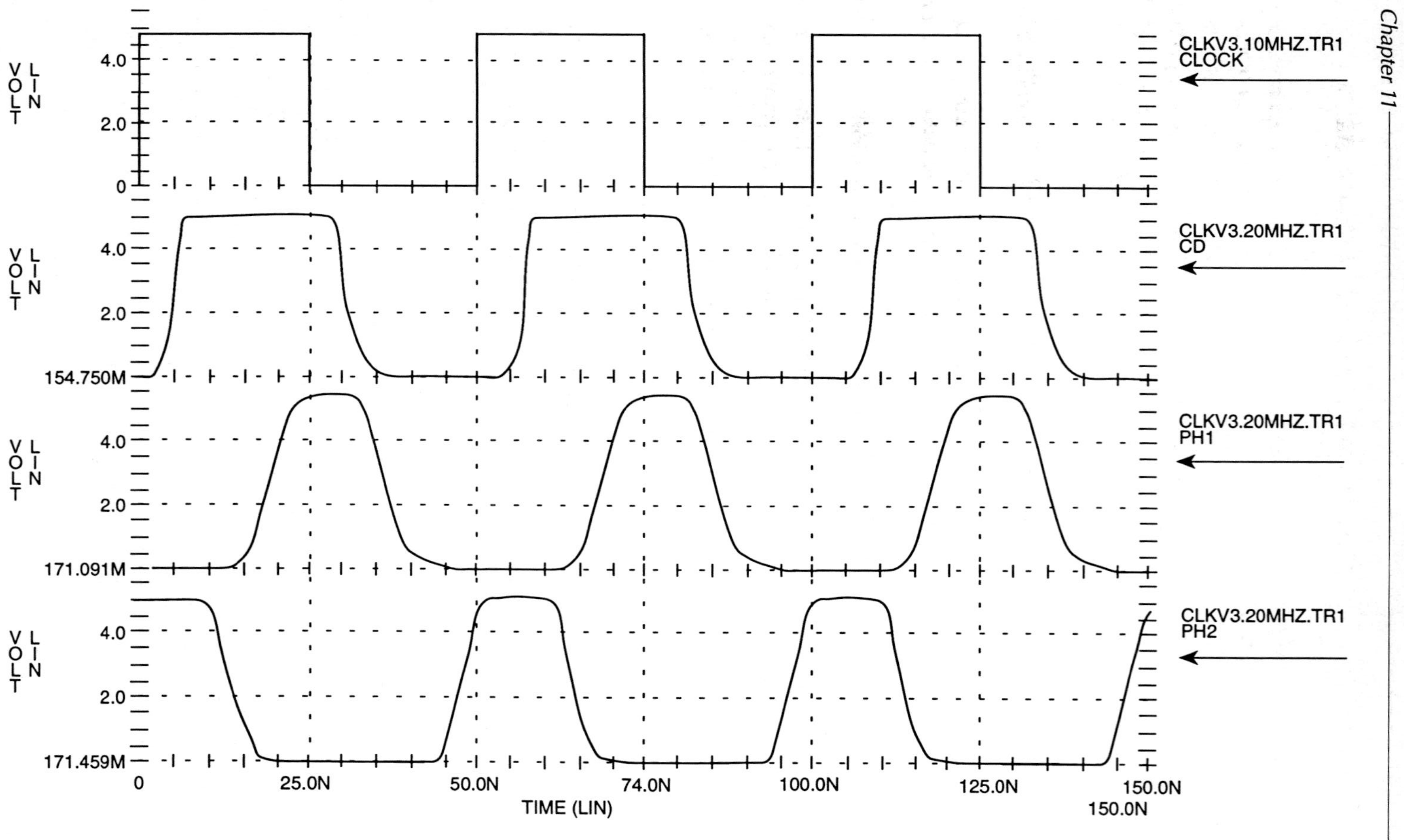

Figure 11–32 20 MHz simulation results for two-phase clock generator (version 3)

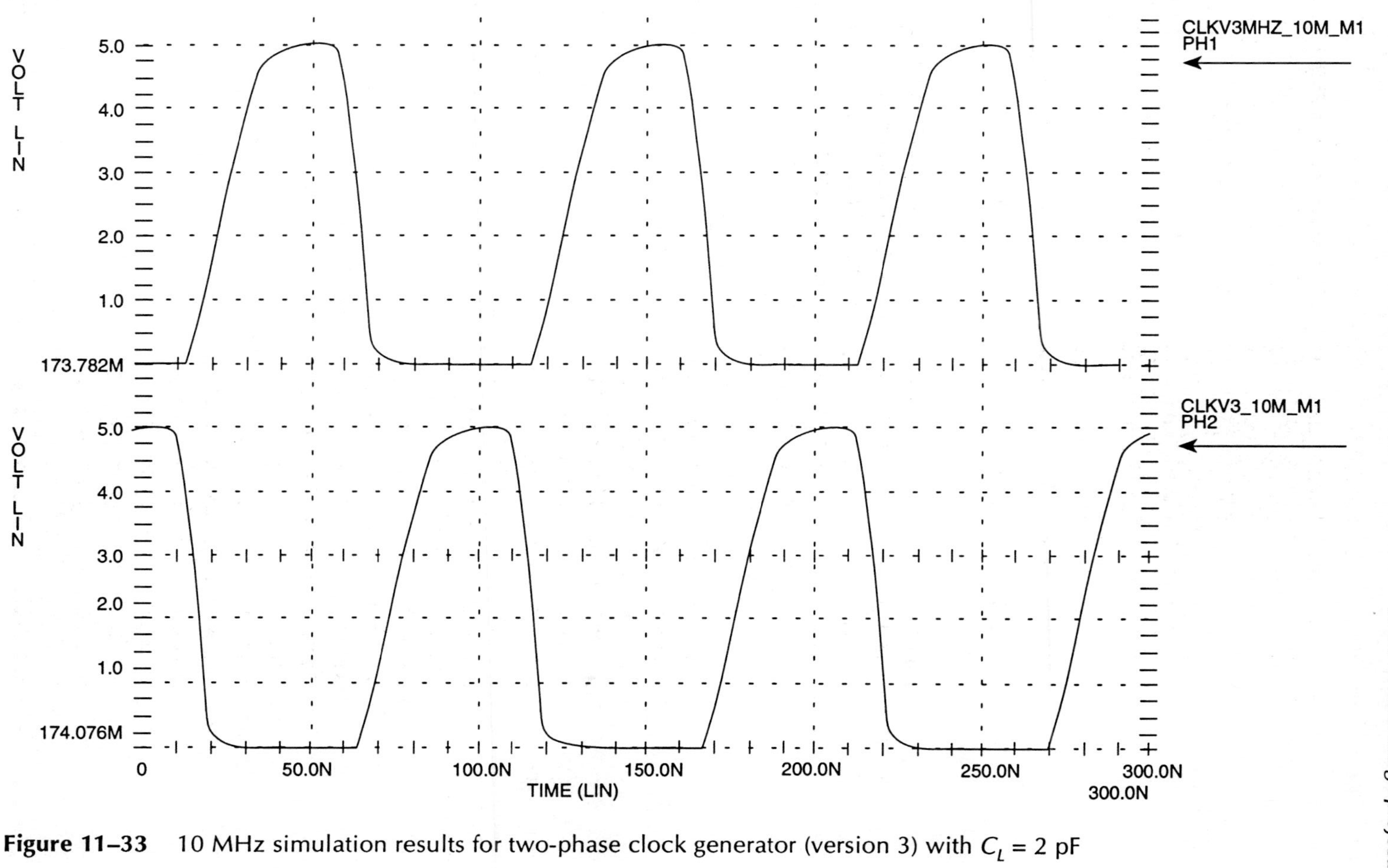

Figure 11–33 10 MHz simulation results for two-phase clock generator (version 3) with $C_L = 2$ pF

1. Increase the number of cascaded inverter buffers at each output.
2. If the technology in use caters for BiCMOS circuits, then redesign the two-phase generator to include BiCMOS output stages for each phase.

A further need may be for buffered complementary outputs, that is ($\overline{\text{phase 1}}$) and ($\overline{\text{phase 2}}$) to be also generated.

The mask layout for an arrangement which employs BiCMOS technology and provides four outputs —(phase 1), ($\overline{\text{phase 1}}$), (phase 2), and ($\overline{\text{phase 2}}$) — is presented as Color plate 12.

11.6 CMOS project 5* — design of a *∂latch* — an event-driven latch element for EDL systems

This project differs from the previous four since it is concerned with the design of an *event-driven* circuit element. In fact the design to be pursued is that of an event-driven latch (*∂latch*) which is part of ongoing developments in the field of event-driven-logic (EDL). Before a behavioral description can be set out, it is necessary to acquaint the reader with some basic aspects of EDL.

11.6.1 A brief overview of event-driven logic (EDL) concepts (Pucknell, 1993)

An alternative way of approaching the representation and design of asynchronous sequential logic is to take an 'event-driven' or 'transition-based' approach. In concept, the approach taken is to define the initial conditions of a system in terms of the logic level assumed by each variable and then describe subsequent system behavior in terms of the transitions (changes in logic level, also called events) of those variables. Clearly, if all events are defined for each variable, then subsequent logic level states are also defined. In order to pursue this approach, let us first examine some of the basic features and factors associated with the concept of 'event-driven' or 'transition-based' logic and logical operations.

11.6.1.1 An event-driven or transition-based approach to logic

In formulating event-driven logic (EDL), it is necessary to adopt special operators which readily express the *transitions or events* which may occur.

Transition operators and some basic relationships. The operators proposed are an extended set of the two originally proposed (Talantsev, 1959). Considering

*The design work on this latch was carried out by postgraduate reseacher Shannon Morton at the University of Adelaide as part of the digital systems group work on the application of EDL concepts to the design of asynchronous processors.

a single line carrying a logic signal denoted 'A', then at any time 't' there are four possibilities:

1. ΔA denoting a change in A from 0 to 1;
2. ∇A denoting a change in A from 1 to 0;
3. $\overline{\Delta}$ denoting no change in A at logic 0;
4. $\overline{\nabla} A$ denoting no change in A at logic 1.

Note the operators $\Delta \nabla \overline{\Delta}\ \overline{\nabla}$ and their significance.

Possibilities 1 and 2 may be defined as *'events'* and we may write:

$$\partial.A = \Delta A + \nabla A$$

where $\partial.A$ indicates *any event* for signal A.

Possibilities 3 and 4 may be defined as *'non-events'* and we may write:

$$\overline{\partial}.A = \overline{\Delta} A + \overline{\nabla} \mathbf{A}$$

the negated ∂ indicating *no event* for signal A.

11.6.1.2 Some bridging rules between EDL and 'conventional logic'

Clearly, there must be some relatively straightforward rules for converting between conventional and event-driven forms of logic and EDL elements may be constructed from conventional combinational logic circuits, as is also the case for clocked sequential elements.

The basic relationships are simple and may be proved quite readily mathematically or through a process of logical reasoning. Requirements are met by the rules given in Table 11–2. To illustrate the use of these rules, we may predict the transition behavior of a simple conventional *two-input And* gate. To do this, we start with the conventional logic equations and then apply the rules of Table 11–2.

Table 11–2 Simple bridging rules

Event-driven		*Conventional*	*In words*
$\nabla A + \overline{\Delta} A$	<=>	$\overline{A}$	A becomes 0 or remains at 0
$\Delta A + \overline{\nabla} A$	<=>	A	A becomes 1 or remains at 1
$(\Delta A + \nabla A + \overline{\Delta} A + \overline{\nabla} A$	<=>	1	All possible events for A
$\Delta A (\nabla A + \overline{\nabla} A + \overline{\Delta} A)$	<=>	0	'Anding' differing events for A

Where '<=>' indicates 'translates to' and should be considered in the context of what is actually meant in conventional logic when we write, say, $A = B.C$ or $\overline{A} = \overline{B} + \overline{C}$ etc.

$$X = A.B \text{ becomes}$$

$$\Delta X + \overline{\nabla} X = (\Delta A + \overline{\nabla} A).(\Delta B + \overline{\nabla} B)$$

$$= \Delta A.\Delta B + \Delta A.\overline{\nabla} B + \overline{\nabla} A.\Delta B + \overline{\nabla} A.\overline{\nabla} B.$$

A little thought will reveal that this equation comprises two parts:

1. the conditions for X to change from 0 to 1

$$\Delta X = \Delta A.\Delta B + \Delta A.\overline{\nabla} B + \overline{\nabla} A.\Delta B$$

2. the conditions for X to remain at logic 1

$$\overline{\nabla} X = \overline{\nabla} A.\overline{\nabla} B$$

Similarly, starting with the complementary form of the expression

$$\overline{X} = \overline{A} + \overline{B}$$

we may arrive at expressions for

3. $$\nabla X = \nabla A + \nabla B$$

and

4. $$\overline{\Delta} X = \overline{\Delta} A + \overline{\Delta} B$$

Taking events only:

$$\Delta X = \Delta A.\Delta B + \Delta A\,\overline{\nabla} B + \overline{\nabla} A.\Delta B$$

and $$\nabla X = \nabla A + \nabla B.$$

These are the EDL equations defining the conditions for X to change from 0 to 1 and from 1 to 0 respectively. EDL equations can be written for any gate. For example, a two input *Nor* gate (inputs *A* and *B*, output *Y*) can be represented by:

$$\nabla Y = \Delta A + \Delta B \text{ and } \Delta Y = \nabla A \nabla B + \nabla A\,\overline{\Delta} B + \overline{\Delta} A \nabla B$$

Clearly, then, the behavior of simple combinational logic gates may be expressed in terms of events. Note, however, that the common combinational logic gates may well not generate simple EDL functions but it is possible to conceive a specifically designed set of EDL gates which perform straightforward EDL functions but which, in turn, may not generate simple combinational logic functions. The exception is the *Exclusive Or* gate which is the point of intersection between the two gate sets.

11.6.1.3 The inverter as an EDL element

The inverter converts one transition of its input variable (e.g. 0 to 1) into the other transition (1 to 0 for the example) at the output. It also quite clearly converts a logic level at the input to its complement at the output. The inverter will *not convert events into non-events or vice versa* unless it is faulty.

11.6.1.4 Other EDL elements

So far, this discussion has covered the EDL aspects of gate logic circuits and we may now turn attention to the application of EDL concepts to the design of storage elements. EDL storage elements will be driven and activated by events on specified control inputs. For example, the event-driven latch to be discussed here is activated by events on *pass* and *capture* control inputs.

11.6.2 Behavioral description of a *∂latch*

The circuit is required to accept a single input, pass this to a single output when any event occurs on a *pass* (*p*) control line and latch this output when any event occurs on a *capture* (*c*) control line. The basic, most general, arrangement is shown in symbolic form in Figure 11–34 and it may be seen that a delayed version of each event control line, namely *pass done* (*pd*) and *capture done* (*cd*), is presented as an output control signal. A *Clear* (*clr*) input is also required. In a particular configuration, the *pd* output provides the *c* input and the delay through the two inverters is sufficient to allow the select line of the latch input switch to go high long enough for data to propagate through the latch from input to output before it is captured. Thus, the whole latch action is controlled by events on the *p* input line. It is that version which is to be implemented here.

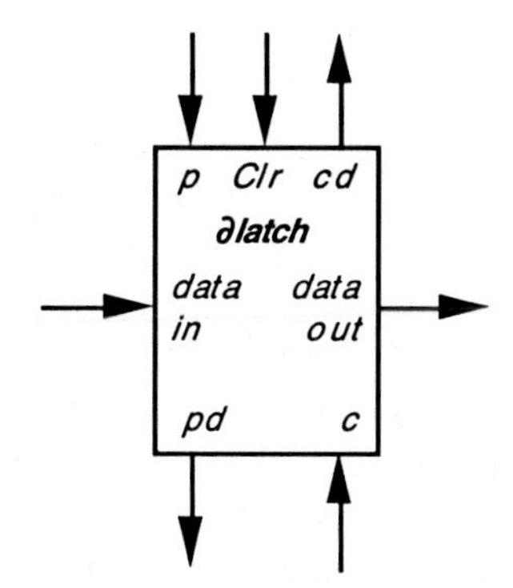

Figure 11–34 Symbolic form of *∂latch* element

11.6.3 Structural description

The structure of a suitable basic latch circuit arrangement is quite simple, comprising three inverter pairs, an *Xor* gate and a switch (multiplexer) as shown in Figure 11–35(a). Note that, in this case, *pd* and *c* will be joined, as in Figure 11–35(b), internally in the mask layout.

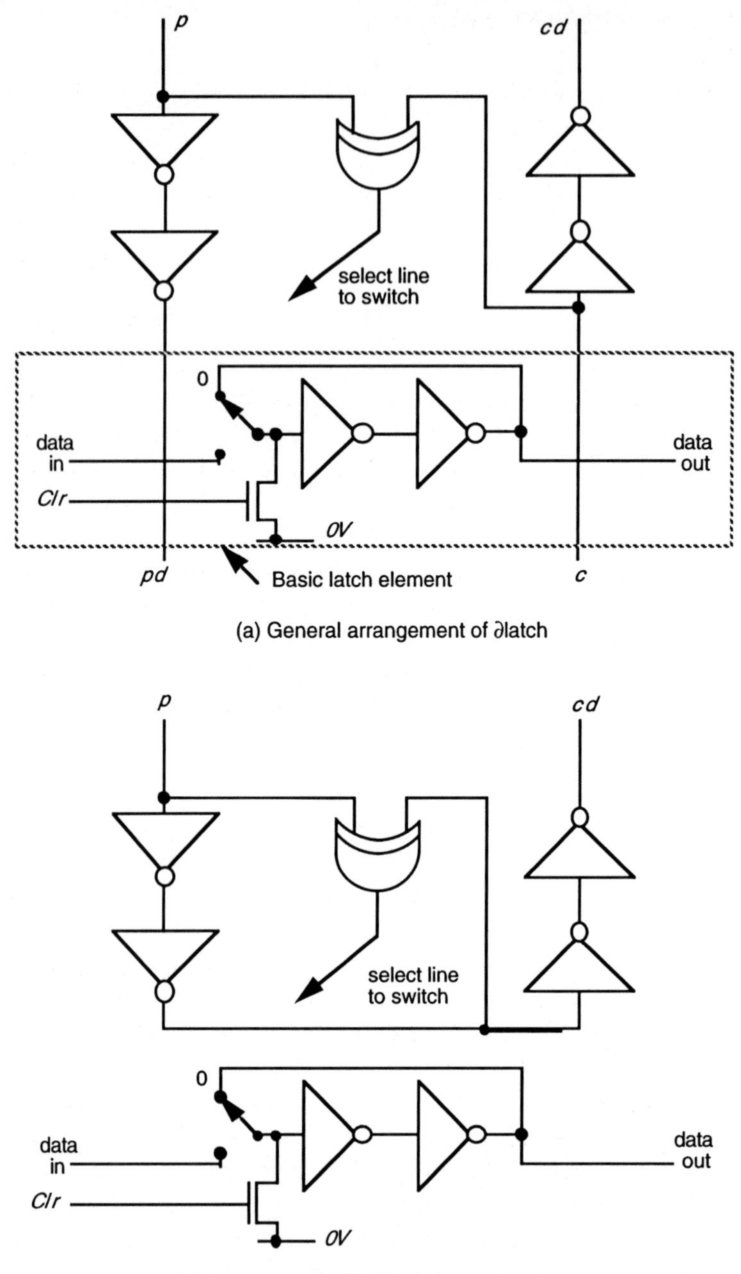

Figure 11–35 Basic arrangement of the *∂latch* element

11.6.4 Circuit action

The select line is generated by an *Xor* gate between inputs *p* and *c*. The *done* events will occur after the select line has reached its new state which will activate the actual latching/storage part of the circuit. This consists of two pass transistor switches with a supporting pull-up transistor, a two inverter buffer/driver, and a clear transistor. If one wishes to latch more than 1-bit of data, then it is this part of the circuit alone which must be replicated, for example 16 times for latching a 16-bit word.

When the select line goes high, the input logic level is connected to the buffer/driver through one of the pass transistor switches and the output of the buffer/driver will assume the same logic level. If this is a logic 1, the logic level would be degraded by the threshold voltage of the pass transistor, but the output (logic 0) of the first inverter of the buffer/driver is used to turn on a p-type pull-up transistor which acts as a pull-up to the output of the pass transistor, thus restoring a good logic 1 level. When the latch enters the *capture* state, the select line goes low, the input pass transistor switch is turned off and the other pass transistor switch is turned on, thus connecting the input of the buffer/driver pair to its output. Thus, the data is latched.

The *clear* line is inactive when low but when enabled with a logic 1, the pass transistor switch output node is forced low and will remain low even if the select line goes high and the logic level at the input is a 1.

11.6.5 Mask layout and performance simulation

The translation of the latch circuit into a mask layout is conveniently achieved using either a symbolic entry editor or a direct mask entry editor. In either case, the technology chosen will determine absolute widths, separations and overlaps and will also determine C and R values for the various layers.

In this case, the geometry of a suitable mask layout is given in Figure 11–36 and network extraction and simulations have been carried out in both 5 μm and 1.2 μm double metal, single poly., p-well CMOS technologies. Simulation results are given in Figure 11–37 (5 μm) and Figure 11–38 (1.2 μm). Noting the differing time scales used to plot the simulation results, it can be seen that the 1.2 μm latch is faster than the 5 μm latch by a factor of approximately 5. This compares favorably with the theoretical speed-up factor = $^{5}/_{1.2} \approx 4.2$. Propagation times through the latch are approximately 5.4 nsec for the 5 μm design and 1.1 nsec for the 1.2 μm design.

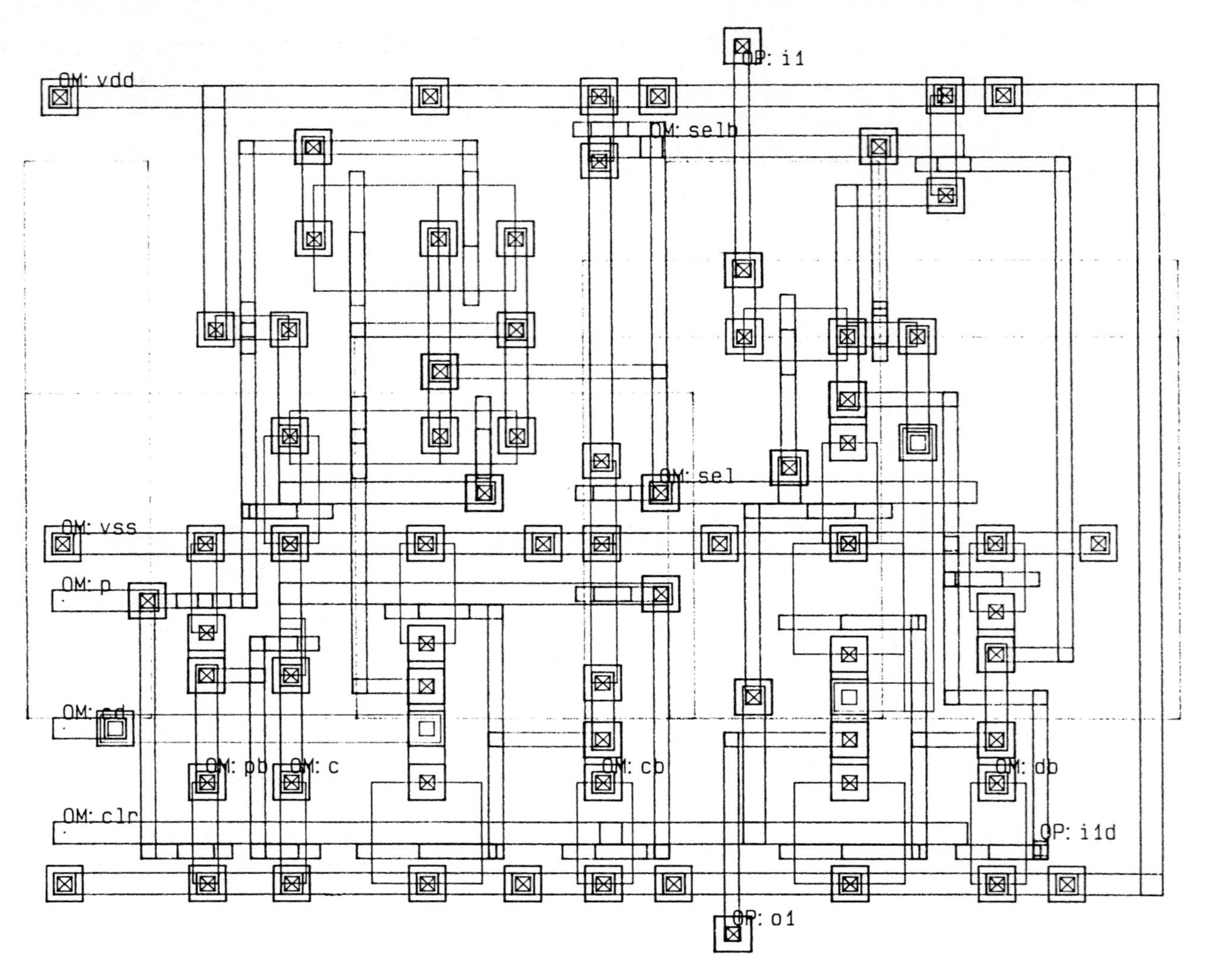

Figure 11–36 A mask layout for the ∂latch

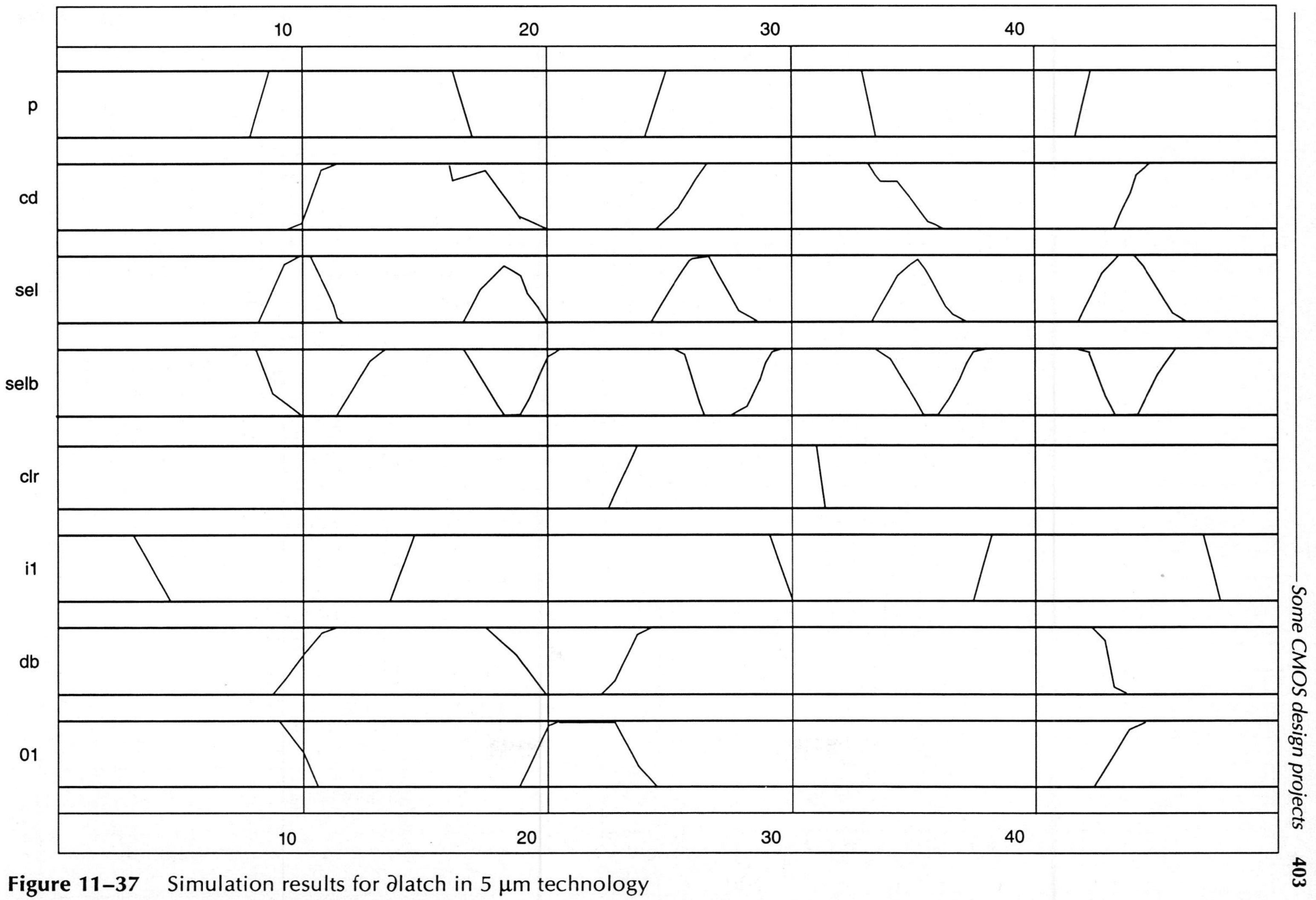

Figure 11–37 Simulation results for ∂latch in 5 μm technology

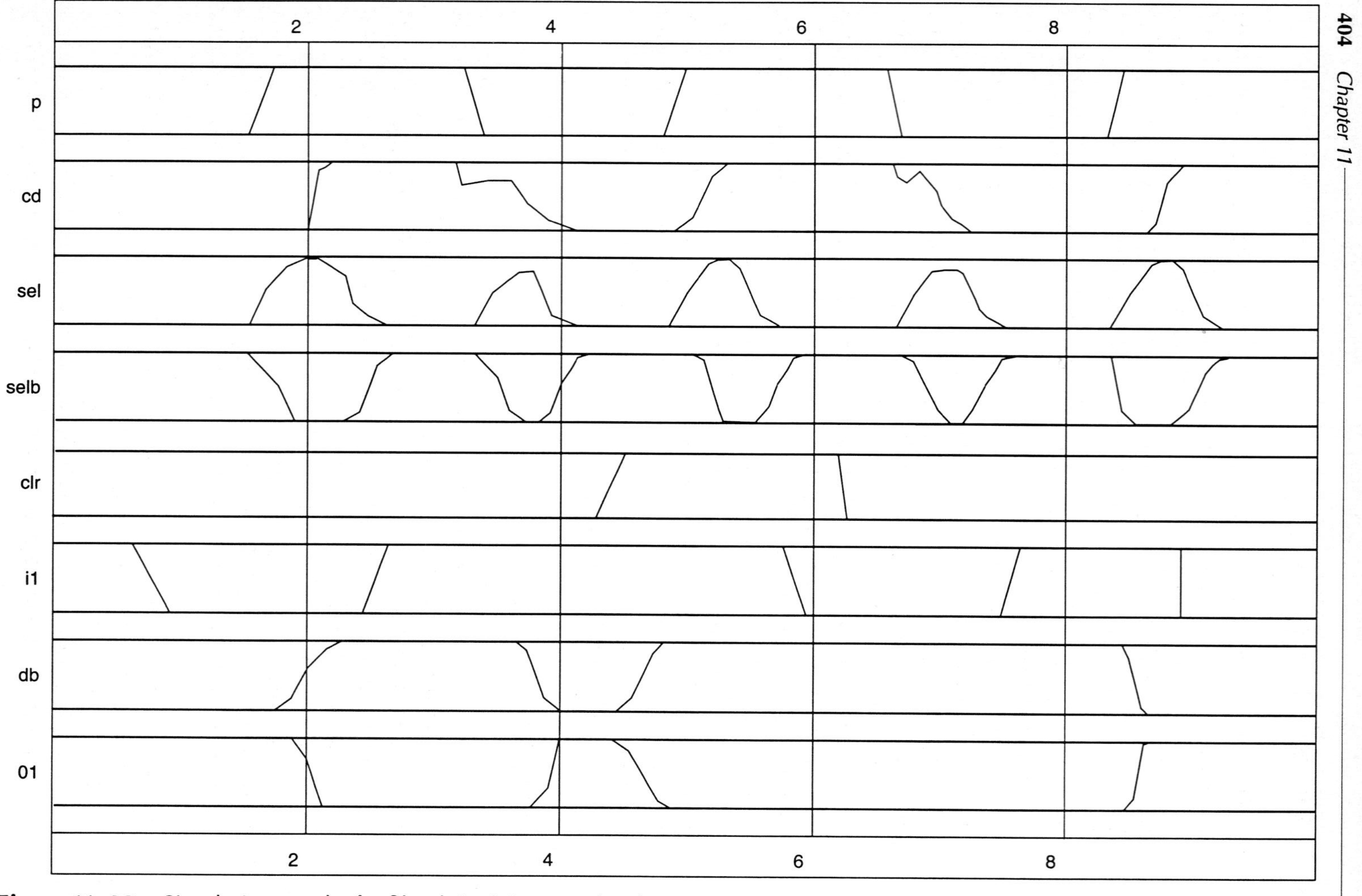

Figure 11–38 Simulation results for ∂latch in 1.2 μm technology

11.7 Observations

We have seen that the design process for the design of digital systems in silicon is a reasonably straightforward proposition, provided that an orderly, structured approach is taken. The tutorials, exercises, and project work in the text have illustrated approaches to design, and readers should by now begin to feel comfortable in their ability to tackle the design of systems of modest size and complexity. An ability to understand the characteristics of the available technologies and the design processes should enable system designers to specify an appropriate technology and, where necessary, design 'custom' digital chips.

This text has not attempted to seriously address the problems of complexity management and the design time associated with the design of large digital systems. We have also largely ignored the ever-growing need for custom-designed analog circuits in MOS technologies, both for pure analog applications and for 'on-chip' interfaces between the analog world and digital systems.

We have seen that there are factors which limit the ultimate scaling of silicon circuits and thus there are ultimate limitations on the speed of silicon circuitry. This will not be a problem in any but the fastest areas of application, but emerging needs in real-time control and in signal processing applications, to name just two, may well impose needs beyond the capability of MOS silicon systems alone. It is in such applications that other technologies, in particular gallium arsenide, will find application as fast 'front-end' processors to silicon systems. To introduce the reader to this important area, the next, and final, chapter introduces gallium arsenide technology.

11.8 References

Pucknell, D. A. (1970) 'Transition equations for the analysis and synthesis of sequential circuits', *IEE. Electron. Lett.*, 6 (23), 731–33.

Pucknell, D. A. (1993) 'An event-driven-logic (EDL) approach to digital system representation and related design processes', *IEE. Proc.-E, Computers and Digital techniques*, Vol. 140, No. 2, 119–26.

Pucknell, D. A. (1973, May) 'Sequential circuit characterisation and synthesis using a transition equation approach', *Proc. IEE.*, 120(5), 551–56.

Pucknell, D. A., and Liebelt, M. J. (1990, July) 'Aspects of event-driven logic', Proc. 9th Australian Microelectronics Conference, Adelaide, South Australia, 171–73.

Smith, J. R., and Roth, C. H. (1971) 'Analysis and synthesis of asynchronous sequential networks using edge-sensitive flip-flops', *IEEE. Trans. Comput.*, C-20, 847–55.

Talantsev, A. D. (1959) 'On the analysis and synthesis of certain electrical circuits by means of special logical operators', *Autom. and Telemech.*, 20, 895–907.

Ultra-fast VLSI circuits and systems — introduction to GaAs technology

There was a young lady named Bright,
Whose speed was far faster than light,
She set out one day
in a relative way,
And returned home the previous night.

Arthur Henry Buller

12.1 Ultra-fast systems

In this final chapter we will briefly review some of the limitations of silicon devices and then look at the emerging alternative for ultra-fast systems — gallium arsenide.

12.1.1 Submicron CMOS technology

Speed and smaller device dimensions are closely interrelated, and we have already touched on the fact that the foreseeable limits on channel length for MOS transistors is in the region of 0.14 μm, after which further scaling down results in unworkable transistor geometry.

In CMOS devices we have also seen that the p-transistors have inherently slower performance than similar n-transistors. This is primarily due to the lower mobility of holes compared with that of electrons. Typically

$$\mu_p \div 240 \text{ cm}^2 / \text{V.sec}$$
$$\mu_n \div 650 \text{ cm}^2 / \text{V.sec}$$

In long-channel devices this means a difference in current drive transition times of about 2.5:1. However, as the channel lengths are scaled down, the influence of mobility starts to diminish as the effects of velocity saturation begin to be felt.

For long-channel MOS transistors, the current/voltage relationship below saturation can be approximated by

$$I_{ds} = \frac{W\mu C_{OX}}{L}[(V_{gs} - V_t)V_{ds} - 0.5V_{ds}^{2}]$$

where

$$C_{OX} = \text{gate/channel capacitance per unit area}$$

$$= \frac{\varepsilon_{ins}\varepsilon_0}{D}$$

This implies that current drive is proportional to mobility and inversely proportional to channel length.

Transconductance g_m is similarly influenced. When velocity saturation occurs along the entire channel length, then the current/voltage relationship is given by

$$I_{dsat} = WC_{OX}v_{sat}(V_{gs} - V_t)$$

where v_{sat} is the *saturation velocity*. Current is now independent of both mobility and channel length but dependent on the saturation velocity. Transconductance is constant and thus independent of channel length.

It should be noted that velocity saturation occurs at lower electric field strengths in n-devices owing to their higher mobility when compared with p-devices. Thus, as dimensions are scaled down, the current drive from n-transistors tends to a constant value independent of channel length while the current drive from p-transistors does not tend to a constant value until, at a shorter channel length, the holes start to run into velocity saturation. We must therefore look to other than silicon-based MOS technology to provide for the faster devices which will undoubtedly be required as the sophistication of our system design capabilities increases. An alternative technology is based on gallium arsenide.

12.1.2 Gallium arsenide VLSI technology

He that will not apply new remedies must expect new evils:
for time is the greatest innovator.

Francis Bacon

Silicon MOS technology has been the main medium for computer and system applications for a number of years and will continue to fill this role in the foreseeable future. However, silicon logic has speed limitations that are already becoming apparent in state-of-the-art fast digital system design. Paralleling developments in silicon technology, some very interesting results have emerged for gallium arsenide (GaAs)-based technology. Gallium arsenide will not displace silicon but is being used in conjunction with silicon to satisfy the need for very high speed integrated (VHSI) technology in many new and innovative systems.

Much of the development work in material technology that has paralleled that in silicon has been related to groups II–VI and groups III–V compounds, with gallium arsenide, a group III–V compound, showing the most promise.

The compound gallium arsenide was discovered in 1926. However, its potential as a high speed semiconductor was not realized until the 1960s. The high speed electron mobility of gallium arsenide with respect to silicon, a semi-insulating substrate with consequent lower parasitics, a 1.4 improvement factor for carrier saturation velocity of GaAs over silicon, its opto-electrical properties, as well as significant improvement in power dissipation and radiation hardness, have promised an ultimate system performance advantage for gallium arsenide products, given similar lithographical processes.

The developments in integrated circuit fabrication technology in the 1970s made such gallium arsenide products a possibility and finally, as the result of significant advances in ion implantation in the 1980s, GaAs VLSI technology is a commercial reality for the 1990s.

Therefore, in the sections to follow we are going to concentrate on this new material and explore the various possibilities that exist to design circuits and systems using an appropriate class of logic together with suitable design methodology for this technology.

12.2 Gallium arsenide crystal structure

Gallium (Ga), a toxic material, is produced as a byproduct in both the zinc and aluminum production processes. Similarly, arsenic (As), which is also very toxic, is produced from ores such as As_2S_3 or As_2S_4. The process entails firstly oxidation of the ores to form As_2O_3 and subsequently, through reduction with carbon, arsenic is produced.

In order to better appreciate the structure and the properties of gallium arsenide crystal, it is appropriate to focus some attention on the characteristics of the individual atoms themselves. Figure 12–1 shows Bohr's model of the atomic structures for gallium and arsenic. Similar representation for silicon is also illustrated for comparison.

Gallium possesses a positively charged nucleus of +31, while the arsenic atom's nucleus has a positive charge of +33. In each case, the total positive charge of the nucleus is equalized by the total effective negative charge of the electrons.

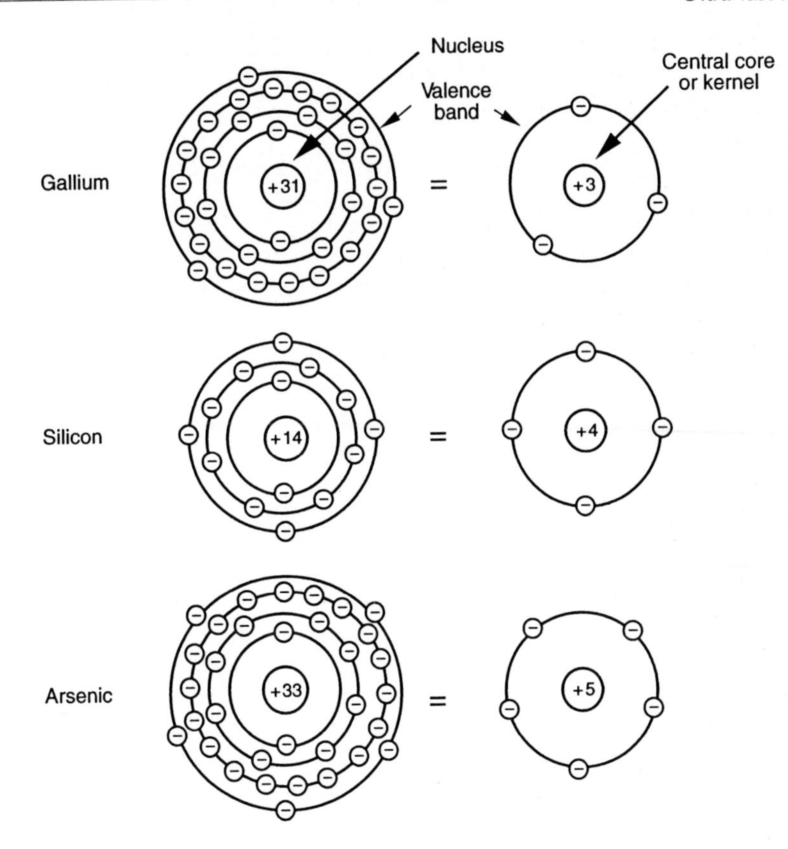

Figure 12–1 Bohr's model for silicon, gallium and arsenic

Electrons, traveling within their respective orbits, possess energy since they are a definite mass in motion (i.e. rest mass of electron is $9.108*10^{-23}$gm). This means each electron in its relationship with its parent nucleus exhibits an energy value and functions at a distinct energy level. This energy level is dictated by the electron's momentum and its physical proximity to the nucleus. The closer the electron is to the nucleus, the greater is the holding influence of the nucleus on the electron and the greater is the energy required for the electron to break loose and become free.

Outer orbit electrons are said to be stronger than inner orbit electrons because of their ability to break loose from the parent atom, and as a result they are referred to as *'valence electrons'*. The outer orbit in which valence electrons exist is called the *'valence band'*. It is the electrons from this band that are being considered in much of the discussions in the section to follow.

Crystal chemical bonds result through sharing of valence electrons. In materials such as Si, Ga and As, the outer-shell valence configuration can be represented by

$$Si \rightarrow 3s^2\ 3p^2$$
$$Ga \rightarrow 4s^2\ 4p^1$$
$$As \rightarrow 4s^2\ 4p^3$$

Here the core is not shown and the superscripts denote the number of electrons in the subshells (i.e. s and p orbitals). With this concept in mind, the structure of the atoms shown in Figure 12–2 can be simplified by representation as in Table 12–1.

Table 12–1 Periodic table

GROUP II	GROUP III	GROUP IV	GROUP V	GROUP VI
+2, 2 × −e	+3, 3 × −e	+4, 4 × −e	+5, 5 × −e	+6, 6 × −e
$Be^{4}_{9.01}$	$B^{5}_{10.82}$	$C^{6}_{12.01}$	$N^{7}_{14.008}$	$O^{8}_{16.0}$
$Mg^{12}_{24.32}$	$Al^{13}_{26.97}$	$Si^{14}_{28.09}$	$P^{15}_{31.02}$	$S^{16}_{32.07}$
$Zn^{30}_{65.38}$	$Ga^{31}_{69.72}$	$Ge^{32}_{72.60}$	$As^{33}_{74.91}$	$Se^{34}_{79.0}$
$Cd^{48}_{112.4}$	$In^{49}_{114.8}$	$Sn^{50}_{118.7}$	$Sb^{51}_{121.8}$	$Te^{52}_{127.6}$

Note: Numbers in the table refer to the atomic number and the atomic weight.

12.2.1 A compound semiconductor

Gallium arsenide is a compound semiconductor which may be defined as a semiconductor made of a compound of two elements (as opposed to silicon, which is a single element semiconductor). From Table 12–1, which shows the materials in a periodic table, it is possible to deduce the manner in which III–V semiconductors can be produced. For example, gallium, having three valence electrons, can be combined with arsenic, which has five valence electrons, to form the compound GaAs.

Figure 12–2 shows the arrangement of atoms in a gallium arsenide substrate material. Note the alternate positioning of gallium and arsenic atoms in their exact crystallographic locations. Since gallium arsenide is a binary semiconductor, special care is required during the processing to avoid high temperatures that could result in dissociation of the surface, this being one of the basic difficulties in the growth of GaAs bulk material.

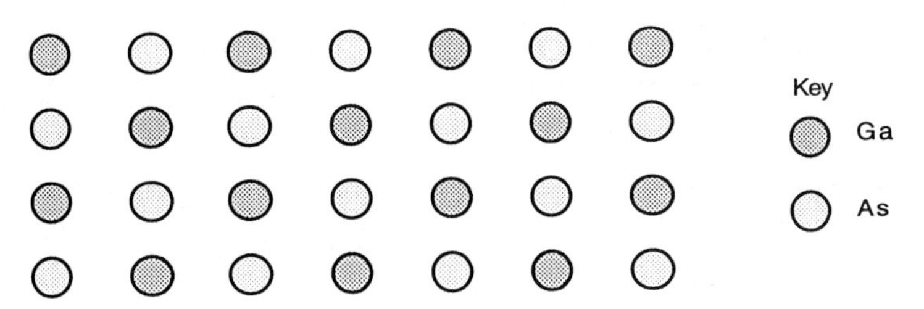

Figure 12–2 Arrangement of atoms in GaAs substrate

12.2.2 Doping process

Much as it is with silicon, it is necessary to introduce impurities into the semi-insulating GaAs material in order to facilitate the creation of switching devices. Selection of the impurity and its concentration density determine the behavior of the switching element. According to the dopant used, both n-type and p-type material can be realized.

12.2.2.1 n-type material

Group IV elements such as silicon can act as either donors (i.e. on Ga sites) or acceptors (i.e. on As sites). Since arsenic is smaller than gallium and silicon (the covalent radius for Ga is 1.26 Å and for As is 1.18 Å), group IV impurities tend to occupy gallium sites. Thus, silicon is used as the dopant for the formation of n-type material as shown in Figure 12–3.

The shrinkage of atomic radii across a given row of the periodic table (Table 12–1) can best be explained by noting that in any given period, electrons are added to s and p orbitals, which are not able to shield each other effectively from the increasing positive nuclear charge. Thus an increase in the positive charge of the nucleus results in an increase in the effective nuclear charge, thereby decreasing the effective atomic radius. This is why, for example, an As atom is smaller than a Ga atom.

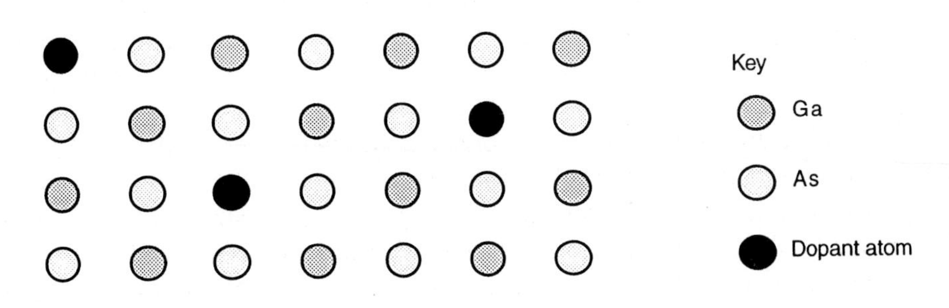

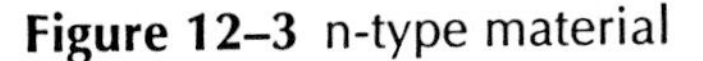
Figure 12–3 n-type material

12.2.2.2 *p-type material*

Beryllium (Be) or magnesium (group II) can be used for the formation of p-type material. Since Be is the lightest p-type dopant for GaAs, deep implantation of the dopant atoms can be accomplished with relatively less lattic damage. Nevertheless, Mg is also finding its way as a suitable dopant in a number of processes. Formation of p-type material is fundamental to both JFET and CE-JFET (i.e. complementary JFET) processes, to be described in the later part of this chapter.

12.2.3 Channeling effect

The whole concept of crystal orientation becomes important during

- the etching of the crystal;
- ion implantation;
- passivation.

This introduces an 'orientation dependency' that influences the properties of GaAs field effect transistor. For example, during implantation, when a high energy ion enters a single crystal lattice at a critical angle to the major axis of the GaAs crystal, the ion is steered down the open directions of the lattice. This steering is called *axial channeling*. This implies that if a random equivalent direction is not used during ion implantation, the depth distribution will be greater than those predicted by range statistics which are used to establish penetration depth.

The channeling effect is not as dramatic in the <100> direction when compared with <110> direction. Many of the current GaAs wafers employ the <100> direction. It should be noted that the profile difference between the aligned <100> direction implant and any other direction of implant has a significant influence upon the threshold voltages of the fabricated devices.

12.2.4 Energy band structure

One of the important characteristics that is attributed to GaAs is its superior electron mobility brought about as the result of its energy band structure as shown in Figure 12–4.

Gallium arsenide is a direct gap material with valence band maximum and conduction band minimum coinciding in k space at the Brillouin zone centers. Valleys in the band structure that are narrow and sharply curved correspond to electrons with low effective mass state, while valleys that are wide with gentle curvature are characterized by larger effective masses.

The curvature of the energy versus electron momentum profile determines the effective mass of electrons traveling through the crystal. The minimum point of gallium arsenide's conduction band is near the zero point of crystal-lattice momentum, as opposed to silicon, where conduction band minimum occurs at high momentum. Now, mobility, μ, depends upon

- concentration of impurity, N;
- temperature, T;

  and is inversely related to

- electron effective mass, m_e.

For GaAs, the effective mass of these electrons is 0.067 times the mass of a free electron (i.e. $0.067m_e$, where m_e is the free electron rest mass). This means electrons travel faster in gallium arsenide than in silicon as the result of their superior electron mobility brought about by the shapes of their conduction bands. Electrons in the higher valleys have high mass and strong intervalley scattering and therefore exhibit very low mobility, which is very similar to conduction electrons in silicon.

Furthermore, gallium arsenide is a direct-gap semiconductor. Its conduction band minimum occurs at the same wave vector as the valence band maximum (Figure 12–4), which means little momentum change is necessary for the transition of an electron from the conduction band to the valence band. Since the probability of photon emission with energy nearly equal to the band gap is somewhat high, GaAs makes an excellent light-emitting diode. Silicon on the other hand, is an indirect-gap semiconductor since the minimum associated with its conduction

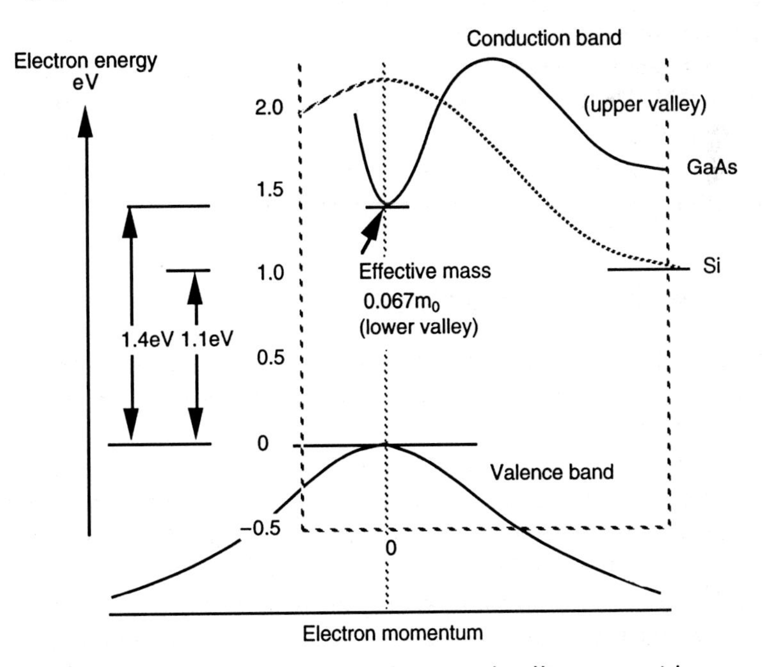

Figure 12–4 Energy band structure of silicon and gallium arsenide

band is separated in momentum from the valence band minimum. Therefore it cannot be a light-emitting device.

12.2.5 Electron velocity-field behavior

As the applied electric field, E, across the GaAs material is increased, the charge carriers, that is electrons in this case, gain energy from the applied field. At the same time, through collisions (i.e. optical phonon scattering) with the lattice, the electrons also lose a small portion of this energy. So long as the resultant balance is positive, the energy and drift velocity of the charge carriers increases with an increase in the applied field. However, at some point, the energy gained from the field becomes equal to the energy lost as the result of collisions. This results in the drift velocity approaching a limiting value referred to as the saturation velocity, v_{sat}.

Since gallium arsenide is a multivalley semiconductor, when the energy of lower valley electrons rises sufficiently, that is at electric fields greater than approximately 3500V/cm, electrons become 'hot'. There is a region in the electron velocity-field characteristics where some of the 'hot' electrons populate an upper conduction band that is characterized by larger electron effective mass. The resultant effect is a reduction in the number of high mobility electrons and hence the drift velocity.

In this region the drift velocity is no longer proportional to the electric field, but instead passes through a maximum of about $2*10^7$ cm/sec with increasing field, and decreases to an electric field independent saturation value of about $1.4*10^7$ cm/sec.

The velocity-field characteristics illustrating the three regions of interest are shown in Figure 12–5. For convenience of comparison, characteristics for silicon are also illustrated. From the figure it can readily be noted that in low electric field regions, silicon has a much lower mobility than gallium arsenide. This increases monotonically until the drift velocity saturates at a value of about $1*10^7$ cm/sec.

12.3 Technology development

Although this technology is confronted with similar technological problems as was silicon in the mid-1970s, during the last few years considerable progress has been made in GaAs integrated circuitry and the technology has progressed to the point where a number of foundries that provide GaAs fabrication are now in operation.

Typically, the current offerings have the following characteristics:

- less than one-micron gate geometry;
- less than two-micron metal pitch;
- up to four-layer metal;

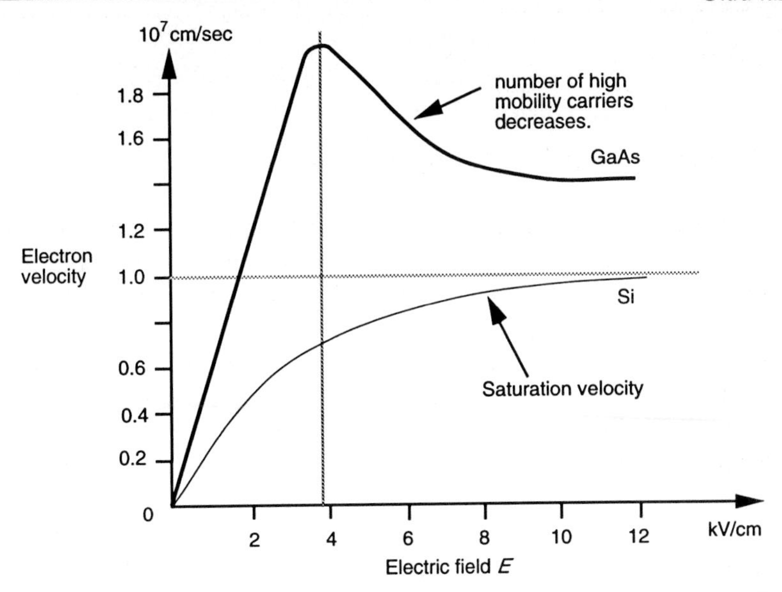

Figure 12–5 Electron velocity versus electric field for silicon and gallium arsenide

- 'ON' and 'OFF' devices;
- four-inch diameter wafers;
- suitability for clock rates in the range 1–2 GHz.

The salient features of this technology include:

- Electron mobility of six to seven times that of silicon, resulting in very fast electron transit times.
- Saturated drift velocity for GaAs and silicon are approximately equal, that is, $1.4*10^7$ cm/sec and $1.0*10^7$ cm/sec respectively. However, what is significant is that for GaAs saturation velocity occurs at a lower threshold field than for silicon.
- Large energy bandgap offers bulk semi-insulating substrate with resistivities in the order of 10^7 to 10^8 ohm.cm. This minimizes parasitic capacitances and allows easy electrical isolation of multiple devices in a single substrate.
- Radiation resistance is stronger due to absence of gate oxide to trap charges.
- A wider operating temperature range is possible due to the larger bandgap. GaAs devices are tolerant of wide temperature variations over the range –200 to +200°C.
- Direct bandgap of GaAs allows efficient radiative recombination of electrons and holes; this means forward-biased pn junctions can be used as light-emitters. Thus, efficient integration of electronics and optics becomes possible.
- Up to 70% reduction in power dissipation can be obtained over the fastest of the silicon technology such as ECL.

Table 12–2 provides an insight into the major differences between silicon and gallium arsenide. Progress in terms of speed/power projections for GaAs and commonly used silicon technologies may be assessed with reference to Figure 12–6.

Table 12–2 Comparisons between silicon and gallium arsenide

Properties	*Si*	*GaAs*	*Units*
Intrinsic mobility			
Electrons	1300	8000	cm^2/V.sec
Holes	500	400	cm^2/V.sec
Intrinsic resistivity	$2.2*10^5$	$1*10^8$	ohm.cm
Dielectric constant	11.9	13.1	
Density	2.33	5.32	gm/cm^3
Energy gap	1.12	1.43	*eV*
Thermal conductivity	1.5	0.46	W/cm° K
Effective electron mass	$0.97m_e$	$0.067m_e$	
Coefficient of thermal expansion	$2.6*10^{-6}$	$5.9*10^{-6}$	/°C
Vapor pressure (900°C)	$7.5*10^{-19}$	$7.5*10^{-3}$	mmHg
Breakdown field	$3*10^5$	$4*10^5$	V/cm
Schottky barrier height ϕ_B	0.4–0.6	0.7–0.8	V

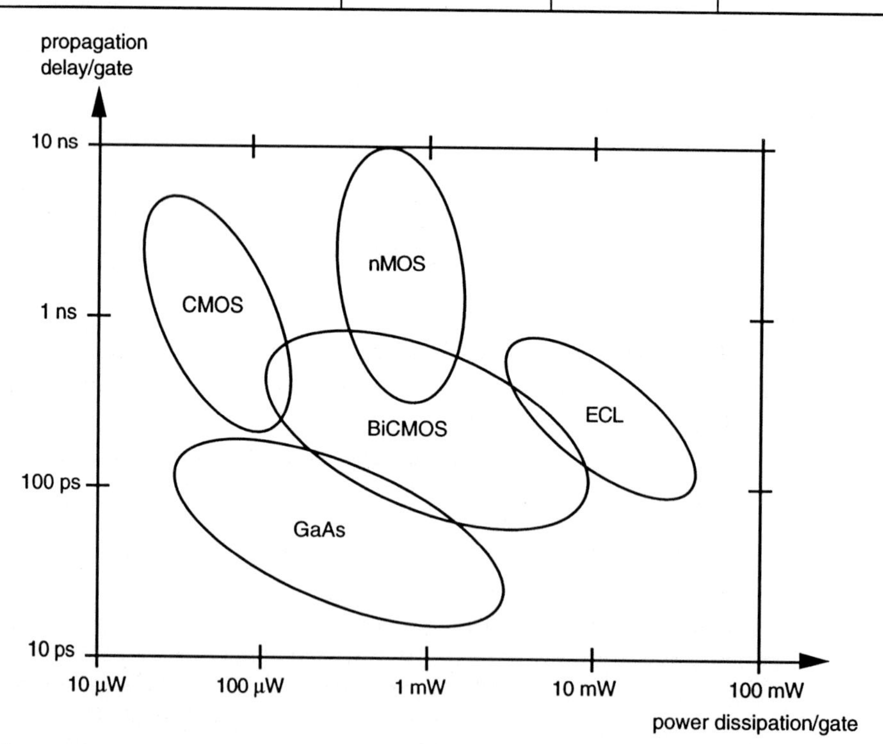

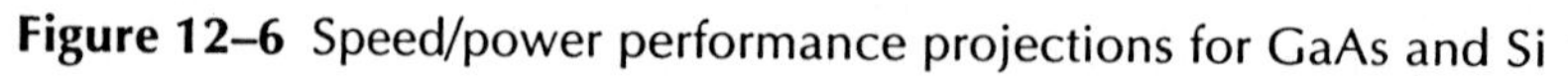
Figure 12–6 Speed/power performance projections for GaAs and Si

Table 12–3 Comparison between CMOS, bipolar and GaAs technologies

CMOS	*Bipolar*	*GaAs*
• Low dissipation	• High dissipation	• Medium dissipation
• High I/P impedance — low drive current	• Low I/P impedance — high drive current	• High I/P impedance — below ϕ_B
• High noise margin	• Medium noise margin	• Low noise margin
• Medium speed — high voltage swing	• High speed — low voltage swing	• Very high speed — low voltage swing
• High packing density	• Low packing density	• High packing density
• High delay sensitivity to load — fan-out	• Low delay sensitivity to load — fan-out	• High delay sensitivity to fan-in and fan-out
• Low output drive	• High output drive	• Low output drive
• $g_m \alpha V_{in}$	• $g_m \alpha\, e^{V_{in}}$	• $g_m \alpha\, V_{in}$
• Bidirectional	• Unidirectional	• Bidirectional possible
• Ideal switching device	• Not ideal switching device	• Reasonable switching device
• Medium f_t	• High f_t at low current	• Very high f_t
• Indirect gap	• Indirect gap	• Direct gap — good light-emitter
• Mask levels 12 to 16	• Mask levels 12 to 20	• Mask levels 6 to 10

In view of rapid developments in silicon technology itself, it is also appropriate to compare gallium arsenide with CMOS and BiCMOS. This comparison is highlighted in Table 12–3.

For very high speed operation in a semiconductor medium, three factors become significant, namely:

- carrier mobility;
- carrier saturation velocity;
- existence of semi-insulating substrate.

Gallium arsenide mostly fulfills the requirements and, together with its moderate power dissipation, provides the technology base for a new generation of circuits and subsystems.

12.3.1 Gallium arsenide devices

During the last few years a number of different devices have been developed. The so-called 'first generation' of GaAs devices includes:

- depletion-mode metal semiconductor field-effect transistor, D–MESFET;
- enhancement-mode metal-semiconductor field-effect transistor, E–MESFET;
- enhancement-mode junction field-effect transistor, E–JFET; and
- complementary enhancement-mode junction field-effect transistor, CE–JFET.

First generation GaAs gates have exhibited switching delays as low as 70 to 80 psec for a power dissipation in the order of 1.5 mW to 150 μW.

There are other more sophisticated 'second generation' devices such as:

- high electron mobility transistor, HEMT;
- heterojunction bipolar transistor, HBT.

Electron mobility in second generation transistors can be up to five times greater than in the first generation. In consequence, very fast devices are possible.

However, in the following sections we will concentrate on establishing some of the fundamental principles of GaAs design methodology for the first generation devices only, particularly the predominant MESFETs, which are now at a stage of development that enables them to be incorporated in very fast VLSI systems.

12.3.2 Metal semiconductor FET (MESFET)

The gallium arsenide field-effect transistor, a bulk-current-conduction majority-carrier device, is fabricated from bulk gallium arsenide by high-resolution photolithography and ion implantation into a semi-insulating GaAs substrate. Processing is relatively simple, requiring no more than six to eight masking stages. For the purpose of comparison, Figure 12–7 shows the evolution of process complexity in terms of mask count as function of time for both silicon and gallium arsenide technologies.

The structure of the basic MESFET as shown in Figure 12–8 is very simple. It consists of a thin n-type active region joining two ohmic contacts with a narrow metal Schottky barrier *gate* that separates the more heavily doped *drain* and *source*.

GaAs MESFETs are similar to silicon MOSFETs. The major difference is the presence of a Schottky diode at the gate region which separates two thin n-type active regions, that is, source and drain, connected by ohmic contacts. It should be noted that both D type and E type MESFETs, that is, 'ON' and 'OFF' devices, operate by the depletion of an existing doped channel. This can be contrasted with silicon MOS devices where the E (enhancement) mode transistor functions by inverting the region below the gate to produce a channel, while the D (depletion)

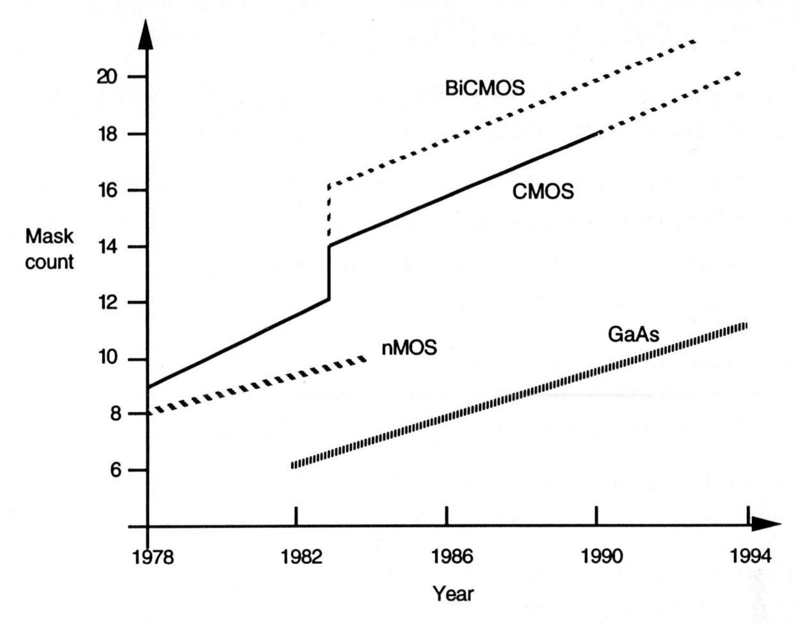

Figure 12–7 Evolution of process complexity for silicon and gallium arsenide technologies

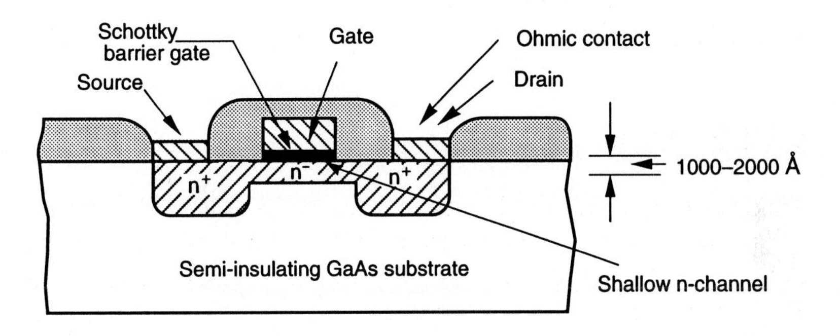

Figure 12–8 Side view for basic MESFET

mode device operates by doping the region under the gate slightly in order to shift the threshold to a normally 'ON' condition.

This similarity provides us with the basis for extending to gallium arsenide the design methodology used so successfully in silicon to simplify circuit and system design and layout issues.

The D–MESFET is normally 'ON' and its threshold voltage, V_{tdep}, is negative. The E–MESFET is normally 'OFF' and its threshold V_{tenh} is positive. The threshold voltage is determined by the channel thickness, a, and concentration density of the implanted impurity, N_D. A highly doped, thick channel exhibits a larger negative threshold voltage. By reducing the channel thickness, and decreasing the

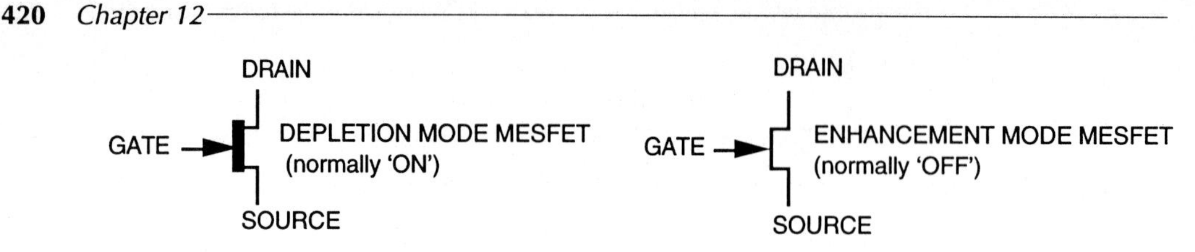

Figure 12–9 MESFET circuit symbols

concentration density a normally 'OFF' enhancement mode MESFET with a positive threshold voltage can be fabricated. Circuit symbols for the depletion and enhancement mode MESFETs are set out in Figure 12–9.

The MESFET has a maximum gate to source voltage V_{gs} of about 0.7–0.8 volt owing to the diode action of the Schottky diode gate. Since the principle underlying the operation of MESFETs is based upon the behavior of metal-semiconductor interface, we will briefly outline some of the features that characterize such an interface.

12.3.2.1 Characteristics of Schottky barriers

When a metal is brought into contact with a semiconductor, an electrostatic potential barrier (refered to as Schottky barrier) is created at the interface as the result of the difference in the work function of the two materials. To appreciate the physical nature of the barrier we can model the interface by visualizing a situation whereby the metal is gradually brought toward the semiconductor surface until the separation becomes zero.

As this separation between the metal-semiconductor surface is reduced, the induced charge in the semiconductor increases, while at the same time the space charge layer widens. A greater part of the contact potential difference begins to appear across the space charge layer within the semiconductor. Because the carrier concentration in the metal is several orders of magnitude larger than that in the semiconductor when the separation is brought to zero, the entire potential drop then appears within the semiconductor itself. This is in the form of a depletion layer situated adjacent to the metal and extending into the semiconductor. A simplified view of such a transistor showing the depletion layer profile is shown in Figure 12–10 for two conditions, one when the drain to source voltage V_{ds} is zero and the other when it is greater than the saturation voltage.

12.3.3 GaAs fabrication

Although there are various approaches that are currently used, high-pressure liquid-encapsulated Czochralski (LEC) growth of gallium arsenide crystals from high purity pyrolytic boron nitride (PBN) crucibles is becoming the primary growth technique over several other methods that have emerged during the last few years.

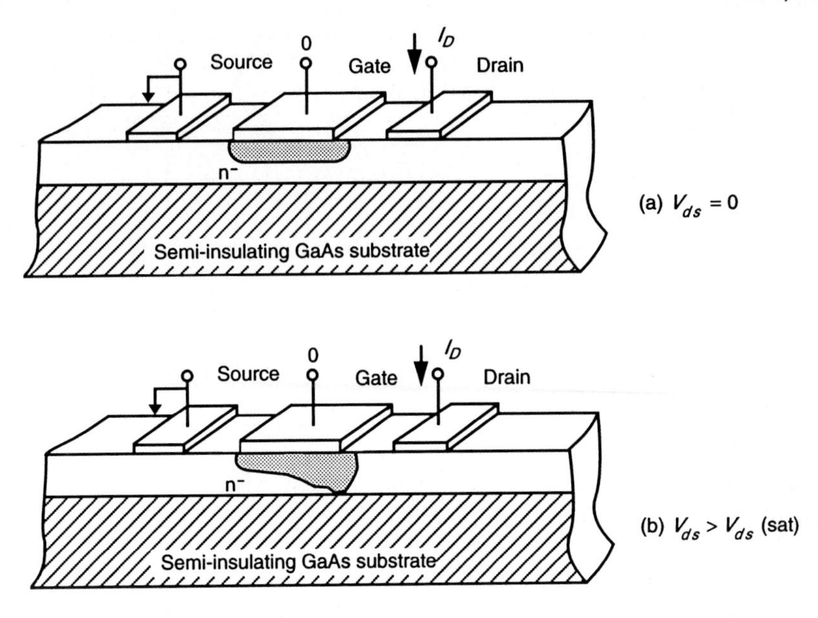

Figure 12–10 Depletion profile of a MESFET

Since preference is usually for wafers grown in the <100> orientation, much of the success of the above method is achieved as the result of the ability to grow LEC material in the <100> direction, which produces relatively large diameter, round (100) wafers that are thermally stable and have superior semi-insulating properties.

Since the <110> cleavage planes are at a right angle, square chips can be obtained with a diamond scribe and break. This means that by adhering to the <100> growth plane many of the problems associated with cutting and subsequent handling can be alleviated.

The sequence for GaAs wafer preparation is very similar to that of silicon wafer preparation technique. The first step involves mechanically grinding the As-grown boules to a precise diameter and incorporating orientation flats. This is followed by

- wafering using a diamond ID saw;
- edge rounding;
- lapping;
- polishing;
- wafer scrubbing.

12.3.3.1 Depletion-mode MESFET

The profile for the metal gate depletion-mode MESFET (D–MESFET), the most mature of the current GaAs technologies, is illustrated in Figure 12–11.

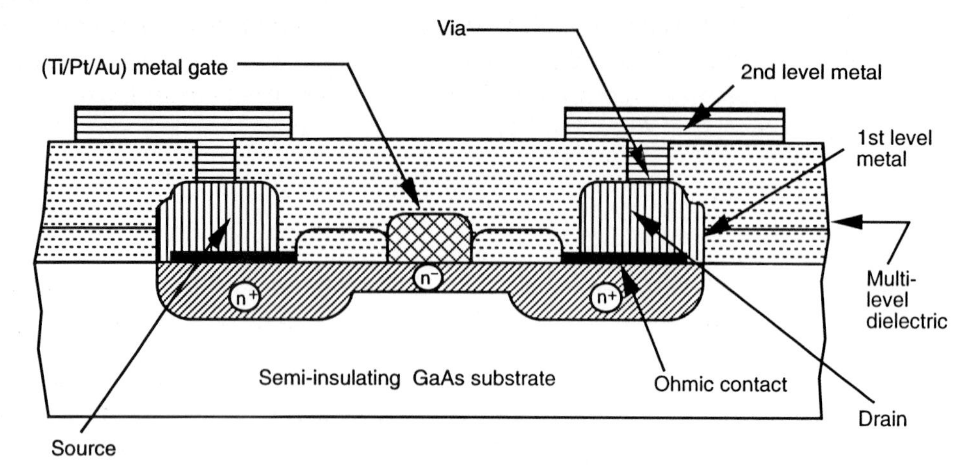

Figure 12–11 Structure of a metal-gate depletion-mode MESFET

Basically, a thin n–type region joins two ohmic contacts with a narrow metal Schottky barrier gate. Usually, the depletion-mode devices are fabricated using the planar process where n-type dopants (having concentration density typically in the range of $1*10^{17}$ cm^{-3} to $2*10^{17}$ cm^{-3}) are directly implanted into the semi-insulating GaAs substrate to form the channel as well as the more heavily doped source and drain regions. The semi-insulating substrate is ideal for all 'ion implantation' planar technology. The gate and first level interconnect metallizations are typically deposited by E-beam evaporation techniques. The gate length and its position relative to the source and drain contacts have a significant influence upon the transconductance of the device and control the performance of the MESFET. The conducting n-channel is confined between the gate depletion region and the semi-insulating GaAs substrate. By varying the channel thickness (usually in the range 1000 Å to 2000 Å) and the doping level of the active region, it is possible to vary the threshold, V_{tdep}, to the desired negative value, that is, in the range – 0.5 V to –2.0 V.

12.3.3.1.1 Depletion-mode planar process flow

The driving force and indeed much of the success associated with silicon technology were brought about as the result of the presence of a stable native oxide which was readily produced through the oxidation of silicon. However, owing to the absence of a stable native oxide, GaAs technology relies on deposited dielectric films for passivation and/or encapsulation.

The fabrication process varies from foundry to foundry. However, one approach is illustrated in Figure 12–12, which entails the use of 3-inch or 4-inch liquid-encapsulated Czochralski (LEC) wafers. Initially, the GaAs substrate is coated with the first level of insulator, that is a thin layer of silicon nitride (Si_3N_4), which is sputtered on the GaAs substrate. This thin film of insulator remains on the wafer throughout the processing steps that follow, allowing the annealing of GaAs at temperatures of up to 900°C. The next step entails the formation of an n^- type active layer. This is achieved by direct ion implantation into the GaAs semi-insulating substrate through the insulating layer where the photoresist is used as the implant mask. Implantation of Si^+ ions takes place at about 220 to 230 keV to a dose of approximately $6*10^{12}/cm^2$. There are only two main implantation steps:

1. a shallow high-resistivity n– layer for formation of the channel layer; and
2. a deep low-resistivity n^+ layer for the formation of source and drain.

The resultant channel resistance is in the order of 1000 to 2500 ohm/square, which is too high for source and drain contacts. Therefore, by keeping the surface concentration at the source, and drain regions relatively high by additional implantation, it is possible to reduce the contact resistances of these contacts.

The wafer is then coated with the interlevel dielectrics, SiO_2 by CVD (chemical vapour deposited) process. SiO_2 layer has a thickness of 400 to 500 nm and is deposited over the Si_3N_4 layer primarily to provide protection against physical damage. This is followed by an anneal in a hydrogen ambient at a temperature of about 800–850°C for approximately 30 minutes. This encapsulation phase is very important as it prevents out-diffusion of arsenic, brought about as the result of high vapour pressure associated with GaAs (Table 12–2) when subjected to temperatures over 600°C or so during the anneal step.

It should be noted that there are only a few capping materials that can be used in the process since the mechanical stability of the thin film encapsulation layer depends upon the stress that is present at the interface.

There are several sources that this stress can originate from:

- lattice mismatch;
- intrinsic stress of the encapsulation layer itself;
- thermal mismatch.

For example, the coefficients of thermal expansion for the commonly used capping materials such as Si_3N_4 and SiO_2 are:

$$Si_3N_4 = 3.2*10^{-6}/°C$$
$$SiO_2 = 0.5*10^{-6}/°C$$

This can be compared with GaAs, which has a thermal expansion coefficient of $5.9*10^{-6}/°C$. Thus, it is readily recognizable that SiO_2 has the greatest mismatch.

Since Si_3N_4 has a dielectric constant of 7, compared to 3.9 for silicon dioxide, a sandwich structure of SiO_2 and Si_3N_4 increases the effective dielectric constant of the insulator layer. Furthermore, SiO_2 was initially employed as the first-level capping material. However, it was found that Ga can diffuse through this layer.

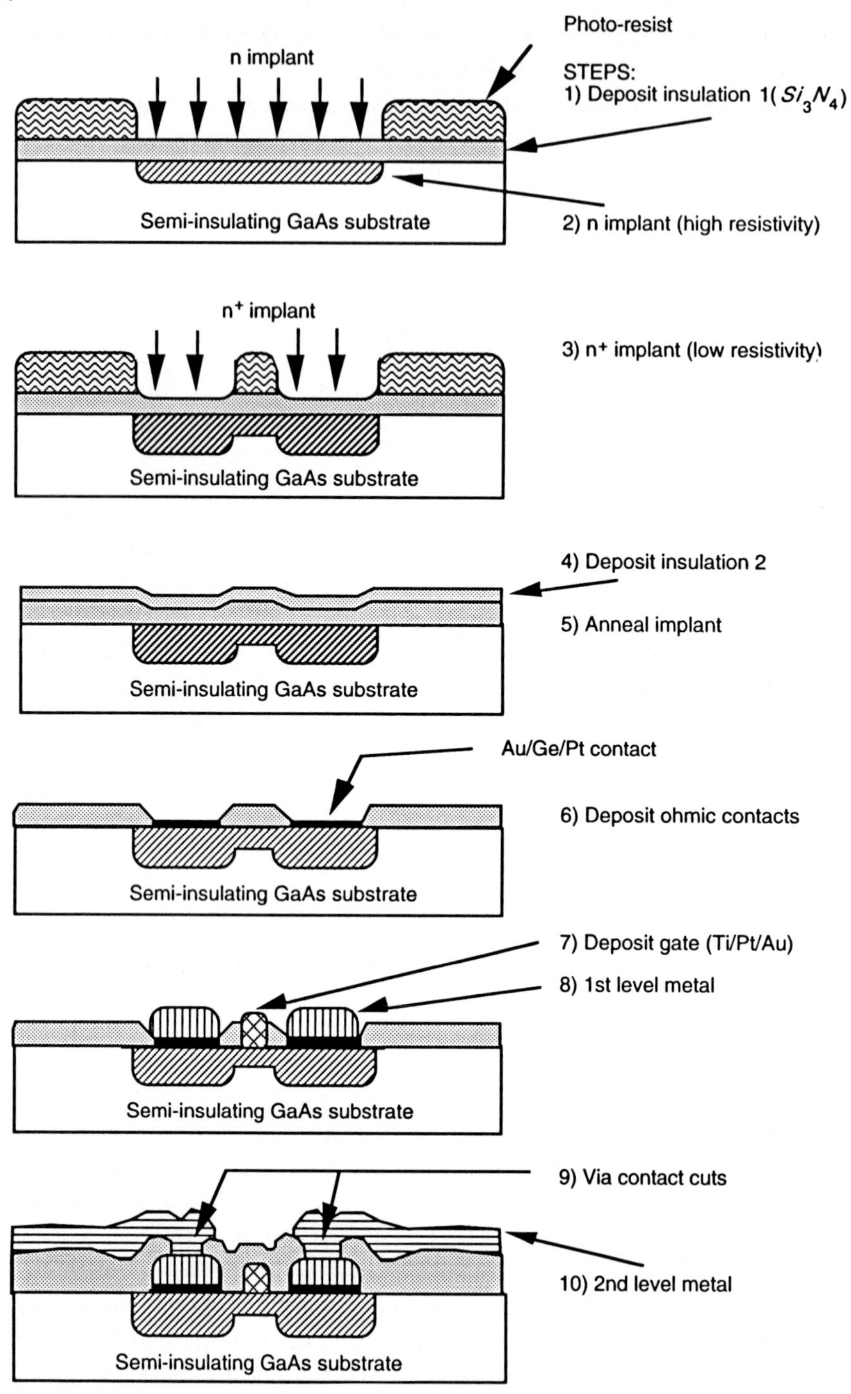

Figure 12–12 A typical gallium arsenide DMESFET metal gate process using planar technology

This problem was subsequently alleviated by using Si_3N_4 as the first-level insulator with SiO_2 as the second-level insulator.

The next step in the process entails defining the MESFET gates, the ohmic contacts and the first-level metal interconnects. There are several points that must be considered during this phase. These are:

- The metals must be carefully alloyed to ensure reliable low resistance contacts, that is less than 10^{-6} Ω-cm^2.
- The ohmic contacts between the metal interconnect and the source and drain are deposited by evaporation using E-beam technology. A thin layer of gold-germanium-nickel (Au/Ge/Ni) or gold-germanium-platinum (Au/Ge/Pt) is alloyed on the wafer at a temperature of about 450°C to 500°C.
- One of the most critical steps in the fabrication process is the gate metallization.
- Schottky gates, together with first-level metal for interconnects, are formed by multilayer gold-refractory thin films such as titanium/platinum/gold (Ti/Pt/Au: 300 Å/400 Å/3000 Å) or alternatively titanium/tungsten/gold (Ti/W/Au) alloys deposited by E-beam evaporation. Titanium provides a good, high barrier, Schottky contact, but it has a high parasitic gate resistance. To reduce the parasitic resistance, gold is used as the top layer with platinum or tungsten as the intermediate layer. In the absence of either Pt or W layers, gold could diffuse into the GaAs surface, thus converting the Schottky contact into an ohmic one.

First-level metallization, which is about 3000 Å to 4000 Å, is accomplished by:

- delineating photoresist patterns;
- plasma etching the underlying insulator;
- deposition of the metal on GaAs wafer either by vacuum evaporation or by sputtering;
- photoresist lift-off.

The metal contacts and interconnects are precisely registered with the plasma-etched insulator windows. By fabricating the first-level metal within windows in the first-level insulator, and by ensuring that the first-level metallization thickness is close to the insulator thickness as in Figure 12–13, a more complex multilevel interconnect structure becomes possible due to the planar nature of the surface. Thus, a third-level metal and a fourth-level metal can readily be implemented.

Second-level metal is not in contact with gallium arsenide substrate; therefore platinum, which is used to prevent the interaction of gold with the GaAs surface (i.e. Au dissolves in GaAs), is usually eliminated from this step. Second-level metallization, which is about 7000 Å to 8000 Å thick, entails magnetron sputtered titanium/gold (Ti/Au: 300 Å/7000 Å) alloy only, which is followed by filling the vias between first-level and second-level metal. The sputtering process entails

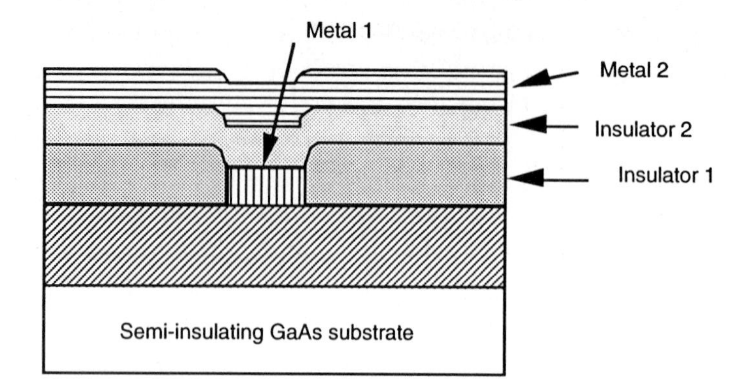

Figure 12–13 Metallization process

the physical deposition of a thin film by ion bombardment of the required material. Usually the deposition rate (i.e. thickness per unit time) depends upon the sticking coefficient of the depositing material and the nature of the sputtering equipment. The main feature of the magnetron itself is that it involves a set of powerful magnets, located behind the target surface, that provides an intense magnetic field for concentrating the plasma in the vicinity of the target.

The final step in the fabrication is passivation, used to protect against contamination and moisture. This entails a 0.4 μm to 0.5 μm thick passivation layer being deposited using a low temperature, plasma-enhanced chemical vapor deposition (PECVD) process. This is a chemical deposition technique used for fabrication of both insulating and conducting films. The method is very similar to the low pressure CVD except that plasma excitation is provided in addition to the usual thermal energy.

Since in D–MESFETs any regions of the source or drain channel that are not under the gate are automatically strongly conducting, one does not require the precise alignments of the gate nor gate recesses to avoid parasitic source and drain resistances. However, in the metal-gate planar technology the position of the gate relative to the source and drain contacts has significant influence upon the performance of the device. Because of the very thin undepleted n^- layer, the source resistance can be rather high, which subsequently causes the degradation of the transconductance, g_m.

Extension of the surface depletion layer cannot be avoided because of the presence of traps localized at the gallium arsenide surface. Subsequently, the extension of the interface depletion layer, owing to traps near the interface between the active layer and the substrate, has an influence on the drain resistance also. Hence process optimization is essential in order to minimize these resistances so that the device performance is not degraded.

It is interesting to note the close similarity between the planar implanted D–MESFET GaAs fabrication process and the Si planar process. This can readily be observed by noting that the GaAs substrate is totally protected by dielectric layers throughout the fabrication process. Cuts are made in the dielectric only where ohmic contacts, Schottky barriers, or interconnect metallizations are required. As far as process technology is concerned, the most difficult layer to control is the shallow, lightly doped high-resistance n^- MESFET channel layer. This implant layer determines the threshold voltage V_t of the MESFETs.

If some reduction in speed can be tolerated, then instead of using the exotic gold process for first-level and second-level metals it is possible to use the less costly aluminum. Thus, significant cost savings could be achieved at the expense of speed. This will be outlined in the following section on the self-aligned gate (SAG) process.

12.3.3.1.2 Ion implantation and annealing

The ion implantation and the subsequent annealing are very significant in this technology. In ion-implantation, doping is achieved by bombarding the semiconductor surface with a high-velocity ion beam. Doping density and dopants distribution in the semi-insulating material are controlled by varying the ion flux and velocity. Using this approach, crystal defects brought about as the result of ion bombardment are annealed at about 800–850°C. The advantages of ion implantation are:

- independent control of doping level;
- independent control of doping profile;
- good reproducibility;
- ease of selective doping of selected areas.

The original arrangement of atoms on the crystal lattice was indicated in Figure 12–1. As implanted ions penetrate the GaAs substrate, they lose energy by several mechanisms, including the displacement of target atoms from the lattice sites. After ion implantation, the dopant atoms come to rest in the crystal and, as a result of interactions and collisions, the crystal lattice is disrupted as indicated in Figure 12–14(a).

Ions now occupying interstitial positions are electrically inactive. Annealing provides energy to the implanted impurities and results in moving the interstitial dopant ions into lattice positions where they become electrically active. Furthermore, the displaced substrate atoms are subsequently moved back to their crystallographic lattice locations (Figure 12–14(b)) which then gives the high electron mobility.

The extent of damage to the crystal depends on several factors, including:

- mass of the implanted ion;
- target mass;
- energy associated with the ion;
- dose;

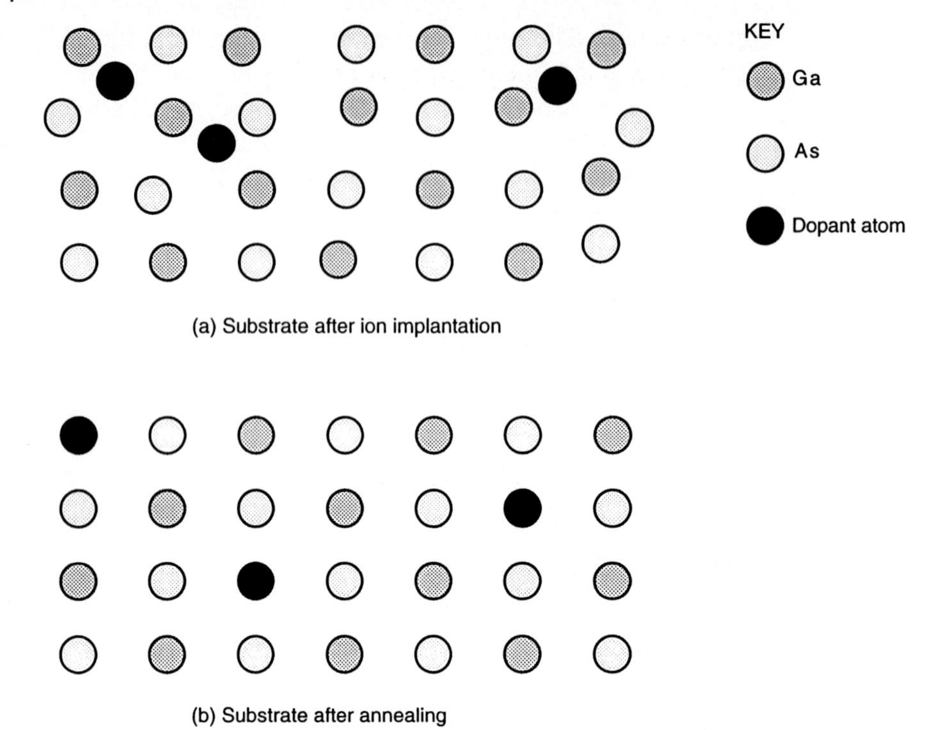

Figure 12–14 Ion implantation before and after annealing

- temperature;
- displacement energies.

12.3.3.2 Enhancement-mode MESFET

The E–MESFET structure is similar to that of the D–MESFET, except for a shallower and more lightly doped channel. This means the channel is in 'pinch-off' at zero gate voltage, due to the built-in potential of the metal Schottky barrier gate. A positive gate voltage is required for the channel to begin conduction. In order to ensure that the depletion layer extends through the channel height at zero gate voltage, the gate is usually recessed into the underlying channel. The steps in the fabrication of the E–MESFET are somewhat similar to those for the D–MESFET.

12.3.3.2.1 Process details

Steps for fabrication of gallium arsenide enhancement/depletion-mode MESFETs are reproduced here to highlight some of the complexities of the process. The details of the process are:

1. Encapsulation phase:
    - wafer preparation;
    - encapsulation (deposition of first-level insulator Si_3N_4);

- alignment mark mask;
- alignment mark metallization and lift-off.

2. Ion implantation:
   - first Si^+ implant (E–MESFET) mask;
   - channel implant;
   - second Si^+ implant (D–MESFET) mask;
   - channel implant;
   - S/D implant (formation of source and drain) mask;
   - n^+ implant;
   - anneal.
3. Schottky junctions and first-level metal:
   - patterning ohmic contact mask;
   - plasma etching contact windows;
   - contact metallization (ohmic-Au/Ge/Ni);
   - contact definition and alloy;
   - H^+ implant mask;
   - H^+ implant (isolation);
   - Schottky gate mask;
   - plasma etch Schottky windows;
   - metallization (Ti/Pt/Au);
   - lift-off.
4. Second-level metal:
   - dielectric (SiO_2 sputter);
   - via cut mask;
   - metallization (Ti/Au) mask;
   - lift-off.
5. Scratch protection:
   - Si_3N_4 plasma deposition;
   - pad/scribe street mask;
   - plasma etch.

12.3.3.3 Self-aligned gate E/D process

An alternative approach in process technology is the self-aligned gate (SAG) process, which is showing promise and is beginning to emerge as a strong contender for silicon in the area of very high speed VLSI systems.

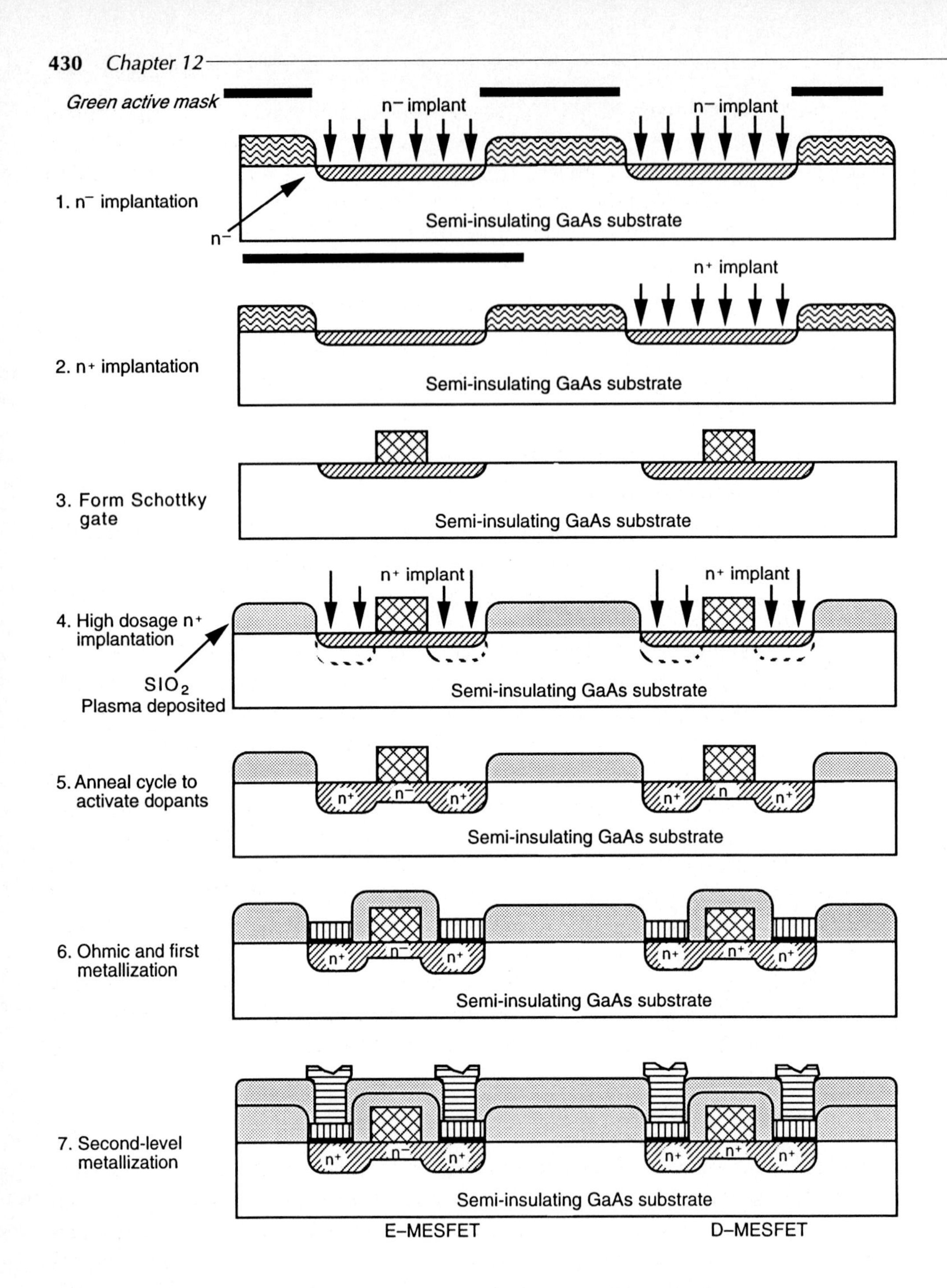

Figure 12–15 Self-aligned processing steps for GaAs E–/D–MESFET

Note that in the self-aligned process as shown in Figure 12–15, the n^+ implant regions prevent the extension of both the surface and the interface depletion layers, thus reducing the effect of the undepleted n^+ layer parasitic resistance which subsequently improves the performance of a device.

Process steps for a GaAs self-aligned gate are as follows:

- a n^- implantation for formation of E–MESFET;
- a second n implantation for formation of D–MESFET;
- formation of Schottky gates on an n-type GaAs layer;
- a third, n^+, implantation for formation of source and drain;
- an anneal cycle at 850°C to activate dopants;
- ohmic metallizations of the source and drain regions;
- interconnect metallizations.

Owing to the anneal cycle that requires a temperature up to 850°C to activate the dopants, it is necessary to choose a high-temperature-stable gate. Tungsten nitride has been found to be satisfactory as gate material. It has film resistivity ρ = 70 $\mu\Omega$-cm and Schottky barrier height $\phi_B = 0.8$ volt to n-type GaAs.

Since Schottky barrier gates on GaAs cannot be forward biased above 0.7 to 0.8 volt without drawing excessive currents, the permissible voltage swing is relatively low. This limits the noise immunity of the gate and places stringent fabrication requirements on threshold voltage control and uniformity.

As can be seen, this technology very closely resembles that of nMOS, which means it is very likely that ratio rule needs to be applied when designing logic circuits using the enhancement/depletion process shown in Figure 12.16.

In summary, the steps in the process entail defining the active areas (Green Mask) that would eventually form the E-type and the D-type MESFETs, followed by two ion implantions, that is, lightly doped for E-type and heavily doped (Yellow Mask) for D-type followed by formation of gate-metal (Red Mask).

The E-type MESFET is defined by intersection of Green and Red masks while the D-type MESFET is defined by intersection of the Green, Red and Yellow masks. This abstraction as an aid to design will be dealt with in more detail in Section 12.5.

12.3.3.4 Enhancement mode junction FET (E–JFET)

The operation of an E–JFET is similar to that of an E–MESFET, in which source and drain regions are formed by n^+ ion implantation while the channel is formed by n^-type implantation. However, in contrast to the E–MESFET, where the metal gate rests above the channel, in the E–JFET the gate is buried below the channel surface by p^+ implantation (using either beryllium or magnesium as the implant). Through this process, a pn junction is formed between the gate and the channel, as illustrated in Figure 12–17. The structure offers lower parasitic source and drain resistances than the E–MESFET owing to the doping of the channel region.

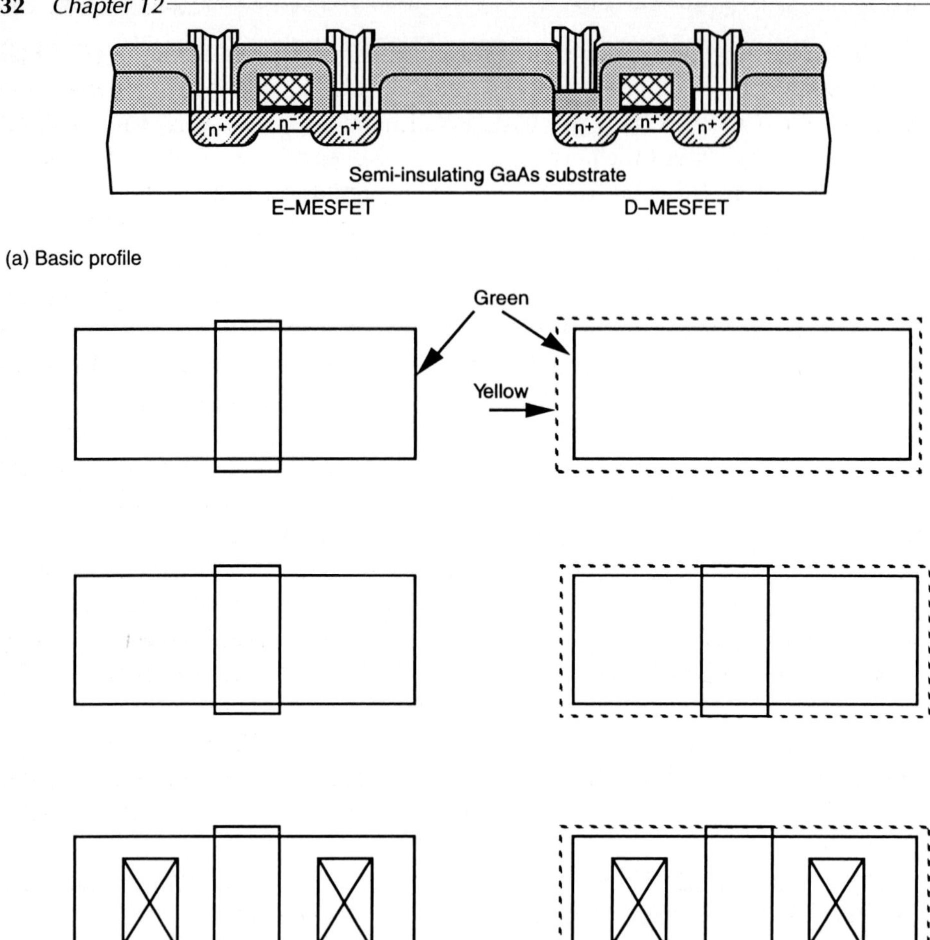

Figure 12–16 Enhancement/depletion self-aligned gate process

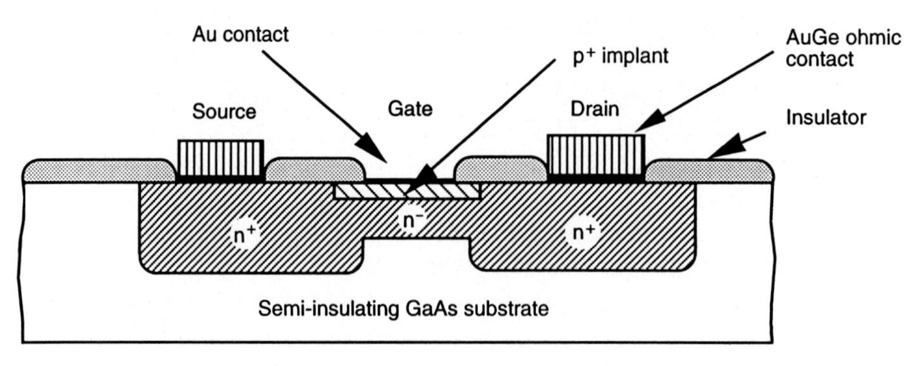

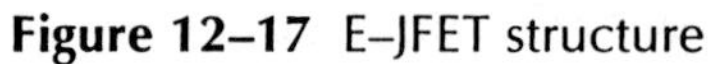
Figure 12–17 E–JFET structure

From previous discussions, the permissible voltage swing for E–MESFETs is rather low since Schottky barrier gates on GaAs cannot be forward biased above 0.7 to 0.8 volt without drawing excessive current. However, with E–JFETs, because of the larger built-in pn junction voltage, the device can be biased to about $V_{gs} = +1$ volt without incurring excessive conduction, thus alleviating some of the problems that are encountered in the control of the threshold voltage. E–JFETs are more difficult to fabricate than MESFETs primarily because of the additional p^+ implant step. It is necessary to have a precise control over the implant thickness to ensure that the desired pinch-off voltage of the device is maintained. However, here we have an additional advantage over the E–MESFET in the control of threshold voltage not only through the implant but also by adjusting the pn junction location.

A significant aspect of E–JFET technology is that complementary devices can be fabricated, whereas in MESFET technology considerable difficulty is encountered in forming Schottky barriers to a p-type implanted channel.

The presence of the additional pn junction sidewall gate capacitance makes the E–JFET slower than an equivalent E–MESFET. However, reduced power requirements together with larger logic voltage swings makes E–JFET technology a possible contender for the emerging ultra high speed VLSI systems.

12.3.3.5 Complementary enhancement-mode junction FET (CE-JFET)

A CE-JFET device, shown in Figure 12–18, is similar to the silicon complementary MOSFET. Here the nMOS depletion mode transistor, or alternatively, a resistive load, is replaced by a p-channel EJFET.

The n-channel and p-channel JFET is fabricated by a series of ion implantations into the semi-insulating GaAs substrate. The sequence entails:

- n^+ implantation;
- n^- channel implantation;

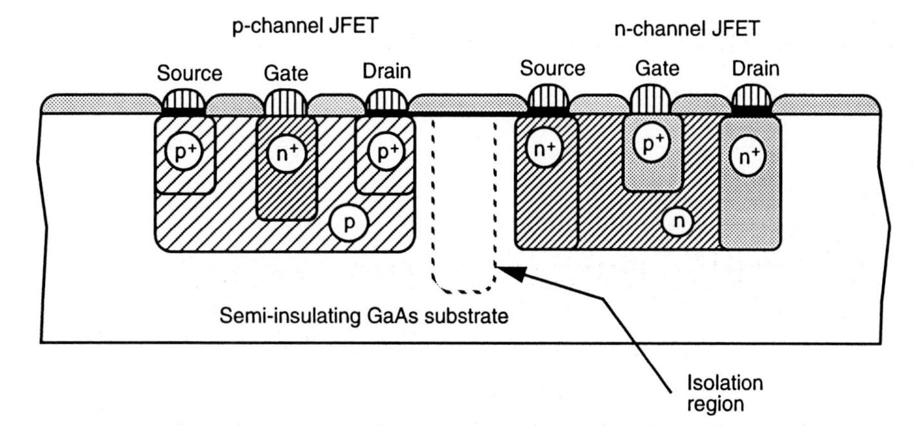

Figure 12–18 Complementary enhancement mode JFET structure

- p^- channel implantation;
- p^+ implantation.

In a CE-JFET the ratio of the effective channel electron mobility μ_n of the n-channel device to that of the hole mobility μ_p of the p-channel device is given by:

$$\frac{\mu_n}{\mu_p} = 10$$

Thus, for a p-channel device requiring the same drain current I_{ds} as that of the complementary n-channel device, it is necessary that

$$W_p = 10W_n$$

where W_p = channel width for p-channel device,
W_n = channel width for n-channel device.

This means that circuits requiring equal numbers of p- and n-devices will consume large areas. Therefore, one must resort to other design methods such as precharge techniques, which require a single pull-up transistor to serve a number of n-transistors performing the logical functions.

12.3.3.6 High electron mobility transistor (HEMT)

Here, multilayered stuctures of very thin (10 Å to 100 Å) alternating layers of GaAs and AlGaAs are used. This multilayered approach is referred to as a multiquantum-well structure. The key in these structures is to place the donor atoms in a wider bandgap GaAlAs layer adjacent to an undoped GaAs channel layer, which receives the free electrons from the ionized donors. Electrons are transferred from the AlGaAs charge control layer to the undoped GaAs layer where they form a two-dimensional electron gas. Since the electrons are spatially separated from the ionized donors, they exhibit high mobility.

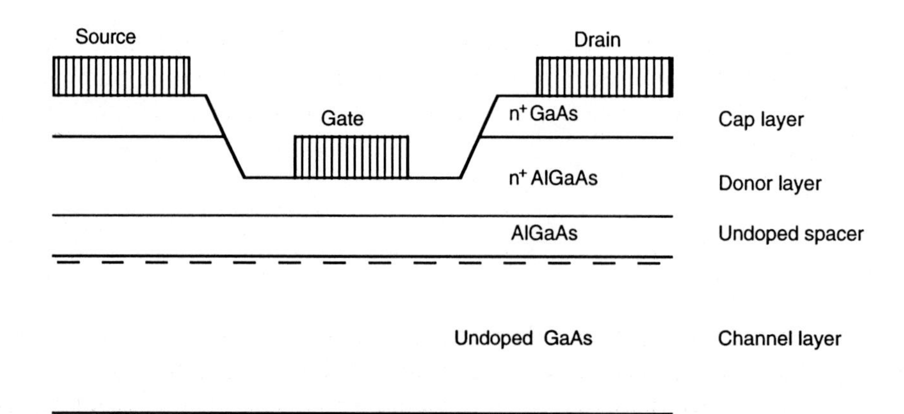

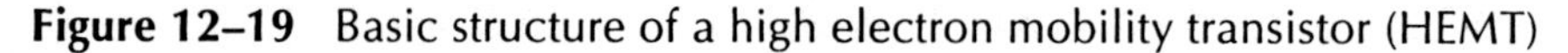

Figure 12–19 Basic structure of a high electron mobility transistor (HEMT)

Although there are variations to the processing steps, the basic structure is as illustrated in Figure 12–19 and will be seen to comprise four distinct layers:

- channel layer — GaAs;
- undoped spacer layer — AlGaAs;
- donor layer;
- cap layer.

12.4 Device modeling and performance estimation

VLSI designers, as a rule, should have a good knowledge of the behavior of the circuits they are designing. Even when large systems are being designed using computer-aided design processes, it is essential that the designs are based on a sound foundation of understanding if the system is to meet a given performance specification.

12.4.1 Device characterization

In order to preserve simplicity, the prime consideration in this section is to provide an approximate model for the MESFET which not only preserves the essential features of the device, but also assists the VLSI systems designer with performance estimations and optimization processes.

As the gallium arsenide transistor and the processes used to produce it have been introduced, it is now possible to gain some insight into the electrical characteristics of the basic GaAs MESFET circuits.

12.4.2 Drain to source current derivation

MESFETs are channel-area modulation devices, that is, they depend upon the capacitance of the Schottky barrier to control the effective charge in the channel. As for silicon MOS transistors, gallium arsenide devices have also three regions of operation:

- cut off;
- linear;
- saturation.

To appreciate some of the features that characterize I_{ds}, in the first instance, we will proceed with deriving a simple model that highlights first order effects only, and then focus attention on the more complex models, without losing our objective of simplicity which is so critical for VLSI systems designers.

Consider a typical structure (see Figure 12–20) where the majority carriers, that is, electrons, flow from source to drain. The current I_{ds}, as the result of this movement, is given by:

$$I_{ds} = -I_{sd} = \frac{\text{Charge induced in channel } (Q_c)}{\text{Electron transit time } (\tau)} \tag{12.1}$$

First, the electron transit time τ can be determined by noting

$$\tau = \frac{\text{Channel length } (L)}{\text{Channel velocity } (v)}$$

Now carrier velocity, that is the movement of electrons, is given by

$$v = \mu_n E_{ds}$$

where E_{ds} is the electric field between the drain and the source, and μ_n is the electron mobility.

The transit time τ can thus be expressed as

$$\tau = \frac{L}{\mu_n E_{ds}} \tag{12.2}$$

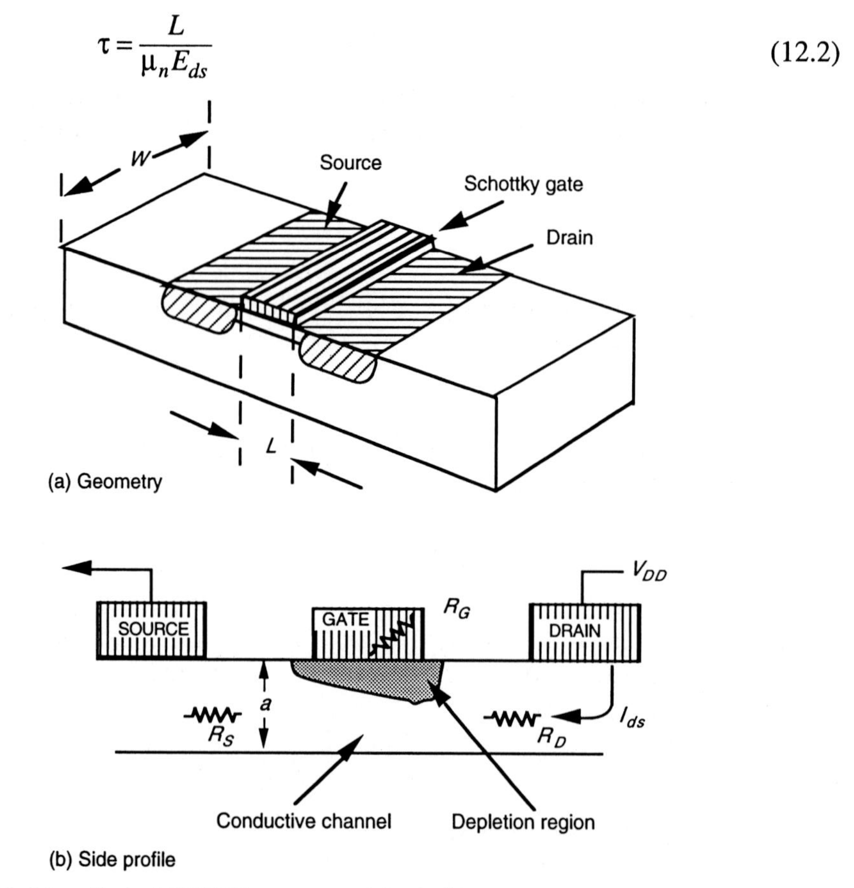

Figure 12–20 GaAs MESFET cross-sectional view

If we denote the average potential difference between the gate and the channel by V_{gb}, then owing to the shape of the depletion layer, this average potential can be written as

$$V_{gb} = 0.5(V_{gs} - V_t) \tag{12.3}$$

where V_{gs} is the gate to source voltage and V_t, is the threshold voltage of the device.

The threshold voltage V_t is defined as the gate voltage at which the depletion layer (Figure 12–17) just pinches off the channel, that is, the gate voltage that extends the depletion layer down to the substrate.

The average electric field E_{ds} along the length of the gate is:

$$E_{ds} = \frac{0.5(V_{gs} - V_t)}{L} \tag{12.4}$$

Upon substitution of equation 12.4 into equation 12.2, the transit time, τ, becomes

$$\tau = \frac{2L^2}{\mu_n(V_{gs} - V_t)} \tag{12.5}$$

The average electric field E_{ave} across the channel can also be approximated in terms of implant depth, a, and the voltage $(V_{gs} - V_t)$ that appears across the channel. Thus

$$E_{ave} = \frac{(V_{gs} - V_t)}{a}$$

The induced charge in terms of the device geometry and the average electric field becomes

$$Q_c = E_{ave}\varepsilon_r\varepsilon_0 WL \tag{12.6}$$

Upon substitution for E_{ave} in equation 12.6, the resultant expression for the induced charge becomes

$$Q_c = (WL)\left(\frac{\varepsilon_r\varepsilon_0}{a}\right)(V_{gs} - V_t) \tag{12.7}$$

Now, by combining equations 12.5 and 12.7 with equation 12.1, we obtain the principal result for the drain to source current I_{ds}

$$I_{ds} = \left(\frac{\mu_n\varepsilon_r\varepsilon_0}{2a}\right)\left(\frac{W}{L}\right)(V_{gs} - V_t)^2 \tag{12.8}$$

which, when rewritten, results in

$$I_{ds} = \beta(V_{gs} - V_t)^2 \tag{12.9}$$

where

$$\beta = \left(\frac{\mu_n \varepsilon_r \varepsilon_0}{2a}\right)\left(\frac{W}{L}\right)$$

β is a common parameter used in the SPICE MESFET model specification, denoted by K_p. For a typical process, K_p is in the order of 0.1 to 0.5 mA/V^2.

β may be seen to consist of a process dependent factor $(\mu_n \varepsilon_r \varepsilon_0/2a)$, which contains all the process terms and a geometry dependent term (W/L), which depends on the actual layout of the transistor. The geometric terms in equation 12.8 are illustrated in Figure 12–20.

Sometimes the channel length of the MESFET is predetermined by the process, which means the designer can control the gain factor through varying the channel width of the MESFET only.

Equation 12.8 describes the behavior of the MESFET in the saturation region only. Now using a similar approach, it is possible to derive a relation for the MESFET to represent the operation in the linear region also. The model in this region becomes:

$$I_{ds} = \beta[2(V_{gs} - V_t)V_{ds} - V_{ds}^2]\ ;\ V_{ds} < (V_{gs} - V_t) \text{ and } V_{gs} \geqslant V_t \qquad (12.10)$$

Special note should be made here that in GaAs the saturation of drain current, I_{ds}, with an increasing drain to source voltage, V_{ds}, is brought about by carrier velocity saturation, whereas in silicon the resultant saturation effect is due to 'channel pinch-off'.

12.4.2.1 More complete device equation

The model described by equation 12.10 unfortunately does not provide for a smooth transition between the saturation and the linear regions of MESFET operation. It is possible to modify equation 12.10 by including a hyperbolic tangent term that will facilitate this smooth transition between the two regions.

The modified model describing the behavior of a GaAs MESFET in the three regions can now be written as

$$I_{ds} = \begin{cases} V_{gs} - V_t < 0 \text{ (Cut off)} \\ \beta\left[(V_{gs} - V_t)^m + \lambda(V_{gs} - V_t)^b V_{ds}\right] \tanh(aV_{ds}) \\ V_{gs} - V_t > 0 \text{ (Linear and saturation)} \end{cases} \qquad (12.11)$$

Where λ is the channel length modulation factor and varies in the range 0.01 to 1.

Parameters λ and $\tanh(aV_{ds})$ are channel length modulation and hyperbolic tangent function respectively, while m and b are constants that are derived empirically. It should be noted that the hyperbolic tangent function $\tanh(aV_{ds})$ is used to describe the channel conductance at low drain-to-source voltage, V_{ds}. This effect is the result of the decrease in magnitude of the depletion region beneath the gate as the gate-to-source voltage, V_{gs}, is increased.

Usually m and b can be adjusted to suit a particular process. For example, with $m = 2$ and $b = 2$, the drain current I_{ds} as described by equation 12.11 reduces to:

$$I_{ds} = \begin{cases} V_{gs} - V_t < 0 \text{ (Cut off)} \\ \beta(V_{gs} - V_t)^2 (1 + \lambda V_{ds}) \tanh(aV_{ds}) \\ V_{gs} - V_t > 0 \text{ (Linear and saturation)} \end{cases} \qquad (12.12)$$

This is referred to as the *Curtice* model.

It is still possible to improve the device model further by considering the influence of velocity saturation. The new relation for the drain to source current referred to the Raytheon model is given by:

$$I_{ds} = \begin{cases} \beta \dfrac{(V_{gs} - V_t)^2}{1 + b(V_{gs} - V_t)} (1 + \lambda V_{ds})(1 - \left(1 - a\dfrac{V_{ds}}{3}\right)^3); \; 0 < V_{ds} < \dfrac{3}{a} \\ \dfrac{\beta(V_{gs} - V_t)^2}{1 + b(V_{gs} - V_t)} (1 + \lambda V_{ds}) \qquad V_{ds} \geq \dfrac{3}{a} \end{cases} \qquad (12.13)$$

Where b, in equation 12.13, is an empirical term and the 'slope factor' a is used to take into consideration the influence of slope in both the linear and saturation regions.

12.4.2.2 V-I characteristics for GaAs MESFET

A typical voltage-current characteristic as described by equation 12.12 is shown in Figure 12–21 for both the depletion and the enhancement devices. When $V_{gs} < V_t$, the increase in the drain-to-source voltage V_{ds} above the saturation voltage $V_{ds\,(sat)}$ leads to current saturation. The saturation of drain current with increasing drain-to-source voltage is caused by velocity saturation in the high electric field in the channel.

The boundary between the linear and saturation regions defined by ($V_{ds} = V_{gs} - V_t$) is referred to as the *'knee voltage'*, and appears as a dashed line in Figure 12–21. Note that the drain current saturates at the same drain-to-source voltage V_{ds} and is independent of the gate-to-source voltage V_{gs}. This behavior can best be explained by noting that the critical electric field $E_{critical}$ in the channel is reached at the same drain-to-source voltage, V_{ds}, given by:

$$V_{ds} = E_{critical} * L$$

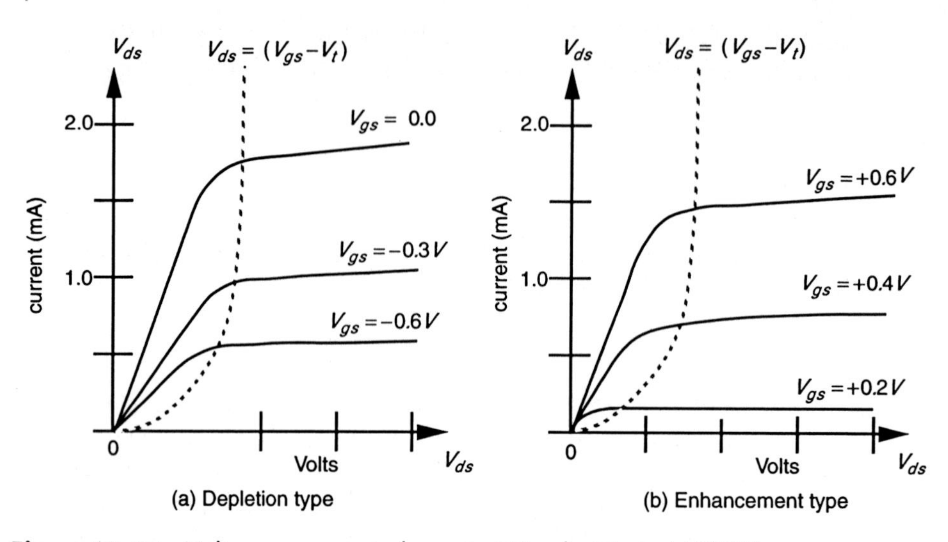

Figure 12–21 Voltage v current characteristics for GaAs MESFET

As a matter of interest the critical electric field is in the order of 3500 V/cm.

As can be seen, the characteristic is similar to that of silicon gate technology, with the exception of the magnitude of the gate-to-source voltage V_{gs}, which is limited to about 0.8 volt. This limit is brought about by the presence of the Schottky diode at the gate region. This is illustrated in Figure 12–22 for both the E type and D type MESFETs.

Depending on whether the MESFET is operating with reverse ($V_t < V_{gs} < 0$) or forward ($0 < V_{gs} < \phi_B$) gate to source bias, the mode of operation is referred to as:

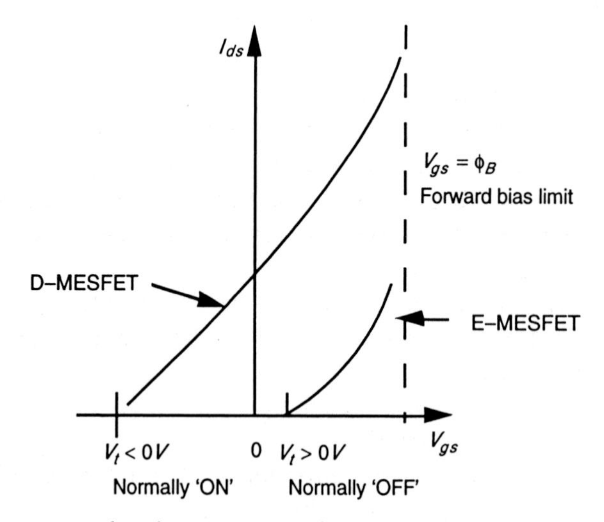

Figure 12–22 Transfer characteristics for MESFET

DEPLETION	→	REVERSE; $V_t < V_{gs} < 0$
ENHANCEMENT	→	FORWARD; $0 < V_{gs} < \phi_B$

12.4.2.3 Threshold voltage definition

The threshold voltage V_t that appears in our model has a significant influence upon the sizing of circuits. V_t is dependent upon the *pinch-off* voltage V_{po} and the barrier potential ϕ_B given by:

$$V_t = \phi_B - V_{po} \tag{12.14}$$

This relation simply means that the pinch-off voltage V_{po} is the total voltage; that is, both built-in potential and applied voltage necessary to completely deplete the channel of mobile charge carries. In other words, it is the gate voltage at which the depletion layer just pinches off the channel; that is, the gate voltage that extends the depletion layer down to the substrate as was illustrated in Figure 12–10(b). The pinch-off voltage is a function of both the channel thickness, a, and concentration density N_d and is always positive. The pinch-off voltage is:

$$V_{po} = \frac{qN_d a^2}{2\varepsilon_0 \varepsilon_r} \tag{12.15}$$

where

a = channel thickness of the n$^-$ implant
N_d = effective channel concentration density
q = electron charge ($1.6 * 10^{-19}$ Coulomb)
ε_0 = permittivity of free space ($8.85 * 10^{-14}$ F.cm^{-1})
ε_r = relative permittivity of GaAs (13.1).

This relation illustrates the difference that exists between the threshold voltage V_t and the pinch-off voltage V_{po}. This difference is somewhat significant and is brought about as the result of the built-in potential ϕ_B which can no longer be neglected as was the case with silicon. Furthermore, the threshold voltage V_t is very sensitive to both the channel thickness a (i.e. the vertical geometry) and the doping of the channel layer.

One significant aspect of the above model is that it illustrates the parameters that influence the transition of a device from being a depletion mode to an enhancement mode.

12.4.2.3.1 Threshold variation

In logic structure, the dynamic switching energy must exceed the energy stored in the load capacitor C_L. This can be written as:

$$P_g \tau_g > 1/2\,(C_L[\Delta V_o]^2) \tag{12.16}$$

where ΔV_o *is* the logic voltage swing, P_g is the gate dynamic dissipation and τ_g is the associated gate delay. To keep the dynamic switching energy small, the logic voltage swing ΔV_o must be kept small also. This requires precise control over the threshold voltages of both the D type and E type MESFETs not only between adjacent devices but also across the whole wafer.

In order to achieve such a control, it is necessary for the standard deviation of the threshold voltage σV_t to be less than 5% of the logic voltage swing ΔV_o. Thus the logic swing can be expressed as:

$$\Delta V_o > 20\,\sigma V_t$$

The logic swing for an E type MESFET is in the order of 500 mV. Above this value one can expect excessive gate current. Thus, the variation of the threshold voltage for the E–MESFETs over the chip must be better than

$$\begin{aligned} \sigma V_t &= \frac{\Delta V_o}{20} \\ &= 25 \text{ mV} \end{aligned} \tag{12.17}$$

This can be compared with the D–MESFET in which the logic voltage swing ΔV_o can be larger than 1 V, which means tolerance to larger threshold voltage variation — that is, at least 50 mV, can be more readily accommodated.

Basically this implies that it becomes necessary to have a high degree of control over the threshold voltage and drain-to-source current to ensure that GaAs circuits with reasonable yields and circuit performance are produced. Also, owing to the almost exponential influence of implant depth on the threshold voltage, there is the need to control the channel thickness within ± 20 Å, and to ensure that change in impurity concentration is < ± 20% to achieve reasonable device yield across the wafer.

12.4.3 Transconductance and output conductance

The two parameters, transconductance g_m and output conductance g_o, are important since they are directly related to the gain of the MESFET. The transconductance describes the relationship between the output current I_{ds} and the input control voltage V_{gs} and is used to measure the gain of the MESFET, while the output conductance determines the slope of the output characteristics.

12.4.3.1 The transconductance parameter g_m

The transconductance g_m is derived by differentiating equation 12.12 with respect to V_{gs}, giving the principal result:

$$g_m = \left.\frac{\Delta I_{ds}}{\Delta V_{gs}}\right|_{V_{ds}} = \text{constant}$$

$$g_m = \begin{cases} = 0;\ \text{for cut off} \\ \text{or} \\ = 2\beta(V_{gs} - V_t)(1 - \lambda V_{ds})\tanh(aV_{ds});\ \text{linear and saturation} \end{cases} \tag{12.18}$$

A major difference between GaAs and Si devices to be noted is the transconductance. For GaAs the transconductance is high with very low gate capacitance. Thus a high gain, bandwidth product can be expected.

Figure 12–23 shows typical transconductances for several types of devices, including those for silicon, primarily for comparison purposes.

It is interesting to compare the transconductance of the silicon bipolar transistor with that of a MESFET. The transconductance of the silicon bipolar transistor is given by:

$$g_m = I_c\left(\frac{q}{kT}\right)$$

where I_c is the collector current.

The expression can be rewritten as

$$g_m \propto A_E e^{V_{be}(q/kT)}$$

Here it is significant that the transconductance is independent of process and it is only slightly influenced by the transistor size. This can be contrasted with GaAs, where the transconductance is both process-dependent and size-dependent.

12.4.3.1.1 Figure of merit f_t

An indication of frequency response may be obtained from the parameter f_t. Thus, it becomes possible to predict the expected intrinsic speed of a GaAs device from a knowledge of the figure of merit commonly referred to as the gain bandwidth product f_t given by:

$$\begin{aligned} f_t &= \frac{g_m}{2\pi(C_{gs} + C_{gd})} \\ &= \frac{\mu_n}{2\pi L^2}(V_{gs} - V_t) \end{aligned} \tag{12.19}$$

where C_{gs} and C_{gd} are the gate-to-source and gate-to-drain capacitances respectively.

The current gain bandwidth product f_t illustrates that switching speed depends upon

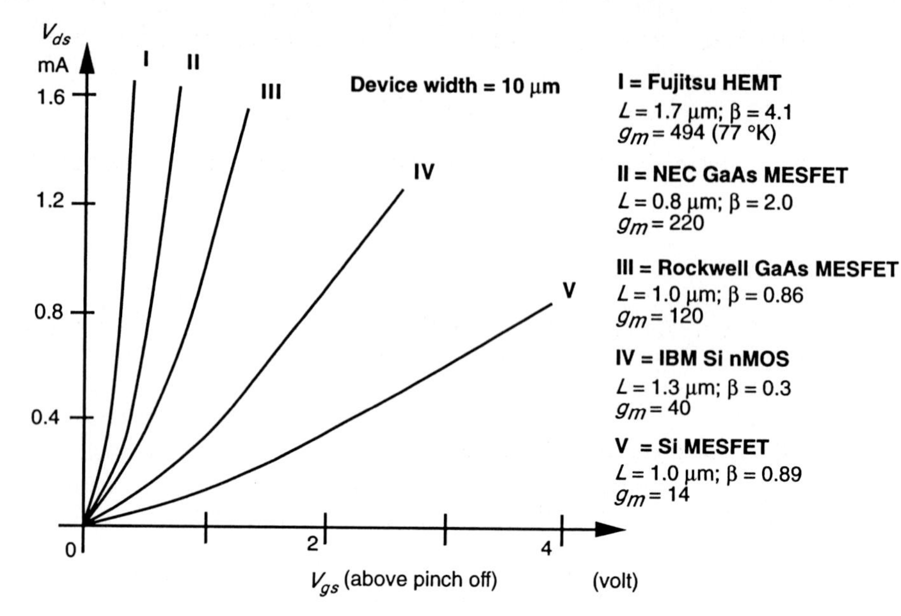

Figure 12–23 Transconductance variations for several devices

- gate length L;
- carrier mobility μ_n in the channel;
- gate voltage.

However, if we consider the limiting condition, that is, velocity saturation, the gain bandwith product may be expressed as

$$f_t = \frac{V_{sat}}{2\pi L} \tag{12.20}$$

where v_{sat} is the saturation velocity. This means for a typical 0.8 μm gate, a f_t of about 28 GHz can be expected.

12.4.3.2 *Output conductance*

The output conductance can also be determined by differentiating equation 12.12 with respect to the drain voltage V_{ds}. Thus

$$g_o = \begin{cases} 0; \text{ for cut off} \\ \lambda\beta(V_{gs} - V_t)^2 \tanh(aV_{ds}) + a\beta[(V_{gs} - V_t)^2(1 + \lambda V_{ds})] \operatorname{sech}^2(aV_{ds}); \text{ linear} \\ \text{and saturation} \end{cases} \tag{12.21}$$

In the saturation region the above relation can be simplified to

$$g_o = \lambda\beta(V_{gs} - V_t)^2 \tanh\ (aV_{ds}) \tag{12.22}$$

from which the drain-to-source resistance R_{ds} can be estimated. Thus

$$R_{ds} = \frac{V_{ds}}{\lambda\beta(V_{gs} - V_t)^2} \tag{12.23}$$

A typical characteristic illustrating the variation of R_{ds} as a function of V_{ds} is shown in Figure 12–24.

12.4.4 Logic voltage swing

In order to improve the switching speed, the options are:

- increase logic voltage swing (logic voltage swing is comparable with the gate voltage above threshold); and
- reduce gate length.

Although the former option is possible, the switching energy in this case is increased, resulting in an increase in dissipation. The dynamic dissipation P_g can be expressed in terms of the logic voltage swing ΔV_o.

Thus

$$P_g = \frac{1}{2} C_L (\Delta V_o)^2 f \tag{12.24}$$

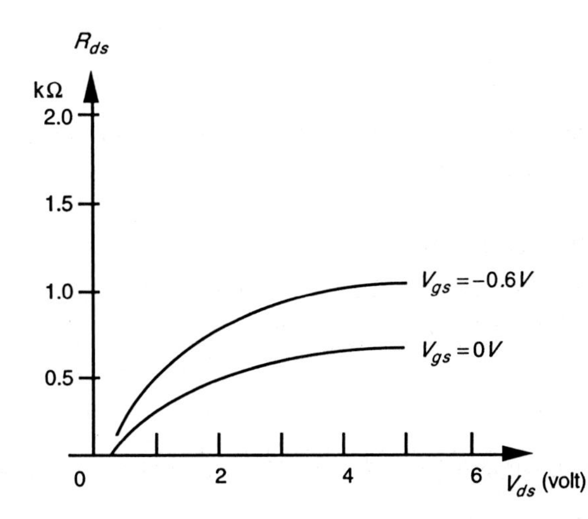

Figure 12–24 Variation of drain-to-source resistance R_{ds} as a function of V_{ds}

where

$$P_g = \text{dynamic dissipation}$$
$$C_L = \text{load capacitance}$$
$$\Delta V_o = \text{logic voltage swing}$$
$$f = \text{frequency of switching}$$

Devices must develop their transconductance at control voltages only a small logic swing above the threshold voltage in order to exhibit small dynamic switching energy.

To establish the logic voltage swing ΔV_o two conditions must be satisfied:

1. The low logic voltage level V_{low} must satisfy

$$V_{low} < V_t$$

which ensures that the device turns off.

2. The gate should not be driven higher than the barrier potential, ϕ_B.

The logic high level V_{high} therefore should satisfy

$$V_{high} < \phi_B$$

Thus, the logic voltage swing can be expressed

$$\Delta V_o = V_{high} - V_{low} \quad (12.25)$$
$$= \phi_B - V_t$$

which is simply the channel pinch-off voltage V_{po}.

12.4.5 Direct-coupled FET logic (DCFL) inverter

A basic requirement for creating a complete range of logic circuits is the inverter. Although several options for the classes of logic are available, direct-coupled FET logic (DCFL) is chosen as this is the only class of logic that shows promise for VLSI implementation.

In this class of logic the inverter uses both the depletion mode and the enhancement mode transistors. The E–MESFET is used as the switching device, while the D–MESFET is used as the load. This basic structure is illustrated in Figure 12–25. From the figure it is evident that the design of the inverter closely resembles that of silicon nMOS circuitry with one exception: the allowable output voltage is limited by the barrier height ϕ_B of the Schottky gate diode.

Now, there several issues that must be highlighted before proceeding with the design of the inverter:

- With no current drawn from the output, the drain to source current I_{ds} for both the E type and D type devices are equal.

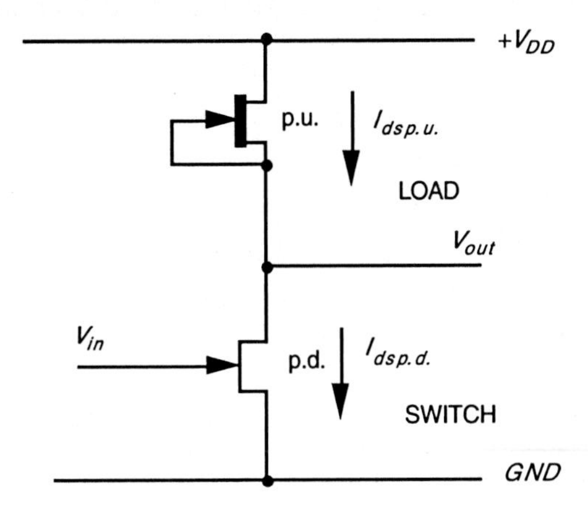

Figure 12–25 Direct coupled FET logic (DCFL)

- For the depletion mode transistor, the gate is connected to the source so it is always on and only the characteristic curve $V_{gs} = 0$ (Figure 12–21(a)) is relevant.
- In this configuration the depletion mode device is called the pull-up (p.u.) and the enhancement mode device is called the pull-down (p.d.) transistor.
- To obtain the inverter transfer characteristic, we superimpose the $V_{gs} = 0$ depletion mode characteristic curve on the family of curves for the enhancement mode device, noting that maximum voltage across the enhancement mode device corresponds to minimum voltage across the depletion mode transistor.
- The point of intersection of the curves as in Figure 12–26 gives points on the transfer characteristic, which is of the form shown in Figure 12–27.
- Note that as V_{in} (= V_{gs} p.d. transistor) exceeds the p.d. threshold voltage, current begins to flow. The output voltage V_{out} thus decreases, and the subsequent increases in V_{in} will cause the p.d. transistor to come out of saturation and become resistive. Note that the p.u. transistor is initially resistive as the p.d. turns on.
- The point at which $V_{out} = V_{in}$ is denoted as V_{inv} (inverter threshold voltage). Note that the transfer characteristics and V_{inv} can be shifted by variation of the ratio of pull-up to pull-down resistances (denoted $Z_{p.u.} / Z_{p.d.}$ where Z is determined by the length to width ratio of the MESFETs).
- During transition, the slope of the transfer characteristic determines the gain

$$Gain = \frac{\Delta V_{out}}{\Delta V_{in}}$$

12.4.5.1 Determination of pull-up to pull-down ratio

Consider the arrangement as shown in Figure 12–28 in which an inverter is driven from the output of another similar inverter.

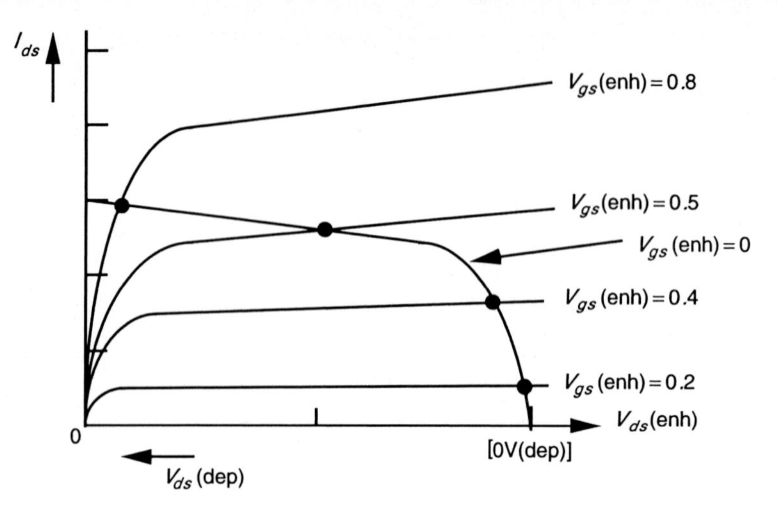

Figure 12–26 Derivation of DCFL inverter transfer characteristics

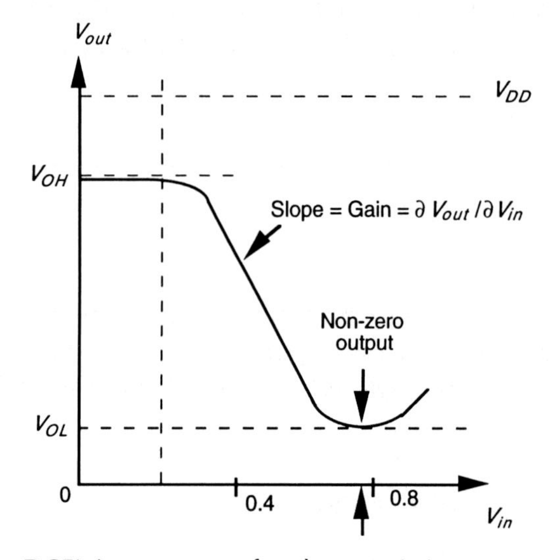

Figure 12–27 DCFL inverter transfer characteristics

In order to cascade inverters without degradation of levels we are aiming to meet the requirement:

$$V_{in} = V_{out} = V_{inv}$$

Since the logic high level is limited by the barrier potential ϕ_B, then for equal margins around the inverter threshold we set V_{inv} equal to half the logic voltage swing.

Thus:

$$V_{inv} = (\phi_B - V_t)/2 = 300 \text{ mV}$$

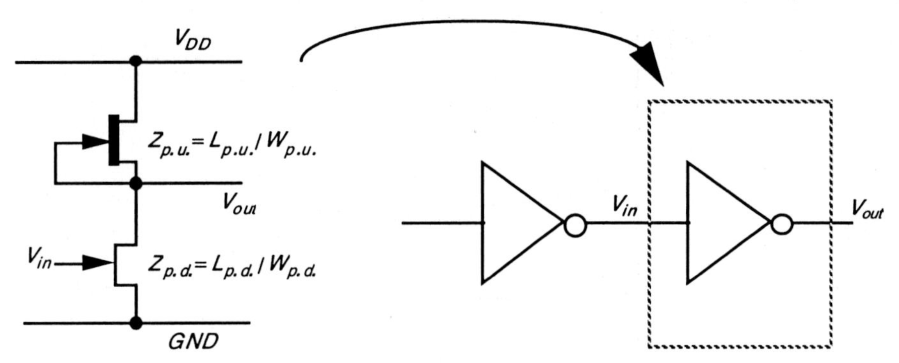

Figure 12–28 DCFL inverter driven directly from another inverter

Now assuming a supply voltage V_{DD} = +2.0 V, and with typical values for threshold voltages V_{tdep} = –700 mV, V_{tenh} = +200 mV, both the pull-up and pull-down transistors are in saturation, that is, $V_{ds} > (V_{gs} - V_t)$ for the D type and E type MESFETs. The pull-up to pull-down ratio ($Z_{p.u.} / Z_{p.d.}$) is defined as:

$$Z_{p.u.} = \frac{L_{p.u.}}{W_{p.u.}}$$

$$Z_{p.d.} = \frac{L_{p.d.}}{W_{p.d.}}$$

where $W_{p.u.}$, $L_{p.u.}$, $W_{p.d.}$ and $L_{p.d.}$ are the widths and lengths of the pull-up and pull-down transistors (i.e. the D–MESFET and E–MESFET) respectively.

The drain to source current for the pull-up transistor (D–MESFET) can be expressed by

$$I_{dsp.u.} = \beta_{p.u.}(V_{gsp.u.} - V_{tdep})^2 \tag{12.26}$$

where

$$\beta_{p.u.} = \left[\frac{\mu_n \varepsilon_0 \varepsilon_r}{2a_{p.u.}}\right]\left[\frac{W_{p.u.}}{L_{p.u.}}\right] \tag{12.27}$$

and

$$a_{p.u.} = \text{implant depth for D–MESFET.}$$

For the pull-down device (E–MESFET), the drain current is

$$I_{dsp.d.} = \beta_{p.d.}(V_{gsp.d.} - V_{tenh})^2 \tag{12.28}$$

where

$$\beta_{p.u.} = \left[\frac{\mu_n \varepsilon_0 \varepsilon_r}{2a_{p.u.}}\right]\left[\frac{W_{p.u.}}{L_{p.u.}}\right] \tag{12.29}$$

and

$$a_{p.d.} = \text{implant depth for E–MESFET}$$

Now equating the two currents, and with $V_{gsp.u.} = 0$, $V_{gsp.d.} = V_{in} = V_{inv}$, we have:

$$\frac{1}{Z_{p.d.}}\left(V_{inv} - V_{tp.d.}\right)^2 = \left(\frac{a_{p.d.}}{a_{p.u.}}\right)\left(\frac{1}{Z_{p.u.}}\right)(-V_{tp.u.})^2 \tag{12.30}$$

and on rearrangement

$$\frac{Z_{p.u.}}{Z_{p.d.}} = \left(\frac{a_{p.d.}}{a_{p.u.}}\right)\left(\frac{-V_{tdep}}{V_{inv} - V_{tp.d.}}\right)^2$$

whence

$$V_{inv} = V_{tenh} - \sqrt{\frac{a_{p.d.}}{a_{p.u.}}}\left(\frac{V_{tdep}}{\sqrt{Z_{p.u.} / Z_{p.d.}}}\right)$$

V_{inv} is set approximately midway between ϕ_B and ground.

Substituting typical values for the threshold voltages $V_{tdep} = -700$ mV, $V_{tenh} = +200$ mV, and with $a_{p.u.}/a_{p.d.} = 4{:}1$ and $\phi_B = 800$ mV, we obtain the principal result

$$\frac{Z_{p.u.}}{Z_{p.d.}} = \frac{10}{1}$$

For MESFETs having $L_{p.u.} = L_{p.d.}$, we have

$$\frac{W_{p.u.}}{W_{p.d.}} = \frac{1}{10}$$

However, in order to improve the packing density, as in the case for VLSI applications, it becomes necessary to use a larger gate length for the pull-up device. This will reduce the drain to source saturation current $I_{ds(sat)}$ but with appropriate optimization this may not be very significant.

It should be noted that such an approach provides us with an approximate method to size up a typical DCFL inverter and therefore it becomes essential to resort to simulation tools such as HSPICE in order to optimize a circuit.

12.5 MESFET-based design

12.5.1 MESFET design methodology

The major aim that a circuit designer is faced with is to turn circuit specifications into masks for processing. However, the physical characteristic of the gallium arsenide processing brings about statistical variations in all process parameters, including those of line width, junction depth and film thickness. The objective of this section is to develop an approach to capture the topology of the actual layout so that through a simple representation both layer information and topology can be described and at the same time interaction between signal and power buses is minimized to guard against degradation of noise margin.

12.5.2 Gallium arsenide layer representations

The advances that are taking place in the gallium arsenide process are very complex and sometimes inhibit the visualization of all the mask levels that are used in the actual fabrication process. Nevertheless, the design process can be abstracted to a manageable number of conceptual levels that represent the physical features one observes in the final GaAs wafer.

We have already seen that MESFET circuits are formed effectively on two layers:

1. green implant layer; and
2. red gate-metal layer.

If the gate-metal layer is in contact with the implant layer a transistor is formed, that is, the implant layer and the gate-metal layer interact to form the Schottky gate where they cross one another. However, if an insulating layer is introduced between the implant and the gate-metal, then there is no interaction between these layers and in this case the gate-metal can be used as an interconnect.

We have also seen that the basic MESFET properties can be modified by varying the implant concentration density. Therefore using a simple color scheme we can capture the topology of the actual layout in gallium arsenide so that simple circuit diagrams which convey both layer information and topology for different layers, including those for the E-type and D-type MESFETs, can be set out.

Through color encoding and symbolic representation of layers it is possible to remove much of the complexity associated with a given design. To convey layer information the encoding used to represent a basic transistor is:

- green (*implant*) for the active implant regions; and
- red (*gate-metal*) for Schottky gate and short interconnections.

Now to facilitate changes to characteristics of the basic transistor and to include representation of other layers, the above encoding is complemented by:

- yellow (*nplus*) for the more heavily doped shallow *n* channel implant;
- blue (*metal 1*) for first level metal; and
- dark blue (*metal* 2) for second level metal.

Transistors are formed by intersection of the green and red masks. The devices that are formed can either be enhancement mode, if no yellow implantation is provided, or depletion mode, if such an implantation is provided. Therefore, the E-type MESFETs are formed whenever the two masks red and green intersect; the D-type MESFETs are formed by intersection of green, red, and yellow masks.

It is essential that one fully understand what set of masks a particular process line uses if an interface format is to be generated. At mask level, some layers can be omitted for clarity while others are derived. The layers for a typical gallium arsenide E–/D–MESFET process are represented in Table 12–4. The following comments should assist with clarifying the color encoding used in Table 12–4.

The *green* layer mask identifies all the active regions, that is, areas that eventually form D and E type devices, active loads, Schottky diodes, and implant resistors.

Green regions that are inside the *yellow* layer mask form the more heavily doped channel of the D–MESFET.

Green regions outside the *yellow* form the lightly doped channel of the E–MESFET.

Table 12–4 Layer representation for E/D GaAs process

Layer	*Color*	*Symbolic*	*CIF*	*Comments*
Implant	Green	E–MESFET	GD	Inside is the active area, outside is the substrate. E–MESFET is formed when crossed by gate-metal.
Depletion implant n^+	Yellow	D–MESFET	GI	Defines the more heavily doped depletion MESFET.
Ohmic contact	Brown	—	GH	Used with source/drain contacts.
Gate-metal	Red	Gate-metal	GP	—
Metal 1	Blue	Metal 1	GM	—
Metal 2	Dark blue	Metal 2	GN	—
Contact	Black	Contact	GC	Source/drain and gate contacts to metal 1.
Via	Gray	Via	GV	Metal 1 to metal 2 contacts.
Passivation	White stipples	—	GG	—

12.5.3 Design methodology and layout style

Having introduced the color and encoding convention for layer representation and device formations, we are now in a position to illustrate the approach to be used to turn MESFET circuits into a mask layout.

12.5.3.1 Ring notation for GaAs MESFETs

Communication paths between cells or group of cells and organization and positioning of power (V_{DD}) and ground (*GND*) buses have significant influence upon the performance of very high and ultra high speed VLSI systems. For example, fast transitions on a signal bus could bring about significant noise on the 'Power Bus'. Thus, both the design methodology and layout will have to address the influence of coupling between buses on performance. This leads to the concept of *'ring notation'* or *'ring diagrams'*, a generic term given to a free form topological symbolic layout in which graphical symbols are placed relative to each other rather than in an absolute manner. These are subsequently interconnected by colored sticks representing mask level interconnection layers, paying particular attention to organizational aspects of 'Power 'and 'Ground' buses in relation to high speed signal carrying paths.

In this text the color coding has been complemented by monochrome encoding of the lines so that black and white copies of circuit representation using *'rings'* do not lose the layer information. The encoding is shown in Figure 12–29.

In the *ring diagram* as shown in Figure 12–30(a), the *'green'* or *'dotted'* line represents the E–MESFET while the *'yellow'* or *'solid'* line represents the D–MESFET. The two 'E type' and 'D type' features are joined together using *'blue'* metal 1. Since this rule is implicit, for simplicity of representation it is possible to remove both the metal and the cut representation at this level of abstraction and include a demarcation line as a reminder, which can be left out after gaining some layout experience. It should be noted that the missing geometries will appear when the *ring diagram* is translated into either symbolic or mask layout form. This simplification is shown in Figure 12–30(b). At this level of abstraction it is important that the length (L) and width (W) for each transistor be included.

12.5.3.2 MESFET design style

Having conveyed layer information and topology by using *ring diagrams,* the rings can then be turned into mask layout either directly or through an intermediate 'symbolic' representation stage of 'grid assignment' where *rings* are converted into circuit elements. This translation phase is illustrated in Figure 12–31. For the mask layouts produced during design to be compatible with the fabrication processes, a set of generic design rules are set out for layouts so that, if obeyed, the rules will produce layouts which will work in practice. Therefore, with the aid of *ring notation* the designer is able to layout the skeleton of a circuit quickly,

Layer	Color	CIF	MONOCHROME ENCODING
Diffusion/ implant	Green	GD	
Gate-metal	Red	GP	
n+	Yellow	GI	
Ohmic contact	Brown	GH	
Metal 1	Blue	GM	
Metal 2	Dark blue	GN	
Contact	Black	GC	
Via 1	Gray	GV	
Overglass/ passivation	White stipples	GG	

Figure 12–29 Layer/feature encoding schemes

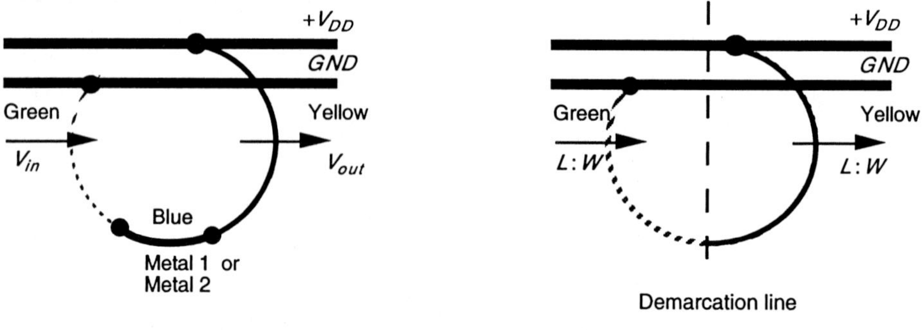

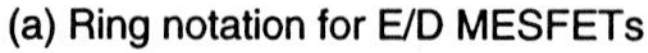
(a) Ring notation for E/D MESFETs

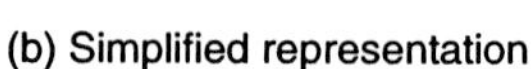
(b) Simplified representation

Figure 12–30 Ring diagram showing the topology of a circuit

paying particular attention to interconnects between adjacent circuitry as well as to the positioning of signal buses in relation to both the Power (+V_{DD}) and Ground (*GND*) buses.

When starting a layout, the first step is normally to draw the Metal 2 (*dark blue*) +V_{DD} and *GND* rails in parallel and in the close proximity of one another at the top. Next, the *green* followed by *yellow* paths are drawn for inverters and inverter-based logic (such as *Nor* gates), as shown in Figure 12–31, not forgetting to make appropriate contacts. Inverters and inverter-based logic comprise a pull-

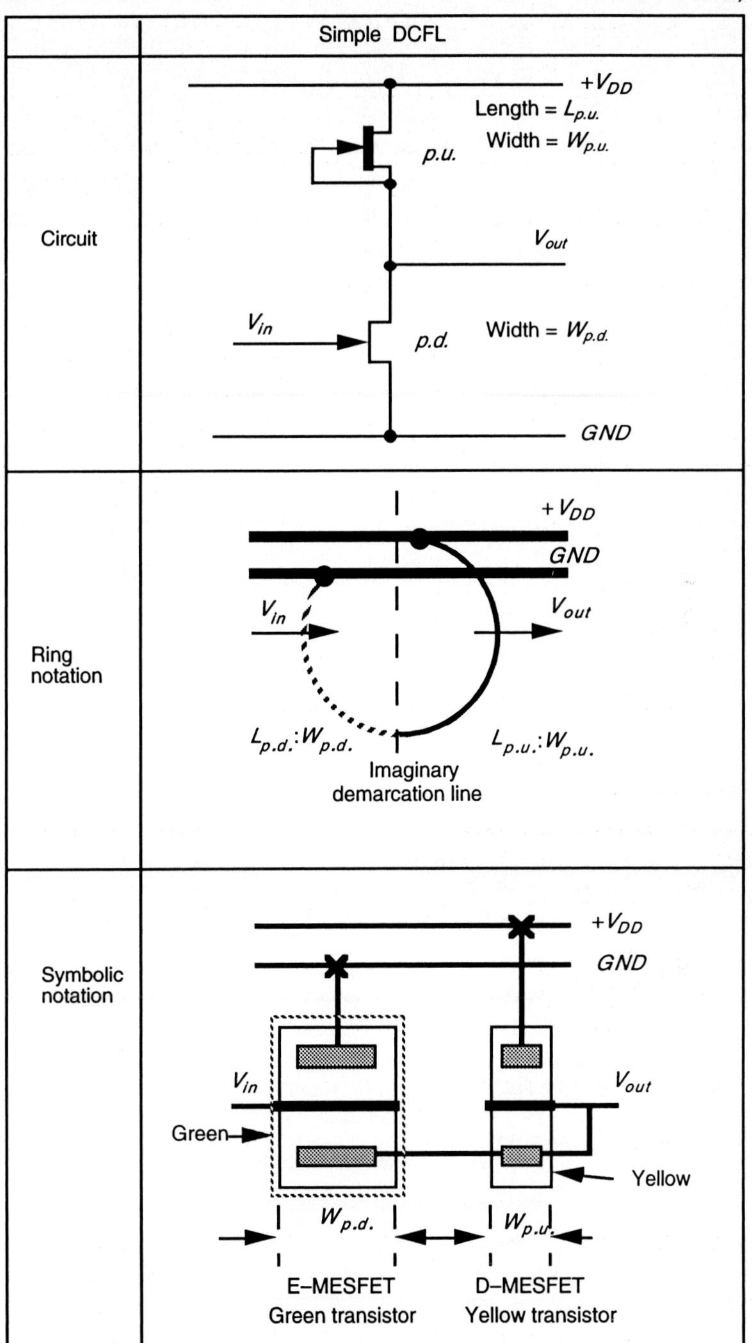

Note: Pull-down always uses minimum size gate length.

Figure 12–31 Translation of DCFL inverter circuit to ring and symbolic forms

up structure, usually a depletion mode transistor, connected from the output point to V_{DD} and a pull-down structure of enhancement mode transistors suitably interconnected between the output point and *GND*. Long signal and global control paths are conveniently run in in metal 2, parallel with the power rails with the GND bus located in between the two to reduce the coupling of fast transients into the power bus. Finally the remaining interconnects are made using either metal 1 (*blue*) or metal 2 (*dark blue*) and the control signals and data inputs added. In some processes it is also possible to use the gate metal (*red*) for very short paths.

Since, in this technology, we restrict ourselves to parallel branches in the input path — that is, *Nor* gates only — then the ring notation for *Nor* gates may be simplified by eliminating the parallel input branches as shown in Figure 12–32. The transformation to symbolic form is then straightforward, as in Figure 12–33.

12.5.3.2 Layer connections

As for nMOS and CMOS, intersections on the same layer form connections, as in Figure 12–34(a). Intersections on different layers do not form connections or transistors as shown in Figure 12–34(b). Different layers may also be connected by a contact or a via as in Figure 12–34(c). Some processes do not support *blue* (metal 1) crossing *green* (diffusion). This is primarily to reduce some of the complexities that emerge during the design phase.

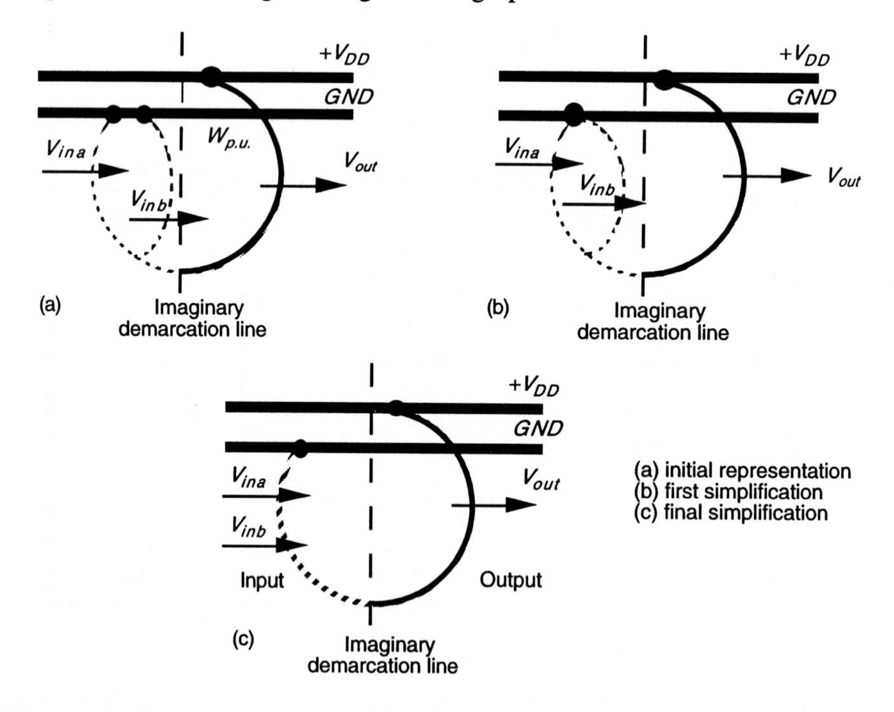

Figure 12–32 Ring notation for 2 input *Nor* gate

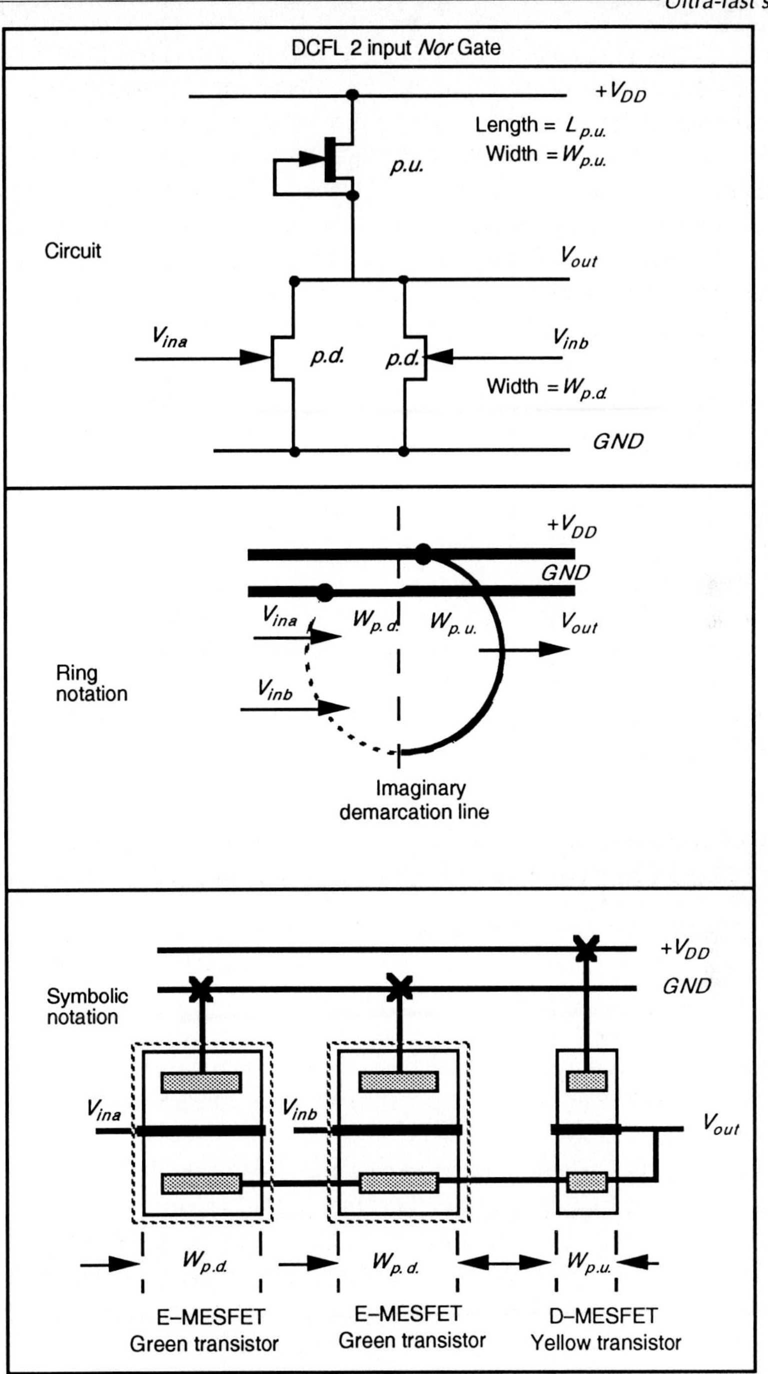

Note: Pull-down MESFETS are always of minimum size gate length.

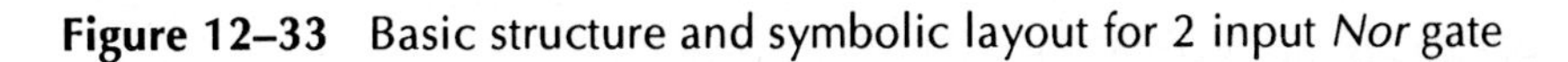

Figure 12–33 Basic structure and symbolic layout for 2 input *Nor* gate

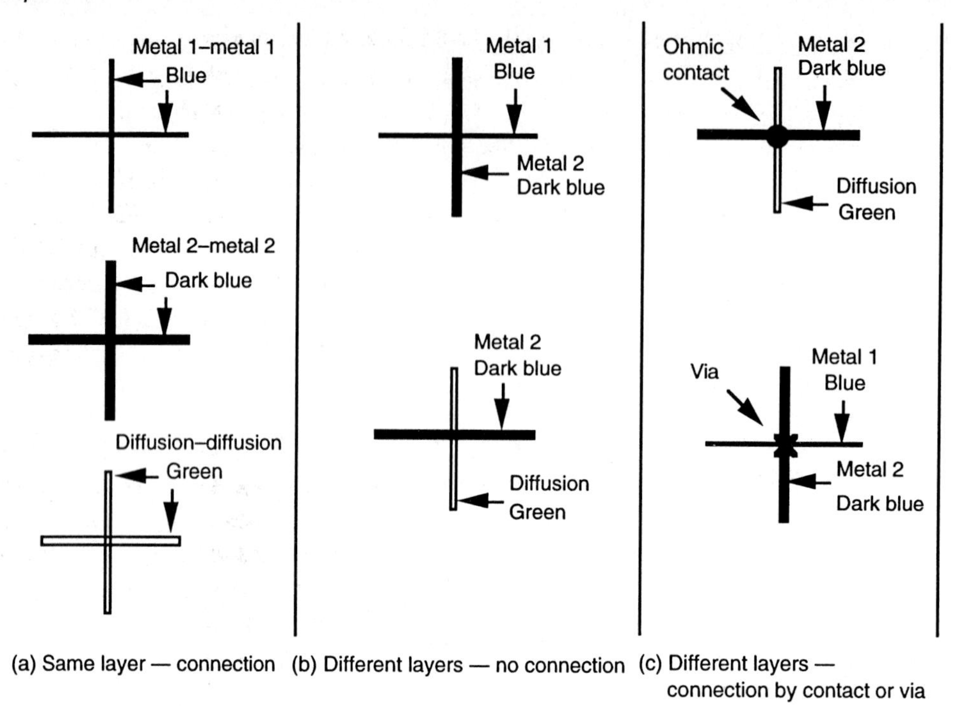

Figure 12–34 Layer connectivity*

* Some restrictions may apply for specific processes.

12.5.4 Layout design rules

Design rules, or *layout rules,* can be considered as a prescription for the preparation of the photomasks that are to be used in the fabrication of integrated circuits. The rule set provides a necessary communication link between circuit designer and process engineer during the manufacturing phase of the integrated circuit. The main objective associated with the design rules is to obtain the circuit with optimum yield in as small a geometry as possible without compromising reliability of the circuit.

Usually, the layout rules represent the best possible compromise between yield and performance. In fact, the more conservative the rules are, the more likely it is that the circuit will function. However, the more aggressive the rules are, the greater the probability of improvements in circuit performance. Such an improvement may be at the expense of yield. Design rules specify to the designer certain geometric constraints on the layout artwork so that the patterns on the processed wafer will preserve the topology and geometry of the design. What is significant is that layout rules do not represent some hard boundary between correct and incorrect fabrication, but a tolerance that ensures very high probability of correct fabrication and subsequent operation.

Circuit designers usually want tighter, smaller layouts for improved performance and decreased area. On the other hand, the process engineer calls for rules that result in a controllable and reproducible process. One important factor associated with design rules is the achievable definition of the process line equipment. Definition is determined by process line equipment and process design. For example, it is found that if a 10:1 wafer stepper is used instead of a 1:1 projection mask aligner, the level-to-level registration will be closer.

Design rules can also be influenced by the maturity of the process line. If the process is mature, then one can be assured of the process line capability allowing tighter design with fewer constraints on the designer. Layout rules address two main issues:

1. geometrical reproduction of features that can be reproduced by the mask-making and lithographical process; and
2. interaction between different layers.

Over the years several approaches have been used to describe the design rules. However, in this text we are going to concentrate on two methods that are appropriate for gallium arsenide technology. These are:

1. The *lambda-based* rule; and
2. The *micron-based* design rule.

The lambda-based design rules used earlier in the text were made popular by Mead and Conway (1980)* for silicon, and are based on a single parameter, *lambda* (λ), which characterizes the linear features as well as the resolution of the complete wafer implementation process.

Note that the degradation in circuit performance could make the lambda-based design approach unsuitable for GaAs processes. However, in this text, for simplicity, initially we will use lambda rules to illustrate principles and to familiarize the designers with the geometric features and the layout process associated with GaAs MESFETs. Then, by adopting symbolic techniques, micron rules can be applied directly.

12.5.4.1 Lambda-based rules for GaAs MESFET

Table 12–5 and Figure 12–35 are a version of a lambda-based rule set. From Figure 12–35 it can be seen that the rule set is defined in terms of:

- feature sizes; and
- separations and overlaps.

Several rule set issues require discussion.

---

* C. A. Mead & L. A. Conway, *Introduction to VLSI Systems*, Addison-Wesley, 1980.

Table 12–5 Lambda-based layout rules for gallium arsenide

Layer	*CIF*	*Rule feature*	*Dimension (lambda)*
Active (Diffusion)	GD	A1 minimum width	5
		A2 minimum spacing	5
		A3 minimum to n^+	5
		A4 minimum E–MESFET width	5
Depletion implant n^+	GI	B1 minimum D–MESFET gate overlap	2
		B2 minimum width	7
		B3 minimum spacing	5
		B4 minimum spacing to E–MESFET	2
Ohmic contact	GH	C1 minimum ohmic contact width	5
		C2 minimum ohmic-metal spacing	5
		C3 minimum cut overlap	2
		C4 minimum ohmic contact size	5×5
Gate metal	GP	D1 min. gate-metal gate extension	2
		D2 min. gate-metal length	3
		D3 min. gate-metal width	3
		D4 minimum cut overlap	2
		D5 min. gate-metal spacing	5
		D6 min. spacing to ohmic contact	3
Contact	GC	E1 minimum cut size	4×4
		E2 minimum cut spacing	4
		E3 minimum spacing to via	4
Metal 1 (Diffusion)	GM	F1 minimum width	4
		F2 minimum spacing	5
		F3 minimum cut overlap	2
		F4 minimum via 1 overlap	2
Via 1	GV	G1 minimum via size	5×5
		G2 minimum via spacing	5
Metal 2	GN	H1 minimum width	5
		H2 minimum spacing	5
		H3 minimum overlap of via 1	2

12.5.4.2 Width and spacing rules

Although diffusion, metal 1, and metal 2 can cross each other without interaction, in some processes metal 1 is not permitted to cross diffusion.

Width and separation rules given in Figure 12–35 are dependent upon the width of the photoresist.

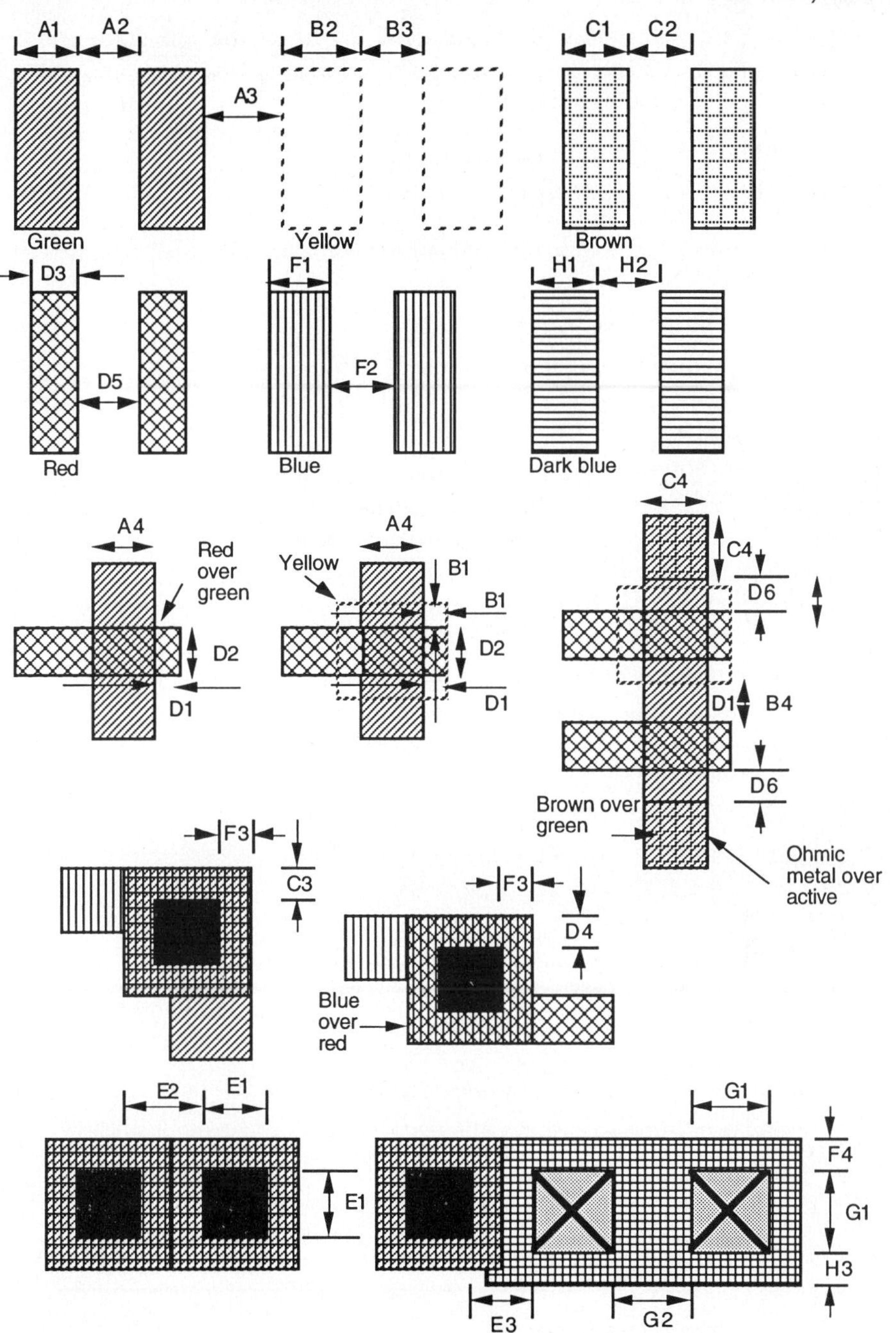

Figure 12–35 Lambda-based rules for GaAs MESFET process

As for nMOS and CMOS, we need to ensure that the depletion regions of two unrelated implants do not contact. The separation between implant is determined from:

(a) width of depletion region; and
(b) width of the photoresist.

Crossing of metal 2 over channel areas of the MESFETs should be avoided.

12.5.4.3 Transistor rules

There are two types of implants used to form the two different MESFETs. A transistor is depletion type if it is inside the n^+ (*yellow*) region, otherwise it is enhancement mode.

It is essential for gate-metal (*red*) to completely cross the implant (*green*) region, otherwise the transistor that has been created will be shorted by a n^- path between source and drain. To ensure this condition is satisfied, 2λ of gate-metal extension is necessary. This is termed the 'Schottky gate extension'.

Orientation is an important consideration during layout. All MESFETs need to be positioned horizontally owing to the anisotropic nature of GaAs, which influences the threshold voltage of the device brought about as a result of variation in both concentration density and channel thickness.

Some processes require isolation between devices to reduce their interaction. This is achieved through lattice damage. The mask is derived from the 'logical' operation of the active layer masks.

12.5.4.4 Contact cut and via rules

Generally the size of a cut is established from the knowledge of the minimum dimensions necessary to give an acceptable resistance. The ohmic contact has a current capability in the range 0.5–1.0 mA/μm long. The rules that one may follow are:

- minimum dimensions of ohmic cut for source/drain are $5\lambda \times 5\lambda$;
- minimum dimensions of a cut are $4\lambda \times 4\lambda$;
- via dimension is $5\lambda \times 5\lambda$;
- metal 1 overlap of via is 2λ; and
- metal 2 overlap of via is 2λ.

12.5.4.5 Process enhancements

There are several enhancements that may be added to the GaAs processes, primarily to provide active load, capacitors and resistors, as well as to increase routability of circuits through a third metal or fourth metal level.

12.5.4.5.1 Saturated resistor

The saturated resistor is simply a MESFET with the Schottky gate removed. The preferred direction for layout is vertical.

12.5.4.5.2 Capacitors

Several of the processes provide for at least two kinds of capacitors. These are:

1. metal-insulator-metal (MIM); and
2. diode-capacitor diode (DCAP).

There is considerable complexity and variation in the approach to realize the DCAP, but the MIM capacitor structure is quite simple using metal 1 and metal 2 as the plates of a parallel plate capacitor.

12.5.4.6 Design rules summary

The approach taken here has been to focus our attention on the main features of typical design rules that the designer must become intimately familiar with. Although the introduction of a lambda-based approach has no real value in terms of creating 'real' circuits, because of its simplicity, much insight can be gained for the important issues that must be considered during the layout phase. With this basic knowledge, it is not too difficult to use actual micron rules that may be obtained from different foundries for an actual design.

12.5.5 Symbolic approach to layout for GaAs MESFETs

Now that the concept of lambda-based rules for GaAs has been introduced, and its limitations have been commented on, it becomes evident that perhaps to implement circuits and systems that synthesize the correct geometry from an intermediate form referred to as *symbolic notation* (Figure 12–31) is more appropriate to this technology. This means symbolic styles of design would provide a solution for creating generic GaAs circuits that can be fabricated in various foundries or processes.

The adoption of symbolic design allows the designer to directly manipulate transistors as well as other circuit features that could be of interest. Figure 12–36 shows the symbolic layout for the D type latch using SDCFL logic (see next section) and based upon the notation of Figures 12–30 and 12–31. Translation from this level into masks once again requires the introduction of geometric details as before, using the micron rules.

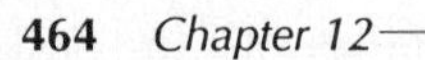

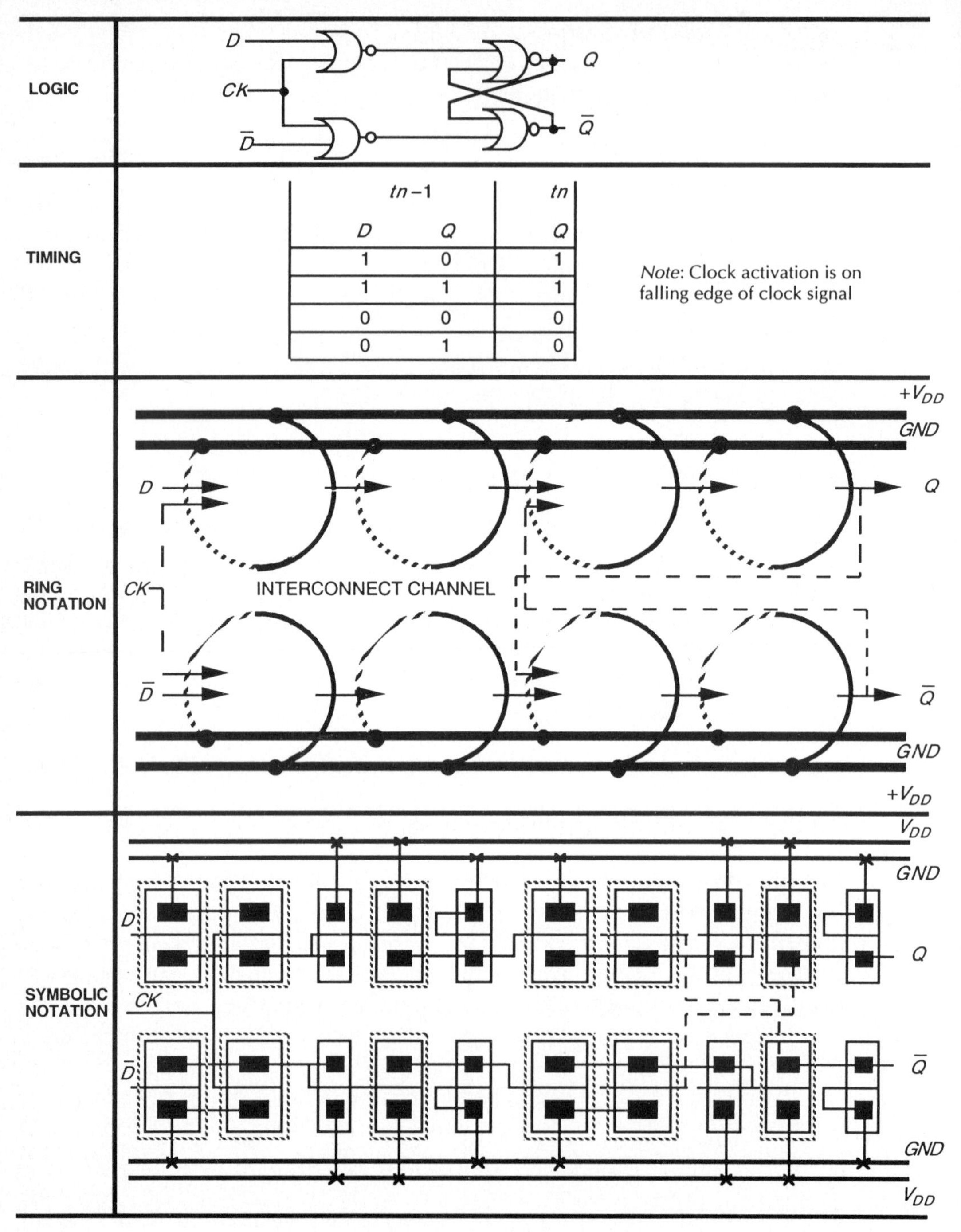

Figure 12–36 Symbolic representation of D type latch using SDCFL

12.6 GaAs MESFET classes of logic

There are two main approaches to logic design:

1. normally-on logic; and
2. normally-off logic.

The normally-on logic uses depletion mode MESFETs which are 'ON' devices and when used as switching elements are required to be turned OFF. Thus, a number of circuit techniques have been developed to facilitate logic turn-off. The approaches in this class of logic include:

- unbuffered FET logic (UFL);
- buffered FET logic (BFL);
- D–MESFET Schottky diode FET logic (SDFL);
- capacitor-coupled FET logic (CCFL); and
- capacitor-diode FET logic (CDFL).

The normally-off logic uses enhancement mode MESFETs as the switching element. Although several approaches have emerged during the last few years, the following structures have shown the most promise:

- direct-coupled FET logic (DCFL);
- buffered DCFL; and
- source-follower DCFL (SDCFL).

12.6.1 Normally-on logic gates

Depletion mode devices are basic switching elements for this class of circuits. Since DMESFETs are ON devices, a negative voltage is needed at the gate to facilitate turn-off. This means that two supply rails together with level shifting networks are necessary for proper circuit operation. Owing to this complexity, this class of logic is unsuitable for VLSI implementation.

12.6.2 Normally-off logic gates

The normally-off logic includes direct coupled FET logic (DCFL), buffered DCFL, and source-follower (SDCFL). The following section provides a brief outline of this particular class of logic families, with particular emphasis on the DCFL and SDCFL being the main contenders for ultra high speed VLSI systems.

12.6.2.1 Direct-coupled FET logic (DCFL)

In this class of logic both the depletion mode and the enhancement mode transistors are used. The enhancement mode FET acts as the switching device, and the depletion type device acts as the load. From the basic structure it is evident that, first, there is no need for level shifting circuitry as was the case for normally-on logic. Secondly, the design of the logic gate closely resembles that of nMOS circuitry; and finally only a single power supply is required. DCFL gate dissipation is typically 100 μW with an associated delay of about 50 ps, which is considerably less than the normally-on logic families. Thus, the logic appears as a suitable contender for very high speed VLSI systems.

The allowable output voltage is limited by the barrier height of the Schottky gate diode, which means only a small voltage swing is possible from DCFL circuits, which in turn implies relatively small noise margins.

12.6.2.2 DCFL with super buffers

DCFL circuits have weak load drive capability. This implies that the delay associated with a gate increases with an increase in both the fan-out and the interconnect line lengths. Introduction of super buffers can alleviate much of the problem at the expense of extra area. Usually the basic DCFL gate is used for light load conditions, while the super buffers are used where larger loads are to be driven.

12.6.2.3 Source-follower DCFL FET logic (SDCFL)

The source-follower FET logic (SFFL) uses both the enhancement and depletion mode devices. The basic structure for a SFFL inverter is shown in Figure 12–37. This logic family has both a power dissipation and switching delay that are comparable with the DCFL family, but with a larger noise margin which is brought

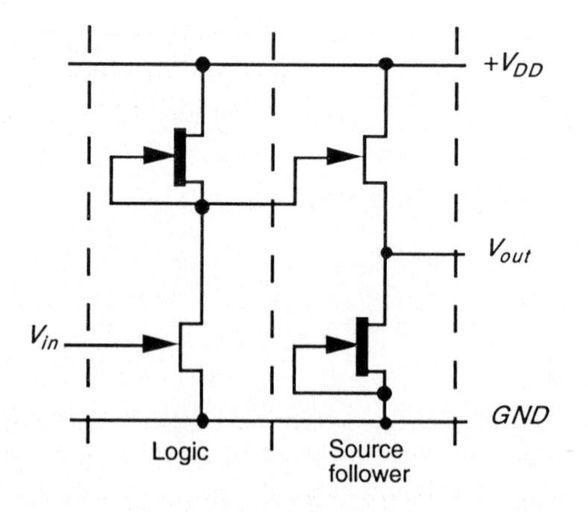

Figure 12–37 Basic inverter structure for source follower DCFL FET logic (SDCFL)

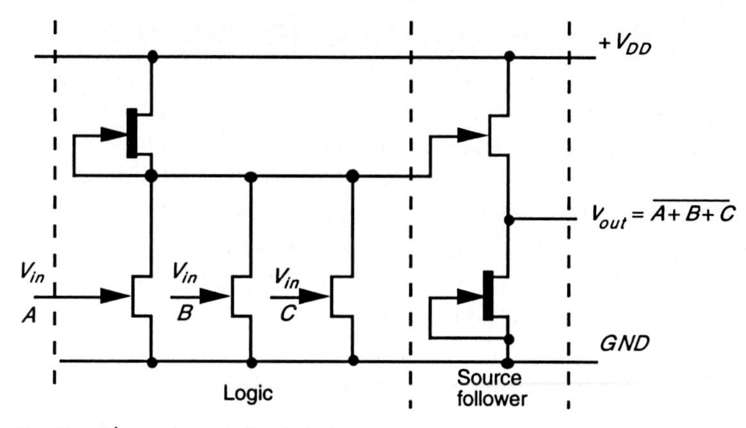

Figure 12–38 Three input SDCFL *Nor* gate

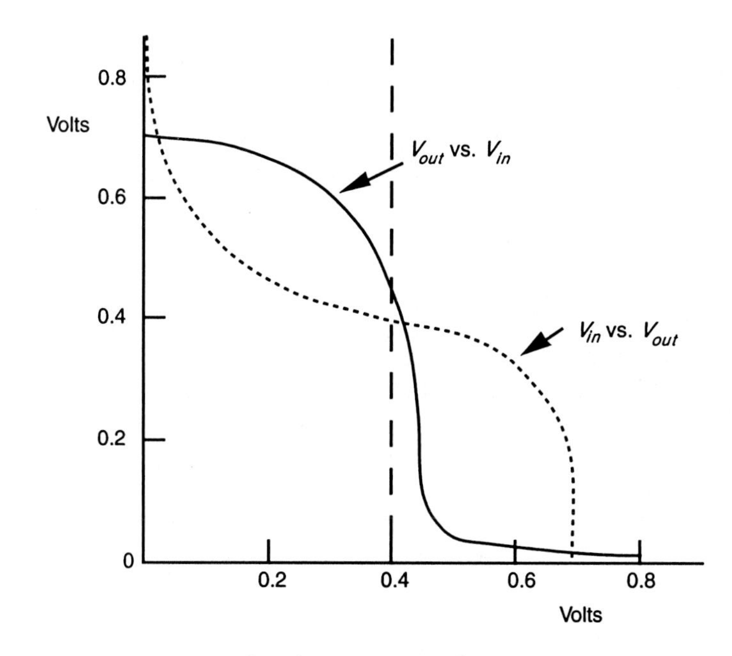

Figure 12–39 DC transfer characteristics for SDCFL

about as the result of the pull-up transistor (enhancement mode) being able to be turned off, thus permitting the source follower output to pull-down all the way toward zero voltage.

This basic structure can be extended to perform logic functions. A typical three input gate is shown in Figure 12–38. The DC transfer characteristics, showing the larger noise margin, are illustrated in Figure 12–39. This class of logic is most suitable for the realization of the *And-Or-Invert* (AOI) function which usually assists in the optimization of logical functions.

12.7 VLSI design — the final ingredients

We are living in an age of unprecedented revolutionary change in engineering, particularly electronic engineering. The digital computer and associated processing revolution of the past two or more decades has been complemented and augmented by the even more dramatic advances in microcircuitry in silicon. We have come to accept world-shattering advances as a matter of course, and predictions such as the computing power of a CRAY 1 computer in one's pocket hardly raise an eyebrow. Further, the full potential of newer technologies, such as GaAs, has yet to be explored.

However, it is a fact that unlike in the situations faced by engineers in the past, we are no longer technology-bound or limited. Indeed, we have solutions to problems that don't even exist yet! This is a situation in which the potential applications of VLSI technology are limited only by the creativity and imagination of those working in engineering or computer science.

VLSI design is also an enjoyable area in which to work. The designer has a great deal of freedom as there are few constraints associated with VLSI system designs. It is also an area which captures the imagination and it is hard not to become highly motivated.

The authors therefore recognize enthusiasm founded on a *knowledgeable base* as the final ingredient. We feel it appropriate to end with three quotations, two from R. W. Emerson and one from Franklin D. Roosevelt:

Nothing great was ever achieved without enthusiasm.

R.W.E.

The reward of a thing well done is to have done it.

R.W.E.

The only limit to our realization of tomorrow will be our doubts of today.

F.D.R.

12.8 Tutorial exercises

1. Using the electron velocity versus electric field characteristics as illustrated in Figure 12–5, compare the carrier velocity behavior of silicon with that of gallium arsenide. For a bipolar device the base region may be considered to be in the order of 0.2 μm–0.25 μm, while in the GaAs technology a typical gate has a dimension of 1.0 μm.
2. Typical values for a D-MESFET are as follows: $\mu_n = 7000$ cm^2/V-sec, $\varepsilon_r = 13.1$, $\varepsilon_0 = 8.85^{*}10^{-14}$ F/cm, $a = 1000$ Å. Using these parameters as the base determine the gain factor for the D–MESFET.

3. For depletion mode devices, typical channel doping is in the order of $1*10^{17}$ cm^{-3} and the channel implant thickness, *a,* is about 1500 Å. Calculate the pinch-off voltage and hence the threshold voltage for this device. What conclusions can you make?
4. Using the *ring notation,* design a simple D-latch. With the aid of color encoding create a layout for this structure.

Appendix A 2.0 micron double poly. double metal n-well CMOS* — Electrical parameters

Process specs: 2.0 micron double poly. double metal n-well CMOS*

	Minimum	Typical	Maximum
Oxide thickness (angstroms)			
Poly. 1 gate oxide	370	400	430
Poly. 2 oxide	470	500	530
Field oxide (poly. 1 & 2 to sub.)	5500	6000	6500
Metal 1 to poly. 1 & 2	8000	8500	9000
Metal 1 to sub.	13500	14500	15500
Metal 1 to n^+/p^+ diff.	8500	9000	9500
Metal 2 to metal 1	6000	6500	7500
Poly. 1 to poly. 2	650	750	850
Conductors			
Poly. 1	3700	4000	4300
Poly. 2	3700	4000	4300
Metal 1	5500	6000	6500
Metal 2	10500	11500	12500

* In all cases, the serious user is advised to contact Orbit for their latest process details.

Device specs: 2.0 micron double poly. metal n-well CMOS

	Minimum	Typical	Maximum
P-channel poly. 1			
Threshold (volts)	–1.0	–0.75	–0.5
Gamma (volts **.5)	0.45	0.55	0.65
K' = μCox/2 (μA/V**2) VDS = 0.1V, VGS = 2 – 3V	6.0	7.5	8.5
Punchthrough for min. length channel (volts)	–16	–14	–10
Subthreshold slope (volts** –3/decade)	90	100	110
Delta length = effective-drawn (microns)	–0.7	–0.4	–0.1

Poly. 2 etch delta from mask C.D. to wafer is 1.1 μm.
For drawn C.D. of 2 μm, and biased to 3 μm, the final wafer dimension is 2.0 μm.
Recommended minimum poly. 2 gate width is 2.5 μm but interconnect can be 2.0 μm.

	Minimum	Typical	Maximum
P-channel poly. 2			
Threshold (volts)	–1.5	–1.15	–0.8
Gamma (volts **.5)	0.5	0.6	0.8
K' = μCox/2 (μA/V**2)	5.0	6.0	7.0
Punchthrough for min. length channel (volts) 2.5 μm	–16	–14	–10
Subthreshold slope (volts** – 3/decade)			
Delta length = effective – drawn (microns)	–0.8	–0.5	–0.2
N-channel poly. 1			
Threshold (volts)	0.5	0.75	1.0
Gamma (volts **.5)	0.15	0.25	0.35
K' = μCox/2 (μA/V**2) VDS = 0.1V, VGS = 2 – 3V	20	23	26
Subthreshold slope (volts** – 3/decade)	90	100	110
Punchthrough for min. length channel (volts)	10	14	16
Delta length = effective-drawn (microns)	–0.7	–0.3	–0.0

Poly. 2 etch delta from mask C.D. to wafer is 1.1 μm.
For drawn C.D. of 2 μm, and biased to 3 μm, the final wafer dimension is 2.0 μm.
Recommended minimum poly. 2 gate width is 2.5μm but interconnect can be 2.0 μm.

	Minimum	Typical	Maximum
N-channel poly. 2			
Threshold (volts)	0.7	1.10	1.40
Gamma (volts**.5)	0.215	0.30	0.40

	Minimum	Typical	Maximum
K' = μCox/2 (μA/V**2)	18	20	22
Subthreshold slope (volts** – 3/decade)			
Punchthrough for min. length channel (volts) 2.5 μm	10	14	16
Delta length = effective – drawn (microns)	−0.8	−0.4	−0.1
CCD channel potential (volts)			
Poly. 1 VG = 0	3.0	5.0	8.0
Poly. 2 VG = 0	3.0	5.0	8.0

NPN transistor in the n-well

Beta = 80 to 200 at $I_B = 1\mu A$

BVEBO = 10V
BVCEO ≥ 10V
BVCES > 10V
BVCBO ≥ 60V

P-base Xj = 0.45 to 0.50 micron
N+ emitter Xj = 0.3 micron

Rcollector = 1.0 +_0.2 kohm/sq
P- base resistance 1.2 + _0.2 kohm/sq

Early voltage > 30 volts

Sheet resistance (ohms per square)

	Minimum	Typical	Maximum
P+ Active	40	57	80
N+ Active	20	28	40
N-well	2000	2500	3000
Poly. 1	15	21	30
Poly. 2	18	25	30
Metal 1	.050	.070	.090
Metal 2	.030	.040	.050

Contact Resistance (ohms)	Minimum (single contact 2 by 2μm)	Maximum
Metal 1 to p^+ Active	35	75
Metal 1 to n^+ Active	20	50
Metal 1 to poly. 1	20	50
Metal 1 to poly. 2	20	50
Metal 1 to metal 2	0.4	0.7

Field inversion and breakdown voltages (volts)

	Minimum	Typical	Maximum
N-channel poly. 1 field inversion	10	14	
N-channel poly. 2 field inversion	10	14	
N-channel metal 1 field inversion	10	14	
P-channel poly. 1 field inversion		−14	−10
P-channel poly. 2 field inversion		−14	−10
P-channel metal 1 field inversion		−14	−10
N-diffusion to substrate junction breakdown		14	16
P-diffusion to substrate junction breakdown		15	18
N-well to P-subjunction breakdown		50	90

Interlayer capacitances (Plate: 10** – 5 pF micron ** – 2)

	Capacitance Min.	Capacitance Max.	Equiv. thickness Min. (angstroms)	Equiv. thickness Max. (angstroms)
Gate oxide plate poly. 1	78	90	370	430
Gate oxide plate poly. 2	64	70	470	530
Poly. 1 to poly. 2 over active	43	55	650	850
Poly. 1 to poly. 2 over field	43	55	650	850
Metal 1 to active plate	3.6	4.0	8500	9500
Metal 1 to subs plate	2.2	2.5	13500	15500
Metal 1 to poly. plate	3.7	4.4	8000	9000
Metal 2 to active plate	1.9	2.4	14500	17500
Metal 2 to subs plate	1.5	1.65	19500	22000
Metal 2 to poly. plate	1.9	2.4	14500	17500
Metal 2 to metal 1 plate	4.6	5.6	6000	7500

Appendix B
1.2 micron single poly. double metal n-well and p-well CMOS*—design rules and process and device specifications

* In all cases, the serious user is advised to contact Orbit for the latest design rules and process details.

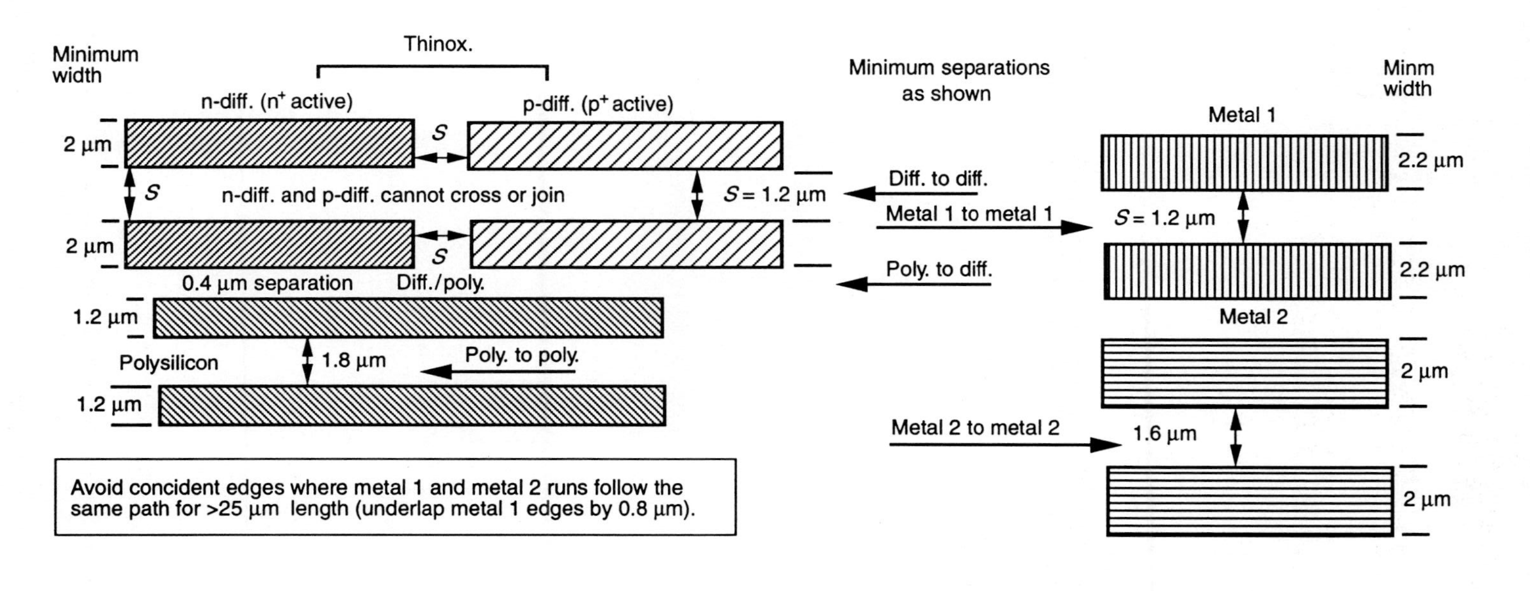

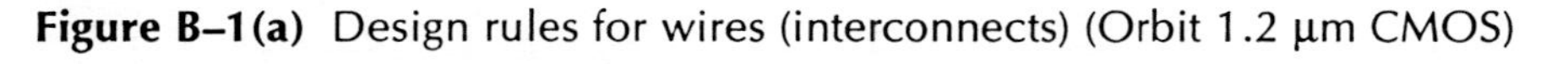

Figure B–1(a) Design rules for wires (interconnects) (Orbit 1.2 μm CMOS)

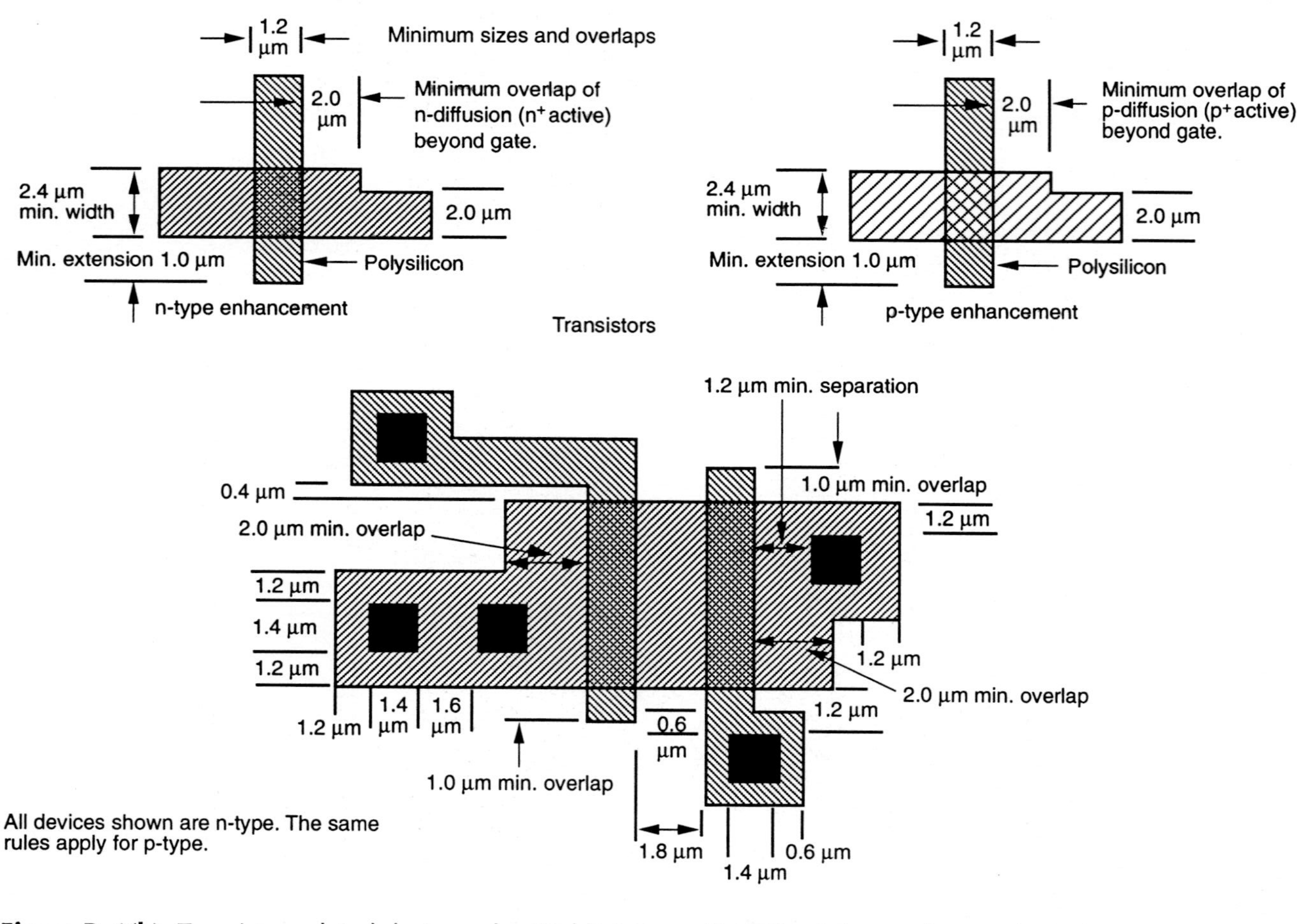

Figure B–1(b) Transistor-related design rules (Orbit 1.2 µm CMOS) minimum sizes and overlaps

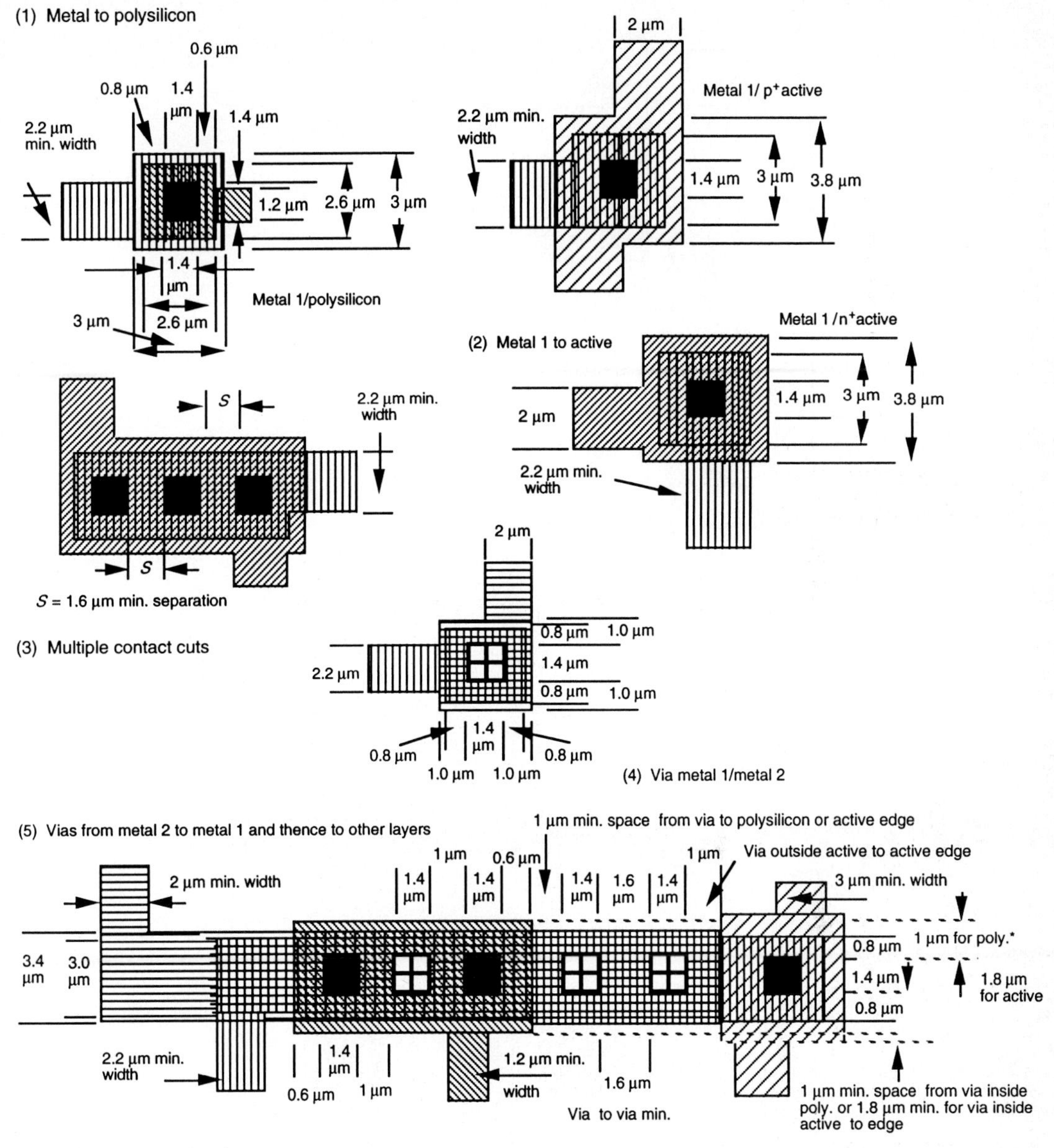

Note: Vias must not be placed over contacts.

Figure B–1(c) Rules for contacts and vias (Orbit 1.2 μm CMOS)

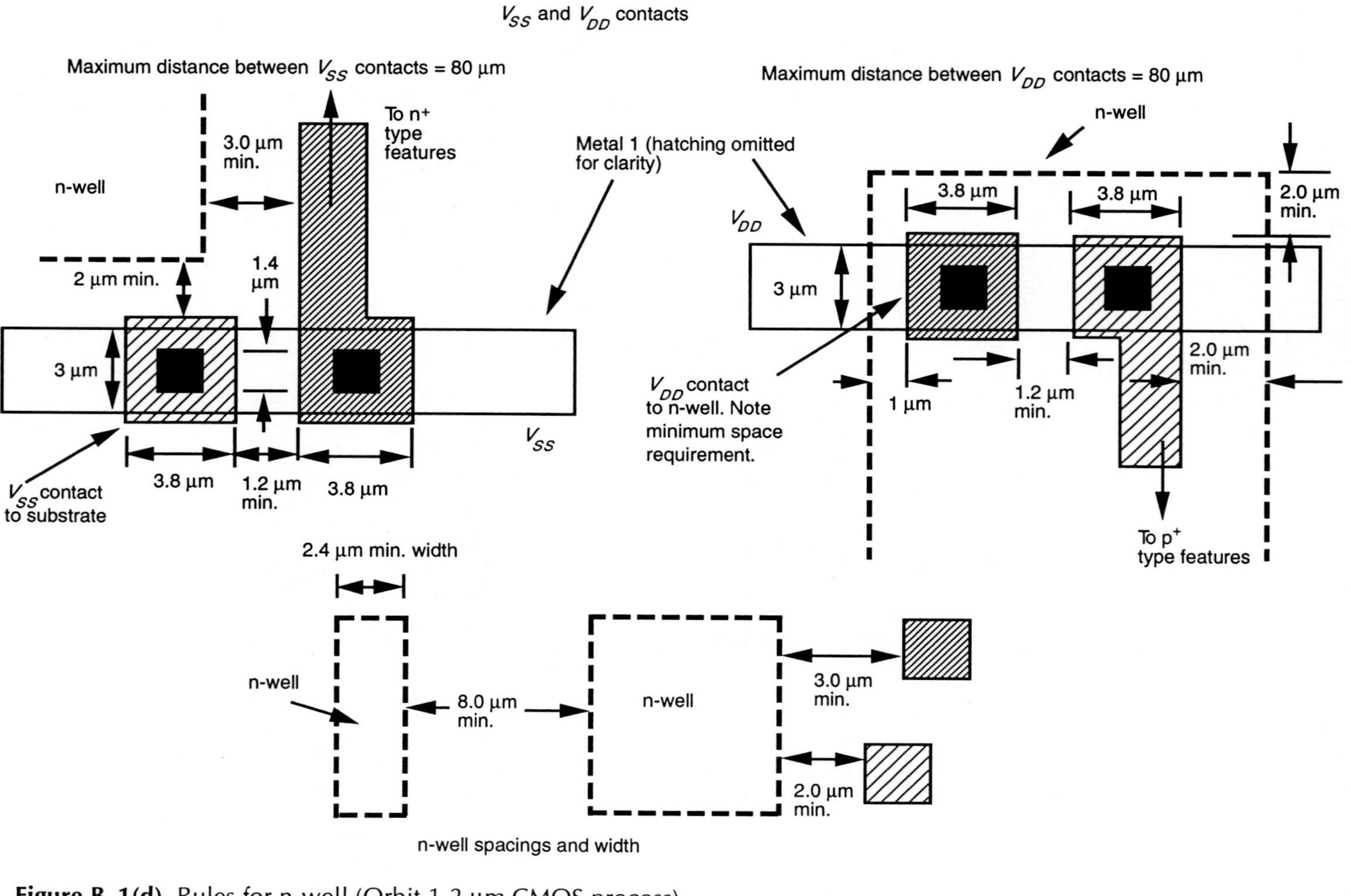

Figure B–1(d) Rules for n-well (Orbit 1.2 µm CMOS process)

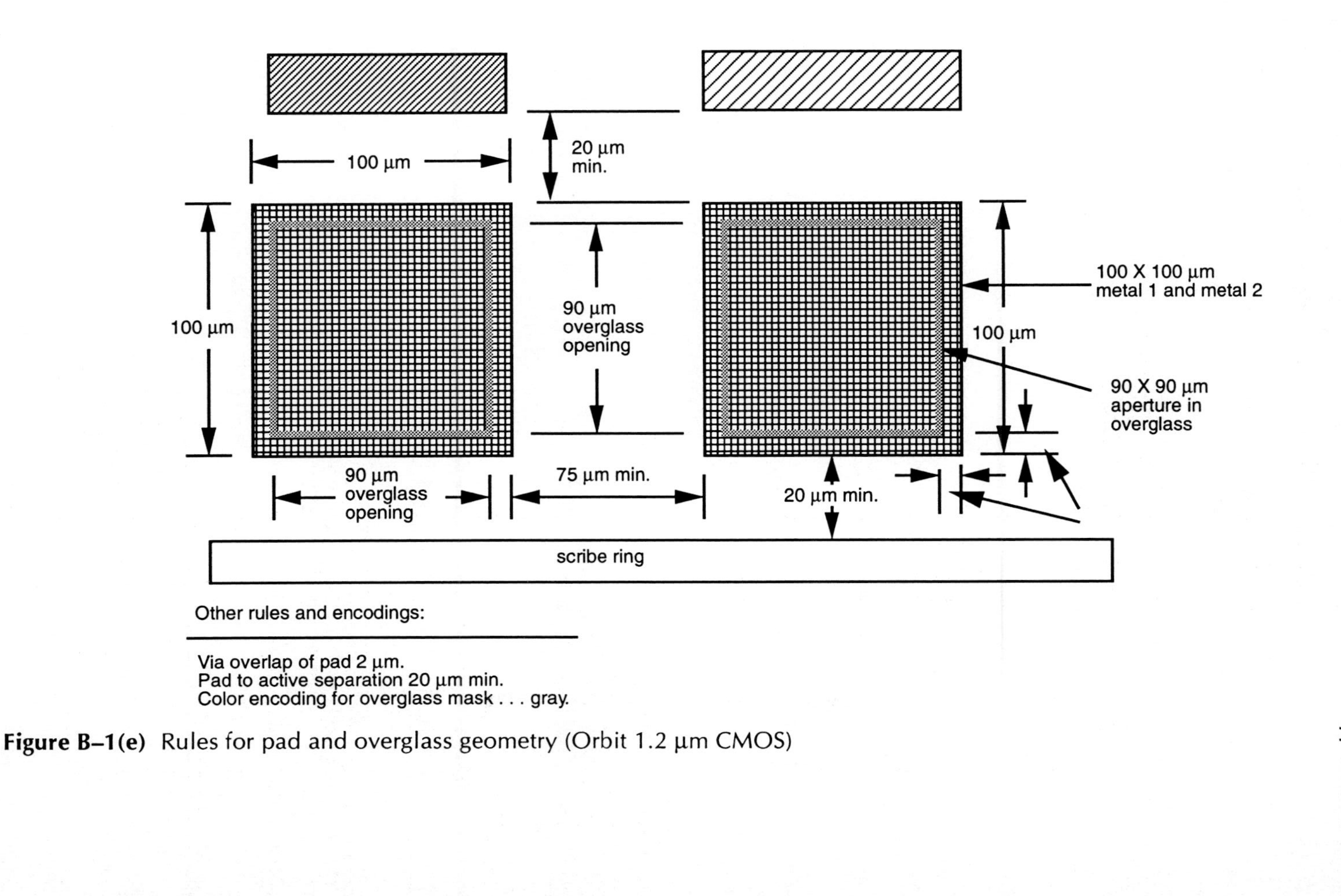

Figure B–1(e) Rules for pad and overglass geometry (Orbit 1.2 µm CMOS)

	Thickness/separation (angstroms)	Capacitance (10^–4 pf/um^2) Min.	Typ.	Max.
Gate oxide	225 +/– 25			
Field oxide	6000 +/– 300 (as grown)			
Poly.	4000 +/– 250			
Intermediate oxide	6000 +/– 600			
Metal 1	6000 +/– 500			
Metal 2	11500 +/– 750			
Metal 1 to polysilicon	6000 +/– 10000	0.56	0.6	0.68
Metal 1 to substrate	11000 +/– 1000	0.31	0.33	0.35
Metal 1 to diffusion	6000 +/– 1000	0.56	0.6	0.68
Metal 2 to poly. 1	13000 +/– 1500	0.25	0.28	0.31
Metal 2 to substrate	20000 +/– 2000	0.15	0.17	0.18
Metal 2 to diffusion	13000 +/– 1500	0.25	0.28	0.31
Metal 2 to metal 1	7000 +/– 10000	0.42	0.50	0.56
N+ to P– JCN		3.1	3.87	4.7
P+ to N – JCN		3.0	3.74	4.4

Device specs: 1.2 micron single poly. double metal n-well and p-well CMOS

	N-channel Min.	Typ.	Max.	P-channel Min.	Typ.	Max.	
VTE (VBS = 0) 30 x 1.2 μm	0.6	0.8	1.0	–1.0	–0.8	–0.6	(volts)
BVDSS (VBS = 0) 30 x 1.2 μm		10	13	–13	–10		(volts)
IDS @ VGS = 5V, VDS = 5V, L = 1.2μm	0.18	0.2	0.22	– 0.11	– 0.093	– 0.083	(mA/micr
K Prime (linear) 30 x 30 μm	30	33	36	7.5	9.5	11.5	(μA/V**2
Leff @ Ldrawn = 1.2 μm	0.8	0.9	1.0	1.0	1.1	1.2	(microns)
Oxide encroachment/side	0.48	0.52	0.56	0.48	0.52	0.56	(microns)

BE (short channel) 1.2 μm = delta VT (VBS = 0,2)	0.4	0.6	0.8	0.2	0.35	0.5	(volts)
BE (long channel) 30 μm = delta VT (VBS = 0,2)	0.5	0.7	0.9	0.3	0.45	0.6	(volts)
VTF polysilicon		10	13	−13	−10		(volts)
Diffusion resistance	25	35	45	50	70	100	(ohm/sq)
Poly. resistance	15	20	30	15	22	30	(ohm/sq)
Substrate resistance		1.3	1.6	1.8			(kohms/sq)
Substrate Cs	1E16	1.5E16	2E16	6E15	7E15	8E15	(/cm)
Diffusion junction	0.25	0.3	0.35	0.25	0.3	0.45	(microns)
Well junction	3.5	4.0	4.5	3.5	4.0	4.5	(microns)
Oxide spacer		0.2			0.2		(microns)
Contact resistance (1.4 x 1.4 μm)		75			150		(ohms)
Junction breakdown voltage		15			15		(volts)
N-well to P-substrate breakdown			45				(volts)
Metal 1 sheet resistance		35	45	55			(mohm/sq)
Metal 2 sheet resistance		20	25	30			(mohm/sq)

The UCB-Mosfet model in TECAP is an exact copy of the model in U. C. Berkeley's 2g.5 and 2g.6 versions of SPICE, except for the parameter WD.

The TECAP model takes oxide encroachment and any biasing between drawn and mask into account through the parameter WD†. The effective channel width Weff is W-2*WD, where W is the drawn channel width. When doing SPICE simulations, use Weff as the device channel width. It is important to take WD into account for devices whose drawn channel widths are small. The other parameters that SPICE needs are L, AS, and AD, which are the drawn channel length, the area of the source and the area of the drain respectively. Do not enter in the effective channel length L. SPICE figures out the effective channel length for you by internally subtracting twice the lateral diffusion from the drawn channel length that you enter.

Because of the different biasing of the drawn active layer for each rule set, there is a different value of WD† associated with each rule set and it may be found in the beginning of each of the rule set descriptions.

† Parameter WD (channel width reduction) = 0.4 μm for Orbit 1.2 μm technology and WD = 0.25 μm for Orbit 2 μm technology.

Corner simulations may be done by using the following fast and slow models:

Fast model — change values Weff, L, and Tox to Weff + 0.25 microns, L – 0.15 microns, and Tox = 21.0 nanometers respectively.
Slow model — change values of Weff, L, and Tox to Weff – 0.25 microns, L + 0.15 microns, and Tox = 24.0 nanometers respectively.

Appendix C
The programmable logic array (PLA)

An elegant solution to the mapping of irregular combinational logic functions into regular structures is provided by the PLA. The PLA provides the designer with a systematic and regular way of implementing multiple output functions of n variables in sum of products (SOP) form. The general arrangement of a PLA is given as Figure C–1 and it may be seen to consist of a programmable two-level *And/Or* structure.

Clearly, the structure is regular and may be expanded in any of its dimensions — the number of input variables v, the number of product (*And*) terms p, and the number of output functions (*Or* terms) z. It will also be noted that if there are v input variables, for complete generality each of the product forming *And* gates must have v inputs, and if there are p product terms, each output *Or* gate must have p inputs.

In practice, a range of 'off-the-shelf' PLAs is available to the TTL-based system designer. Typically, PLAs with 14 variable inputs, 96 product terms, and eight output functions are readily obtained, and much larger PLAs (e.g. with more than 200 product terms) are also available. Such elements are programmed by the manufacturer or field programmed by the user to meet requirements.

In VLSI design, however, custom PLAs can be readily designed and must be 'programmed' during the design process. Thus for the VLSI designer, PLAs are tailored to specific tasks with little wastage of functions or space. However, the PLA structure is regular and readily expanded, contracted, or modified during design. This contrasts sharply with the attributes of random logic.

In VLSI design our objective is to map circuits onto silicon to meet particular specifications. The way in which a PLA maps onto the chip may be indicated by a 'floor plan' which gives the notional areas and relative disposition of the particular circuits and subsystems. A floor plan layout for a PLA is given in Figure C–2(a).

For MOS fabrication, *And* and *Or* gates are neither as simple nor as suitable as the *Nor* gate. Thus, we look to De Morgan's theorem to manipulate *And-Or* combinational logic requirements into *Nor* form.

For an *n* input *Nor* gate, we may write

$$X' = A + B + C + \ldots\ldots + N$$

where X is the output and A to N the inputs.

By De Morgan's theorem

$$X = A'.B'.C' \ldots\ldots N'$$

In other words, the *Nor* gate is an *And* gate to inverted input levels.

Obviously, the output *Or* functions of the PLA can be realized with *Nor* gates each followed by an inverter. Thus, the requirements and floor plan of the PLA may be adapted to *Nor* gate form as in Figure C–2(b). A MOS *Nor* gate-based PLA realization for the multiple output functions used as an example in Figure C–1 is presented in circuit form as Figure C–3.

It will be noted that Figure C–3 is a PLA, tailored to meet the particular needs and drawn in mixed circuit and logic symbol notation. Although not in mask layout form, it can be clearly seen how the factors v, p, and z affect the PLA dimensions. A PLA circuit is readily turned into a stick diagram and then to mask layout form. A similar $4 \times 8 \times 4$ programmed PLA is given in stick diagram form as Figure C–4 and the regular nature of the topology is clearly apparent. The reader is left to determine the functions implemented by this PLA.

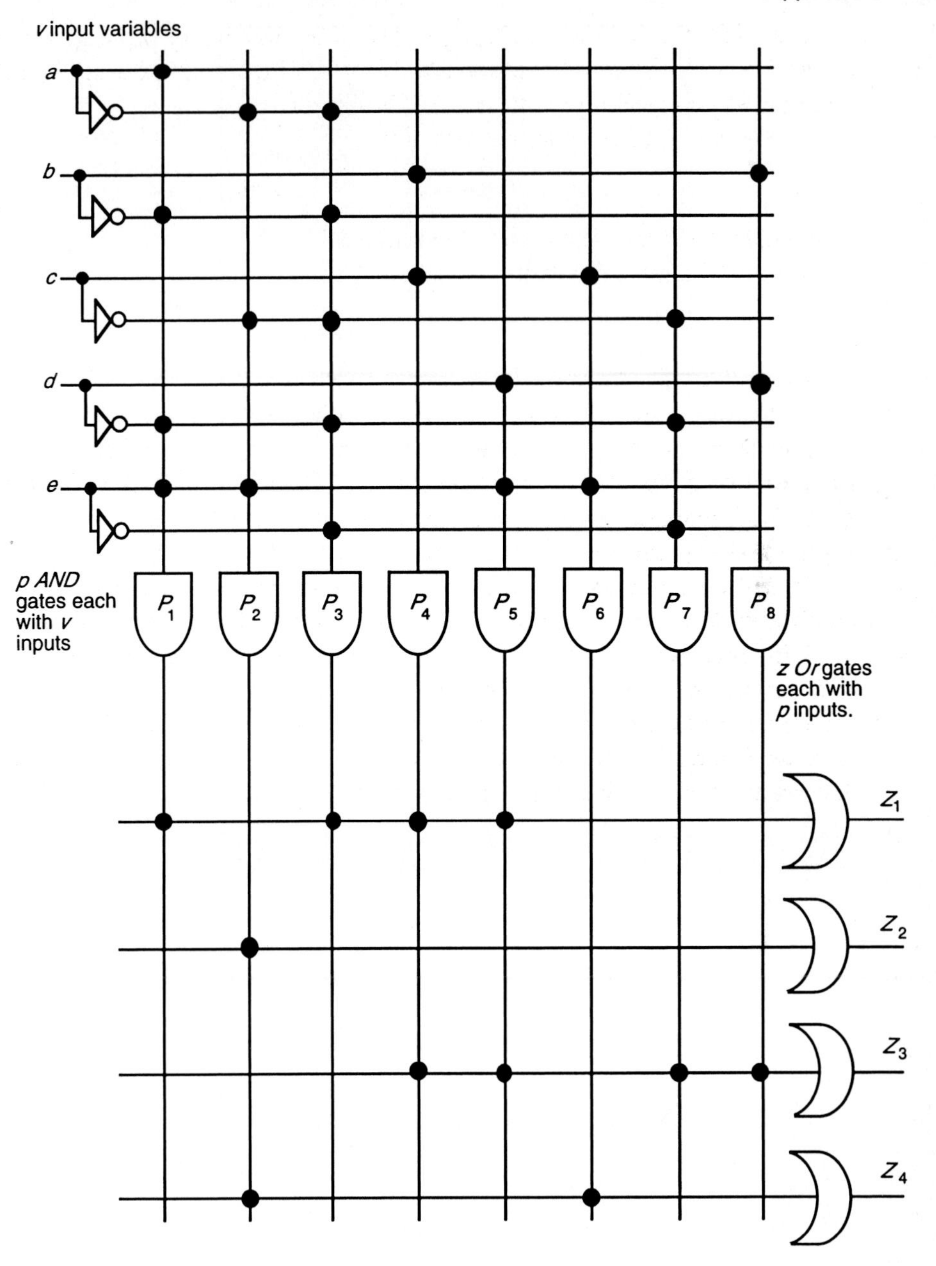

Note: 5 × 8 × 4 PLA shown symbolically and programmed for:

$Z_1 = p_1 + p_3 + p_4 + p_5 \therefore Z_1 = a\bar{b}\bar{d}e + \bar{a}\bar{b}\bar{c}\bar{d}\bar{e} + bc + de$

$Z_2 = p_2 \qquad \therefore Z_2 = \bar{a}\bar{c}e$

$Z_3 = p_4 + p_5 + p_7 + p_8 \therefore Z_3 = bc + de + \bar{c}\bar{d}\bar{e} + bd$

$Z_4 = p_2 + p_6 \qquad \therefore Z_4 = \bar{a}\bar{c}e + ce$

Figure C–1 $v \times p \times z$ PLA

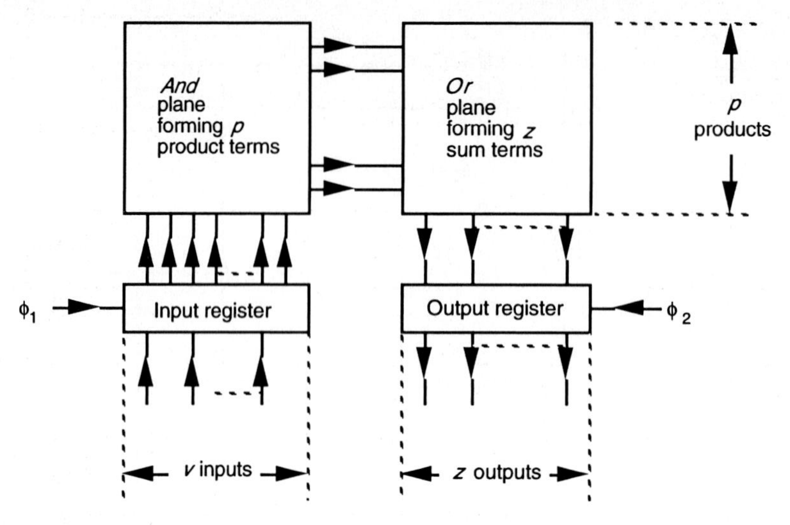

(a) *And/Or* based

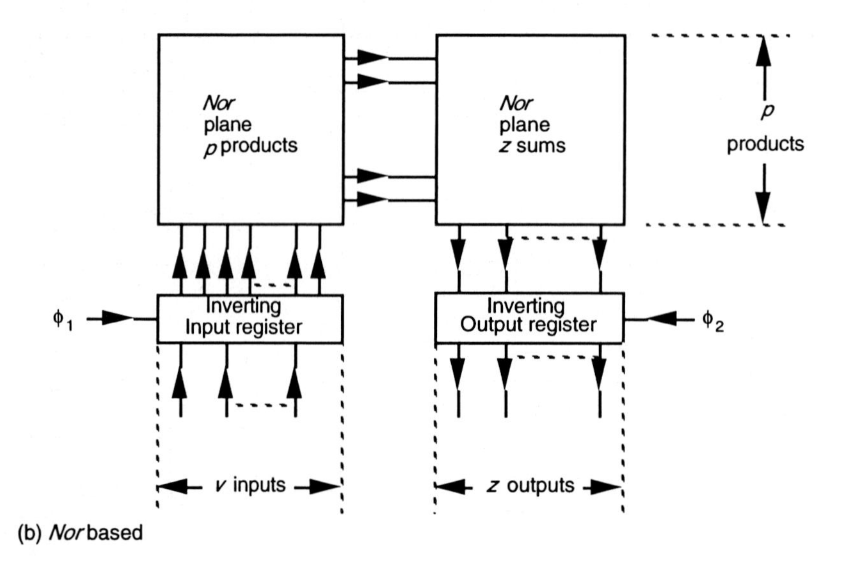

(b) *Nor* based

Figure C–2 PLA floor plans

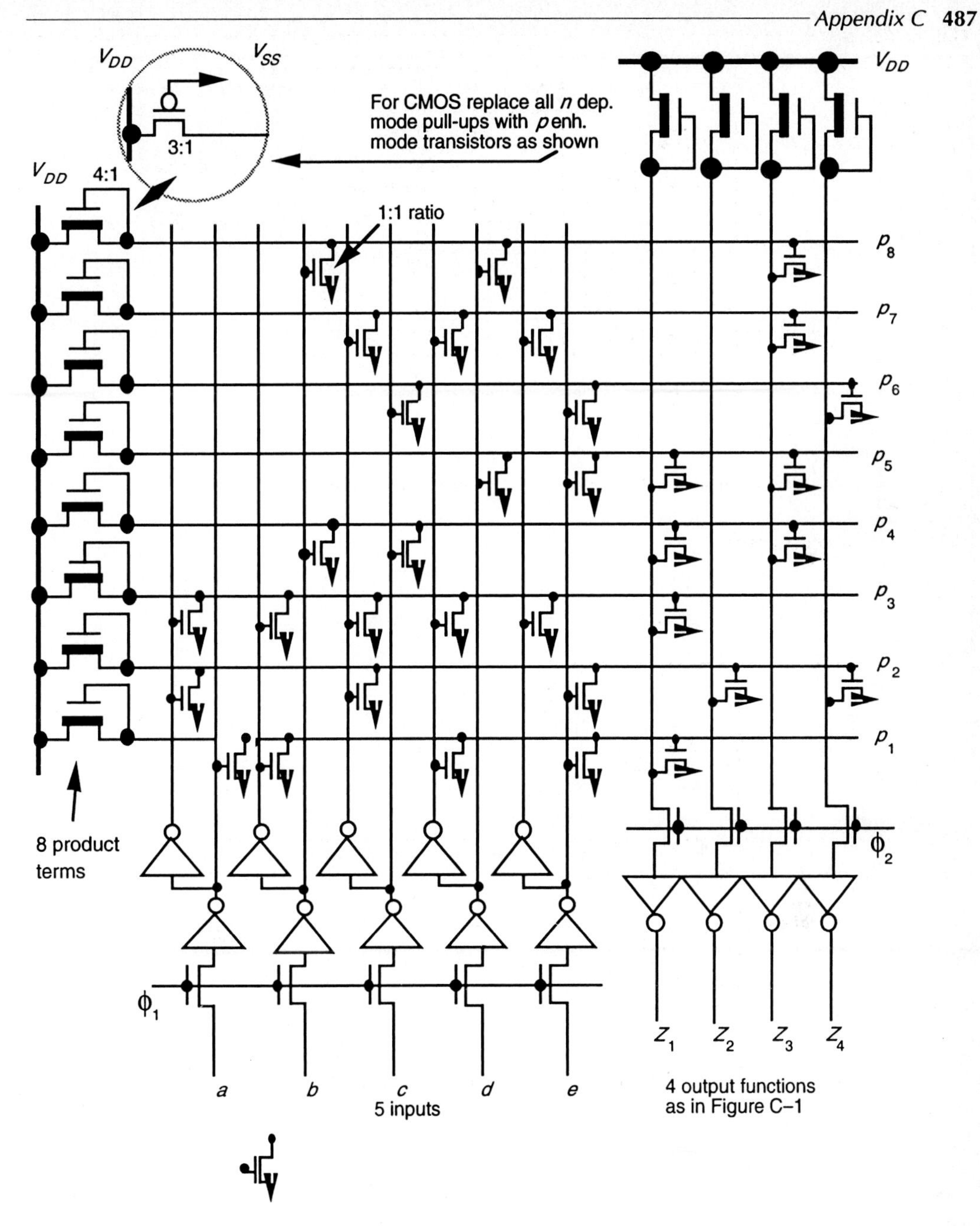

Figure C–3 PLA arrangement for multiple output function

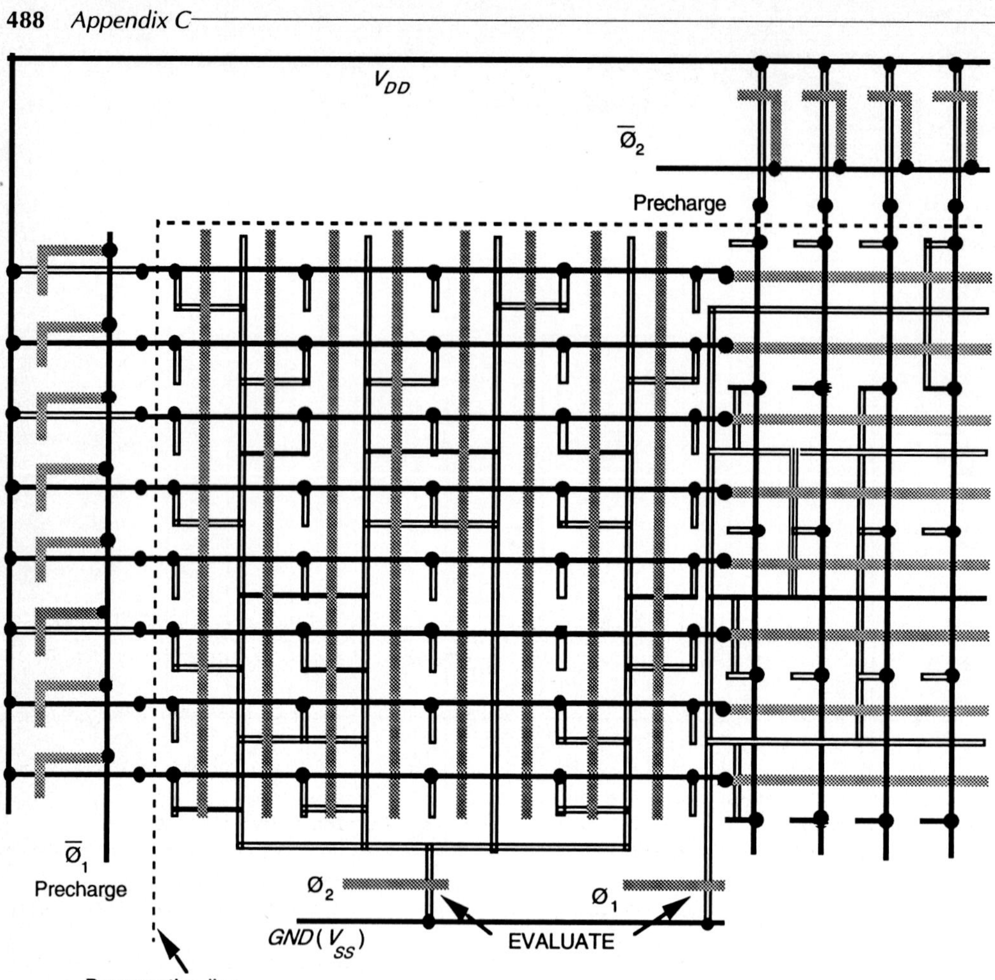

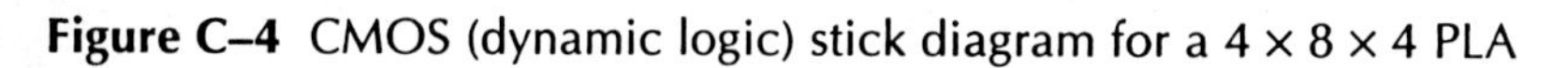
Figure C–4 CMOS (dynamic logic) stick diagram for a 4 × 8 × 4 PLA

Further reading

Allison, J. (1975) *Electronic Integrated Circuits—Their Technology and Design,* McGraw-Hill.
Ayers, R. F. (1983) *VLSI—Silicon Compilation and the Art of Automatic Microchip Design,* Prentice-Hall, USA.
Barbe, D. F. (ed.) (1982) *Very Large Scale Integration — VLSI — Fundamentals and Applications,* Springer-Verlag, West Germany/USA.
Barna, A. (1981) *VHSIC (Very High Speed Integrated Circuits) — Technologies and Trade Offs,* Wiley, USA and Canada.
Camenzind, H. R. (1968) *Circuit Design for Integrated Electronics,* Addison-Wesley, USA.
Cobbold, R. S. (1970) *Theory and Application of Field-Effect Transistors,* Wiley, USA.
Colclaser, R. A. (1981) *Microelectronics: Processing and Device Design,* Wiley, USA.
Denyer P. & Renshaw D. (1985) *VLSI Signal Processing: A Bit-Serial Approach,* Addison-Wesley, UK.
Eichelberger, E. B. & Williams, T. W. (1978, May) 'A logic design structure for LSI testability', *Journal of Design Automation and Fault-Tolerant Computing*, Vol. 2, No. 2, pp. 165–78.
Einspruch, N. G. & Wisseman, W. R. (ed.) (1985) *VLSI Electronics, Microstructure Science, Vol. 11, Ga As Microelectronics,* Academy Press.
Fortino, A. (1983) *Fundamentals of Computer Aided Analysis and Design of Integrated Circuits,* Reston, USA.
Glasser, L. A. & Dobberpuhl, D. W. (1985) *The Design and Analysis of VLSI Circuits,* Addison-Wesley.
Gray, J. P. (1981) *VLSI 81: Very Large Scale Integration,* Academic Press, UK, 1981.
Grove, A. S. (1981) *Physics and Technology of Semiconductor Devices,* Wiley, USA.
Haskard, M. & May, I. (1987) *Analog VLSI Design, nMOS and CMOS,* Prentice-Hall, USA.
Hicks, P. J. (1983) *Semi-Custom IC Design and VLSI,* Peter Peregrinus Ltd, UK.
Hon, R. W. & Sequin, C. M. (1980) *A Guide to LSI Implementation,* 2nd edn, Xerox, USA.
Lindmayer, J. & Butner S.E. (1965) *Gallium Arsenide Digital Integrated Circuit Design,* McGraw-Hill, USA.
Long, S. I. & Wrigley, C. Y. (1990) *Fundamentals of Semiconductor Devices,* Van Nostrand, USA.
McCarthy, O. J. (1982) *MOS Device and Circuit Design,* Wiley, USA.
Maly, W. (1987) *Atlas of I. C. Technologies: An Introduction to VLSI Processes,* Benjamin/ Cummings Publishing, USA.
Marcus M. (1967) *Switching Circuits for Engineers,* 2nd edn, Prentice-Hall, USA.

Mavor, J., Jack, M. A. & Denyer, P. B. (1983) *Introduction to MOS LSI Design,* Addison-Wesley, UK.

Mead, C. A. & Conway, L. A. (1980) *Introduction to VLSI Systems,* Addison-Wesley, USA.

Mukherjee, A. (1986) *Introduction to nMOS and CMOS Systems Design,* Prentice-Hall, USA.

Muroga, S. (1982) *VLSI System Design,* Wiley, USA.

Nadig, H. J. (1977, May) 'Signature analysis — Concepts, examples, and guidelines', *Hewlett-Packard Journal*, USA, pp. 15–21.

Newkirk, J. A. & Mathews, R. G. (1984) *The VLSI Designer's Library,* Addison-Wesley, USA and Canada.

Pucknell, D. A. (1990) *Fundamentals of Digital Logic Design with VLSI Circuit Applications,* Prentice Hall, Australia.

Rene Segers, M. T. M. (1982, June) 'The impact of testing on VLSI design methods', *IEEE Journal of Solid-State Circuits*, USA, Vol. SC-17, No. 3, pp. 481–86.

Richman, P. (1967) *Characteristics and Operation of MOS Field-Effect Devices,* McGraw-Hill, USA, 1967.

Rubin, S. M. (1987) *Computer Aids for VLSI Design,* Addison-Wesley, USA.

Streetman, B. G. (1980) *Solid State Electronic Devices,* Prentice-Hall, USA.

Sze, S. M. (ed.) (1983) *VLSI Technology,* McGraw-Hill, USA.

Till, C. W. and Luxon, J. T. (1982) *Integrated Circuits: Materials, Devices, and Fabrications,* Prentice-Hall, USA.

Weste, N. H. E. (1982, July–August) 'Mulga — An interactive symbolic system for the design of integrated circuits', *Bell System Technical Journal*, 60, USA, pp. 823–57.

Weste, N. H. E. & Eshraghian, K. (1984) *Principles of CMOS VLSI Design — A Systems Perspective,* Addison-Wesley, USA.

Westinghouse Defense and Space Center (1970) *Integrated Electronic Systems,* Prentice-Hall, USA.

Index

active bus 186
adder
  block diagram 212
  bounding box 221
  complete stick diagram CMOS 217
    bounding box 221
  element 214–23
  enhancement techniques 228 *ff*
    carry look-ahead 233–36
    carry select 229
    carry skip 230–31
    comparison of 236–40
    Manchester carry-chain 226–38
  4-bit design *see* 4-bit adder 213 *ff*
  implementation of ALU with 224 *ff*
  multiplexer-based 215 *ff*
  requirements 214
  standard equations 214, 226
  truth table 213
allocation of layers
  considerations 119
ALU (Arithmetic and Logic Unit) 224
arbitration logic example 167 *ff*
architecture
  nature of design for in VLSI 198
area capacitance (layers) 99 *ff*
arithmetic processor, 4-bit 212 *ff*
array
  forming from memory cells 273 *ff*
  4 × 4-bit register 274
  RAM 278
ASM (algorithmic state machine) design 269
  chart for JK flip-flop 270
asymmetric two-phase clock 305–6

barrel shifter 205 *ff*
  bounding box for 4 × 4 208
  4 × 4 circuit 205
    stick 206
  standard cell for 206
beta (β) 31 *ff*
BiCMOS
  comparison with CMOS 22
  drivers 111–14
  fabrication 24 *ff*
  technology 21 *ff*
  transistor, npn 23
BILBO 355–57
bipolar npn transistor
  comparison with MOS transistors 53
  transconductance 52
body effect 34, 35
Boltzmann's constant *k* 34
bounding box concept 205, 207 *ff*
buffer, super 110–11
bus arrangements 186 *ff*
  precharged arrangement 186

CAD (Computer-Aided Design) 324–32
Caltech Intermediate Form 319–23
capacitance
  area 99, 308
  calculations 100–2
  MOS circuit model 50

peripheral 117–18, 308
standard unit $\square C_g$ 100, 309
wiring 116
capacitive load driving 107–14
carry look-ahead adders 233 *ff*
cell, design *see* mask layouts
channel length modulation 331
charge
electron, on 34
gate to channel Q_c 30
precharged bus 186–87
storage-based dynamic shift register 183
storage-based register 182
storage on C_g 181
CIF (Caltech Intermediate Form) 319–23
circuit extractor *see* CAD
circuit simulator *see* CAD
clock, two-phase 176 *ff*
clock asymmetric 305, 306
clock generator 179, 180
CMOS
design projects 362–405
design style 68–72
fabrication 15–19
inverter 47 *ff*
latch-up 57
submicron technology 406, 407
color
layers, coding of 62–67 Color plates 1 (a)–(d)
crossbar switch 204
current
limitations 189–90
I_{ds} versus V_{ds} relationship
(Ga As) 435–440
(MOS) in saturation 32
(MOS) non-saturated 30–32
(Si) 29 *ff*

D flip-flop 273
data path, 4-bit 199 *ff*
data selectors (multiplexers) 171 *ff*
decoder 277
delay
in long polysilicon wires 115–16
pass transistor chains, in 114–15, 309–11
through inverters 104–7
delay unit τ 102–4
design
ALU 212–25
bus arbitration logic 167–71
∂*latch* 396 *ff*
4-bit arithmetic processor 199 *ff*
4-bit shifter, of 203–7
4–line Gray/binary code converter 175–76
ground rules for 307–315
guidelines 311 *ff*
incrementer/decrementer 363–66
L/R shift register, serial/parallel 367–71
n-bit comparator 372–81
parameters 307–9
parity generator 165–67
process 211 *ff*, 313–15
observations on 211
regularity 211–12
style 316–18
testability, for 341–49
2-phase clock generator 381 *ff*
design rules 72 *ff*, Appendix B 475–79
checkers *see* CAD
CMOS 73–83
GaAs 458–63
nMOS 73–78
Orbit
2 μm double metal/double poly. CMOS/ BiCMOS 83–88 and Colorplates 3–6
1.2 μm double metal/single poly. CMOS 84, Appendix B 475-79
design style
CMOS 68–72
GaAs 453 *ff*
symbolic 463–64
nMOS 67–68
design tools 324–32
device parameters
silicon 96–104, 307–9
diffusion
area capacitance 99
peripheral capacitance 117
sheet resistance R_s 95–96
dissipation
power, BiCMOS 188–89
power, CMOS 188–89
double metal process
design rules 83–88, Appendix B 475–79
double polysilicon process 83–88
drivers for large capacitive load 107–14
dynamic storage
elements 182
4–bit shift register 183
on C_g 181 *ff*

electrical
MOS parameters 96–106, 307–9
Orbit parameters
1.2 μm process Appendix B 474–81
2 μm process Appendix A 470–73

electron
charge 34
Exclusive-Or gate 177
exercises 60, 92, 120–22, 192–95, 209, 254, 283–84, 468–69

fabrication
CMOS 15–17
GaAs MESFET 428–31
nMOS 10–15
figure of merit 37
flip-flop
D type 273
JK type 269
ASM chart for 270
floorplan
4–bit processor 202, 294, 297–99
4-bit adder 213 *ff*
design of 213–223
element for, *see* adder, element
4-bit shifter
design of 203–7
4 × 4-bit register array 274 *ff*
selection and control 276–78

gallium arsenide (GaAs) 406–69
CE-JFET 433–34
comparison with other technologies 416
crystal structure 408–10
DCFL 446–47
device modeling and performance estimation 435 *ff*
E-JFET 431–33
HEMT 434–35
MESFET 418 *ff*
design methodology 451
design rules 458–63
logic voltage swing 445–46
ring notation 453 *ff*
transconductance and output conductance 442
transfer characteristics 440
general logic function block 174
ground rules for design 307–15

HEMT 434–35
Hochet, Dr B. 228

I/O pads 298–302
I_{ds} versus V_{ds} relationship
(Si) 29 *ff*
(GaAs) 435
incrementer/decrementer design (CMOS) 363–67
Integrated Silicon Design Pty Ltd (ISD) software 326 *ff*
interlayer capacitance 117
inverter
alternative pull-ups 45–47
BiCMOS 54 *ff*
CMOS 47 *ff*
nMOS 38 *ff*
noise margins 292–94
optimization (nMOS and CMOS) 287–92
pseudo-nMOS 159–61
p.u./p.d. ratio nMOS 40–45
p.u./p.d. ratio pseudo-nMOS 160
threshold voltage V_{inv} 40 *ff*
transfer characteristic
(CMOS) 51
(nMOS) 40

JK flip-flops 269

lambda 73 *ff*
latch-up
BiCMOS 59
CMOS 57
layer representation (GaAs) 451–53
layers (Si) 62 *ff*
choice of 118–19
encoding 63–66
Color plates 1(a)–(d)
layout diagrams *see* Mask layouts
layout style (GaAs) 453
length to width ratio 40 *ff*
logic
other forms of CMOS 159 *ff*
pseudo-nMOS 159–61
switch arrangements 148–49

mask encoding 63 *ff*
mask layouts
barrel-shifter cell 206, 296
clock generator 386, 389, 392
comparator cell 380
∂latch 402
4-way multiplexer (Transmission gate) Color plate 11
inverters
CMOS 220, Color plate 2
nMOS 152, 219
memory cell
CMOS static 280
nMOS pseudo-static 265
one-transistor 262
three-transistor 259–61

multiplexer
cells 218
four-way n-type 172, Color plate 10
Nand gate, 2I/P BiCMOS 154, Color plate 8(a)
Nor gate
3I/P nMOS Color plate 9(a)
2I/P BiCMOS 158, Color plate 8(b)
2I/P CMOS Color plate 9(b)
2I/P nMOS 89
shift register cells 89, 91, 185, Color plate 7
two-phase clock generator (BiCMOS) Color plate 12
Xor gate 177
memory arrays 273
4 × 4-bit register 274–78
floor plan 275–76
RAM arrays 278–82
memory cells *see* mask layouts
CMOS dynamic and static 266–69
refresh 306
MESFET-based design 451 *ff*
layer connections 456–58
layer representation 451–53
layout style 453
Moore's law 3
multiplexers 171 *ff*
general logic function block 174
multipliers 240

Nand gates 150 *ff*
Nor gates 156 *ff*

observability 334
optimization of inverters 287–292
Osseiran, Dr A. 334

pads I/O 298–301
geometry 87, Color plate 5
parallel multiplier 242 *ff*
parity generator design example 165–67
pass transistor 38 *ff*
cascaded delay 114–15
logic 149
properties 150
periodic table groups 410
peripheral capacitance 117–18
permittivity ε 30 *ff*
pinch off
pMOS transistor 8 *ff*
PLA 176, 482–86
polysilicon long wire delay 115–16
power dissipation
CMOS and BiCMOS 188–89
nMOS 128
power rails, current limitations 189–90
priority encoder, example 167–71
propagation delay, basic unit τ 102 *ff*
pull-up, alternative forms 45
pull-up/pull-down ratios 42 *ff*

RAM *see* memory arrays
ratio calculation, inverters 40–45, 159–60
Recursive decomposition, multiplication 251–53
regularity 211–12
resistance, of layers 96 *ff*
ring notation (GaAs design) 453 *ff*

saturated region, I_{ds} versus V_{ds} 32 *ff*
saturation 10 *ff*
Schottky barrier diode 420 *ff*
sheet resistance (R_s) 95–98
typical values 96, 308
shift register 183, 257, 367, *see also* mask layouts
shifter, 4-bit, barrel 203–7
signature analysis 354 *ff*
silicide layer 96
simulations
clock generator 387–95
comparator cells 382–83
∂*latch* 403–4
4-bit shift register 374
2-bit incrementer/decrementer 368
Xor gate 178
simulators 324 *ff*
speed, power product 4 *ff*
standard units
$\square C_g$ *see* capacitance, standard unit
R_s *see* sheet resistance
τ *see* delay unit
stick diagrams 62 *ff*
storage *see* memory arrays
strategy, interconnection 200 *ff*
structured design 198 *ff*
submicron CMOS 406–7
super buffers 110–11
switch crossbar 204
symbolic design (as in GaAs) 463
system
timing considerations 257

thermal aspects of processing 19
threshold voltage V_t 33–35
transconductance g_m 35–37, 52–53

transfer characteristics
  inverter 40, 51
  MESFET (GaAs) 440
transistor
  BiCMOS 22 *ff*
  MOS 6 *ff*
transmission gate 148 *ff*
  properties 150
Trivedi, Prof. K. S. 287
tutorial exercises 60, 92–93, 120–22, 192–95, 209, 254, 283–84, 468–69
two-phase clock 176 *ff*

velocity of electrons 30 *ff*
velocity saturation 332
V_{inv} 40 *ff*
V_t 9, 33 *ff*

Wallace tree multiplier 251 *ff*
wiring
  capacitance 116 *ff*
  choice of layer 118
    rules for 119

yield 317